Student's Solutions Manual

Algebra and
Trigonometry

Student's Solutions Manual

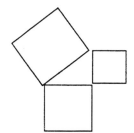

Carol A. Aaronson

Fourth Edition

Algebra and Trigonometry
A PROBLEM-SOLVING APPROACH

Walter Fleming
Hamline University

Dale Varberg
Hamline University

Herbert Kasube
Bradley University

Prentice Hall
Upper Saddle River, NJ 07458

Special thanks to Shmuel Gerber for his work on the Review Exercises.

Editorial production/supervision: *Joanne E. Jimenez*
Pre-press buyer: *Paula Massenaro*
Manufacturing buyer: *Lori Bulwin*
Supplement Acquisitions Editor: *Susan Black*
Acquisitions Editor: *Priscilla McGeehon*

Printed in the United States of America

10 9 8 7 6 5 4 3 2

ISBN   0-13-028937-X

Prentice-Hall International (UK) Limited,London
Prentice-Hall of Australia Pty. Limited, Sydney
Prentice-Hall Canada Inc., Toronto
Prentice-Hall Hispanoamericana, S.A., Mexico
Prentice-Hall of India Private Limited, New Delhi
Prentice-Hall of Japan, Inc., Tokyo
Pearson Education Asia Pte. Ltd., Singapore
Editora Prentice-Hall do Brasil, Ltda., Rio de Janeiro

CONTENTS

Student's Solutions Manual

Algebra and Trigonometry

Chapter 1: Numbers and their Properties

Problem Set 1.1

1. If $x = $ first number and $y = $ second number, then we want $x + \frac{y}{3}$.

3. If $x = $ first number and $y = $ second number, then we want $\frac{2x}{3y}$.

5. If $x = $ the number, then $(.10)x = 10\%$ of the number. We want $x + (.10)x$, or $1.10x$.

7. If the sides of the triangle are s and t, we want $s^2 + t^2$.

9. Using the formula $d = rt$, Distance $= xy$.

11. Using the formula $r = \frac{d}{t}$, Rate $= \frac{y}{x}$.

13. Using the formula $t = \frac{d}{r}$, Time $= \frac{30}{x} + \frac{30}{y}$.

15. Using the formula $A = bh$, let $b = x$ and $h = x$:
$$A = x \cdot x = x^2$$

17. Using the formula $A = 2lw + 2wh + 2hl$,
let $l = x, w = x$, and $h = x$:
$$A = 2x \cdot x + 2x \cdot x + 2x \cdot x$$
$$= 6x^2$$

19. Since the diameter of the sphere is x, the radius is $\frac{x}{2}$. Using the formula
$A = 4\pi r^2$ with $r = \frac{x}{2}$:
$$A = 4\pi \left(\frac{x}{2}\right)^2$$
$$= 4\pi \frac{x^2}{4}$$
$$= \pi x^2$$

21. Using the formula $V = lwh$,
let $l = x$, $w = x$, and $h = 10$:
$$V = 10(x)(x) = 10x^2$$

23. Since the diameter of the sphere is x, the radius is $\frac{x}{2}$. Using the formula
$V = \frac{4}{3}\pi r^3$ with $r = \frac{x}{2}$:
$$V = \frac{4}{3}\pi \left(\frac{x}{2}\right)^3$$
$$= \frac{4}{3}\pi \frac{x^3}{8}$$
$$= \frac{\pi x^3}{6}$$

25. $V = $ volume of cube $-$ volume of cylindrical hole
$$= x^3 - \pi \cdot 2^2 \cdot x = x^3 - 4\pi x$$

27.

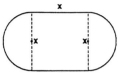

Note that the radius of each semicircular end is $\frac{x}{2}$. The circumference of each semicircle is $\pi \cdot \frac{x}{2}$.
Therefore,
$$\text{Perimeter} = x + \frac{\pi x}{2} + x + \frac{\pi x}{2}$$
$$= 2x + \pi x$$

The area of each semicircular end is $\frac{1}{2}\pi\left(\frac{x}{2}\right)^2$ and the area of the square is x^2.
Therefore,
$$\text{Area} = x^2 + \frac{1}{2}\pi\left(\frac{x^2}{4}\right) + \frac{1}{2}\pi\left(\frac{x^2}{4}\right)$$
$$= x^2 + \frac{\pi x^2}{4}$$

29. Let x represent the number.
Then $\frac{x}{2}$ represents one-half the number.
$$x + \frac{x}{2} = 45$$
$$\frac{3x}{2} = 45$$
$$3x = 90$$
$$x = 30$$
The numbers are 30 and 15.

31. Let $x = $ one number.
Then $x + 2 = $ the other number.
$$x + (x + 2) = 168$$
$$2x + 2 = 168$$
$$2x = 166$$
$$x = 83$$
The numbers are 83 and 85.

33. If the car is traveling at x mph, then the distance the car travels in $4\frac{1}{2}$ hours is $4.5x$. Since the distance is 252 miles,
$$4.5x = 252$$
$$x = \frac{252}{4.5}$$
$$x = 56 \text{ mph}$$

35. If x represents the width of the rectangle, in meters, then $x + 3$ represents the length of the rectangle, in meters.
$$\text{Perimeter} = 2x + 2(x + 3)$$
Since the perimeter is 48 meters,
$$2x + 2(x + 3) = 48$$
$$2x + 2x + 6 = 48$$
$$4x + 6 = 48$$
$$4x = 42$$
$$x = 10.5 \text{ meters}$$

37. We want the area of the rectangle minus the area of the two circles, or

$$20 \cdot 40 - 2 \cdot \pi \cdot 10^2 = 800 - 200\pi$$
$$= 200(4 - \pi)$$
$$\approx 171.68 \text{ square feet}$$

39. Let x represent the radius of the sphere, in centimeters. Using the formula $A = 4\pi r^2$, $4\pi x^2$ represents the surface area of the sphere. The area of a circle of radius 6 cm is $\pi(6)^2$, or 36π square centimeters.

$$4\pi x^2 = 36\pi$$
$$x^2 = \frac{36\pi}{4\pi}$$
$$x^2 = 9$$
$$x = +3 \text{ or } x = -3$$

Since the radius of a geometric figure cannot be a negative number, the radius of the sphere is 3 cm.

41. Use the formula $V = \pi r^2 h$ to compute the volume of oil each can can hold.

$$V_{\text{can 1}} = \pi\left(\tfrac{5}{2}\right)^2(8)$$
$$= \pi \cdot \tfrac{25}{4} \cdot 8$$
$$= 50\pi \text{ in}^3$$
$$V_{\text{can 2}} = \pi(3)^2 10$$
$$= 90\pi \text{ in}^3$$

Assume that the price of each can of oil is directly proportional to the volume of oil the can can hold. Let x represent the price of can 2. Then,

$$\frac{x}{90\pi} = \frac{1.80}{50\pi}$$
$$x = 90\pi\left(\frac{1.80}{50\pi}\right)$$
$$x = 3.24$$

A fair price for can 2 is $3.24.

43. Let r = radius of the ball
Height of can = $12r$
Radius of base of can = r
Volume of air = volume of can − volume of balls
$$V = \pi r^2 \cdot 12r - 8\pi r^3$$
$$= 4\pi r^3$$

45. Let x represent the number of miles Orville can travel on his outward flight.
Then x also represents the number of miles Orville can travel on his return flight.
When Orville flies against the wind, the speed of the plane will be $150 - 50$, or 100 mph.
When Orville makes the return trip, with the wind behind him, the speed of the plane will be $150 + 50$, or 200 mph.

Using the formula $t = \frac{d}{r}$,
$\frac{x}{100}$ represents the length of time Orville can spend on his outward flight, in hours.
$\frac{x}{200}$ represents the number of hours it takes Orville to make the return flight.
Since the total time of his trip must be 6 hours,
$$\frac{x}{100} + \frac{x}{200} = 6$$
multiply by 200 to clear fractions:
$$2x + x = 1200$$
$$3x = 1200$$
$$x = 400$$
Orville can fly 400 miles on his outward flight.

47. We can assume that the rotations of tires are equally spaced through the 60,000 miles. Each of the 5 tires must remain on the car for 4 out of every 5 rotations, or $\frac{4}{5}$ of the 60,000 miles.

$$\tfrac{4}{5}(60{,}000) = 48{,}000$$

Each tire will be on the car for 48,000 miles.

49. Let r = radius of the earth in feet
Distance Arnold's feet travel = $2\pi r$
Distance Arnold's nose travels = $2\pi(r + 6)$
His nose travels $2\pi(r + 6) - 2\pi r = 12\pi$ feet further—no matter whether the radius of the circle is 6 feet or 4000 miles!

Problem Set 1.2

1. $4 - 2(8 - 12) = 4 - 2(-4)$
$$= 4 + 8$$
$$= 12$$

3. $-3 + 2[-5 - (12 - 3)] = -3 + 2[-5 - 9]$
$$= -3 + 2[-14]$$
$$= -3 - 28$$
$$= -31$$

5. $-4[3(-6 + 13) - 2(7 - 5)] + 1$
$$= -4[3(7) - 2(2)] + 1$$
$$= -4[21 - 4] + 1$$
$$= -4[17] + 1$$
$$= -68 + 1$$
$$= -67$$

7. $-3[4(5 - x) + 2x] + 3x$
$$= -3[20 - 4x + 2x] + 3x$$
$$= -3[20 - 2x] + 3x$$
$$= -60 + 6x + 3x$$
$$= 9x - 60$$

9. $2[-t(3+5-11)+4t]$
$= 2[-t(-3)+4t]$
$= 2[3t+4t]$
$= 2[7t]$
$= 14t$

11. $\dfrac{24}{27} = \dfrac{3\cdot 8}{3\cdot 9} = \dfrac{8}{9}$

13. $\dfrac{45}{-60} = \dfrac{3\cdot 15}{-4\cdot 15} = -\dfrac{3}{4}$

15. $\dfrac{3-9x}{6} = \dfrac{3(1-3x)}{3(2)} = \dfrac{1-3x}{2}$

17. $\dfrac{4x-6}{4-8} = \dfrac{2(2x-3)}{-4}$
$= \dfrac{2(2x-3)}{2(-2)}$
$= \dfrac{2x-3}{-2}$
$= -\dfrac{2x-3}{2}$
$= \dfrac{-2x+3}{2}$

19. $\dfrac{5}{6}+\dfrac{11}{12} = \dfrac{10}{12}+\dfrac{11}{12} = \dfrac{21}{12} = \dfrac{7\cdot 3}{4\cdot 3} = \dfrac{7}{4}$

21. $\dfrac{4}{5}-\dfrac{3}{20}+\dfrac{3}{10} = \dfrac{16}{20}-\dfrac{3}{20}+\dfrac{6}{20} = \dfrac{19}{20}$

23. $\dfrac{5}{12}+\dfrac{7}{18}-\dfrac{1}{6} = \dfrac{15}{36}+\dfrac{14}{36}-\dfrac{6}{36} = \dfrac{23}{36}$

25. $\dfrac{-5}{27}+\dfrac{5}{12}+\dfrac{3}{4} = \dfrac{-20}{108}+\dfrac{45}{108}+\dfrac{81}{108} = \dfrac{106}{108} = \dfrac{53}{54}$

27. $\dfrac{5}{6}\cdot\dfrac{9}{15} = \dfrac{5\cdot 9}{6\cdot 15} = \dfrac{\cancel{5}\cdot\cancel{3}\cdot\cancel{3}}{\cancel{3}\cdot 2\cdot\cancel{5}\cdot\cancel{3}} = \dfrac{1}{2}$

29. $\dfrac{3}{4}\cdot\dfrac{6}{15}\cdot\dfrac{5}{2} = \dfrac{3\cdot 6\cdot 5}{4\cdot 15\cdot 2} = \dfrac{3\cdot\cancel{2}\cdot\cancel{3}\cdot\cancel{5}}{4\cdot\cancel{3}\cdot\cancel{5}\cdot\cancel{2}} = \dfrac{3}{4}$

31. $\dfrac{\frac{5}{6}}{\frac{8}{12}} = \dfrac{5}{6}\cdot\dfrac{12}{8} = \dfrac{5}{4}$

33. $\dfrac{\frac{3}{4}}{2} = \dfrac{3}{4}\cdot\dfrac{1}{2} = \dfrac{3\cdot 1}{4\cdot 2} = \dfrac{3}{8}$

35. $6\div\dfrac{7}{9} = \dfrac{6}{1}\cdot\dfrac{9}{7} = \dfrac{6\cdot 9}{1\cdot 7} = \dfrac{54}{7}$

37. $\dfrac{\frac{2}{3}+\frac{3}{4}}{\frac{7}{12}} = \dfrac{12\left(\frac{2}{3}+\frac{3}{4}\right)}{12\left(\frac{7}{12}\right)} = \dfrac{8+9}{7} = \dfrac{17}{7}$

39. $\dfrac{\frac{2}{3}+\frac{3}{4}}{\frac{2}{3}-\frac{3}{4}} = \dfrac{12\left(\frac{2}{3}+\frac{3}{4}\right)}{12\left(\frac{2}{3}-\frac{3}{4}\right)} = \dfrac{8+9}{8-9} = \dfrac{17}{-1} = -17$

41. $\dfrac{\frac{5}{6}-\frac{1}{12}}{\frac{3}{4}+\frac{2}{3}} = \dfrac{12\left(\frac{5}{6}-\frac{1}{12}\right)}{12\left(\frac{3}{4}+\frac{2}{3}\right)} = \dfrac{10-1}{9+8} = \dfrac{9}{17}$

43. $\dfrac{5}{6}+\dfrac{1}{9}-\dfrac{2}{3}+\dfrac{5}{18} = \dfrac{15}{18}+\dfrac{2}{18}-\dfrac{12}{18}+\dfrac{5}{18} = \dfrac{10}{18} = \dfrac{5}{9}$

45. $-\dfrac{2}{3}\left(\dfrac{7}{6}-\dfrac{5}{4}\right) = -\dfrac{2}{3}\left(\dfrac{14}{12}-\dfrac{15}{12}\right)$
$= -\dfrac{2}{3}\left(-\dfrac{1}{12}\right) = \dfrac{1}{18}$

47. $\dfrac{2}{3}-\dfrac{1}{2}\left(\dfrac{2}{3}-\dfrac{1}{2}+\dfrac{1}{12}\right) = \dfrac{2}{3}-\dfrac{1}{2}\left(\dfrac{8}{12}-\dfrac{6}{12}+\dfrac{1}{12}\right)$
$= \dfrac{2}{3}-\dfrac{1}{2}\left(\dfrac{3}{12}\right)$
$= \dfrac{2}{3}-\dfrac{3}{24}$
$= \dfrac{16}{24}-\dfrac{3}{24}$
$= \dfrac{13}{24}$

49. $\left(\dfrac{5}{7}+\dfrac{7}{5}\right)\div\dfrac{2}{5} = \left(\dfrac{25+49}{35}\right)\cdot\dfrac{5}{2} = \dfrac{74}{35}\cdot\dfrac{5}{2} = \dfrac{37}{7}$

51. $\left[2-\left(3\div\dfrac{4}{5}\right)\right]\dfrac{2}{7} = \left[2-\left(\dfrac{3}{1}\cdot\dfrac{5}{4}\right)\right]\dfrac{2}{7}$
$= \left[2-\dfrac{15}{4}\right]\dfrac{2}{7}$
$= \left[\dfrac{8}{4}-\dfrac{15}{4}\right]\dfrac{2}{7}$
$= \left[-\dfrac{7}{4}\right]\dfrac{2}{7}$
$= -\dfrac{1}{2}$

53. $\dfrac{\frac{1}{2}+\frac{3}{4}-\frac{7}{8}}{\frac{1}{2}-\frac{3}{4}+\frac{7}{8}}+\dfrac{2}{5} = \dfrac{8\left(\frac{1}{2}+\frac{3}{4}-\frac{7}{8}\right)}{8\left(\frac{1}{2}-\frac{3}{4}+\frac{7}{8}\right)}+\dfrac{2}{5}$
$= \dfrac{4+6-7}{4-6+7}+\dfrac{2}{5}$
$= \dfrac{3}{5}+\dfrac{2}{5}$
$= \dfrac{5}{5}$
$= 1$

55.
$$\left[3+\frac{2}{3+\frac{1}{3}}\right]\div\frac{6}{11}=\left[3+\frac{2}{\frac{10}{3}}\right]\div\frac{6}{11}$$
$$=\left(3+\frac{6}{10}\right)\cdot\frac{11}{6}$$
$$=\left(3+\frac{3}{5}\right)\frac{11}{6}=\frac{18}{5}\cdot\frac{11}{6}=\frac{33}{5}$$

57. $\left(1-\frac{1}{2}\right)\left(1-\frac{1}{3}\right)\left(1-\frac{1}{4}\right)\left(1-\frac{1}{5}\right)\cdots\left(1-\frac{1}{18}\right)\left(1-\frac{1}{19}\right)$
$$=\left(\frac{1}{2}\right)\left(\frac{2}{3}\right)\left(\frac{3}{4}\right)\left(\frac{4}{5}\right)\cdots\left(\frac{17}{18}\right)\left(\frac{18}{19}\right)$$
$$=\frac{1}{19}\ \text{(Notice the pattern of cancellations.)}$$

59. Let x represent the needed test grade.
$\dfrac{x+80+82+98}{4}$ represents the average of the 4 grades.
Since we want the average to equal 90,
$$\frac{x+80+82+98}{4}=90$$
$$\frac{x+260}{4}=90$$
$$x+260=360$$
$$x=100$$
A test grade of 100 will bring the average up to 90.

61. The first class scored $20\cdot70=1400$ points and the second class scored $30\cdot60=1800$ points. The average for the two classes
$$=\frac{1400+1800}{50}=\frac{3200}{50}=64\text{ points}$$

63. Let x represent the number of meters Ronald walked.
Then $(2\frac{1}{2})x$, or $\frac{5}{2}x$, represents the number of meters Ronald jogged.
$(2\frac{1}{3})(\frac{5}{2}x)$, or $\frac{35}{6}x$, represents the number of meters Ronald sprinted.
Since he covered a total of 2352 meters,
$$x+\frac{5}{2}x+\frac{35}{6}x=2352$$
$$\frac{6}{6}x+\frac{15}{6}x+\frac{35}{6}x=2352$$
$$\frac{56}{6}x=2352$$
$$x=2352\cdot\frac{6}{56}$$
$$x=252$$
Ronald walked 252 meters.

Problem Set 1.3

1. $250=2\cdot125=2\cdot5\cdot25=2\cdot5\cdot5\cdot5$

3. $200=2\cdot100=2\cdot2\cdot50=2\cdot2\cdot2\cdot25=2\cdot2\cdot2\cdot5\cdot5$

5. $2100=2\cdot1050=2\cdot2\cdot525=2\cdot2\cdot3\cdot175$
$=2\cdot2\cdot3\cdot5\cdot35=2\cdot2\cdot3\cdot5\cdot5\cdot7$

7. lcm$(250,200)=$ lcm$(2\cdot5\cdot5\cdot5,2\cdot2\cdot2\cdot5\cdot5)$
$=2\cdot2\cdot2\cdot5\cdot5\cdot5=1000$

9. lcm$(250,2100)=$ lcm$(2\cdot5\cdot5\cdot5,2\cdot2\cdot3\cdot5\cdot5\cdot7)$
$=2\cdot2\cdot3\cdot5\cdot5\cdot5\cdot7$
$=10,500$

11. lcm$(250,200,2100)$
$=$ lcm$(2\cdot5\cdot5\cdot5,2\cdot2\cdot2\cdot5\cdot5,2\cdot2\cdot3\cdot5\cdot5\cdot7)$
$=2\cdot2\cdot2\cdot3\cdot5\cdot5\cdot5\cdot7$
$=21,000$

13. $\frac{3}{250}+\frac{17}{200}=\frac{3\cdot4}{1000}+\frac{17\cdot5}{1000}=\frac{12+85}{1000}=\frac{97}{1000}$

15. $\frac{7}{250}-\frac{1}{2100}=\frac{7\cdot42}{10,500}-\frac{1\cdot5}{10,500}$
$$=\frac{294-5}{10,500}$$
$$=\frac{289}{10,500}$$

17. $\frac{3}{250}-\frac{17}{200}+\frac{11}{2100}=\frac{3\cdot84}{21,000}-\frac{17\cdot105}{21,000}+\frac{11\cdot10}{21,000}$
$$=\frac{252-1785+110}{21,000}$$
$$=\frac{-1423}{21,000}$$

19. $\frac{2}{3}$:
$$3\overline{\smash)2.00}\quad.66\cdots$$
$1\ 8$
20
18
$2\cdots$
$\frac{2}{3}=.\overline{6}$

21. $\frac{5}{8}$:
$$8\overline{\smash)5.000}\quad.625$$
$4\ 8$
20
16
40
40
0
$\frac{5}{8}=.625\overline{0}$

4

23. $\frac{6}{13}$:

$$
\begin{array}{r}
.4615384\cdots \\
13\overline{\smash)6.0000000} \\
5\;2 \\
\overline{80} \\
78 \\
\overline{20} \\
13 \\
\overline{70} \\
65 \\
\overline{50} \\
39 \\
\overline{110} \\
104 \\
\overline{60} \\
52 \\
\overline{8\cdots}
\end{array}
$$

$\frac{6}{13} = .\overline{461538}$

25.
$$x = .\overline{7} = .777777\ldots$$
$$10x = 7.777777\ldots$$
Subtracting the upper line from the lower,
$$9x = 7$$
$$x = \tfrac{7}{9}$$

27. Let $x = .235235\ldots$
$$1000x = 235.235235\ldots$$
Subtracting the upper line from the lower,
$$999x = 235$$
$$x = \tfrac{235}{999}$$

29. $.325 = \frac{325}{1000} = \frac{13}{40}$

31. $.3\overline{21} = .321212121\ldots = x$
$$1000x = 321.212121\ldots$$
$$10x = 3.212121\ldots$$
Subtracting, $990x = 318$
$$x = \frac{318}{990} = \frac{106}{330} = \frac{53}{165}$$

33.
$$
\begin{aligned}
\frac{11}{420}\cdot\frac{11}{3} - \left(\frac{13}{420}+\frac{11}{630}\right) &= \frac{121}{1260} - \frac{13}{420} - \frac{11}{630} \\
&= \frac{121}{1260} - \frac{13\cdot 3}{1260} - \frac{11\cdot 2}{1260} \\
&= \frac{121-39-22}{1260} \\
&= \frac{60}{1260} \\
&= \frac{1}{21}
\end{aligned}
$$

35.
$$
\begin{array}{lll}
\text{Let } x = .\overline{23} & \text{and} & y = .\overline{27} \\
100x = 23.\overline{23} & \text{and} & 100y = 27.\overline{27} \\
\text{Then } 99x = 23 & \text{and} & 99y = 27 \\
x = \frac{23}{99} & \text{and} & y = \frac{27}{99}
\end{array}
$$
$$x + y = \frac{23}{99} + \frac{27}{99} = \frac{50}{99}$$

37. Assume $\sqrt{3} = \frac{a}{b}$, where a and b are positive integers.

Squaring, $3 = \frac{a^2}{b^2}$ and then $3b^2 = a^2$. Any factor of a is found twice in a^2, and similarly, any factor of b is found twice in b^2. Thus there is an even number of 3's in a^2, and an even number plus 1, or an odd number of 3's in $3b^2$. But this is impossible, and so the assumption that $\sqrt{3}$ is rational is impossible.

39. Let $r = \frac{a}{b}$ and $s = \frac{c}{d}$, where a, b, c, and d are integers, and $b \neq 0$ and $d \neq 0$.

Then $r + s = \frac{a}{b} + \frac{c}{d} = \frac{ad+bc}{bd}$, a rational number, and $r \cdot s = \frac{a}{b} \cdot \frac{c}{d} = \frac{ac}{bd}$, a rational number.

41. Let $r = \sqrt{2} + \frac{2}{3}$

Then $r + \left(\frac{-2}{3}\right) = \sqrt{2}$

If r is rational, then $r + \left(\frac{-2}{3}\right)$ must be a rational number. But this is impossible, since we know that $\sqrt{2}$ is irrational. Hence, r, or $\sqrt{2} + \frac{2}{3}$, must be an irrational number.

43. Assume $\frac{2}{3}\sqrt{2}$ is rational. Let $r = \frac{2}{3}\sqrt{2}$. Then $\sqrt{2} = \frac{3}{2}r$. But an irrational number cannot be the product of rationals, so our assumption that r is rational is impossible.

45. d: $\sqrt{2}\sqrt{8} = \sqrt{16} = 4$
e: $(1+\sqrt{2})(1-\sqrt{2}) = 1 + \sqrt{2} - \sqrt{2} - 2 = -1$
f: $\sqrt{\frac{27}{75}} = \sqrt{\frac{9}{25}} = \frac{3}{5}$

i: $(.12)(.\overline{12})$ is the product of two rational numbers, which must be a rational number.

5

47.

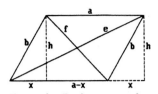

Using the Pythagorean theorem,
$b^2 = x^2 + h^2$
$e^2 = (a+x)^2 + h^2 = a^2 + 2ax + x^2 + h^2$
$f^2 = (a-x)^2 + h^2 = a^2 - 2ax + x^2 + h^2$
The sum of the two diagonals is $e^2 + f^2$, and the sum of the squares of the four sides is $2(a^2 + b^2)$.
$e^2 + f^2 = 2a^2 + 2x^2 + 2h^2$
Since $2x^2 + 2h^2 = 2b^2$,
$e^2 + f^2 = 2a^2 + 2b^2 = 2(a^2 + b^2)$

49. See the rectangular parallelepiped in Figure 15. Let the diagonals of sides with edges a and b be e, edges a and c be f, and with edges b and c be g. Then
$e^2 = a^2 + b^2$
$f^2 = a^2 + c^2$
$g^2 = b^2 + c^2$
Adding,
$e^2 + f^2 + g^2 = 2(a^2 + b^2 + c^2)$

$a^2 + b^2 + c^2 = \dfrac{e^2 + f^2 + g^2}{2}$

From Exercise 48,

$d = \sqrt{a^2 + b^2 + c^2}$

$d = \dfrac{\sqrt{e^2 + f^2 + g^2}}{\sqrt{2}}$

Problem Set 1.4

1. $420 + 431 + 580 = 420 + 580 + 431 = 1431$
Uses associative and commutative properties of addition.

3. $938 + 400 + 300 + 17 = 938 + 717 = 1655$
Uses associative property of addition.

5. $\frac{11}{13} + 43 + \frac{2}{13} + 17 = \frac{11}{13} + \frac{2}{13} + 43 + 17 = 1 + 60 = 61$
Uses commutative and associative properties of addition.

7. $6 \cdot \frac{3}{4} \cdot \frac{1}{6} \cdot 4 = 6 \cdot \frac{1}{6} \cdot \frac{3}{4} \cdot 4 = 1 \cdot 3 = 3$

Uses commutative and associative properties of multiplication.

9. $5 \cdot \frac{1}{3} \cdot \frac{2}{5} \cdot 6 \cdot \frac{1}{2} = 5 \cdot \frac{2}{5} \cdot 6 \cdot \frac{1}{3} \cdot \frac{1}{2} = 2$
Uses associative and commutative properties of multiplication.

11. $\frac{5}{6}\left(\frac{3}{4} \cdot 12\right) = \frac{5}{6}\left(12 \cdot \frac{3}{4}\right)$ Commutative property of multiplication

$= \left(\frac{5}{6} \cdot 12\right)\frac{3}{4}$ Associative property of multiplication

$= \left(12 \cdot \frac{5}{6}\right)\frac{3}{4}$ Commutative property of multiplication

13. $(2 + 3) + 4 = 4 + (2 + 3)$
                    Commutative property of addition

15. $6 + [-6 + 5] = [6 + (-6)] + 5$ Associative property of addition

          $= 0 + 5$ Additive inverse

          $= 5$ Zero is the additive identity element.

17. $-(\sqrt{5} + \sqrt{3} - 5) = -\sqrt{5} - \sqrt{3} + 5$
                    Distributive property

19. $(x+4)(x+2) = (x+4)x + (x+4)2$
                    Distributive property

21. True: $a - (b-c) = a + (-1)(b-c)$
                $= a + (-1)[b + (-c)]$
                $= a + (-1)b + (-1)(-c)$
                $= a - b + c$

23. False. Let $a = 12$, $b = 3$, and $c = 4$.
Then $a \div (b + c) = 12 \div (3 + 4) = 12 \div 7 = \frac{12}{7}$
But $a \div b + a \div c = 12 \div 3 + 12 \div 4 = 4 + 3 = 7$

25. True. $ab(a^{-1} + b^{-1}) = (ab)a^{-1} + (ab)b^{-1}$
$= (ba) \cdot a^{-1} + a \cdot (bb^{-1})$
$= b(aa^{-1}) + a(bb^{-1}) = b \cdot 1 + a \cdot 1$
$= b + a$

27. False. Let $a = 1$ and $b = 1$.
Then $a^{-1} = 1$ and $b^{-1} = 1$
$(a + b)(a^{-1} + b^{-1}) = (1 + 1)(1 + 1) = (2)(2) = 4$

29. False. Let $a = 2$ and $b = 3$.
Then $(a + b)(a + b) = (2 + 3)(2 + 3) = (5)(5) = 25$
But $a^2 + b^2 = 2^2 + 3^2 = 4 + 9 = 13$

31. False. $a \div (b \div c) \neq (a \div b) \div c$
Let $a = 3$, $b = 5$, and $c = 6$. Then

$3 \div (5 \div 6) = 3 \div \frac{5}{6} = 3 \cdot \frac{6}{5} = \frac{18}{5}$

But $(3 \div 5) \div 6 = \frac{3}{5} \div 6 = \frac{3}{5} \cdot \frac{1}{6} = \frac{1}{10}$

33. a) $4 \# 3 = 4^3 = 64$
$2(3 \# 2) = 2(3^2) = 2(9) = 18$
$2 \# (3 \# 2) = 2 \# 3^2 = 2 \# 9 = 2^9 = 512$

b) $\#$ is not commutative:
$3 \# 2 = 3^2 = 9$, but $2 \# 3 = 2^3 = 8$

c) $\#$ is not associative:
As shown in part (a),
$2 \# (3 \# 2) = 512$, but
$(2 \# 3) \# 2 = (2^3) \# 2 = 8 \# 2 = 8^2 = 64$

35. Show that $(ab)^{-1} = b^{-1}a^{-1}$.
$(ab)(b^{-1}a^{-1}) = a(bb^{-1})a^{-1}$
$= aa^{-1}$
$= 1$
So, $b^{-1}a^{-1}$ is the multiplicative inverse of ab, or
$b^{-1}a^{-1} = (ab)^{-1}$.

Show that $(a^{-1})^{-1} = a$.
Since $a^{-1} \cdot a = 1$, a is the multiplicative inverse of a^{-1}. Translated into a mathematical statement, that becomes $a = (a^{-1})^{-1}$

Show that $\left(\frac{a}{b}\right)^{-1} = \frac{b}{a}$
$\frac{a}{b} \cdot \frac{b}{a} = (ab^{-1})(ba^{-1})$
$= a(b^{-1}b)a^{-1}$
$= aa^{-1}$
$= 1$
So, $\frac{b}{a}$ is the multiplicative inverse of $\frac{a}{b}$, or $\frac{b}{a} = \left(\frac{a}{b}\right)^{-1}$

37. $0 = -(a \cdot 0) + (a \cdot 0)$ Additive inverse
$= -(a \cdot 0) + a \cdot (0 + 0)$
         Neutral element for addition
$= -(a \cdot 0) + (a \cdot 0 + a \cdot 0)$
         Distributive property
$= [-(a \cdot 0) + a \cdot 0] + a \cdot 0$
         Associative Property of addition
$= 0 + a \cdot 0$ Additive inverse property
$= a \cdot 0$ Neutral element of addition

39. $(-a)(-b) + [-(ab)] = (-a)(-b) + (-a)b$
         Problem 38
$= -a(-b + b)$ Distributive property
$= -a \cdot 0$ Additive inverse
$= 0$ Problem 37
Thus, $(-a)(-b)$ is the additive inverse of $-(ab)$. But, by definition, the additive inverse of $-(ab)$ is ab. Therefore, $(-a)(-b) = ab$.

Problem Set 1.5

1. $1.5 > -1.6$

3. Since $\sqrt{2} \approx 1.414$,
$\sqrt{2} > 1.4$

5. $\frac{1}{5} = \frac{6}{30}; \frac{1}{6} = \frac{5}{30}$
Since $\frac{6}{30} > \frac{5}{30}, \frac{1}{5} > \frac{1}{6}$

7. Since $\sqrt{2} < \sqrt{3}$,
$-\sqrt{2} > -\sqrt{3}$
$5 - \sqrt{2} > 5 - \sqrt{3}$

9. Since $\frac{1}{16} > \frac{1}{17}$,
$\frac{3\pi}{16} > \frac{3\pi}{17}$
$\frac{-3\pi}{16} < \frac{-3\pi}{17}$

11. $\left| -\pi + (-2) \right| = \left| -(\pi + 2) \right|$
$= |\pi + 2|$
$= \pi + 2$
$|-\pi| + |-2| = \pi + 2$
Thus $\left| -\pi + (-2) \right| = |-\pi| + |-2|$

13. $\frac{-3\sqrt{2}}{2} = -2.12132$
$-2 = -2$
$\frac{-\pi}{2} = -1.570796$
$\frac{3}{4} = .75$
$\sqrt{2} = 1.4142135$
$\frac{43}{24} = 1.791667$

Comparing the decimal approximations, we see that the numbers as listed above are in ascending order of size.
$\frac{-3\sqrt{2}}{2} < -2 < \frac{-\pi}{2} < \frac{3}{4} < \sqrt{2} < \frac{43}{24}$

15.
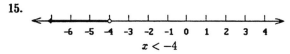
$x < -4$

17.
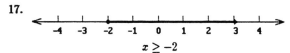
$x \geq -2$

19.
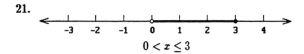
$-1 < x < 3$

21.
$0 < x \leq 3$

23.

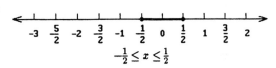

$$-\tfrac{1}{2} \le x \le \tfrac{1}{2}$$

25. $-2 \le x \le 3$

27. $x \ge 2$

29. $-2 < x \le 3$

31. $-1 < x < 2$

33. $|x| \le 4$
$$-4 \le x \le 4$$

35. $|x - 3| < 2$
$$-2 < x - 3 < 2$$
$$1 < x < 5$$

37. $|x + 1| \le 3$
$$-3 \le x + 1 \le 3$$
$$-4 \le x \le 2$$

39. $|x - 5| > 5$
$$x - 5 > 5 \quad \text{or} \quad x - 5 < -5$$
$$x > 10 \quad \text{or} \quad x < 0$$

41. a) x is less than or equal to 12:
$$x \le 12$$

b) $-11 \le x < 3$

c) The distance between x and y is less than or equal to 4 units:
$$|x - y| \le 4$$

d) The distance between x and 7 is greater than 3 units:
$$|x - 7| > 3$$

e) $|x - 5|$ represents the distance between x and 5; $|x - y|$ represents the distance between x and y
$$|x - 5| < |x - y|$$

43. $1.414 = 1.414$
$$\frac{\sqrt{2} + 1.414}{2} = 1.4141067\ldots$$
$$1.4\overline{14} = 1.4141414\ldots$$
$$\sqrt{2} = 1.414213\ldots$$
$$1.\overline{414} = 1.414414414\ldots$$
$$1.41\overline{4} = 1.414444444\ldots$$

We see from comparison of the decimal expansions that the numbers are listed above in increasing size.

45. Use a calculator to convert each fraction to a decimal. You need only enough decimal places to enable you to realize which number is larger.

a) $\frac{11}{46} = .23\ldots$ $\quad \frac{6}{25} = .24$ $\quad \frac{6}{25}$ is larger.

b) $\frac{4}{17} = .235\ldots$ $\quad \frac{7}{29.8} = .234\ldots$ $\quad \frac{4}{17}$ is larger.

c) $\frac{17.1}{85} = .201\ldots$ $\quad \frac{33}{165} = .2$ $\quad \frac{17.1}{85}$ is larger.

d) $\frac{11}{13} = .\overline{846153}$ $\quad$ The numbers are equal.

47. Since $A = \frac{1}{2}ab$, we can say
$$50 < \tfrac{1}{2}ab < 60$$
$$100 < ab < 120$$
$$100\left(\tfrac{1}{b}\right) < ab\left(\tfrac{1}{b}\right) < 120\left(\tfrac{1}{b}\right)$$
$$100\left(\tfrac{1}{b}\right) < a < 120\left(\tfrac{1}{b}\right)$$

We know that $10 < b$, making $\frac{1}{b} < \frac{1}{10}$, and
$$12 > b, \text{ making } \tfrac{1}{b} > \tfrac{1}{12}.$$

Therefore,
$$100\left(\tfrac{1}{12}\right) < 100\left(\tfrac{1}{b}\right) < a < 120\left(\tfrac{1}{b}\right) < 120\left(\tfrac{1}{10}\right)$$
$$\frac{100}{12} < a < 12$$
$$\frac{25}{3} < a < 12$$

49. If $|x - 4| < 2$,
$$-2 < x - 4 < 2$$
$$2 < x < 6$$
$$\tfrac{1}{6} < y < \tfrac{1}{2}$$

51. $|a + b| = |a| + |b|$ if a and b have the same sign.

53. $(|a| + |b|)^2 = |a|^2 + 2|a\|b| + |b|^2$
$$\ge |a|^2 + |b|^2, \text{ since } 2|a\|b| \ge 0.$$

Since $|a|^2 = a^2$ and $|b|^2 = b^2$,
$$(|a| + |b|)^2 \ge a^2 + b^2$$

Taking the principal square root of both sides of the inequality,
$$|a| + |b| \ge \sqrt{a^2 + b^2} \text{ or}$$
$$\sqrt{a^2 + b^2} \le |a| + |b|$$

Problem Set 1.6

1. $(2 + 3i) + (-4 + 5i) = (2 - 4) + (3i + 5i) = -2 + 8i$

3. $5i - (4 + 6i) = -4 + (5i - 6i)$
$$= -4 - i$$

5. $(3i - 6) + (3i + 6) = (-6 + 6) + (3i + 3i)$
$= 6i$

7. $4i^2 + 7i = 4(-1) + 7i$
$= -4 + 7i$

9. $i(4 - 11i) = 4i - 11i^2$
$= 4i - 11(-1)$
$= 11 + 4i$

11. $(3i + 5)(2i + 4) = (5 + 3i)(4 + 2i)$
$= 20 + 6i^2 + 12i + 10i$
$= 20 + 6(-1) + 22i$
$= 14 + 22i$

13. $(3i + 5)^2 = 9i^2 + 30i + 25$
$= (9)(-1) + 30i + 25$
$= 25 - 9 + 30i$
$= 16 + 30i$

15. $(5 + 6i)(5 - 6i) = 25 - 36i^2 + 30i - 30i$
$= 25 - 36(-1)$
$= 25 + 36$
$= 61$

17. $\dfrac{5 + 2i}{1 - i} = \dfrac{(5 + 2i)(1 + i)}{(1 - i)(1 + i)}$
$= \dfrac{5 + 2i^2 + 2i + 5i}{1 - i^2 - i + i}$
$= \dfrac{5 + 2(-1) + 7i}{1 - (-1)}$
$= \dfrac{3 + 7i}{2}$
$= \dfrac{3}{2} + \dfrac{7}{2}i$

19. $\dfrac{5 + 2i}{i} = \dfrac{(5 + 2i)i}{i \cdot i}$
$= \dfrac{5i + 2i^2}{-1}$
$= \dfrac{-2 + 5i}{-1}$
$= 2 - 5i$

21. $\dfrac{(2 + i)(3 + 2i)}{1 + i} = \dfrac{6 + 2i^2 + 3i + 4i}{1 + i}$
$= \dfrac{6 - 2 + 7i}{1 + i}$
$= \dfrac{4 + 7i}{1 + i}$
$= \dfrac{(4 + 7i)(1 - i)}{(1 + i)(1 - i)}$
$= \dfrac{4 - 7i^2 + 7i - 4i}{1 - i^2}$
$= \dfrac{4 + 7 + 3i}{1 - (-1)}$
$= \dfrac{11 + 3i}{2}$
$= \dfrac{11}{2} + \dfrac{3}{2}i$

23. $(2 - i)^{-1} = \dfrac{1}{2 - i}$
$= \dfrac{(1)(2 + i)}{(2 - i)(2 + i)}$
$= \dfrac{2 + i}{4 - i^2} = \dfrac{2 + i}{4 + 1}$
$= \dfrac{2 + i}{5} = \dfrac{2}{5} + \dfrac{1}{5}i$

25. $(\sqrt{3} + i)^{-1} = \dfrac{1}{\sqrt{3} + i}$
$= \dfrac{\sqrt{3} - i}{(\sqrt{3} + i)(\sqrt{3} - i)}$
$= \dfrac{\sqrt{3} - i}{3 - i^2}$
$= \dfrac{\sqrt{3} - i}{4}$
$= \dfrac{\sqrt{3}}{4} - \dfrac{1}{4}i$

27. $\dfrac{2 + 3i}{2 - i} = (2 + 3i)(2 - i)^{-1}$
$= (2 + 3i)(\tfrac{2}{5} + \tfrac{1}{5}i)$
$= \tfrac{4}{5} + \tfrac{3}{5}i^2 + \tfrac{6}{5}i + \tfrac{2}{5}i$
$= \tfrac{4}{5} - \tfrac{3}{5} + \tfrac{8}{5}i$
$= \tfrac{1}{5} + \tfrac{8}{5}i$

29. $\dfrac{4 - i}{\sqrt{3} + i} = (4 - i)(\sqrt{3} + i)^{-1}$
$= (4 - i)\left(\dfrac{\sqrt{3}}{4} - \dfrac{1}{4}i\right)$
$= \sqrt{3} + \tfrac{1}{4}i^2 - \dfrac{\sqrt{3}}{4}i - i$
$= \left(\sqrt{3} - \tfrac{1}{4}\right) + \left(\dfrac{-\sqrt{3}}{4} - 1\right)i$

31. $i^{94} = i^{92} \cdot i^2 = (i^4)^{23} \cdot (-1) = 1^{23}(-1) = -1$

33. $(-i)^{17} = (-1)^{17}i^{17} = -i^{17} = -(i^{16}i) = -(1 \cdot i) = -i$

35. $\dfrac{(3i)^{16}}{(9i)^5} = \dfrac{3^{16}i^{16}}{9^5i^5}$

$\qquad = \dfrac{3^{16}i^{16}}{(3^2)^5i^5}$

$\qquad = \dfrac{3^{16}i^{16}}{3^{10}i^5}$

$\qquad = 3^6 i^{11}$ (Subtract exponents)

$\qquad = 3^6 \cdot i^8 \cdot i^3$

$\qquad = 3^6(1)(-i)$

$\qquad = -729i$

37. $(1+i)^3 = (1+i)(1+i)(1+i)$

$\qquad = (1+i)(1+i^2+i+i)$

$\qquad = (1+i)(1-1+2i)$

$\qquad = (1+i)(2i)$

$\qquad = 2i + 2i^2$

$\qquad = 2i - 2$

$\qquad = -2 + 2i$

39. The fourth roots of 1 are:

$\quad i: \ i^4 = i^2 \cdot i^2 = (-1)(-1) = 1$

$\quad -i: \ (-i)^4 = (-i)(-i)(-i)(-i) = i^4 = 1$

$\quad 1: \ (1)^4 = 1$

$\quad -1: \ (-1)^4 = (-1)(-1)(-1)(-1) = 1$

41. $(1-i)^4 = (1-i)^2(1-i)^2$

$\qquad = (1-2i+i^2)(1-2i+i^2)$

$\qquad = (1-1-2i)(1-1-2i)$

$\qquad = (-2i)(-2i)$

$\qquad = 4i^2$

$\qquad = 4(-1)$

$\qquad = -4$

43. a) $\quad 2+3i-i(4-3i) = 2+3i-4i+3i^2$

$\qquad\qquad\qquad\qquad\quad = 2+3i-4i-3$

$\qquad\qquad\qquad\qquad\quad = -1-i$

b) $\qquad \dfrac{(i^7+2i^2)}{i^3} = \dfrac{-i-2}{-i}$

$\qquad\qquad\qquad = 1 + \dfrac{2}{i}$

$\qquad\qquad\qquad = 1 + \dfrac{2i}{i^2}$

$\qquad\qquad\qquad = 1 - 2i$

c) $3+4i+(3+4i)(-1+2i) = 3+4i-3+8i^2-4i+6i$

$\qquad\qquad\qquad\qquad\qquad\quad = -8+6i$

d) $\quad i^{14} + \dfrac{5+2i}{5-2i} = i^{12} \cdot i^2 + \dfrac{(5+2i)(5+2i)}{(5-2i)(5+2i)}$

$\qquad\qquad\qquad = (1)(-1) + \dfrac{25+4i^2+10i+10i}{25-4i^2}$

$= -1 + \dfrac{25-4+20i}{29}$

$= \left(-1 + \dfrac{21}{29}\right) + \dfrac{20}{29}i$

$= \dfrac{-8}{29} + \dfrac{20}{29}i$

e) $\dfrac{5-2i}{3+4i} + \dfrac{5+2i}{3-4i} = \dfrac{(5-2i)(3-4i)}{(3+4i)(3-4i)} + \dfrac{(5+2i)(3+4i)}{(3-4i)(3+4i)}$

$\qquad = \dfrac{15+8i^2-6i-20i}{9-16i^2} + \dfrac{15+8i^2+6i+20i}{9-16i^2}$

$\qquad = \dfrac{15-8-26i}{25} + \dfrac{15-8+26i}{25}$

$\qquad = \dfrac{7}{25} - \dfrac{26}{25}i + \dfrac{7}{25} + \dfrac{26}{25}i = \dfrac{14}{25}$

f) $\qquad \dfrac{(3+2i)(3-2i)}{2\sqrt{3}+i} = \dfrac{9-4i^2}{2\sqrt{3}+i}$

$\qquad\qquad\qquad = \dfrac{9+4}{2\sqrt{3}+i}$

$\qquad\qquad\qquad = \dfrac{13(2\sqrt{3}-i)}{(2\sqrt{3}+i)(2\sqrt{3}-i)}$

$\qquad\qquad\qquad = \dfrac{26\sqrt{3}-13i}{12-i^2}$

$\qquad\qquad\qquad = \dfrac{26\sqrt{3}-13i}{13}$

$\qquad\qquad\qquad = 2\sqrt{3}-i$

45. a) $\quad (2+i)(2-i)(a+bi) = (4-i^2)(a+bi)$

$\qquad\qquad\qquad\qquad\qquad = 5(a+bi) = 5a+5bi$

Since $5a+5bi = 10-4i$,

$\quad 5a = 10$ and $\quad 5b = -4$

$\qquad a = 2 \qquad\qquad b = \dfrac{-4}{5}$

b) $\qquad (2+i)(a-bi) = 2a - bi^2 + ai - 2bi$

$\qquad\qquad\qquad\qquad = (2a+b) + (a-2b)i$

Since $(2a+b) + (a-2b) = 8-i$,

$\quad 2a+b = 8$ and $a-2b = -1$

Solve these two equations simultaneously:

$\quad 4a + 2b = 16$

$\quad \underline{a - 2b = -1}$

$\quad 5a \qquad\ = 15$

$\qquad a = 3$

Let $a = 3$ in $2a+b = 8$:

$\quad 2(3) + b = 8$

$\quad 6 + b = 8$

$\qquad b = 2$

47. a) $\left(-\frac{1}{2}+\frac{\sqrt{3}}{2}i\right)^3 = \left(-\frac{1}{2}+\frac{\sqrt{3}}{2}i\right)\left(-\frac{1}{2}+\frac{\sqrt{3}}{2}i\right)\left(-\frac{1}{2}+\frac{\sqrt{3}}{2}i\right)$

$\quad\quad = \left(\frac{1}{2}\right)\left(\frac{1}{2}\right)\left(\frac{1}{2}\right)(-1+\sqrt{3}i)(-1+\sqrt{3}i)(-1+\sqrt{3}i)$

$\quad\quad = \frac{1}{8}(1+3i^2-\sqrt{3}i-\sqrt{3}i)(-1+\sqrt{3}i)$

$\quad\quad = \frac{-2}{8}(1+\sqrt{3}i)(-1+\sqrt{3}i)$

$\quad\quad = -\frac{1}{4}(-1+3i^2-\sqrt{3}i+\sqrt{3}i)$

$\quad\quad = -\frac{1}{4}(-4)$

$\quad\quad = 1$

b) $1+\left(-\frac{1}{2}+\frac{\sqrt{3}}{2}i\right)+\left(-\frac{1}{2}+\frac{\sqrt{3}}{2}i\right)^2$

$\quad = \frac{1}{2}+\frac{\sqrt{3}}{2}i+\left(-\frac{1}{2}+\frac{\sqrt{3}}{2}i\right)\left(-\frac{1}{2}+\frac{\sqrt{3}}{2}i\right)$

$\quad = \frac{1}{2}+\frac{\sqrt{3}}{2}i+\left(\frac{1}{2}\right)\left(\frac{1}{2}\right)(-1+\sqrt{3}i)(-1+\sqrt{3}i)$

$\quad = \frac{1}{2}+\frac{\sqrt{3}}{2}i+\frac{1}{4}(1+3i^2-\sqrt{3}i-\sqrt{3}i)$

$\quad = \frac{1}{2}+\frac{\sqrt{3}}{2}i+\frac{1}{4}(-2-2\sqrt{3}i)$

$\quad = \frac{1}{2}+\frac{\sqrt{3}}{2}i-\frac{1}{2}-\frac{\sqrt{3}}{2}i$

$\quad = 0$

c) $(1-x)(1-x^2) = 1-x-x^2+x^3$

In part (b) we found that

$\quad 1+x+x^2 = 0$

So,

$\quad\quad x+x^2 = -1$

$\quad\quad -x-x^2 = 1$

$\quad\quad 1-x-x^2 = 2$

In part (a) we found that $x^3 = 1$.

Using these results, we see that

$\quad (1-x)(1-x^2) = 1-x-x^2+x^3$

$\quad\quad\quad\quad\quad\quad\quad = 2+1 = 3$

49. Assume that the square roots of i have the form $a+bi$, where a and b are real numbers. Then

$$(a+bi)^2 = i$$
$$a^2+2abi+b^2i^2 = i$$
$$(a^2-b^2)+2abi = i$$

Equating the real and imaginary parts of the complex numbers on each side of the equation, we get

$\quad a^2-b^2 = 0$ and $2ab = 1$

$\quad\quad a^2 = b^2 \quad\quad ab = \frac{1}{2}$

$\quad\quad a = b$ or $a = -b$

Replace a with b in $ab = \frac{1}{2}$:

$\quad\quad b^2 = \frac{1}{2}$

$\quad\quad b = \pm\sqrt{\frac{1}{2}} = \pm\frac{\sqrt{2}}{2}$

$\quad\quad a = b = \pm\frac{\sqrt{2}}{2}$

Replacing a with $-b$ in $ab = \frac{1}{2}$ does not yield a real value for b:

$\quad\quad (-b)b = \frac{1}{2}$

$\quad\quad -b^2 = \frac{1}{2}$

$\quad\quad b^2 = -\frac{1}{2}$

$\quad\quad b = \pm\sqrt{\frac{1}{2}}i$

Therefore, the two square roots of i are

$$\frac{\sqrt{2}}{2}+\frac{\sqrt{2}}{2}i \text{ and } -\frac{\sqrt{2}}{2}-\frac{\sqrt{2}}{2}i.$$

51. Let $x = a+bi$ and $y = c+di$. Then $\bar{x} = a-bi$ and $\bar{y} = c-di$.

a) $\quad x+y = (a+c)+(b+d)i$

$\quad\quad \overline{x+y} = (a+c)-(b+d)i$

$\quad\quad\quad\quad\quad = a+c-bi-di$

$\quad\quad\quad\quad\quad = (a-bi)+(c-di)$

$\quad\quad\quad\quad\quad = \bar{x}+\bar{y}$

b) $\quad xy = (a+bi)(c+di)$

$\quad\quad\quad = ac+bdi^2+adi+bci$

$\quad\quad\quad = (ac-bd)+(ad+bc)i$

$\quad\quad \overline{xy} = (ac-bd)-(ad+bc)i$

$\quad\quad \bar{x}\,\bar{y} = (a-bi)(c-di)$

$\quad\quad\quad\quad = ac+bdi^2-adi-bci$

$\quad\quad\quad\quad = (ac-bd)-(ad+bc)i$

$\quad\quad\quad\quad = \overline{xy}$

c) $x^{-1} = \dfrac{1}{a+bi} = \dfrac{(1)(a-bi)}{(a+bi)(a-bi)} = \dfrac{a-bi}{a^2-b^2i^2}$

$\quad\quad\quad = \dfrac{a}{a^2+b^2}-\dfrac{b}{a^2+b^2}i$

$\quad \overline{(x^{-1})} = \dfrac{a}{a^2+b^2}+\dfrac{b}{a^2+b^2}i$

$\quad (\bar{x})^{-1} = \dfrac{1}{a-bi} = \dfrac{(1)(a+bi)}{(a-bi)(a+bi)} = \dfrac{a+bi}{a^2-b^2i^2}$

$\quad\quad\quad = \dfrac{a}{a^2+b^2}+\dfrac{b}{a^2+b^2}i = \overline{(x^{-1})}$

d) $\quad \overline{\dfrac{x}{y}} = \overline{x(\bar{y})^{-1}}$, by the definition of division

$\quad\quad = \overline{\bar{x}(y^{-1})}$, by the result of part c

$\quad\quad = \overline{(xy^{-1})}$, by the result of part b

$\quad\quad = \overline{\left(\dfrac{x}{y}\right)}$

11

Chapter 1 Review Problem Set

1. True, by the distributive law.

2. False. Try $x = 1$: $\frac{2 \cdot 1 + 1}{2} = \frac{3}{2} \neq 1 + 1 = 2$

3. False. By definition, $\sqrt{x^2} = |x|$.
   Notice, if $x = -2$, $\sqrt{x^2} = \sqrt{(-2)^2} = \sqrt{4}$, not -2.

4. True

5. False. Let $a = -3$ and $b = -1$.
   Then $a < b$, but
   $|a| = 3$ and $|b| = 1$, making
   $|a| > |b|$

6. True

7. False. Let $a = 1$ and $b = 1$.
   Then $(a + b)^2 = (1 + 1)^2$
   $ = 2^2 = 4$
   But $\quad a^2 + b^2 = \; = 1^2 + 1^2 = 2$.

8. False. Let $x = -1$. Then $|-x| = |-(-1)| = 1 \neq x$.

9. True. $\frac{a}{b} \cdot c = \frac{a}{b} \cdot \frac{c}{1} = \frac{ac}{b}$

10. False. Try $a = b = 1$: $(1+1)^{-1} = \frac{1}{2} \neq \frac{1}{1} + \frac{1}{1} = 2$

11. $a \cdot (b \cdot c) = (a \cdot b) \cdot c$

12. If a and b are real numbers, then $a + b = b + a$.

13. A rational number is a number that can be expressed as the ratio of two integers $\frac{p}{q}$, where $q \neq 0$.

14. a) $\frac{11}{24} - \frac{7}{12} + \frac{4}{3} = \frac{11}{24} - \frac{14}{24} + \frac{32}{24} = \frac{43 - 14}{24} = \frac{29}{24}$

b) $\frac{9}{28} \cdot \frac{7}{18} \cdot \frac{8}{5} = \frac{1}{4} \cdot \frac{1}{2} \cdot \frac{8}{5} = \frac{1}{5}$

c) $\frac{2}{3}\left[\frac{1}{4} - \left(\frac{5}{6} - \frac{2}{3}\right)\right] = \frac{2}{3}\left[\frac{1}{4} - \frac{5}{6} + \frac{2}{3}\right] = \frac{2}{3}\left[\frac{3}{12} - \frac{10}{12} + \frac{8}{12}\right]$
   $ = \frac{2}{3}\left[\frac{1}{12}\right] = \frac{1}{18}$

d) $3\left[\frac{5}{4}\left(\frac{2}{3} + \frac{8}{15} - \frac{1}{5}\right) + \frac{1}{6}\right] = 3\left[\frac{5}{6} + \frac{2}{3} - \frac{1}{4} + \frac{1}{6}\right]$
   $ = 3\left[\frac{10}{12} + \frac{8}{12} - \frac{3}{12} + \frac{2}{12}\right] = \frac{17}{4}$

15. a) $\frac{\frac{11}{30}}{\frac{33}{25}} = \frac{11}{30} \cdot \frac{25}{33} = \frac{11 \cdot 25}{30 \cdot 33} = \frac{11 \cdot 5 \cdot 5}{5 \cdot 6 \cdot 11 \cdot 3} = \frac{5}{18}$

b) $\frac{\frac{3}{4} - \frac{1}{12} + \frac{3}{8}}{\frac{3}{4} + \frac{5}{12} - \frac{7}{8}} = \frac{24\left(\frac{3}{4} - \frac{1}{12} + \frac{3}{8}\right)}{24\left(\frac{3}{4} + \frac{5}{12} - \frac{7}{8}\right)}$
   $ = \frac{18 - 2 + 9}{18 + 10 - 21} = \frac{25}{7}$

c) $3 + \frac{\frac{3}{4} - \frac{7}{8}}{\frac{5}{12}} = 3 + \frac{12}{5}\left(\frac{3}{4} - \frac{7}{8}\right)$
   $ = 3 + \frac{12}{5} \cdot \frac{3}{4} - \frac{12}{5} \cdot \frac{7}{8}$
   $ = 3 + \frac{9}{5} - \frac{21}{10}$
   $ = \frac{30}{10} + \frac{18}{10} - \frac{21}{10}$
   $ = \frac{27}{10}$

d) $\frac{72}{25}\left(\frac{9}{4} - \frac{11}{6}\right)^3 = \frac{72}{25}\left(\frac{27}{12} - \frac{22}{12}\right)^3$
   $ = \frac{72}{25}\left(\frac{5}{12}\right)^3$
   $ = \frac{72 \cdot 5 \cdot 5 \cdot 5}{25 \cdot 12 \cdot 12 \cdot 12}$
   $ = \frac{12 \cdot 6 \cdot 5 \cdot 5 \cdot 5}{5 \cdot 5 \cdot 12 \cdot 6 \cdot 2 \cdot 12} = \frac{5}{24}$

16. Let $x = 2 - \sqrt{3}$. If x were rational, there would exist integers m and n ($n \neq 0$) such that $x = \frac{m}{n}$ which would imply that $\quad 2 - \sqrt{3} = \frac{m}{n}$
   $$\sqrt{3} = 2 - \frac{m}{n}$$
   $$\sqrt{3} = \frac{2n - m}{n}$$
   $$\sqrt{3} = \text{a rational number}$$
   Since we know that $\sqrt{3}$ is irrational, $2 - \sqrt{3}$ must not be rational.

17. Let x represent the length of the base in centimeters.
   Then $x - 5$ represents the length of the altitude.
   Using the formula Area $= \frac{1}{2}bh$
   with $b = x$ and $h = x - 5$,
   Area$= \frac{1}{2}x(x - 5)$

18. Let $y = $ length of the other side of the rectangle.
   Since the perimeter is 20 feet,
   $$2x + 2y = 20$$
   $$2y = 20 - 2x$$
   $$y = 10 - x$$
   Area $= xy = x(10 - x)$

19. A diagram will help.

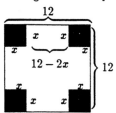

The rectangular box will have length $12 - 2x$, width $12 - 2x$, and height x.

Using the formula Volume $= lwh$,
$$\text{Volume} = (12 - 2x)(12 - 2x)(x)$$
$$= x(12 - 2x)^2$$

20. From the Pythagorean Theorem, the altitude a satisfies $x^2 + \left(\frac{x}{2}\right)^2 = a^2$, $a = \sqrt{x^2 + \left(\frac{x}{2}\right)^2}$. Therefore, the area of the triangle equals $\frac{1}{2}xa = \frac{1}{2}x\sqrt{x^2 + \left(\frac{x}{2}\right)^2}$.

21. Use the formula $S = 2\pi r^2 + 2\pi rh$ replacing r with x and h with y:
$$S = 2\pi x^2 + 2\pi xy$$
$$= 2\pi x(x + y)$$

22. The area of the base is given by xy. The height is given as $\frac{1}{2}x$. The area of each of the two ends of the box is $\frac{1}{2}x^2$ and the area of each of the two sides is given by $\frac{1}{2}xy$. Adding these areas gives the total surface area: $xy + x^2 + xy = x^2 + 2xy$.

23. $x - y$ represents the speed of the plane during its eastbound flight.

$x + y$ represents the speed of the plane during its westbound flight.

Using the formula $t = \frac{d}{r}$,

$\frac{100}{x - y}$ represents the number of hours it will take for the plane to fly 100 miles east.

$\frac{100}{x + y}$ represents the number of hours it will take for the plane to fly 100 miles west.

Total time $= \dfrac{100}{x - y} + \dfrac{100}{x + y}$

24. The area of the bottom rectangular portion is xy. The top semicircle has a radius of $\frac{1}{2}x$ inches and its area is then equal to $\frac{1}{2}\pi(\frac{1}{2}x)^2 = \frac{\pi x^2}{8}$.

Total area $= xy + \dfrac{\pi x^2}{8}$.

25. The area of the rectangle is xy, but to compute the area of the triangle, we need to find its altitude. Dropping its altitude to the base, x, of the triangle we see that it is one leg of a right triangle whose other leg is $\frac{x}{2}$ and whose hypotenuse is z.

Using the Pythagorean theorem,

$$\text{altitude} = \sqrt{z^2 - \left(\frac{x}{2}\right)^2}$$
$$= \sqrt{\frac{4z^2 - x^2}{4}}$$
$$= \frac{1}{2}\sqrt{4z^2 - x^2}$$

The area of the triangle is $\frac{1}{2}$ the base times the altitude:

$$\frac{1}{2}x\left(\frac{1}{2}\right)\sqrt{4z^2 - x^2} = \frac{1}{4}x\sqrt{4z^2 - x^2}$$

The total area of the figure is
$$xy + \frac{1}{4}x\sqrt{4z^2 - x^2}$$

26. $500 = 5^3 \cdot 2^2$; $360 = 2^3 \cdot 3^2 \cdot 5$

27. $500 = 2 \cdot 250 = 2 \cdot 2 \cdot 125 = 2 \cdot 2 \cdot 5 \cdot 25 = 2 \cdot 2 \cdot 5 \cdot 5 \cdot 5$
$360 = 2 \cdot 180 = 2 \cdot 2 \cdot 90 = 2 \cdot 2 \cdot 2 \cdot 45$
$\qquad = 2 \cdot 2 \cdot 2 \cdot 3 \cdot 15 = 2 \cdot 2 \cdot 2 \cdot 3 \cdot 3 \cdot 5$
lcm $(500, 360) = 2 \cdot 2 \cdot 2 \cdot 3 \cdot 3 \cdot 5 \cdot 5 \cdot 5 = 9000$

28. $\dfrac{7}{500} - \dfrac{7}{360} = 7 \cdot \left(\dfrac{1}{500} - \dfrac{1}{360}\right) = 7 \cdot \left(\dfrac{1}{2^2 \cdot 5^3} - \dfrac{1}{2^3 \cdot 3^2 \cdot 5}\right)$
$\qquad = 7 \cdot \left(\dfrac{2 \cdot 3^2}{2^3 \cdot 3^2 \cdot 5^3} - \dfrac{5^2}{2^3 \cdot 3^2 \cdot 5^3}\right) = 7 \cdot \left(\dfrac{18 - 25}{2^3 \cdot 3^2 \cdot 5^3}\right)$
$\qquad = \dfrac{7 \cdot (-7)}{2^3 \cdot 3^2 \cdot 5^3} = -\dfrac{49}{9000}$

29. $15 \cdot 40 \cdot 18 = 3 \cdot 5 \cdot 2 \cdot 2 \cdot 2 \cdot 5 \cdot 2 \cdot 3 \cdot 3$
$\qquad\qquad\quad = 2 \cdot 2 \cdot 2 \cdot 2 \cdot 3 \cdot 3 \cdot 3 \cdot 5 \cdot 5$
$63 \cdot 72 = 3 \cdot 3 \cdot 7 \cdot 2 \cdot 2 \cdot 2 \cdot 3 \cdot 3 = 2 \cdot 2 \cdot 2 \cdot 3 \cdot 3 \cdot 3 \cdot 3 \cdot 7$
lcm$(15 \cdot 40 \cdot 18, 63 \cdot 72) = 2 \cdot 2 \cdot 2 \cdot 2 \cdot 3 \cdot 3 \cdot 3 \cdot 3 \cdot 5 \cdot 5 \cdot 7$
$\qquad\qquad\qquad\qquad = 2^4 \cdot 3^4 \cdot 5^2 \cdot 7$

30. $15^2 \cdot 40^3 \cdot 18^2 = (3 \cdot 5)^2 \cdot (2^3 \cdot 5)^3 \cdot (2 \cdot 3^2)^2$
$\qquad\qquad\qquad = 2^{11} \cdot 3^6 \cdot 5^5$
$63^4 \cdot 72^2 = (3^2 \cdot 7)^4 \cdot (2^3 \cdot 3^2)^2 = 2^6 \cdot 3^{11} \cdot 7^4$
lcm$(15^2 \cdot 40^3 \cdot 18^2, 63^4 \cdot 72^2) = 2^{11} \cdot 3^{11} \cdot 5^5 \cdot 7^4$

31. $\dfrac{5}{11}$:

$$11 \overline{\smash{\big)}\,5.000} \quad\begin{array}{r} .454\cdots \\ \hline \end{array}$$
$$\begin{array}{r} 4\ 4 \\ \hline 60 \\ 55 \\ \hline 50 \\ 44 \\ \hline 6\cdots \end{array}$$

$\dfrac{5}{11} = .\overline{45}$

$\frac{5}{13}$:

$$
\begin{array}{r}
.3846153\cdots \\
13\overline{)5.000000} \\
\underline{3\ 9} \\
1\ 10 \\
\underline{1\ 04} \\
60 \\
\underline{52} \\
80 \\
\underline{78} \\
20 \\
\underline{13} \\
70 \\
\underline{65} \\
50 \\
\underline{39} \\
11\cdots
\end{array}
$$

$\frac{5}{13} = .\overline{384615}$

32. Let $x = 0.\overline{468}$. Then $1000x = 468.\overline{468}$
$$x = 0.\overline{468}$$
$$999x = 468$$
$$x = \frac{468}{999} = \frac{52}{111}$$

Let $y = 3.2\overline{45}$. Then $1000y = 3245.\overline{45}$
$$10y = 32.\overline{45}$$
$$990y = 3213$$
$$y = \frac{3213}{990} = \frac{357}{110}$$

33. $\frac{16}{5} = \frac{16}{5} \cdot \frac{4}{4} = \frac{64}{20}$

$\frac{13}{4} = \frac{13}{4} \cdot \frac{5}{5} = \frac{65}{20}$

Since $\frac{65}{20} > \frac{64}{20}, \frac{13}{4} > \frac{16}{5}$

34. $1.4, \sqrt{2}, 1.\overline{4}, \frac{29}{20}, \frac{13}{8}$

35.
$$a < b$$
$$a + 2a < 2a + b$$
$$3a < 2a + b$$
$$\frac{3a}{3} < \frac{2a+b}{3}$$
$$a < \frac{2a+b}{3}$$

36.
$$|x - 9| \le 2.5$$
$$-2.5 \le x - 9 \le 2.5$$
$$6.5 \le x \le 11.5$$

$$|x + 3| < 5$$
$$-5 < x + 3 < 5$$
$$-8 < x < 2$$

$$|x - 2| > 6$$
$$x - 2 > 6 \text{ or } x - 2 < -6$$
$$x > 8 \text{ or } x < -4$$

37. 8 is midway between 4 and 12.
12 is 4 units to the right of 8, and 4 is 4 units to the left of 8. x is less than 4 units away from 8.
$$|x - 8| < 4$$

38. If a and b are both positive, then $a < b$ would imply $\frac{1}{a} > \frac{1}{b}$.

39. $5x$ represents the cost of x units of product A.
$12y$ represents the cost of y units of product B.
$5x + 12y$ represents the total cost of production.
$$5x + 12y \le 10,000$$

▷ For Problems 40–42, recall that
$i^2 = -1, i^3 = -i, i^4 = 1, i^5 = i, i^6 = -1, i^7 = -i$

40. a) $(4 - 5i) + (-8 + 3i) - (6 - 4i)$
$= (4 - 8 - 6) + (-5 + 3 + 4)i = -10 + 2i$

b) $(5 + 2i)(5 - 2i) - 2i^3 + (2i)^5$
$= (25 + 4) + 2i + 32i^5 = 29 + 2i + 32i = 29 + 34i$

c) $3 + 2i + \frac{2 - 5i}{3 + 2i} = 3 + 2i + \frac{(2 - 5i)(3 - 2i)}{9 + 4}$
$= 3 + 2i + \frac{6 - 4i - 10 - 15i}{13} = 3 + 2i + \frac{-4 - 19i}{13}$
$= \left(3 - \frac{4}{13}\right) + \left(2 - \frac{19}{13}\right)i = \frac{35}{13} + \frac{7}{13}i$

d) $(3 - 3i)^3 = 3^3(1 - i)^3 = 27(1 - i)^2(1 - i)$
$= 27(1 - 2i + i^2)(1 - i) = 27(-2i)(1 - i)$
$= 27(-2i + 2i^2) = 27(-2i - 2)$
$= -54i - 54 = -54 - 54i$

e) $(2 + i)^{-2} = \frac{1}{(2 + i)^2} = \frac{1}{4 + 4i + i^2} = \frac{1}{3 + 4i}$
$= \frac{1(3 - 4i)}{(3 + 4i)(3 - 4i)} = \frac{3 - 4i}{9 + 16} = \frac{3}{25} - \frac{4}{25}i$

41. Making the appropriate substitutions, we find
$1 + 2i + 3i^2 + 4i^3 + 5i^4 - 4i^5 - 3i^6 - 4i^7$
$= 1 + 2i - 3 - 4i + 5 - 4i + 3 + 4i$
$= 6 - 2i$

42. To show that $3 + 2i$ is a solution to the equation $2x^2 - 3x - 1 - 18i = 0$, we substitute $x = 3 + 2i$ into the left-hand side, and we obtain
$2(3 + 2i)^2 - 3(3 + 2i) - 1 - 18i$
$= 2(9 + 12i - 4) - 9 - 6i - 1 - 18i$
$= 10 + 24i - 10 - 6i - 18i = 0\checkmark$

43. Let x, $x+1$, and $x+2$ represent the three consecutive integers. Since the sum of the integers is 225,
$$x + (x+1) + (x+2) = 225$$
$$3x + 3 = 225$$
$$3x = 222$$
$$x = 74, \text{ the smallest integer}$$

44. Let w denote the width of the rectangle. Consequently, the length equals $w+4$ and the perimeter is $2w + 2(w+4) = \quad 4w + 8 = 72$
$$4w = 64$$
$$w = 16$$
The length would then equal 20 meters, and the area equals $16 \cdot 20 = 320$ square meters.

45. Let x represent the width of the rectangle, in feet. Then $2x$ represents the length of the rectangle.
Since the diagonal is the hypotenuse of a right triangle, whose legs are of length x and $2x$,
$$x^2 + (2x)^2 = (\sqrt{125})^2$$
$$x^2 + 4x^2 = 125$$
$$5x^2 = 125$$
$$x^2 = 25$$
$$x = 5 \text{ or } x = -5$$
Since a negative length makes no sense, reject $x = -5$.
Width $= x = 5$ feet
Length $= 2x = 10$ feet

46. Let $s =$ length of the side of the smaller square. The larger square then has a side of length $2s$. Their respective areas are s^2 and $(2s)^2 = 4s^2$. If their areas differ by 108 square inches,
$$4s^2 - s^2 = 3s^2 = 108$$
$$s^2 = 36$$
$$s = 6 \text{ inches}$$

47. The original cylinder has $V = \pi r^2 h$.
Since r is halved, replace r with $\frac{r}{2}$;
since h is doubled, replace h with $2h$.
The new cylinder has $V = \pi \left(\frac{r}{2}\right)^2 (2h)$
$$= \frac{\pi r^2 (2h)}{4}$$
$$= \frac{\pi r^2 h}{2}$$

We see that the new volume is half the original volume.

48. Let $m =$ no. of coins Mary received,
$h =$ no. of coins Helen received, and
$e =$ no. of coins Eloise received. We are given that
$$m = h+1, \quad h = e+1 \text{ and } \frac{m+h+e+57}{4} = 30.$$
Since we are looking for m, we solve the above equation for m. Hence $h = m-1$, $e = h-1 = m-2$
and $\quad \dfrac{m + (m-1) + (m-2) + 57}{4} = 30$
$$3m + 54 = 120$$
$$3m = 66$$
$$m = 22$$

Mary got 22 coins.

49. Let x represent the speed at which Jenny should walk the second two miles.
Using $t = \frac{d}{r}$,
$\frac{2}{x}$ represents the number of hours it takes Jenny to walk the second two miles.
Since 40 minutes $= \frac{2}{3}$ hrs,
$\frac{2}{x} + \frac{2}{3}$ represents the total time of Jenny's trip, in hours.
The average rate for the entire 4-mile trip is to be 3.5 mph. Using $d = rt$,
$$4 = 3.5 \left(\frac{2}{x} + \frac{2}{3}\right)$$
$$4 = \frac{7}{x} + \frac{7}{3}$$
$$\frac{5}{3} = \frac{7}{x}$$
$$5x = 21$$
$$x = \frac{21}{5} = 4.2$$
Jenny should walk the second two miles at a rate of 4.2 mph.

50. If a car averages 60 mph for two miles, then it has used two minutes. However, if the car traveled only 30 mph for the first mile, it has already used up its 2 minutes. Therefore, it is impossible to average 60 mph on a 2-mile stretch if it averaged only 30 mph for the first mile.

Chapter 2: Exponents and Polynomials

Problem Set 2.1

1. $\dfrac{3^2 \cdot 3^5}{3^4} = \dfrac{3^7}{3^4} = 3^3 = 27$

3. $\dfrac{(2^2)^4}{2^6} = \dfrac{2^8}{2^6} = 2^2 = 4$

5. $\dfrac{(3^2 \cdot 2^3)^3}{6^6} = \dfrac{3^6 \cdot 2^9}{(3 \cdot 2)^6}$

$\qquad = \dfrac{3^6 \cdot 2^9}{3^6 \cdot 2^6}$

$\qquad = 2^3$

$\qquad = 8$

7. $5^{-2} = \dfrac{1}{5^2} = \dfrac{1}{25}$

9. $-5^{-2} = -\dfrac{1}{5^2} = -\dfrac{1}{25}$

11. $(-2)^{-5} = \dfrac{1}{(-2)^5} = \dfrac{1}{-32} = -\dfrac{1}{32}$

13. $\left(\dfrac{-2}{3}\right)^{-3} = \dfrac{(-2)^{-3}}{3^{-3}} = \dfrac{3^3}{(-2)^3} = \dfrac{27}{-8} = -\dfrac{27}{8}$

15. $\dfrac{2^{-2}}{3^{-3}} = \dfrac{3^3}{2^2} = \dfrac{27}{4}$

17. $\left[\left(\dfrac{3}{2}\right)^{-2}\right]^{-2} = \left(\dfrac{3}{2}\right)^4$

$\qquad = \dfrac{3^4}{2^4}$

$\qquad = \dfrac{81}{16}$

19. $\dfrac{2^{-2} - 4^{-3}}{(-2)^2 + (-4)^0} = \dfrac{\dfrac{1}{2^2} - \dfrac{1}{4^3}}{4 + 1}$

$\qquad = \dfrac{\dfrac{1}{4} - \dfrac{1}{64}}{5}$

$\qquad = \dfrac{\dfrac{16}{64} - \dfrac{1}{64}}{5}$

$\qquad = \dfrac{\dfrac{15}{64}}{5}$

$\qquad = \dfrac{15}{64} \cdot \dfrac{1}{5}$

$\qquad = \dfrac{3}{64}$

21. $3^3 \cdot 2^{-3} \cdot 3^{-5} = 3^{-2} \cdot 2^{-3}$

$\qquad = \dfrac{1}{3^2 \cdot 2^3}$

$\qquad = \dfrac{1}{9 \cdot 8}$

$\qquad = \dfrac{1}{72}$

23. $(3x)^4 = 3^4 x^4 = 81x^4$

25. $(xy^2)^6 = x^6 y^{12}$

27. $\left(\dfrac{2x^2 y}{w^3}\right)^4 = \dfrac{2^4 x^8 y^4}{w^{12}}$

$\qquad = \dfrac{16x^8 y^4}{w^{12}}$

29. $\left(\dfrac{3x^{-1}y^2}{z^2}\right)^3 = \dfrac{3^3 x^{-3} y^6}{z^6}$

$\qquad = \dfrac{27y^6}{x^3 z^6}$

31. $\left(\dfrac{\sqrt{5}i}{x^{-2}}\right)^4 = \dfrac{(\sqrt{5})^4 i^4}{x^{-8}} = +25x^8$

33. $(4y^3)^{-2} = 4^{-2} y^{-6}$

$\qquad = \dfrac{1}{4^2 y^6}$

$\qquad = \dfrac{1}{16y^6}$

35. $\left(\dfrac{5x^2}{ab^{-2}}\right)^{-1} = \dfrac{5^{-1} x^{-2}}{a^{-1} b^2}$

$\qquad = \dfrac{a}{5x^2 b^2}$

37. $\dfrac{2x^{-3} y^2 z}{x^3 y^4 z^{-2}} = \dfrac{2y^2 z z^2}{x^3 x^3 y^4}$

$\qquad = \dfrac{2y^2 z^3}{x^6 y^4}$

$\qquad = \dfrac{2z^3}{x^6 y^2}$

39. $\left(\dfrac{-2xy}{z^2}\right)^{-1} (x^2 y^{-3})^2 = \left(\dfrac{z^2}{-2xy}\right)(x^4 y^{-6})$

$\qquad = \dfrac{z^2 x^4}{-2xyy^6}$

$\qquad = -\dfrac{z^2 x^3}{2y^7}$

41. $\dfrac{ab^{-1}}{(ab)^{-1}} \cdot \dfrac{a^2 b}{b^{-2}} = ab^{-1}(ab) \cdot a^2 bb^2 = a^4 b^3$

43. $\left[\dfrac{(3b^{-2}d)(2)(bd^3)^2}{12b^3d^{-1}}\right]^5 = \left[\dfrac{6b^{-2}db^2d^6}{12b^3d^{-1}}\right]^5$

$\qquad\qquad = \left[\dfrac{db^2d^6d}{2b^3b^2}\right]^5$

$\qquad\qquad = \left[\dfrac{d^8}{2b^3}\right]^5$

$\qquad\qquad = \dfrac{d^{40}}{2^5b^{15}}$

$\qquad\qquad = \dfrac{d^{40}}{32b^{15}}$

45. $(a^{-2} + a^{-3})^{-1} = \left(\dfrac{1}{a^2} + \dfrac{1}{a^3}\right)^{-1}$

$\qquad\qquad = \left(\dfrac{a}{a^3} + \dfrac{1}{a^3}\right)^{-1}$

$\qquad\qquad = \left(\dfrac{a+1}{a^3}\right)^{-1}$

$\qquad\qquad = \dfrac{a^3}{a+1}$

47. $\dfrac{x^{-1}}{y^{-1}} - \left(\dfrac{x}{y}\right)^{-1} = \dfrac{x^{-1}}{y^{-1}} - \dfrac{x^{-1}}{y^{-1}} = 0$

49. $\left(\dfrac{1}{2}x^{-1}y^2\right)^{-3} = \dfrac{1}{2^{-3}}x^3y^{-6}$

$\qquad\qquad = \dfrac{2^3x^3}{y^6}$

$\qquad\qquad = \dfrac{8x^3}{y^6}$

51. $\dfrac{2^{-2}}{1+\dfrac{3^{-1}}{1+3^{-1}}} = \dfrac{\dfrac{1}{2^2}}{1+\dfrac{\frac{1}{3}}{1+\frac{1}{3}}} = \dfrac{\frac{1}{4}}{1+\dfrac{\frac{1}{3}}{\frac{4}{3}}}$

$\qquad\qquad = \dfrac{\frac{1}{4}}{1+\frac{1}{3}\cdot\frac{3}{4}} = \dfrac{\frac{1}{4}}{1+\frac{1}{4}} = \dfrac{\frac{1}{4}}{\frac{5}{4}} = \dfrac{1}{4}\cdot\dfrac{4}{5} = \dfrac{1}{5}$

53. $\dfrac{(2x^{-1}y^2)^2}{2xy} \cdot \dfrac{x^{-3}}{y^3} = \dfrac{2^2x^{-2}y^4x^{-3}}{2xyy^3}$

$\qquad\qquad = \dfrac{2^2y^4}{2xx^2x^3y^4}$

$\qquad\qquad = \dfrac{2}{x^6}$

55. $\left[\left(\dfrac{1}{2}x^{-2}\right)^3(4xy^{-1})^2\right]^2 = \left(\dfrac{x^{-6}}{2^3}16x^2y^{-2}\right)^2$

$\qquad\qquad = \left(\dfrac{16x^2}{8x^6y^2}\right)^2$

$\qquad\qquad = \left(\dfrac{2}{x^4y^2}\right)^2$

$\qquad\qquad = \dfrac{4}{x^8y^4}$

57. $(x^{-1}+y^{-1})^{-1}(x+y) = \left(\dfrac{1}{x}+\dfrac{1}{y}\right)^{-1}(x+y)$

$\qquad\qquad = \left(\dfrac{y}{xy}+\dfrac{x}{xy}\right)^{-1}(x+y)$

$\qquad\qquad = \left(\dfrac{y+x}{xy}\right)^{-1}(x+y)$

$\qquad\qquad = \left(\dfrac{xy}{y+x}\right)(x+y)$

$\qquad\qquad = xy$

59. a) $\dfrac{1}{2}\cdot\dfrac{1}{4}\cdot\dfrac{1}{8}\cdot\dfrac{1}{16}\cdot\dfrac{1}{32} = \dfrac{1}{2\cdot2^2\cdot2^3\cdot2^4\cdot2^5}$

$\qquad\qquad = \dfrac{1}{2^{15}}$

$\qquad\qquad = 2^{-15}$

b) $\dfrac{1}{2}+\dfrac{1}{4}+\dfrac{1}{8}+\dfrac{1}{16}+\dfrac{1}{32}+\dfrac{1}{32} = \dfrac{16}{32}+\dfrac{8}{32}+\dfrac{4}{32}+\dfrac{2}{32}+\dfrac{1}{32}+\dfrac{1}{32}$

$\qquad\qquad = \dfrac{32}{32}$

$\qquad\qquad = 1$

$\qquad\qquad = 2^0$

61. When we try to calculate 2^{1000} or 10^{300} on a calculator, we get an error message; the calculator can handle only up to 100 digits (in scientific notation). However, if we take the fifth root of each number (i.e., raise each to the 1/5 power,) we have 2^{200} and 10^{60}, which the calculator can handle. $2^{200} \approx 1.606 \times 10^{60}$, which is of course larger than 10^{60}. Since positive numbers raised to positive powers retain their relative inequality, we have the conclusion that

$$2^{1000} > 10^{300}$$

This could even be done without a calculator by raising the numbers to the one-one hundredth power. Then we must compare $2^{10} = 1024$ and $10^3 = 1000$. Since $1024 > 1000$, we again have the conclusion

$$2^{1000} > 10^{300}$$

63. a)

Day #	Amount earned on this day	Total amount earned to date
1	1¢	1¢
2	2¢	3¢
3	4¢	7¢
4	8¢	15¢
5	16¢	31¢
6	32¢	63¢

b) Noticing the pattern that $31 = 2^5 - 1$ and $63 = 2^6 - 1$, we can say that by the nth day the secretary will have earned $2^n - 1$ cents.

c) 2 billion dollars is equal to 2×10^{11} cents. We want to know the smallest value of n such that $2^n - 1 > 2 \times 10^{11}$.

One approach is to use a calculator, replacing n with various values, until $2^n - 1$ becomes greater than 2×10^{11}.

A more sophisticated method is to try to approximate $2 \cdot 10^{11}$ as a power of 2:
$2(10^{11}) = 200(10^3)^3 < 200(2^{10})^3 = 200(2^{30}) < 2^{38}$
He will go broke on the 38th day, or February 7.

Problem Set 2.2

1. $341{,}000{,}000 = 3.41 \times 10^8$

3. $.0000000513 = 5.13 \times 10^{-8}$

5. $.0000000001245 = 1.245 \times 10^{-10}$

7. $(1.2 \times 10^5)(7 \times 10^{-9}) = 8.4 \times 10^{-4}$

9.
$$\frac{(.000021)(240000)}{7000} = \frac{(2.1 \times 10^{-5})(2.4 \times 10^5)}{7.0 \times 10^3}$$
$$= \frac{(2.1)(2.4)}{7} \times 10^{-5+5-3}$$
$$= .72 \times 10^{-3}$$
$$= 7.2 \times 10^{-4}$$

11. $(54)(.00005)(2000000)^2$
$$= (5.4 \times 10^1)(5.0 \times 10^{-5})(2.0 \times 10^6)^2$$
$$= (5.4)(5.0)(2.0)^2(10^1 \times 10^{-5} \times 10^{12})$$
$$= (5.4)(5.0)(4) \times 10^{1-5+12}$$
$$= 108 \times 10^8$$
$$= 1.08 \times 10^{10}$$

13. 413.2 meters $= 4.132 \times 10^2$ meters
$$= 4.132 \times 10^2 \times 10^2 \text{ centimeters}$$
$$= 4.132 \times 10^4 \text{ cm}$$

15. 4×10^{15} mm $= 4 \times 10^{15}$ mm $\times \dfrac{10^{-6} \text{ km}}{\text{mm}}$
$$= 4 \times 10^9 \text{ km}$$

17. 1 yard $= 36$ inches $\times \dfrac{2.54 \text{ cm}}{\text{inch}} \times \dfrac{10 \text{ mm}}{\text{cm}}$
$$= 91.44 \times 10^1 \text{ mm}$$
$$= 9.144 \times 10^2 \text{ mm}$$

19. $34.1 \boxed{-} 49.95 \boxed{+} 64.2 \boxed{=} 48.35$

21. $\boxed{(}\, 3.42 \boxed{-} 6.71 \,\boxed{)} \boxed{\times} \boxed{(}\, 14.3 \boxed{\times} 51.9 \,\boxed{)}$
$\boxed{=} -2441.7393$

23. $\boxed{(}\, 514 \boxed{+} 31.9 \,\boxed{)} \boxed{\div} \boxed{(}\, 52.6 \boxed{-} 50.8 \,\boxed{)}$
$\boxed{=} 303.27778$

25. $6.34 \boxed{EE} 7 \boxed{\times} 537.8 \boxed{\div} 1.23 \boxed{EE} 5 \boxed{+/-}$
$\boxed{=} 2.7721 \times 10^{15}$

27. $6.34 \boxed{EE} 7 \boxed{\div} \boxed{(}\, .00152 \boxed{+} .00341 \,\boxed{)}$
$\boxed{=} 1.2860 \times 10^{10}$

29. $\boxed{(}\, 532 \boxed{+} 1.346 \,\boxed{)} \boxed{\times} \boxed{(}\, 1.75 \boxed{-} 2.61 \,\boxed{)} \boxed{\div} 34.91$
$\boxed{=} -13.138859$

31. $1.214 \boxed{y^x} 3 \boxed{=} 1.7891883$

33. $1.215 \boxed{\sqrt[3]{}} 1.067068$

35. $1.34 \boxed{\times} 2.345 \boxed{y^x} 3 \boxed{\div} 364 \boxed{\sqrt{}} \boxed{=} .90569641$

37. $\boxed{(}\, 130 \boxed{\sqrt{}} \boxed{-} 5 \boxed{\sqrt{}} \,\boxed{)} \boxed{\div} \boxed{(}\, 15 \boxed{y^x} 6 \boxed{-} 4 \boxed{y^x} 8 \,\boxed{)}$
$\boxed{=} .00000081$

39. Time
$$= 2.39 \times 10^5 \text{ miles} \times \frac{1.609 \text{ km}}{\text{mile}} \times \frac{10^5 \text{ cm}}{\text{km}}$$
$$\times \frac{1 \quad \text{sec}}{2.9979 \times 10^{10} \text{ cm}}$$
$$= \frac{2.39 \times 10^5 \times 1.609 \times 10^5}{2.9979 \times 10^{10}} \text{ seconds}$$
$$= 1.2827 \text{ seconds}$$

41. Light travels
$$\frac{2.9979 \times 10^{10} \text{ cm}}{\text{sec}} \times \frac{60 \text{ sec}}{\text{min}} \times \frac{60 \text{ min}}{\text{hr}} \times \frac{24 \text{ hr}}{\text{day}}$$
$$\times \frac{365.24 \text{ day}}{\text{year}} \times \frac{1 \text{ meters}}{10^2 \text{ cm}} = \frac{9.46 \times 10^{15} \text{ m}}{\text{year}}$$
4.300 light years $= 4.300 \times 9.46 \times 10^{15}$
$$= 4.068 \times 10^{16} \text{ meters}$$

43. As found in problem 41,
1 light year $= 9.46 \times 10^{15}$ meters
Area $= (2.3 \times 9.46 \times 10^{15})(4.5 \times 9.46 \times 10^{15})$ m^2
$$= 9.26 \times 10^{32} \text{ square meters}$$

45. a) $3.151 \boxed{EE} 2 \boxed{y^x} 4 \boxed{\times} 32400 \boxed{\div} 21300 \boxed{y^x} 2$
$\boxed{=} 7.04 \times 10^5$

b) $\boxed{(}\, .433 \boxed{y^x} 3 \boxed{-} 2.31 \boxed{y^x} 4 \boxed{+/-} +.0932 \boxed{\sqrt{}} \,\boxed{)}$
$\boxed{\div} 5.23 \boxed{EE} 3 \boxed{=} 6.72 \times 10^{-5}$

47. Area of ring = Area of outer circle − area of inner circle
$$= \pi(26.25)^2 - \pi(14.42)^2 \text{ square cm}$$
$$= 1511.5 \text{ square cm}$$

49.
$$\bar{x} = \frac{121 + 132 + 155 + 161 + 133 + 175}{6}$$
$$= \frac{877}{6} = 146.17$$
$$s =$$
$$\sqrt{\frac{121^2 + 132^2 + 155^2 + 161^2 + 133^2 + 175^2}{6} - \left(\frac{877}{6}\right)^2}$$
$$= \sqrt{\frac{130,325}{6} - \left(\frac{877}{6}\right)^2}$$
$$= \sqrt{21720.833 - 21364.694}$$
$$= \sqrt{356.138558} = 18.87$$

51. Number of heartbeats = 86 years
$$\times \frac{365.24 \text{ days}}{\text{year}} \times \frac{24 \text{ hours}}{\text{day}} \times \frac{60 \text{ min}}{\text{hr}} \times \frac{75 \text{ heartbeats}}{\text{min}}$$
$$= 3.3923 \times 10^9$$

53. Hilda heard the speaker after
$$200 \text{ ft} \times \frac{1 \text{ sec}}{1100 \text{ feet}} = .182 \text{ seconds}$$
Hans heard the speaker after
$$2000 \text{ miles} \times \frac{1 \text{ sec}}{186,000 \text{ miles}} = .011 \text{ seconds}$$
Hans heard the speaker sooner:
$$(.182 - .011) = .171 \text{ seconds sooner}$$

55. Area of circle = $\pi r^2 = \pi(6.25)^2 = 122.718 \text{ cm}^2$
To find the area of the triangle, use Heron's formula:
Area of triangle = $\sqrt{s(s-a)(s-b)(s-c)}$ where a, b, c are the lengths of the sides of the triangle, and s is half the perimeter.
$$s = \frac{a+b+c}{2} = \frac{6.25 + 6.25 + 10.64}{2} = 11.57$$
Area of triangle =
$$\sqrt{11.57(11.57-6.25)(11.57-6.25)(11.57-10.64)}$$
$$= 17.451 \text{ square centimeters}$$
The area we seek = Area of circle − Area of triangle
$$= 122.718 - 17.451$$
$$= 105.27 \text{ square centimeters}$$

57. a) Number of
molecules = $20 \text{ gm} \times \frac{1 \text{ mole}}{18 \text{ gm}} \times \frac{6.02 \times 10^{23} \text{ molecules}}{\text{mole}}$
$$= 6.69 \times 10^{23} \text{ molecules}$$

b) Number of
molecules = $30 \text{ gm} \times \frac{1 \text{ mole}}{44 \text{ gm}} \times \frac{6.02 \times 10^{23} \text{ molecules}}{\text{mole}}$
$$= 4.10 \times 10^{23} \text{ molecules}$$

59. a) $x_1 = 3$
$x_2 = 3^2 = 9$
$x_3 = (x_2)^2 = (3^2)^2 = 81$
$x_4 = (x_3)^2 = [(3^2)^2]^2 = 6561$
$\vdots$

An easy way to calculate x_8 is to enter 3 in the calculator, and then press the $\boxed{x^2}$ button 7 times:
$x_8 = 1.179 \times 10^{61}$

b) x_n will grow infinitely large.

c) x_n will tend towards zero.

Problem Set 2.3
_____

1. Polynomial of degree 2

3. Polynomial of degree 5

5. Polynomial of degree 0

7. Not a polynomial

9. Not a polynomial

11. $(2x - 7) + (-4x + 8) = 2x - 7 - 4x + 8$
$$= 2x - 4x - 7 + 8$$
$$= -2x + 1$$

13. $(2x^2 - 5x + 6) + (2x^2 + 5x - 6)$
$$= 2x^2 - 5x + 6 + 2x^2 + 5x - 6$$
$$= 4x^2$$

15. $(5 - 11x^2 + 4x) + (x - 4 + 9x^2)$
$$= 5 - 11x^2 + 4x + x - 4 + 9x^2$$
$$= -11x^2 + 9x^2 + 4x + x + 5 - 4$$
$$= -2x^2 + 5x + 1$$

17. $(2x - 7) - (-4x + 8) = 2x - 7 + 4x - 8$
$$= 2x + 4x - 7 - 8$$
$$= 6x - 15$$

19. $(2x^2 - 5x + 6) - (2x^2 + 5x - 6)$
$$= 2x^2 - 5x + 6 - 2x^2 - 5x + 6$$
$$= -10x + 12$$

21. $5x(7x - 11) + 19 = (5x)(7x) - (5x)(11) + 19$
$$= 35x^2 - 55x + 19$$

23. $(t + 5)(t + 11) = t^2 + 5t + 11t + 55$
$$= t^2 + 16t + 55$$

25. $(x + 9)(x - 10) = x^2 + 9x - 10x - 90$
$$= x^2 - x - 90$$

27. $(2t - 1)(t + 7) = 2t^2 - t + 14t - 7$
$$= 2t^2 + 13t - 7$$

29. $(4 + y)(y - 2) = 4y + y^2 - 8 - 2y$
$$= y^2 + 2y - 8$$

19

31. $(2x-5)(3x^2-2x+4)$
$= 2x(3x^2-2x+4) - 5(3x^2-2x+4)$
$= 6x^3 - 4x^2 + 8x - 15x^2 + 10x - 20$
$= 6x^3 - 19x^2 + 18x - 20$

33. $(x+10)^2 = x^2 + 2(10)x + 10^2$
$= x^2 + 20x + 100$

35. $(x+8)(x-8) = x^2 - 8^2$
$= x^2 - 64$

37. $(2t-5)^2 = (2t)^2 - 2(2t)(5) + 5^2$
$= 4t^2 - 20t + 25$

39. $(2x^4 + 5x)(2x^4 - 5x) = (2x^4)^2 - (5x)^2$
$= 4x^8 - 25x^2$

41. $[(t+2)+t^3]^2 = (t+2)^2 + 2(t+2)t^3 + (t^3)^2$
$= t^2 + 4t + 4 + 2t^4 + 4t^3 + t^6$
$= t^6 + 2t^4 + 4t^3 + t^2 + 4t + 4$

43. $[(t+2)+t^3][(t+2)-t^3] = (t+2)^2 - (t^3)^2$
$= -t^6 + t^2 + 4t + 4$

45. $(2.3x - 1.4)^2 = (2.3x)^2 - 2(1.4)(2.3x) + (1.4)^2$
$= 5.29x^2 - 6.44x + 1.96$

47. $(x+2)^3 = x^3 + 3(2)x^2 + 3(2)^2 x + 2^3$
$= x^3 + 6x^2 + 12x + 8$

49. $(2t-3)^3 = (2t)^3 - 3(3)(2t)^2 + 3(3)^2(2t) - 3^3$
$= 8t^3 - 36t^2 + 54t - 27$

51. $(2t + t^2)^3 = (2t)^3 + 3(t^2)(2t)^2 + 3(t^2)^2(2t) + (t^2)^3$
$= 8t^3 + 12t^4 + 6t^5 + t^6$
$= t^6 + 6t^5 + 12t^4 + 8t^3$

53. $[(2t+1)+t^2]^3$
$= (2t+1)^3 + 3(t^2)(2t+1)^2 + 3(t^2)^2(2t+1) + (t^2)^3$
$= (2t)^3 + 3(1)(2t)^2 + 3(1)^2(2t) + 1^3$
$\qquad\qquad + 3t^2(4t^2 + 4t + 1) + 3t^4(2t+1) + t^6$
$= 8t^3 + 12t^2 + 6t + 1 + 12t^4 + 12t^3 + 3t^2 + 6t^5$
$\qquad\qquad\qquad\qquad\qquad\qquad + 3t^4 + t^6$
$= t^6 + 6t^5 + 15t^4 + 20t^3 + 15t^2 + 6t + 1$

55. $(x-3y)^2 = x^2 - 2(x)(3y) + (3y)^2$
$= x^2 - 6xy + 9y^2$

57. $(3x-2y)(3x+2y) = (3x)^2 - (2y)^2$
$= 9x^2 - 4y^2$

59. $(3x-y)(4x+5y) = 12x^2 - 4xy + 15xy - 5y^2$
$= 12x^2 + 11xy - 5y^2$

61. $(2x^2 y + z)(x^2 y - z)$
$= (2x^2 y)(x^2 y) - 2x^2 y(z) + x^2 y(z) - (z)(z)$
$= 2x^4 y^2 - x^2 yz - z^2$

63. $(t+1+s)(t+1-s) = [(t+1)+s][(t+1)-s]$
$= (t+1)^2 - s^2$
$= t^2 + 2t + 1 - s^2$

65. $(2t-3s)^3 = (2t)^3 - 3(3s)(2t)^2 + 3(3s)^2(2t) - (3s)^3$
$= 8t^3 - 36st^2 + 54s^2 t - 27s^3$

67. $(2x^2 - 3y)(2x^2 + 3y) = (2x^2)^2 - (3y)^2$
$= 4x^4 - 9y^2$

69. $(2s^3 + 3t)(s^3 - 4t) = 2s^6 + 3ts^3 - 8ts^3 - 12t^2$
$= 2s^6 - 5ts^3 - 12t^2$

71. $(2u - v^2)^3 = (2u)^3 - 3(v^2)(2u)^2 + 3(v^2)^2(2u) - (v^2)^3$
$= 8u^3 - 12v^2 u^2 + 6v^4 u - v^6$

73. $2x(3x^2 - 6x + 4) - 3x[2x^2 - 4(x-1)]$
$= 6x^3 - 12x^2 + 8x - 3x[2x^2 - 4x + 4]$
$= 6x^3 - 12x^2 + 8x - 6x^3 + 12x^2 - 12x$
$= -4x$

75. $(2s+3)^2 - (2s+3)(2s-3)$
$= (2s)^2 + 2(3)(2s) + 3^2 - [(2s)^2 - 3^2]$
$= 4s^2 + 12s + 9 - 4s^2 + 9$
$= 12s + 18$

77. $(x^2 + 2x - 3)(x^2 + 2x + 3)$
$= [(x^2 + 2x) + 3][(x^2 + 2x) - 3]$
$= (x^2 + 2x)^2 - 3^2$
$= (x^2)^2 + 2(2x)(x^2) + (2x)^2 - 9$
$= x^4 + 4x^3 + 4x^2 - 9$

79. $(2x^2 + x - 1)(x + 2)$
$= x(2x^2 + x - 1) + 2(2x^2 + x - 1)$
$= 2x^3 + x^2 - x + 4x^2 + 2x - 2$
$= 2x^3 + 5x^2 + x - 2$

81. $(x^2 - 2xy)(x^2 + 2xy)(x+y)$
$= [(x^2)^2 - (2xy)^2](x+y)$
$= (x^4 - 4x^2 y^2)(x+y)$
$= x^4(x) + x^4(y) - 4x^2 y^2(x) - 4x^2 y^2(y)$
$= x^5 + x^4 y - 4x^3 y^2 - 4x^2 y^3$

83. $(x^2 + 2xy + 4y^2)(x - 2y)$
$= x(x^2 + 2xy + 4y^2) - 2y(x^2 + 2xy + 4y^2)$
$= x^3 + 2x^2 y + 4xy^2 - 2x^2 y - 4xy^2 - 8y^3$
$= x^3 - 8y^3$

85. $(x^2 + xy + y^2)(x^2 - xy + y^2)$
$= [(x^2 + y^2) + xy][(x^2 + y^2) - xy]$
$= (x^2 + y^2)^2 - (xy)^2$
$= x^4 + 2x^2 y^2 + y^4 - x^2 y^2$
$= x^4 + x^2 y^2 + y^4$

87. Each term of the first factor must be multiplied by each term of the second factor. Pick out those combinations that lead to x^3 terms. They are $(x^2)(2x)$, $(2x)(-3x^2)$, and $(3)(x^3)$.

$$2x^3 - 6x^3 + 3x^3 = -x^3$$

The coefficient is -1.

89. We want to show that

$$(2m)^2 + (m^2 - 1)^2 = (m^2 + 1)^2.$$

$$\begin{aligned}(2m)^2 + (m^2 - 1)^2 &= 4m^2 + m^4 - 2m^2 + 1 \\ &= m^4 + 2m^2 + 1 \\ &= (m^2 + 1)^2\end{aligned}$$

m must be greater than one so that the second number in the Pythagorean triple, $m^2 - 1$, will be positive.

91. Notice that

$$\begin{aligned}(r^2 + r^{-2})(r + r^{-1}) &= r^3 + r^2 r^{-1} + r^{-2}r + r^{-3} \\ &= (r^3 + r^{-3}) + (r + r^{-1})\end{aligned}$$

or,

$$r^3 + r^{-3} = (r^2 + r^{-2})(r + r^{-1}) - (r + r^{-1})$$

If we can find the value of $(r + r^{-1})$ and $(r^2 + r^{-2})$, we will know the value of $r^3 + r^{-3}$.

Since $(r + r^{-1})^2 = 5$, $r + r^{-1} = \sqrt{5}$.

Also,

$$\begin{aligned}(r + r^{-1})^2 &= r^2 + 2rr^{-1} + r^{-2} \\ &= r^2 + r^{-2} + 2\end{aligned}$$

or,

$$\begin{aligned}r^2 + r^{-2} &= (r + r^{-1})^2 - 2 \\ &= 5 - 2 \\ &= 3\end{aligned}$$

Substituting in

$$r^3 + r^{-3} = (r^2 + r^{-2})(r + r^{-1}) - (r + r^{-1}),$$

$$\begin{aligned}r^3 + r^{-3} &= 3(\sqrt{5}) - \sqrt{5} \\ &= 2\sqrt{5}\end{aligned}$$

93. Since $(x + y)^4 = x^4 + 4x^3y + 6x^2y^2 + 4xy^3 + y^4$,

$$\begin{aligned}x^4 + y^4 &= (x + y)^4 - 4x^3y - 6x^2y^2 - 4xy^3 \\ &= (x + y)^4 - 4xy(x^2 + y^2) - 6x^2y^2\end{aligned}$$

We know that $x + y = \sqrt{11}$, so

$$(x + y)^4 = (\sqrt{11})^4 = 121$$

We are given that $x^2 + y^2 = 16$.

If we can find a value for xy, we will be able to find the value of $x^4 + y^4$.

Since $(x + y)^2 = x^2 + 2xy + y^2$,

$$\begin{aligned}2xy &= (x + y)^2 - (x^2 + y^2) \\ 2xy &= (\sqrt{11})^2 - 16 \\ 2xy &= -5 \\ xy &= \frac{-5}{2}\end{aligned}$$

Therefore,

$$\begin{aligned}x^4 + y^4 &= 121 - 4\left(\frac{-5}{2}\right)(16) - 6\left(\frac{-5}{2}\right)^2 \\ &= 121 + 160 - \frac{150}{4} \\ &= 243.5\end{aligned}$$

1. $x^2 + 5x = x(x + 5)$

3. $x^2 + 5x - 6 = (x + 6)(x - 1)$

5. $y^4 - 6y^3 = y^3(y - 6)$

7. $y^2 + 4y - 12 = (y + 6)(y - 2)$

9. $y^2 + 8y + 16 = (y + 4)^2$

11. $4x^2 - 12xy + 9y^2 = (2x - 3y)^2$

13. $y^2 - 64 = (y + 8)(y - 8)$

15. $1 - 25b^2 = (1 - 5b)(1 + 5b)$

17. $4z^2 - 4z - 3 = (2z + 1)(2z - 3)$

19. $20x^2 + 3xy - 2y^2 = (4x - y)(5x + 2y)$

21. $\begin{aligned}x^3 + 27 &= x^3 + 3^3 \\ &= (x + 3)(x^2 - 3x + 9)\end{aligned}$

23. $\begin{aligned}a^3 - 8b^3 &= a^3 - (2b)^3 \\ &= (a - 2b)(a^2 + 2ab + (2b)^2) \\ &= (a - 2b)(a^2 + 2ab + 4b^2)\end{aligned}$

25. $\begin{aligned}x^3 - x^3y^3 &= x^3(1 - y^3) \\ &= x^3(1 - y)(1 + y + y^2)\end{aligned}$

27. $x^2 - 3$ cannot be factored over the integers; however, it does factor over the reals.

29. $3x^2 - 4$ does not factor over the integers; however, it does factor over the reals.

31. $y^2 - 5 = (y + \sqrt{5})(y - \sqrt{5})$

33. $\begin{aligned}5z^2 - 4 &= (\sqrt{5}z)^2 - (2)^2 \\ &= (\sqrt{5}z + 2)(\sqrt{5}z - 2)\end{aligned}$

35. $\begin{aligned}t^4 - 2t^2 &= t^2(t^2 - 2) \\ &= t^2(t + \sqrt{2})(t - \sqrt{2})\end{aligned}$

37. $y^2 - 2\sqrt{3}y + 3 = (y - \sqrt{3})^2$

39. $x^2 + 9$ does not factor over the real numbers.

41. $\begin{aligned}x^2 + 9 &= x^2 - 9i^2 \\ &= (x + 3i)(x - 3i)\end{aligned}$

43. $x^6 + 9x^3 + 14 = (x^3 + 7)(x^3 + 2)$

45. $\begin{aligned}4x^4 - 37x^2 + 9 &= (4x^2 - 1)(x^2 - 9) \\ &= (2x + 1)(2x - 1)(x + 3)(x - 3)\end{aligned}$

47. $(x + 4y)^2 + 6(x + 4y) + 9 = (x + 4y + 3)^2$

49. $x^4 - x^2y^2 - 6y^4 = (x^2 - 3y^2)(x^2 + 2y^2)$

51. $\begin{aligned}x^6 - 64 &= (x^3 + 8)(x^3 - 8) \\ &= (x + 2)(x^2 - 2x + 4)(x - 2)(x^2 + 2x + 4)\end{aligned}$

53. $x^8 - x^4y^4 = x^4(x^4 - y^4)$
$= x^4(x^2 + y^2)(x^2 - y^2)$
$= x^4(x^2 + y^2)(x + y)(x - y)$

55. $x^6 + y^6 = (x^2)^3 + (y^2)^3$
$= (x^2 + y^2)(x^4 - x^2y^2 + y^4)$

57. $x^3 - 4x^2 + x - 4 = x^2(x - 4) + 1(x - 4)$
$= (x^2 + 1)(x - 4)$

59. $4x^2 - 4x + 1 - y^2 = (2x - 1)^2 - y^2$
$= (2x - 1 + y)(2x - 1 - y)$

61. $3x + 3y - x^2 - xy = 3(x + y) - x(x + y)$
$= (x + y)(3 - x)$

63. $x^2 + 6xy + 9y^2 + 2x + 6y = (x + 3y)^2 + 2(x + 3y)$
$= (x + 3y)(x + 3y + 2)$

65. $x^2 + 2xy + y^2 + 3x + 3y + 2$
$= (x + y)^2 + 3(x + y) + 2$
$= (x + y + 2)(x + y + 1)$

67. $x^4 + 64 = x^4 + 16x^2 + 64 - 16x^2$
$= (x^2 + 8)^2 - 16x^2$
$= (x^2 + 8 - 4x)(x^2 + 8 + 4x)$
$= (x^2 - 4x + 8)(x^2 + 4x + 8)$

69. $x^4 + x^2 + 1 = x^4 + 2x^2 + 1 - x^2$
$= (x^2 + 1)^2 - x^2$
$= (x^2 + 1 - x)(x^2 + 1 + x)$
$= (x^2 - x + 1)(x^2 + x + 1)$

71. $4 - 9m^2 = (2 + 3m)(2 - 3m)$

73. $6x^2 - 5x + 1 = (3x - 1)(2x - 1)$

75. $5x^2 - 20x = 5x(x^2 - 4)$
$= 5x(x - 2)(x + 2)$

77. $6x^3 - 5x^2 + x = x(6x^2 - 5x + 1)$
$= x(3x - 1)(2x - 1)$

79. $2u^4 - 7u^2 + 5 = (2u^2 - 5)(u^2 - 1)$
$= (2u^2 - 5)(u - 1)(u + 1)$

81. $(a + 2b)^2 - 3(a + 2b) - 28 = (a + 2b - 7)(a + 2b + 4)$

83. $(a + 3b)^4 - 1 = [(a + 3b)^2 + 1][(a + 3b)^2 - 1]$
$= [(a + 3b)^2 + 1][a + 3b + 1][a + 3b - 1]$
$= (a^2 + 6ab + 9b^2 + 1)(a + 3b + 1)((a + 3b - 1)$

85. $x^2 - 6xy + 9y^2 + 4x - 12y = (x - 3y)^2 + 4(x - 3y)$
$= (x - 3y)(x - 3y + 4)$

87. $9x^4 - 24x^2y^2 + 16y^4 - y^2 = (3x^2 - 4y^2)^2 - y^2$
$= (3x^2 - 4y^2 - y)(3x^2 - 4y^2 + y)$

89. $x^4 - 3x^2y^2 + y^4 = x^4 - 2x^2y^2 + y^4 - x^2y^2$
$= (x^2 - y^2)^2 - x^2y^2$
$= (x^2 - y^2 - xy)(x^2 - y^2 + xy)$

91. $(x + 3)^2(x + 2)^3 - 20(x + 3)(x + 2)^2$
$= (x + 3)(x + 2)^2[(x + 3)(x + 2) - 20]$
$= (x + 3)(x + 2)^2(x^2 + 5x + 6 - 20)$
$= (x + 3)(x + 2)^2(x^2 + 5x - 14)$
$= (x + 3)(x + 2)^2(x + 7)(x - 2)$

93. $x^{2n} + 3x^n + 2 = (x^n + 2)(x^n + 1)$

95. a) $(547)^2 - (453)^2 = (547 + 453)(547 - 453)$
$= (1000)(94)$
$= 94,000$

b) $\dfrac{2^{20} - 2^{17} + 7}{2^{17} + 1}$

$= \dfrac{2^{17}(2^3 - 1) + 7}{2^{17} + 1} = \dfrac{2^{17}(7) + 7}{2^{17} + 1}$

$= \dfrac{7(2^{17} + 1)}{2^{17} + 1} = 7$

c) $\left(1 - \dfrac{1}{2^2}\right)\left(1 - \dfrac{1}{3^2}\right)\left(1 - \dfrac{1}{4^2}\right)\cdots\left(1 - \dfrac{1}{29^2}\right)$

$= \left(1 + \tfrac{1}{2}\right)\left(1 - \tfrac{1}{2}\right)\left(1 + \tfrac{1}{3}\right)\left(1 - \tfrac{1}{3}\right)\left(1 + \tfrac{1}{4}\right)\left(1 - \tfrac{1}{4}\right)\cdots$
$\cdots\left(1 + \tfrac{1}{29}\right)\left(1 - \tfrac{1}{29}\right)$

$= \left(\tfrac{3}{2} \cdot \tfrac{1}{2}\right) \cdot \left(\tfrac{4}{3} \cdot \tfrac{2}{3}\right) \cdot \left(\tfrac{5}{4} \cdot \tfrac{3}{4}\right) \cdots \left(\tfrac{30}{29} \cdot \tfrac{28}{29}\right)$

(Notice the pattern of cancellations:) $= \dfrac{1}{2} \cdot \dfrac{30}{29} = \dfrac{15}{29}$

97. Let n = the integer to be cubed.

Let $\dfrac{n(n + 1)}{2}$ be one of the integers to be squared (as suggested in the statement of the problem). [The expression is an integer because either n or $(n + 1)$ is even.] Then

$n^3 = \left(\dfrac{n(n + 1)}{2}\right)^2 - k^2$ where k is the other integer.

$n^3 = \left(\dfrac{n^2 + n}{2}\right)^2 - k^2$

$n^3 = \dfrac{n^4 + 2n^3 + n^2}{4} - k^2$

$k^2 = \dfrac{n^4 + 2n^3 + n^2}{4} - n^3$

$k^2 = \dfrac{n^4 + 2n^3 + n^2 - 4n^3}{4}$

$k^2 = \dfrac{n^4 - 2n^3 + n^2}{4}$

$k^2 = \dfrac{n^2(n^2 - 2n + 1)}{4} = \dfrac{n^2(n - 1)^2}{4}$

Hence $k = \dfrac{n(n - 1)}{2}$, which is also an integer since either n or $(n - 1)$ must be even. Thus,

$n^3 = \left(\dfrac{n(n + 1)}{2}\right)^2 - \left(\dfrac{n(n - 1)}{2}\right)^2$

22

Problem Set 2.5

1. $\dfrac{x+6}{x^2-36} = \dfrac{x+6}{(x-6)(x+6)} = \dfrac{1}{x-6}$

3. $\dfrac{y^2+y}{5y+5} = \dfrac{y(y+1)}{5(y+1)} = \dfrac{y}{5}$

5. $\dfrac{(x+2)^3}{x^2-4} = \dfrac{(x+2)^3}{(x+2)(x-2)}$

$\qquad = \dfrac{(x+2)^2}{x-2}$

7. $\dfrac{zx^2+4xyz+4y^2z}{x^2+3xy+y^2} = \dfrac{z(x^2+4xy+4y^2)}{x^2+3xy+2y^2}$

$\qquad\qquad = \dfrac{z(x+2y)^2}{(x+2y)(x+y)}$

$\qquad\qquad = \dfrac{z(x+2y)}{x+y}$

9. $\dfrac{5}{x-2} + \dfrac{4}{x+2} = \dfrac{5(x+2)}{(x-2)(x+2)} + \dfrac{4(x-2)}{(x+2)(x-2)}$

$\qquad = \dfrac{5x+10+4x-8}{(x+2)(x-2)} = \dfrac{9x+2}{(x+2)(x-2)}$

11. $\dfrac{5x}{x^2-4} + \dfrac{3}{x+2} = \dfrac{5x}{(x+2)(x-2)} + \dfrac{3(x-2)}{(x+2)(x-2)}$

$\qquad = \dfrac{5x+3x-6}{(x+2)(x-2)}$

$\qquad = \dfrac{8x-6}{(x+2)(x-2)}$

$\qquad = \dfrac{2(4x-3)}{(x+2)(x-2)}$

13. $\dfrac{2}{xy} + \dfrac{3}{xy^2} - \dfrac{1}{x^2y^2} = \dfrac{2(xy)}{xy(xy)} + \dfrac{3(x)}{xy^2(x)} - \dfrac{1}{x^2y^2}$

$\qquad = \dfrac{2xy+3x-1}{x^2y^2}$

15. $\dfrac{x+1}{x^2-4x+4} + \dfrac{4}{x^2+3x-10}$

$\qquad = \dfrac{x+1}{(x-2)^2} + \dfrac{4}{(x+5)(x-2)}$

$\qquad = \dfrac{(x+1)(x+5)}{(x-2)^2(x+5)} + \dfrac{4(x-2)}{(x+5)(x-2)^2}$

$\qquad = \dfrac{x^2+6x+5+4x-8}{(x-2)^2(x+5)}$

$\qquad = \dfrac{x^2+10x-3}{(x-2)^2(x+5)}$

17. $\dfrac{4}{2x-1} + \dfrac{x}{1-2x} = \dfrac{4}{2x-1} - \dfrac{x}{2x-1}$

$\qquad = \dfrac{4-x}{2x-1}$

19. $\dfrac{2}{6y-2} + \dfrac{y}{9y^2-1} - \dfrac{2y+1}{1-3y}$

$\qquad = \dfrac{1}{3y-1} + \dfrac{y}{(3y-1)(3y+1)} + \dfrac{2y+1}{3y-1}$

$\qquad = \dfrac{1(3y+1)}{(3y-1)(3y+1)} + \dfrac{y}{(3y-1)(3y+1)}$

$\qquad\quad + \dfrac{(2y+1)(3y+1)}{(3y-1)(3y+1)}$

$\qquad = \dfrac{3y+1+y+6y^2+5y+1}{(3y-1)(3y+1)}$

$\qquad = \dfrac{6y^2+9y+2}{(3y-1)(3y+1)}$

21. $\dfrac{m^2}{m^2-2m+1} - \dfrac{1}{3-3m} = \dfrac{m^2}{(m-1)^2} + \dfrac{1}{3(m-1)}$

$\qquad\qquad = \dfrac{3m^2}{3(m-1)^2} + \dfrac{(m-1)}{3(m-1)^2}$

$\qquad\qquad = \dfrac{3m^2+m-1}{3(m-1)^2}$

23. $\dfrac{5}{2x-1} \cdot \dfrac{x}{x+1} = \dfrac{5x}{(2x-1)(x+1)}$

25. $\dfrac{x+2}{x^2-9} \cdot \dfrac{x+3}{x^2-4} = \dfrac{x+2}{(x-3)(x+3)} \cdot \dfrac{x+3}{(x-2)(x+2)}$

$\qquad = \dfrac{1}{(x-3)(x-2)}$

27. $x^2y^4\left(\dfrac{x}{y^2} - \dfrac{y}{x^2}\right) = \dfrac{x^2y^4 \cdot x}{y^2} - \dfrac{x^2y^4 \cdot y}{x^2}$

$\qquad = x^3y^2 - y^5$

$\qquad = y^2(x^3-y^3)$

29. $\dfrac{x^2+5x}{x^2-16} \cdot \dfrac{x^2-2x-24}{x^2-x-30}$

$\qquad = \dfrac{x(x+5)}{(x+4)(x-4)} \cdot \dfrac{(x-6)(x+4)}{(x-6)(x+5)}$

$\qquad = \dfrac{x}{x-4}$

31. $\dfrac{\dfrac{5}{2x-1}}{\dfrac{x}{x+1}} = \dfrac{5}{2x-1} \cdot \dfrac{x+1}{x} = \dfrac{5(x+1)}{x(2x-1)}$

33. $\dfrac{x+2}{\dfrac{x^2-4}{x}} = (x+2) \cdot \dfrac{x}{(x+2)(x-2)} = \dfrac{x}{x-2}$

23

35. $\dfrac{\dfrac{x^2+a^2}{x^3-a^3}}{\dfrac{x+2a}{(x-a)^2}} = \dfrac{x^2+a^2}{(x-a)(x^2+ax+a^2)} \cdot \dfrac{(x-a)^2}{x+2a}$

$= \dfrac{(x^2+a^2)(x-a)}{(x+2a)(x^2+ax+a^2)}$

37. $\dfrac{\dfrac{y^2+y-2}{y^2+4y}}{\dfrac{2y^2-8}{y^2+2y-8}} = \dfrac{\dfrac{(y+2)(y-1)}{y(y+4)}}{\dfrac{2(y^2-4)}{(y+4)(y-2)}}$

$= \dfrac{(y+2)(y-1)}{y(y+4)} \cdot \dfrac{(y+4)(y-2)}{2(y-2)(y+2)}$

$= \dfrac{y-1}{2y}$

39. $\dfrac{\dfrac{1}{2x+2h+3} - \dfrac{1}{2x+3}}{h} = \dfrac{\dfrac{(2x+3)-(2x+2h+3)}{(2x+3)(2x+2h+3)}}{h}$

$= \dfrac{\dfrac{2x+3-2x-2h-3}{(2x+3)(2x+2h+3)}}{\dfrac{h}{1}}$

$= \dfrac{-2h}{(2x+3)(2x+2h+3)} \cdot \dfrac{1}{h}$

$= \dfrac{-2}{(2x+3)(2x+2h+3)}$

41. $\dfrac{\dfrac{1}{(x+h)^2} - \dfrac{1}{x^2}}{h} = \dfrac{\dfrac{x^2-(x+h)^2}{x^2(x+h)^2}}{h}$

$= \dfrac{\dfrac{x^2-(x^2+2xh+h^2)}{x^2(x+h)^2}}{\dfrac{h}{1}}$

$= \dfrac{-2xh-h^2}{x^2(x+h)^2} \cdot \dfrac{1}{h}$

$= \dfrac{h(-2x-h)}{x^2(x+h)^2} \cdot \dfrac{1}{h}$

$= \dfrac{-2x-h}{x^2(x+h)^2}$

43. $\dfrac{\dfrac{y}{y+4} - \dfrac{2}{y^2+5y+4}}{\dfrac{4}{y+1} + \dfrac{3}{y+4}}$

$= \dfrac{\left(\dfrac{y}{y+4} - \dfrac{2}{(y+4)(y+1)}\right)(y+4)(y+1)}{\left(\dfrac{4}{y+1} + \dfrac{3}{y+4}\right)(y+4)(y+1)}$

$= \dfrac{y(y+1)-2}{4(y+4)+3(y+1)}$

$= \dfrac{y^2+y-2}{4y+16+3y+3}$

$= \dfrac{(y+2)(y-1)}{7y+19}$

45. $\dfrac{\dfrac{a^2}{b^2} - \dfrac{b^2}{a^2}}{\dfrac{a}{b} - \dfrac{b}{a}} = \dfrac{\left(\dfrac{a^2}{b^2} - \dfrac{b^2}{a^2}\right)(a^2b^2)}{\left(\dfrac{a}{b} - \dfrac{b}{a}\right)(a^2b^2)}$

$= \dfrac{a^2 \cdot a^2 - b^2 \cdot b^2}{a \cdot a^2 b - b \cdot ab^2}$

$= \dfrac{a^4-b^4}{a^3b-ab^3}$

$= \dfrac{(a^2+b^2)(a^2-b^2)}{ab(a^2-b^2)}$

$= \dfrac{a^2+b^2}{ab}$

47. $\dfrac{\dfrac{x^2}{x-y} - x}{\dfrac{y^2}{x-y} + y} = \dfrac{\dfrac{x^2-x(x-y)}{x-y}}{\dfrac{y^2+y(x-y)}{x-y}}$

$= \dfrac{\dfrac{x^2-x^2+xy}{x-y}}{\dfrac{y^2+xy-y^2}{x-y}}$

$= \dfrac{\dfrac{xy}{x-y}}{\dfrac{xy}{x-y}} = 1$

49.
$$\frac{y - \dfrac{1}{1+\frac{1}{y}}}{y + \dfrac{1}{y-\frac{1}{y}}} = \frac{y - \dfrac{(1)(y)}{\left(1+\frac{1}{y}\right)(y)}}{y + \dfrac{(1)(y)}{\left(y-\frac{1}{y}\right)(y)}}$$

$$= \frac{y - \dfrac{y}{y+1}}{y + \dfrac{y}{y^2-1}}$$

$$= \frac{\left(y - \dfrac{y}{y+1}\right)(y+1)(y-1)}{\left(y + \dfrac{y}{(y+1)(y-1)}\right)(y+1)(y-1)}$$

$$= \frac{y(y+1)(y-1) - y(y-1)}{y(y+1)(y-1) + y}$$

$$= \frac{y[(y+1)(y-1) - (y-1)]}{y[(y+1)(y-1) + 1]}$$

$$= \frac{y^2 - 1 - y + 1}{y^2 - 1 + 1}$$

$$= \frac{y^2 - y}{y^2}$$

$$= \frac{y(y-1)}{y^2}$$

$$= \frac{y-1}{y}$$

51. $x + y + \dfrac{y^2}{x-y} = \dfrac{(x+y)(x-y) + y^2}{x-y}$

$$= \frac{x^2 - y^2 + y^2}{x-y}$$

$$= \frac{x^2}{x-y}$$

53. $\left(\dfrac{1}{x} + \dfrac{1}{y}\right)\left(x + y - \dfrac{x^2+y^2}{x+y}\right)$

$$= \left(\frac{y+x}{xy}\right)\left[\frac{(x+y)^2 - (x^2+y^2)}{x+y}\right]$$

$$= \frac{x^2 + 2xy + y^2 - x^2 - y^2}{xy}$$

$$= \frac{2xy}{xy}$$

$$= 2$$

55. $\dfrac{x}{x^2 + 11x + 30} - \dfrac{5}{x^2 + 9x + 20}$

$$= \frac{x}{(x+5)(x+6)} - \frac{5}{(x+5)(x+4)}$$

$$= \frac{x(x+4) - 5(x+6)}{(x+4)(x+5)(x+6)}$$

$$= \frac{x^2 + 4x - 5x - 30}{(x+4)(x+5)(x+6)}$$

$$= \frac{x^2 - x - 30}{(x+4)(x+5)(x+6)}$$

$$= \frac{(x-6)(x+5)}{(x+4)(x+5)(x+6)}$$

$$= \frac{x-6}{(x+4)(x+6)}$$

57. $\dfrac{\dfrac{a^2+4a+3}{a} - \dfrac{2a+2}{a-1}}{\dfrac{a+1}{a^2-a}}$

$$= \left[\frac{(a+3)(a+1)}{a} - \frac{2(a+1)}{a-1}\right]\cdot\left[\frac{a(a-1)}{a+1}\right]$$

$$= \frac{(a+3)(a+1)}{a}\cdot\frac{a(a-1)}{(a+1)} - \frac{2(a+1)}{(a-1)}\cdot\frac{a(a-1)}{(a+1)}$$

$$= (a+3)(a-1) - 2a$$

$$= a^2 + 2a - 3 - 2a$$

$$= a^2 - 3$$

59. $\dfrac{18x^2y - 27x^2 - 8y + 12}{6xy - 4y - 9x + 6} = \dfrac{9x^2(2y-3) - 4(2y-3)}{2y(3x-2) - 3(3x-2)}$

$$= \frac{(2y-3)(9x^2-4)}{(3x-2)(2y-3)}$$

$$= \frac{(2y-3)(3x+2)(3x-2)}{(3x-2)(2y-3)}$$

$$= 3x + 2$$

61. $\dfrac{x^3 + y^3}{x^3 + (x-y)^3}$

$$= \frac{(x+y)(x^2 - xy + y^2)}{(x+x-y)[x^2 - x(x-y) + (x-y)^2]}$$

$$= \frac{(x+y)(x^2 - xy + y^2)}{(2x-y)(x^2 - x^2 + xy + x^2 - 2xy + y^2)}$$

$$= \frac{(x+y)(x^2 - xy + y^2)}{(2x-y)(x^2 - xy + y^2)}$$

$$= \frac{x+y}{2x-y}$$

63. a) $\quad \frac{13}{5} = 2 + \frac{3}{5}$

$$= 2 + \frac{1}{\frac{5}{3}}$$

$$= 2 + \frac{1}{1 + \frac{2}{3}}$$

$$= 2 + \frac{1}{1 + \frac{1}{\frac{3}{2}}}$$

$$= 2 + \frac{1}{1 + \frac{1}{1 + \frac{1}{2}}}$$

b) $\quad \frac{29}{11} = 2 + \frac{7}{11}$

$$= 2 + \frac{1}{\frac{11}{7}}$$

$$= 2 + \frac{1}{1 + \frac{4}{7}}$$

$$= 2 + \frac{1}{1 + \frac{1}{\frac{7}{4}}}$$

$$= 2 + \frac{1}{1 + \frac{1}{1 + \frac{3}{4}}}$$

$$= 2 + \frac{1}{1 + \frac{1}{1 + \frac{1}{\frac{4}{3}}}}$$

$$= 2 + \frac{1}{1 + \frac{1}{1 + \frac{1}{1 + \frac{1}{3}}}}$$

c) $\quad -\frac{5}{4} = -2 + \frac{3}{4}$

$$= -2 + \frac{1}{\frac{4}{3}}$$

$$= -2 + \frac{1}{1 + \frac{1}{3}}$$

Chapter 2 Review Problem Set

1. False: $2^m 2^n = 2^{(m+n)}$

2. True

3. False: The rule to remember is $4^{-n} = \frac{1}{4^n}$.

4. False, since 62.345 is not less than 10.

5. False: A polynomial is of the form
$$a_n x^n + a_{n-1} x^{n-1} + \cdots + a_1 x + a_0$$
where n is a nonnegative integer.

6. False, since pq will be a polynomial of degree 7.

7. True: $\begin{aligned}a^6 - 4b^4 &= (a^3)^2 - (2b^2)^2 \\ &= (a^3 + 2b^2)(a^3 - 2b^2)\end{aligned}$

8. True

9. True: $x - i$ is a factor of $x^4 + x^2$ over the complex numbers.
$$\begin{aligned} x^4 + x^2 &= x^2(x^2 + 1) \\ &= x^2[x^2 - (-1)] \\ &= x^2(x^2 - i^2) \\ &= x^2(x + i)(x - i)\end{aligned}$$

10. False: The negation is distributed over either the numerator or the denominator, but not both.

11. $\begin{aligned}\left(\frac{4}{3}\right)^{-3} &= \frac{4^{-3}}{3^{-3}} \\ &= \frac{3^3}{4^3} \\ &= \frac{27}{64}\end{aligned}$

12. $\left(\frac{4}{7}\right)^4 \left(\frac{4}{7}\right)^{-2} = \left(\frac{4}{7}\right)^2 = \frac{16}{49}$

13. $\begin{aligned}\left(2 + \frac{2}{3} - \frac{1}{2}\right)^{-2} &= \left(\frac{12}{6} + \frac{4}{6} - \frac{3}{6}\right)^{-2} \\ &= \left(\frac{13}{6}\right)^{-2} \\ &= \frac{13^{-2}}{6^{-2}} \\ &= \frac{6^2}{13^2} \\ &= \frac{36}{169}\end{aligned}$

14. $\frac{3a^{-1}b^2}{(2a^{-1})^{-3}b^4} = \frac{3a^{-1}b^2}{2^{-3}a^3 b^4} = \frac{8 \cdot 3}{a^4 b^2} = \frac{24}{a^4 b^2}$

15. $\left(\frac{4x^{-3}}{y^2}\right)^{-3} = \frac{4^{-3}x^9}{y^{-6}} = \frac{x^9 y^6}{4^3} = \frac{x^9 y^6}{64}$

16. $\begin{aligned}\frac{(x^3 y^{-1})^3 (2x^{-1}y^2)^{-2}}{(x^4 y^{-2})^4} &= \frac{x^9 y^{-3} 2^{-2} x^2 y^{-4}}{x^{16} y^{-8}} \\ &= \frac{2^{-2} x^{11} y^{-7}}{x^{16} y^{-8}} \\ &= \frac{y}{4x^5}\end{aligned}$

17. $215{,}000{,}000 = 2.15 \times 10^8$

26

18. $0.000107 = 1.07 \times 10^{-4}$

19. $(402{,}000)(2)10^{-8} = 4.02 \times 10^5 \times 2 \times 10^{-8}$
$$= 8.04 \times 10^{-3}$$

20. $\dfrac{1.44 \times 10^4}{4.8 \times 10^{-5}} = 0.3 \times 10^9 = 3.0 \times 10^8$

21. Not a polynomial

22. Polynomial of degree 2

23. Not a polynomial

24. Not a polynomial

25. $(2x^2 + 7) + (-3x^2 + x - 14)$
$$= 2x^2 - 3x^2 + x + 7 - 14$$
$$= -x^2 + x - 7$$

26. $4 + x^4 - (2x - 3x^4 + x^2 - 11) = 4x^4 - x^2 - 2x + 15$

27. $(2y + 1)(3y - 2) = 6y^2 - 4y + 3y - 2$
$$= 6y^2 - y - 2$$

28. $(4x + 3)^2 + (4x - 3)^2$
$$= (16x^2 + 24x + 9) + (16x^2 - 24x + 9)$$
$$= 32x^2 + 18$$

29. $(3xy - 2z^2)^2 = (3xy)^2 - 2(3xy)(2z^2) + (2z^2)^2$
$$= 9x^2y^2 - 12xyz^2 + 4z^4$$

30. $(5a^3 + 2b)(2a^3 - 7b) = 10a^6 - 35a^3b + 4a^3b - 14b^2$
$$= 10a^6 - 31a^3b - 14b^2$$

31. $(x^2 - 3x + 2)(x^2 + 3x - 2)$
$$= [x^2 - (3x - 2)][x^2 + (3x - 2)]$$
$$= (x^2)^2 - (3x - 2)^2$$
$$= x^4 - [(3x)^2 - 2(2)(3x) + 2^2]$$
$$= x^4 - (9x^2 - 12x + 4)$$
$$= x^4 - 9x^2 + 12x - 4$$

32. $(x^2 - 3y)(x^4 + 3x^2y + 9y^2)$
$$= x^6 + 3x^4y + 9x^2y^2 - 3x^4y - 9x^2y^2 - 27y^3$$
$$= x^6 - 27y^3$$

33. $(2s^2 - 5s + 3)^2$
$$= (2s^2 - 5s + 3)(2s^2 - 5s + 3)$$
$$= 2s^2(2s^2 - 5s + 3) - 5s(2s^2 - 5s + 3)$$
$$\qquad\qquad\qquad\qquad + 3(2s^2 - 5s + 3)$$
$$= 4s^4 - 10s^3 + 6s^2 - 10s^3 + 25s^2 - 15s$$
$$\qquad\qquad\qquad\qquad\qquad + 6s^2 - 15s + 9$$
$$= 4s^4 - 20s^3 + 37s^2 - 30s + 9$$

34. $(5x^2 - 3yz)(5x^2 + 3yz) = (5x^2)^2 - (3yz)^2$
$$= 25x^4 - 9y^2z^2$$

35. $x^2 + 3x - 10 = (x + 5)(x - 2)$

36. $x^2 - 7x - 30 = (x - 10)(x + 3)$

37. $6x^2 + 13x - 5 = (3x - 1)(2x + 5)$

38. $2x^2 + 21x - 11 = (2x - 1)(x + 11)$

39. $3x^6 + 2x^3 - 8 = (3x^3 - 4)(x^3 + 2)$

40. $4x^6 - 9 = (2x^3)^2 - 3^2 = (2x^3 - 3)(2x^3 + 3)$

41. $8x^3 - 125 = (2x)^3 - (5)^3$
$$= (2x - 5)[(2x)^2 + 5(2x) + 5^2]$$
$$= (2x - 5)(4x^2 + 10x + 25)$$

42. $25x^2 - 20xy + 4y^2 = (5x - 2y)^2$

43. $9c^2d^4 - 6bcd^2 + b^2 = (3cd^2)^2 - 6bcd^2 + (b)^2$
$$= (3cd^2 - b)^2$$

44. $x^4 - 16 = (x^2)^2 - 4^2$
$$= (x^2 - 4)(x^2 + 4)$$
$$= (x - 2)(x + 2)(x^2 + 4)$$

45. $4x^2 - y^2 + 6y - 9 = 4x^2 - (y^2 - 6y + 9)$
$$= (2x)^2 - (y - 3)^2$$
$$= [2x + (y - 3)][2x - (y - 3)]$$
$$= (2x + y - 3)(2x - y + 3)$$

46. $25x^2 - 4y^2 - 15x - 6y$
$$= (5x - 2y)(5x + 2y) - 3(5x + 2y)$$
$$= (5x + 2y)(5x - 2y - 3)$$

47. $4x^2 - 29 = (2x)^2 - (\sqrt{29})^2$
$$= (2x + \sqrt{29})(2x - \sqrt{29})$$

48. $x^2 + 2x + 2 = (x + 1)^2 + 1$
$$= [(x + 1) - i][(x + 1) + i]$$
$$= (x + 1 - i)(x + 1 + i)$$

49. $\dfrac{3x^2 - 6x}{x^2 + x - 6} = \dfrac{3x(x - 2)}{(x - 2)(x + 3)}$
$$= \dfrac{3x}{x + 3}$$

50. $\dfrac{(x^2 - 4)^3}{(x^3 + 8)^2} = \dfrac{[(x - 2)(x + 2)]^3}{[(x + 2)(x^2 - 2x + 4)]^2}$
$$= \dfrac{(x - 2)^3(x + 2)^3}{(x + 2)^2(x^2 - 2x + 4)^2}$$
$$= \dfrac{(x - 2)^3(x + 2)}{(x^2 - 2x + 4)^2}$$

51. $\dfrac{4}{x + 4} - \dfrac{2}{x - 4} + \dfrac{3x - 1}{x^2 - 16}$

$$= \dfrac{4(x - 4)}{(x + 4)(x - 4)} - \dfrac{2(x + 4)}{(x + 4)(x - 4)} + \dfrac{3x - 1}{(x + 4)(x - 4)}$$
$$= \dfrac{4x - 16 - 2x - 8 + 3x - 1}{(x + 4)(x - 4)}$$
$$= \dfrac{5x - 25}{(x + 4)(x - 4)}$$
$$= \dfrac{5x - 25}{x^2 - 16}$$

52. $\dfrac{2x^2 + 5x - 3}{x^2 - 4} \cdot \dfrac{x^2 - 5x - 14}{2x^2 - 15x + 7}$

$= \dfrac{(2x - 1)(x + 3)}{(x - 2)(x + 2)} \cdot \dfrac{(x - 7)(x + 2)}{(2x - 1)(x - 7)}$

$= \dfrac{x + 3}{x - 2}$

53. $\dfrac{(1 - a^{-1})(1 + a^{-3})}{1 - a^{-2}} = \dfrac{\left(1 - \frac{1}{a}\right)\left(1 + \frac{1}{a^3}\right)}{1 - \frac{1}{a^2}}$

$= \dfrac{\left(\frac{a}{a} - \frac{1}{a}\right)\left(\frac{a^3}{a^3} + \frac{1}{a^3}\right)}{\frac{a^2}{a^2} - \frac{1}{a^2}}$

$= \dfrac{\left(\frac{a - 1}{a}\right)\left(\frac{a^3 + 1}{a^3}\right)}{\frac{a^2 - 1}{a^2}}$

$= \left(\dfrac{a - 1}{a}\right)\left(\dfrac{a^3 + 1}{a^3}\right)\left(\dfrac{a^2}{a^2 - 1}\right)$

$= \dfrac{(a - 1)(a + 1)(a^2 - a + 1)(a^2)}{a^4(a + 1)(a - 1)}$

$= \dfrac{a^2 - a + 1}{a^2}$

54. $1 - \left(\dfrac{x^3 - x^{-2}}{1 - x^{-2}}\right)^2$

$= 1 - \left(\dfrac{x^3 - x^{-2}}{1 - x^{-2}} \cdot \dfrac{x^2}{x^2}\right)^2$

$= 1 - \left(\dfrac{x^5 - 1}{x^2 - 1}\right)^2$

$= 1 - \left(\dfrac{(x - 1)(x^4 + x^3 + x^2 + x + 1)}{(x - 1)(x + 1)}\right)^2$

$= 1 - \left(\dfrac{x^4 + x^3 + x^2 + x + 1}{x + 1}\right)^2$

$= \left(1 - \dfrac{x^4 + x^3 + x^2 + x + 1}{x + 1}\right)\left(1 + \dfrac{x^4 + x^3 + x^2 + x + 1}{x + 1}\right)$

$= \dfrac{x + 1 - x^4 - x^3 - x^2 - x - 1}{x + 1} \cdot$

$\quad\quad\quad\quad \dfrac{x + 1 + x^4 + x^3 + x^2 + x + 1}{x + 1}$

$= \dfrac{-x^4 - x^3 - x^2}{x + 1} \cdot \dfrac{x^4 + x^3 + x^2 + 2x + 2}{x + 1}$

$= \dfrac{-x^2(x^2 + x + 1)(x^4 + x^3 + x^2 + 2x + 2)}{(x + 1)^2}$

28

Chapter 3: Equations and Inequalities

Problem Set 3.1

1. $2(x+4) = 8$
$2x + 8 = 8$
$2x = 0$
$x = 0$

This is a conditional equation, since it is true only if $x = 0$.

3. $3(2x - \frac{2}{3}) = 6x - 2$
$6x - 2 = 6x - 2$
This equation is an identity.

5. $\frac{2}{3}x + 4 = \frac{1}{2}x - 1$
$\frac{2}{3}x - \frac{1}{2}x = -1 - 4$
$\frac{1}{6}x = -5$
$x = -30$

This is a conditional equation, since it is true only if $x = -30$.

7. $(x+2)^2 = x^2 + 4$
$x^2 + 4x + 4 = x^2 + 4$
$x^2 - x^2 + 4x = 4 - 4$
$4x = 0$
$x = 0$

This is a conditional equation, since it is true only if $x = 0$.

9. $x^2 - 9 = (x+3)(x-3)$
$x^2 - 9 = x^2 - 9$
This is an identity.

11. $4x - 3 = 3x - 1$
$4x - 3x = -1 + 3$
$x = 2$

13. $2t + \frac{1}{2} = 4t - \frac{7}{2} + 8t$
$2t - 4t - 8t = -\frac{7}{2} - \frac{1}{2}$
$-10t = -\frac{8}{2} = -4$
$t = \frac{-4}{-10} = +\frac{2}{5}$

15. $3(x-2) = 5(x-3)$
$3x - 6 = 5x - 15$
$3x - 5x = -15 + 6$
$-2x = -9$
$x = \frac{-9}{-2} = \frac{9}{2}$

17. $\sqrt{3}z + 4 = -\sqrt{3}z + 8$
$\sqrt{3}z + \sqrt{3}z = 8 - 4$
$2\sqrt{3}z = 4$
$z = \frac{4}{2\sqrt{3}} = \frac{2}{\sqrt{3}}$

19. $3.23x - 6.15 = 1.41x + 7.63$
$3.23x - 1.41x = 7.63 + 6.15$
$(3.23 - 1.41)x = 7.63 + 6.15$
$x = \frac{7.63 + 6.15}{3.23 - 1.41}$
$x = 7.57$

21. $(6.13 \times 10^{-8})x + (5.34 \times 10^{-6}) = 0$
$(6.13 \times 10^{-8})x = -5.34 \times 10^{-6}$
$x = \frac{-5.24 \times 10^{-6}}{6.13 \times 10^{-8}}$
$x = -0.871 \times 10^2$
$x = -8.71 \times 10^1$
$x = -87.1$

23. $\frac{2}{3}x + 4 = \frac{1}{2}x$
$6\left(\frac{2}{3}x + 4\right) = 6\left(\frac{1}{2}x\right)$
$4x + 24 = 3x$
$4x - 3x = -24$
$x = -24$

25. $\frac{9}{10}x + \frac{5}{8} = \frac{1}{5}x + \frac{9}{20}$
$40\left(\frac{9}{10}x + \frac{5}{8}\right) = 40\left(\frac{1}{5}x + \frac{9}{20}\right)$
$36x + 25 = 8x + 18$
$36x - 8x = 18 - 25$
$28x = -7$
$x = \frac{-7}{28} = -\frac{1}{4}$

27. $\frac{3}{4}(x-2) = \frac{9}{5}$
$20\left[\frac{3}{4}(x-2)\right] = 20\left(\frac{9}{5}\right)$
$15(x-2) = 36$
$15x - 30 = 36$
$15x = 36 + 30$
$15x = 66$
$x = \frac{66}{15} = \frac{22}{5}$

29. $\frac{5}{x+2} = \frac{2}{x-1}$
$(x+2)(x-1)\left(\frac{5}{x+2}\right) = (x+2)(x-1)\left(\frac{2}{x-1}\right)$
$5(x-1) = 2(x+2)$
$5x - 5 = 2x + 4$
$5x - 2x = 4 + 5$
$3x = 9$
$x = 3$

Check:
$\frac{5}{3+2} \stackrel{?}{=} \frac{2}{3-1}$
$\frac{5}{5} \stackrel{?}{=} \frac{2}{2}$
$1 = 1$

31.
$$\frac{2}{x-3}+\frac{3}{x-7}=\frac{7}{(x-3)(x-7)}$$
$$(x-3)(x-7)\left(\frac{2}{x-3}+\frac{3}{x-7}\right)$$
$$=\frac{7}{(x-3)(x-7)}(x-3)(x-7)$$
$$2(x-7)+3(x-3)=7$$
$$2x-14+3x-9=7$$
$$5x=7+14+9$$
$$5x=30$$
$$x=\frac{30}{5}=6$$
Check:
$$\frac{2}{6-3}+\frac{3}{6-7}\overset{?}{=}\frac{7}{(6-3)(6-7)}$$
$$\frac{2}{3}+\frac{3}{-1}\overset{?}{=}\frac{7}{(3)(-1)}$$
$$\frac{2}{3}-3\overset{?}{=}\frac{-7}{3}$$
$$\frac{-7}{3}=\frac{-7}{3}$$

33.
$$\frac{x}{x-2}=2+\frac{2}{x-2}$$
$$(x-2)\left(\frac{x}{x-2}\right)=\left[2+\frac{2}{x-2}\right](x-2)$$
$$x=2(x-2)+2$$
$$x=2x-4+2$$
$$x-2x=-2$$
$$-x=-2$$
$$x=2$$
By inspection, we see that $x=2$ causes division by 0 in the original equation. $x=2$ is not a solution; the equation has no solution.

35.
$$x^2+4x=x^2-3$$
$$x^2-x^2+4x=-3$$
$$4x=-3$$
$$x=\frac{-3}{4}=-\frac{3}{4}$$

37. $(x-4)(x+5)=(x+2)(x+3)$
$$x^2+x-20=x^2+5x+6$$
$$x^2-x^2+x-5x=6+20$$
$$-4x=26$$
$$x=\frac{26}{-4}=-\frac{13}{2}$$

39.
$$\sqrt{5-2x}=5$$
$$(\sqrt{5-2x})^2=(5)^2$$
$$5-2x=25$$
$$-2x=25-5$$
$$-2x=20$$
$$x=\frac{20}{-2}=-10$$
Check:
$$\sqrt{5-2(-10)}\overset{?}{=}5$$
$$\sqrt{25}\overset{?}{=}5,\ 5=5$$

41.
$$\sqrt[3]{1-3x}=4$$
$$(\sqrt[3]{1-3x})^3=(4)^3$$
$$1-3x=64$$
$$-3x=64-1$$
$$-3x=63$$
$$x=\frac{63}{-3}=-21$$
Check:
$$\sqrt[3]{1-3(-21)}\overset{?}{=}4$$
$$\sqrt[3]{64}\overset{?}{=}4$$
$$4=4$$

43. $A=P+prt$ [Solve for P]
$$A=p(1+rt)$$
$$P=\frac{A}{1+rt}$$

45. $I=\frac{nE}{R+nr}$ [Solve for r]
$$I(R+nr)=nE$$
$$IR+Inr=nE$$
$$Inr=nE-IR$$
$$r=\frac{nE-IR}{In}$$

47. $A=2\pi r^2+2\pi rh$ [Solve for h]
$$A-2\pi r^2=2\pi rh$$
$$\frac{A-2\pi r^2}{2\pi r}=h$$

49. $R=\frac{R_1R_2}{R_1+R_2}$ [Solve for R_1]
$$R(R_1+R_2)=R_1R_2$$
$$RR_1+RR_2=R_1R_2$$
$$RR_1-R_1R_2=-RR_2$$
$$R_1(R-R_2)=-RR_2$$
$$R_1=\frac{-RR_2}{R-R_2}=\frac{RR_2}{R_2-R}$$

51.
$$\tfrac{3}{4}x-\tfrac{4}{3}=\tfrac{1}{3}x+\tfrac{5}{6}$$
$$12\left(\tfrac{3}{4}x-\tfrac{4}{3}\right)=12\left(\tfrac{1}{3}x+\tfrac{5}{6}\right)$$
$$9x-16=4x+10$$
$$9x-4x=10+16$$
$$5x=26$$
$$x=\tfrac{26}{5}$$

53.
$$4\left(3x-\tfrac{1}{2}\right)=5x+\tfrac{1}{2}$$
$$12x-2=5x+\tfrac{1}{2}$$
$$2(12x-2)=2(5x+\tfrac{1}{2})$$
$$24x-4=10x+1$$
$$24x-10x=1+4$$
$$14x=5$$
$$x=\tfrac{5}{14}$$

30

55. $(x-2)(3x+1) = (x-2)(3x+5)$
$3x^2 - 5x - 2 = 3x^2 - x - 10$
$-5x + x = 2 - 10$
$-4x = -8$
$x = 2$ which checks since $0 = 0$.

Notice that if we had divided each side originally by $(x-2)$, we would have obtained $3x+1 = 3x+5$, or $1 = 5$, which is never true. But $x = 2$ makes the *original* equation true; division by the factor $(x-2)$ would "throw out" the solution.

57.
$$\frac{2x}{4x+2} = \frac{x+1}{2x-1}$$
$$\frac{2x}{2(2x+1)} = \frac{x+1}{2x-1}$$
$$\frac{x}{2x+1} = \frac{x+1}{2x-1}$$
$$(2x+1)(2x-1)\left(\frac{x}{2x+1}\right) = (2x+1)(2x-1)\left(\frac{x+1}{2x-1}\right)$$
$$(2x-1)(x) = (2x+1)(x+1)$$
$$2x^2 - x = 2x^2 + 3x + 1$$
$$2x^2 - 2x^2 - 1 = 3x + x$$
$$-1 = 4x$$
$$-\tfrac{1}{4} = x$$

Check:
$$\frac{2\left(-\frac{1}{4}\right)}{4\left(-\frac{1}{4}\right)+2} \stackrel{?}{=} \frac{-\frac{1}{4}+1}{2\left(-\frac{1}{4}\right)-1}$$
$$\frac{-\frac{1}{2}}{1} \stackrel{?}{=} \frac{\frac{3}{4}}{-\frac{3}{2}}$$
$$-\tfrac{1}{2} \stackrel{?}{=} \left(\tfrac{3}{4}\right)\left(-\tfrac{2}{3}\right)$$
$$-\tfrac{1}{2} = -\tfrac{1}{2}$$

59.
$$1 + \frac{x}{x+3} = \frac{-3}{x+3}$$
$$(x+3)\left(1 + \frac{x}{x+3}\right) = (x+3)\left(\frac{-3}{x+3}\right)$$
$$(x+3) + x = -3$$
$$2x + 3 = -3$$
$$2x = -3 - 3$$
$$2x = -6$$
$$x = -3$$

By inspection, we see that $x = -3$ causes division by 0 in the original equation. $x = -3$ is not a solution; the equation has no solution.

61. $1 + \dfrac{1}{1 + \dfrac{1}{a + \frac{1}{x}}} = \dfrac{1}{a}$

Left side $= 1 + \dfrac{1}{1 + \dfrac{1}{\frac{ax+1}{x}}}$
$= 1 + \dfrac{1}{1 + \frac{x}{ax+1}} = 1 + \dfrac{1}{\frac{ax+1+x}{ax+1}}$
$= 1 + \dfrac{ax+1}{ax+1+x} = \dfrac{ax+1+x+ax+1}{ax+1+x}$
$= \dfrac{x+2ax+2}{x+ax+1}$

The original equation is then equivalent to
$$\frac{x+2ax+2}{x+ax+1} = \frac{1}{a}$$
$$a(x+ax+1)\left(\frac{x+2ax+2}{x+ax+1}\right) = a(x+ax+1)\left(\frac{1}{a}\right)$$
$$a(x+2ax+2) = x+ax+1$$
$$ax + 2a^2x + 2a = x + ax + 1$$
$$2a^2x - x = 1 - 2a$$
$$x(2a^2 - 1) = 1 - 2a$$
$$x = \frac{1-2a}{2a^2-1}$$

63. a) Let $C = 30$:
$$30 = \tfrac{5}{9}(F-32)$$
$$9(30) = 9\left(\tfrac{5}{9}\right)(F-32)$$
$$270 = 5(F-32)$$
$$270 = 5F - 160$$
$$270 + 160 = 5F$$
$$430 = 5F$$
$$\tfrac{430}{5} = F$$
$$F = 86°$$

b) Let $x =$ degrees F
Then $x =$ degrees C
$$x = \tfrac{5}{9}(x-32)$$
$$9x = 5(x-32)$$
$$9x = 5x - 160$$
$$9x - 5x = -160$$
$$4x = -160$$
$$x = \tfrac{-160}{4} = -40$$
At $-40°$F, a Celsius thermometer would also register -40.

c) Let $x =$ degrees F.
Then $\tfrac{x}{2} =$ degrees C
$$\tfrac{x}{2} = \tfrac{5}{9}(x-32)$$
$$18\left(\tfrac{x}{2}\right) = 18\left(\tfrac{5}{9}\right)(x-32)$$
$$9x = 10(x-32)$$
$$9x = 10x - 320$$
$$320 = 10x - 9x$$
$$320 = x, \ 160 = \tfrac{x}{2}$$

31

At 160° Celsius, the Celsius reading will be one-half the Fahrenheit's reading.

65. $A = P + Prt$, solve for P.
$$A = P(1 + rt)$$
$$\frac{A}{1 + rt} = P$$
With $A = 3813.75$, $r = .075$ and $t = 5.5$,
$$P = \frac{3813.75}{1 + (.075)(5.5)}$$
$$P = \$2700$$

67. a) If S represents an employee's salary this year, then $.10S$ represents 10% of the salary,
and $S - .10S$, or $.90S$, represents the salary after the pay cut.
Let x represent the percent (written as a decimal) increase necessary to restore the salary back to the value S.
Then $x(.90S)$ represents the amount of the increase and $.90S + x(.90S)$ represents the restored salary.
$$.90S + .90xS = S$$
$$S(.90 + .90x) = S$$
$$.90 + .90x = 1$$
$$.90x = 1 - .90$$
$$.90x = .10$$
$$x = \frac{.10}{.90} = \frac{1}{9} = .11\overline{1}$$
$$x = 11\tfrac{1}{9}\%$$

b) Let S represent an employee's salary this year.
Then $\left(\frac{p}{100}\right)S$ represents $p\%$ of the salary,
and $S - \frac{pS}{100}$ represents the salary after the pay cut.
Let x represent the percent increase necessary to restore the salary back to the value S.
Then $\frac{x}{100}\left(S - \frac{pS}{100}\right)$ represents the amount of the increase
and $\left(S - \frac{pS}{100}\right) + \frac{x}{100}\left(S - \frac{pS}{100}\right)$ represents the restored salary.
$$\left(S - \frac{pS}{100}\right) + \frac{x}{100}\left(S - \frac{pS}{100}\right) = S$$
$$S - \frac{pS}{100} + \frac{xS}{100} - \frac{xpS}{10000} = S$$
$$1 - \frac{p}{100} + \frac{x}{100} - \frac{xp}{10000} = 1$$
$$\frac{x}{100} - \frac{xp}{10000} = \frac{p}{100}$$
$$10000\left[\frac{x}{100} - \frac{xp}{10000}\right] = 10000\left[\frac{p}{100}\right]$$
$$100x - xp = 100p$$

$$x(100 - p) = 100p$$
$$x = \frac{100p}{100 - p}$$

1. Let x represent the number.
Then $2x$ represents twice the number.
$$15 + 2x = 33$$
$$2x = 33 - 15 = 18$$
$$x = \frac{18}{2} = 9$$
The number we seek is 9.

3. Let x represent the number.
Then $2x$ represents twice the number and $3x - 12$ represents 12 less than 3 times the number.
$$2x = 3x - 12$$
$$12 = 3x - 2x$$
$$12 = x$$
The number we seek is 12.

5. Let x, $x + 1$, and $x + 2$ represent three consecutive integers.
$$x + (x + 1) + (x + 2) = 72$$
$$3x + 3 = 72$$
$$3x = 72 - 3 = 69$$
$$x = 23$$
The smallest of the integers is 23.

7. Let w represent the width of the rectangle in centimeters. Then $w + 3$ represents the length in centimeters. Since the perimeter of a rectangle is twice the length plus twice the width,
$$2w + 2(w + 3) = 130$$
$$2w + 2w + 6 = 130$$
$$4w = 124$$
$$w = 31$$
The width of the rectangle is 31 cm.

9. Let x represent the score Mary needs on her second test.
Then $\frac{61 + x}{2}$ represents the average of Mary's two tests. Since we want the average to be 75,
$$\frac{61 + x}{2} = 75$$
$$61 + x = 150$$
$$x = 150 - 61$$
$$x = 89$$
Mary needs an 89 on her second test.

11. Let x represent the number of quarters necessary.
Then $25x$ represents the amount of money those quarters generate, in cents.
The 21 dimes amount to 210¢. Since we want a total of 385¢ to be in the change box,

$$25x + 210 = 385$$
$$25x = 385 - 210$$
$$25x = 175$$
$$x = 7$$

7 quarters must be added to the change box.

13. Let D represent the number of dimes the woman has. Then $25 - D$ represents the number of quarters she has.

$10D$ represents the monetary value of the dimes, in cents. $25(25 - D)$ represents the monetary value of the quarters, in cents.

Since the woman has a total of 445¢,

$$10D + 25(25 - D) = 445$$
$$10D + 625 - 25D = 445$$
$$-15D = 445 - 625 = -180$$
$$D = \frac{-180}{-15} = 12$$

Thus there are 12 dimes and 13 quarters.

15. Let t be the number of hours after 2:00 P.M. when Speedy will be 100 miles ahead of Slowpoke.
Slowpoke drove t hours.
Speedy, who left 1 hour later, drove $t - 1$ hours.
Using the formula $D = RT$,
$45t$ represents the distance Slowpoke covered and $60(t - 1)$ represents the distance Speedy covered.
Since Speedy drove 100 miles past Slowpoke,

$$60(t - 1) = 45t + 100$$
$$60t - 60 = 45t + 100$$
$$60t - 45t = 100 + 60$$
$$15t = 160$$
$$t = \frac{160}{15} = 10\frac{2}{3}$$

Speedy will be 100 miles ahead of Slowpoke $10\frac{2}{3}$ hours, or 10 hours and 40 minutes, after 2:00 P.M., that is, at 12:40 A.M. the next day.

17. Let t be the number of hours after 12:00 noon when the two men meet.
Then t represents the number of hours Paul travels and t also represents the number of hours Nick travels. Using $D = RT$,
$60t$ represents the number of miles Paul travels and $45t$ represents the number of miles Nick travels.
When they meet, their total mileage will amount to 455 miles:

$$60t + 45t = 455$$
$$105t = 455$$
$$t = \frac{455}{105} = 4\frac{1}{3}$$

The two men will meet $4\frac{1}{3}$ hours, or 4 hours and 20 minutes, after 12:00 noon; that is, at 4:20 P.M.

19. Let r represent the rate at which Luella can row in still water, measured in miles per hour.
Then $r - 3$ represents the rate at which the boat goes upstream and $r + 3$ represents the rate at which the boat goes downstream. Using $T = \frac{D}{R}$,
$\frac{1}{r - 3}$ represents the number of hours it takes Luella to row 1 mile upstream.

$\frac{2}{r + 3}$ represents the number of hours it takes Luella to row 2 miles downstream.

$$\frac{1}{r - 3} = \frac{2}{r + 3}$$
$$(r + 3)(r - 3)\left(\frac{1}{r - 3}\right) = (r + 3)(r - 3)\left(\frac{2}{r + 3}\right)$$
$$r + 3 = 2(r - 3)$$
$$r + 3 = 2r - 6$$
$$3 + 6 = 2r - r$$
$$9 = r$$

Luella can row at the rate of 9 mph in still water.

21. Let x represent the son's age today.
Then $3x$ represents the father's age today.
$x + 15$ represents how old the son will be in 15 years. $3x + 15$ represents how old the father will be in 15 years. In 15 years the father will be twice as old as the son:

$$3x + 15 = 2(x + 15)$$
$$3x + 15 = 2x + 30$$
$$3x - 2x = 30 - 15$$
$$x = 15$$

The son is 15 years old today.

23. Let x represent the number of cc of the 40% solution that should be added. Then $.40x$ represents the cc of pure HCl in the 40% solution.
$x + 2000$ represents the number of cc of the new solution.
$.35(x + 2000)$ represents the cc of pure HCl in the new solution.
The cc of pure HCl contributed by the 40% solution plus the cc of pure HCl contributed by the 20% solution must equal the cc of pure HCl in the new solution:

$$.40x + .20(2000) = .35(x + 2000)$$
$$.40x + 400 = .35x + 700$$
$$.40x - .35x = 700 - 400$$
$$.05x = 300$$
$$x = \frac{300}{.05} = 6000$$

6000 cc of the 40% solution must be added.

25. The 1000 liters of the original 30% brine solution contain $.30(1000) = 300$ liters of salt.
Let x represent the number of liters of water to be boiled off. Then $1000 - x$ represents the number of liters of solution remaining.
$.35(1000 - x)$ represents the number of liters of salt in the new solution.
There are 300 liters of salt in the new solution:

$$300 = .35(1000 - x)$$
$$300 = 350 - .35x$$
$$.35x = 350 - 300$$
$$.35x = 50$$
$$x = \frac{50}{.35} \approx 142.86$$

Approximately 142.86 liters of water should be boiled off.

27. Let x represent the original price of the suit.
Then $.15x$ represents the amount of the markdown, in dollars.
$x - .15x$ represents the sale price of the suit.
Since the sale price of the suit is $123.25,

$$x - .15x = 123.25$$
$$.85x = 123.25$$
$$x = \frac{123.25}{.85} = 145$$

The original price of the suit was $145.

29. Let x represent the width of the driveway, in *yards*.
Convert all dimensions to yards:

$$36 \text{ feet} = \frac{36 \text{ feet}}{\frac{3 \text{ feet}}{\text{yard}}} = 12 \text{ yards}$$

$$4 \text{ inches} = \frac{4 \text{ inches}}{\frac{36 \text{ inches}}{\text{yard}}} = \frac{1}{9} \text{ yards}$$

The volume of the driveway must be 4 cubic yards:

$$(12)\left(\tfrac{1}{9}\right)x = 4$$
$$\frac{12x}{9} = 4$$
$$x = 4\left(\tfrac{9}{12}\right) = 3$$

The driveway must be 3 yards, or 9 feet, wide.

31. Let x represent the number of days it would take Jill to hoe the garden by herself.
Then $\frac{1}{x}$ represents the fraction of the garden she can hoe by herself in 1 day.
Jack can hoe $\frac{1}{5}$ of the garden by himself in 1 day.
Together, Jack and Jill could complete $\frac{1}{3}$ of the garden in 1 day:

$$\frac{1}{x} + \frac{1}{5} = \frac{1}{3}$$
$$15x\left(\frac{1}{x} + \frac{1}{5}\right) = 15x\left(\frac{1}{3}\right)$$
$$15 + 3x = 5x$$
$$15 = 5x - 3x$$
$$15 = 2x$$
$$\frac{15}{2} = x$$

Jill would take $\frac{15}{2}$, or $7\frac{1}{2}$, days to hoe the garden by herself.

33. Let x be the number of feet from the fulcrum to the point where Susan's dad sits.
$200x$ represents weight $\cdot$ distance for Susan's dad.
$92(10)$ represents weight $\cdot$ distance for Susan and her puppy.

$$200x = 92(10)$$
$$200x = 920$$
$$x = 4.6$$

Susan's dad should sit 4.6 feet from the fulcrum.

35.
$$4x = 50(10)$$
$$4x = 500$$
$$x = \frac{500}{4} = 125 \text{ pounds}$$

37.
$$100 \cdot 10 = \tfrac{5}{2}x + 2x(10)$$
$$1000 = \tfrac{5}{2}x + 20x$$
$$1000 = 22.5x$$
$$x = \frac{1000}{22.5} = 44.4 \text{ pounds}$$

39. Let x represent the length of each edge of the garden, in feet. Then $x + 6$ represents the length of each outer edge of the sidewalk border.
x^2 represents the area of the garden.
$(x + 6)^2$ represents the area of the garden and sidewalk together.
Since the area of the sidewalk alone is 249 square ft

$$(x + 6)^2 - x^2 = 249$$
$$x^2 + 12x + 36 - x^2 = 249$$
$$12x = 249 - 36$$
$$12x = 213$$
$$x = \frac{213}{12} = 17\frac{3}{4}$$

The garden is $17\frac{3}{4}$ feet wide.

41. Let x be Matilda's age on her 50th anniversary.
Then $x + 4$ represents her husband's age on their 50th anniversary. $x - 25$ is Matilda's age on her 25th anniversary. $(x + 4) - 25 = x - 21$ is her husband's age on their 25th anniversary. $x - 50$ is Matilda's age on her wedding day. $(x + 4) - 50 = x - 46$ is her husband's age on their wedding day.

The sum of their ages on their 25th anniversary is double the sum of their ages on their wedding day:

$$(x-25)+(x-21)=2[(x-50)+(x-46)]$$
$$2x-46=2(2x-96)$$
$$2x-46=4x-192$$
$$-46+192=4x-2x$$
$$146=2x$$
$$73=x$$

Matilda will be 73 years old on her 50th anniversary.

43. Let b represent the number of boys in the class.

Total weight of girls $= 14(128)$

Total weight of boys $= b(160)$

Total weight of the class $= (b+14)146$

$$14(128)+160b=146(b+14)$$
$$14(128)+160b=146b+14(146)$$
$$160b-146b=14(146)-14(128)$$
$$14b=14(146-128)=14(18)$$
$$b=\frac{14(18)}{14}=18 \text{ boys}$$

There are $14+18=32$ students in the class.

45. Let x represent Amos's contribution.

Since he is one of the givers, the average contribution of all the givers is $\frac{44000+x}{89}$. Amos's contribution must be $1000 more than the above average:

$$x=\frac{44000+x}{89}+1000$$
$$89x=44000+x+1000(89)$$
$$89x-x=44000+89000$$
$$88x=133000$$
$$x=\frac{133000}{88}\approx 1511.360$$

Amos must donate $1511.36.

47. Let x represent the interest rate (in decimal) at which Susan invested her $7400.

Then $2x$ represents the interest rate (in decimal) at which she invested her $4600.

$7400x$ represents the amount of money she earned on the $7400.

$4600(2x)$ represents the amount of money she earned on the $4600.

$.072625(7400+4600)$ represents her total earnings.

$$7400x+4600(2x)=(.072625)(12000)$$
$$16600x=871.5$$
$$x=\frac{871.5}{16600}=.0525$$

Susan invested the $7400 at 5.25%, and the $4600 at $2(5.25\%)=10.5\%$.

49. Let x represent the length of Diophantus' life.

Draw a time line:

From the time line, we see that

$x-\left(\frac{1}{6}x+\frac{1}{12}x+\frac{1}{7}x+5+4\right)$ represents the length of the son's life.

Since Diophantus lived twice as long as his son,

$$x=2\left[x-\left(\frac{1}{6}x+\frac{1}{12}x+\frac{1}{7}x+5+4\right)\right]$$
$$x=2\left(x-\frac{1}{6}x-\frac{1}{12}x-\frac{1}{7}x-9\right)$$
$$x=2x-\frac{1}{3}x-\frac{1}{6}x-\frac{2}{7}x-18$$
$$18=2x-\frac{1}{3}x-\frac{1}{6}x-\frac{2}{7}x-x$$
$$18=\frac{9}{42}x$$
$$\frac{42}{9}(18)=x$$
$$84=x$$

Diophantus was 84 years old when he died.

51. a) Let $t=$ number of hours after noon before scout A turns around. By the time A returns (2 hours), the column will have marched $2\cdot 5=10$ miles. He rides $20t$ miles before turning around, and rides $20(2-t)$ miles back to the rear of the column. At that point he is 10 miles ahead of the start of his ride. So

$$20t=20(2-t)+10$$
$$20t=40-20t+10$$
$$40t=50$$
$$t=\frac{50}{40}=\frac{5}{4} \text{ hours}=1 \text{ hour and } 15 \text{ minutes}$$

(He rides 25 miles, turns, and rides back 15 miles to the general.)

b) Let $T_1=$ time in hours after noon for rider B to get to the head of the column. In time T_1 the column will have advanced $5T_1$ miles and the head of the column is 2 miles $+5T_1$ miles from the place the rider started. The rider has gone $20T_1$ miles. From these facts we write

$$20T_1=5T_1+2$$
$$15T_1=2$$
$$T_1=\frac{2}{15} \text{ hours}$$

We now let $T_2=$ time in hours from the turnaround until rider B gets back to the rear of the column. B has to go 2 miles less the distance $5T_2$ than the rear of the column has moved forward. He also rides $20T_2$ miles in this time, so

$$20T_2 = 2 - 5T_2$$
$$25T_2 = 2$$
$$T_2 = \tfrac{2}{25} \text{ hours.}$$

Thus the total time for rider B to ride to the head of the column and return is

$$T_1 + T_2 = \tfrac{2}{15} + \tfrac{2}{25} = \tfrac{10}{75} + \tfrac{6}{75} = \tfrac{16}{75} \text{ hours}$$

$$= \tfrac{16}{75} \cdot \frac{60 \text{ minutes}}{\text{hour}} = 12\tfrac{4}{5} \text{ minutes} = 12 \text{ min, } 48 \text{ sec}$$

Problem Set 3.3

1. $\begin{cases} 2x + 3y = 13 \\ y = 13 \end{cases}$

Substituting:
$$2x + 3(13) = 13$$
$$2x = 13 - 39$$
$$2x = -26$$
$$x = -13$$
Solution: $x = -13$, $y = 13$

3. $\begin{cases} 2u - 5v = 23 \\ 2u = 3 \end{cases}$

From the second equation, we see that $u = \tfrac{3}{2}$.
Substituting $u = \tfrac{3}{2}$ in the first equation:
$$2\left(\tfrac{3}{2}\right) - 5v = 23$$
$$3 - 5v = 23$$
$$-5v = 20$$
$$v = \frac{20}{-5} = -4$$
Solution: $u = \tfrac{3}{2}$, $v = -4$

5. $\begin{cases} 7x + 2y = -1 \\ y = 4x + 7 \end{cases}$

Substituting:
$$7x + 2(4x + 7) = -1$$
$$7x + 8x + 14 = -1$$
$$15x = -15$$
$$x = -1$$
Substituting $x = -1$ in the second equation:
$$y = 4(-1) + 7$$
$$y = 3$$
Solution: $x = -1$, $y = 3$

7. $\begin{cases} y = -2x + 11 \\ y = 3x - 9 \end{cases}$

Equating values of y:
$$3x - 9 = -2x + 11$$
$$5x = 20$$
$$x = 4$$
Substituting $x = 4$ in the second equation:
$$y = 3(4) - 9 = 3$$
Solution: $x = 4$, $y = 3$

9. $\begin{cases} x - y = 14 \\ x + y = -2 \end{cases}$

Adding these two equations:
$$2x = 12$$
$$x = 6$$
Substituting $x = 6$ in the second equation:
$$6 + y = -2$$
$$y = -8$$
Solution: $x = 6$, $y = -8$

11. $\begin{cases} 2s - 3t = -10 \\ 5s + 6t = 29 \end{cases}$

Multiplying the first equation by 2:
$$4s - 6t = -20$$
$$5s + 6t = 29$$
Adding these last two equations:
$$9s = 9$$
$$s = 1$$
Substituting $s = 1$ in the second original equation:
$$5(1) + 6t = 29$$
$$6t = 24$$
$$t = 4$$
Solution: $s = 1$, $t = 4$

13. $\begin{cases} 5x - 4y = 19 \\ 7x + 3y = 18 \end{cases}$

Multiplying the first equation by 3 and the second equation by 4:
$$15x - 12y = 57$$
$$28x + 12y = 72$$
Adding these two equations:
$$43x = 129$$
$$x = 3$$
Substituting $x = 3$ in the second original equation:
$$7(3) + 3y = 18$$
$$3y = -3$$
$$y = -1$$
Solution: $x = 3$, $y = -1$

15. $\begin{cases} 7x - 4y = 0 \\ 2x + 7y = 57 \end{cases}$

Multiplying the first equation by 7 and the second equation by 4:
$$49x - 28y = 0$$
$$8x + 28y = 228$$
Adding these last two equations:
$$57x = 228$$
$$x = \frac{228}{57} = 4$$
Substituting $x = 4$ in the first original equation:
$$7(4) - 4y = 0$$
$$-4y = -28$$
$$y = \frac{-28}{-4} = 7$$
Solution: $x = 4$, $y = 7$

17. $\begin{cases} \frac{2}{3}x + y = 4 \\ x + 2y = 5 \end{cases}$

Multiplying the first equation by 3 and the second equation by -2:

$2x + 3y = 12$
$-2x - 4y = -10$

Adding these last two equations:

$-y = 2$
$y = -2$

Substituting $y = -2$ in the second original equation:

$x + 2(-2) = 5$
$x - 4 = 5$
$x = 9$

Solution: $x = 9$, $y = -2$

19. $\begin{cases} .125x - .2y = 3 \\ .75x + .3y = 10.5 \end{cases}$

Multiplying the first equation by 3 and the second equation by 2:

$.375x - .6y = 9$
$\underline{1.5x + .6y = 21}$
$1.875x \quad\quad = 30$

$x = \dfrac{30}{1.875} = \dfrac{30{,}000}{1875} = \dfrac{1200}{75} = \dfrac{48}{3} = 16$

Substituting $x = 16$ in the second original equation:

$.75(16) + .3y = 10.5$
$12 + .3y = 10.5$
$.3y = -1.5$
$y = \dfrac{-1.5}{.3} = -5$

Solution: $x = 16$, $y = -5$

21. $\begin{cases} \frac{4}{x} + \frac{3}{y} = 17 \\ \frac{1}{x} - \frac{3}{y} = -7 \end{cases}$

Let $u = \dfrac{1}{x}$ and $v = \dfrac{1}{y}$

$4u + 3v = 17$
$u - 3v = -7$

Adding these two equations:

$5u = 10$
$u = 2$

Substituting $u = 2$ in $u - 3v = -7$,

$2 - 3v = -7$
$-3v = -9$
$v = 3$

Since $u = \dfrac{1}{x}$ Since $v = \dfrac{1}{y}$

$2 = \dfrac{1}{x}$ $3 = \dfrac{1}{y}$

$x = \dfrac{1}{2}$ $y = \dfrac{1}{3}$

Solution: $x = \frac{1}{2}$, $y = \frac{1}{3}$

23. $\begin{cases} \frac{2}{\sqrt{x}} - \frac{1}{\sqrt{y}} = \frac{2}{3} \\ \frac{1}{\sqrt{x}} + \frac{2}{\sqrt{y}} = \frac{7}{6} \end{cases}$

Let $u = \dfrac{1}{\sqrt{x}}$ and $v = \dfrac{1}{\sqrt{y}}$

$2u - v = \dfrac{2}{3}$
$u + 2v = \dfrac{7}{6}$

Multiplying the first of these two equations by 2:

$4u - 2v = \dfrac{4}{3}$
$\underline{u + 2v = \dfrac{7}{6}}$

Adding these last two equations:

$5u = \dfrac{15}{6}$
$u = \dfrac{15}{6} \cdot \dfrac{1}{5} = \dfrac{1}{2}$

Substituting $u = \frac{1}{2}$ in $u + 2v = \frac{7}{6}$:

$\dfrac{1}{2} + 2v = \dfrac{7}{6}$
$2v = \dfrac{4}{6}$
$v = \dfrac{4}{6} \cdot \dfrac{1}{2} = \dfrac{1}{3}$

Since $u = \dfrac{1}{\sqrt{x}}$ Since $v = \dfrac{1}{\sqrt{y}}$

$\dfrac{1}{2} = \dfrac{1}{\sqrt{x}}$ $\dfrac{1}{3} = \dfrac{1}{\sqrt{y}}$

$\sqrt{x} = 2$ $\sqrt{y} = 3$
$x = 4$ $y = 9$

Solution: $x = 4$, $y = 9$

25. $\begin{cases} 4x - y + 2z = 2 \\ -3x + y - 4z = -1 \\ x \quad\; + 5z = 1 \end{cases}$

From the third equation, $x = 1 - 5z$, which we substitute in each of the other two equations:

$\begin{cases} 4(1 - 5z) - y + 2z = 2 \\ -3(1 - 5z) + y - 4z = -1 \end{cases}$

$\begin{cases} 4 - 20z - y + 2z = 2 \\ -3 + 15z + y - 4z = -1 \end{cases}$

$\begin{cases} -y - 18z = -2 \\ y + 11z = 2 \end{cases}$

Adding these two equations:

$-7z = 0$
$z = 0$

Substituting in $y + 11z = 2$:

$y + 11(0) = 2$
$y = 2$

Substituting in the third original equation:

$x + 5(0) = 1$
$x = 1$

Solution: $x = 1$, $y = 2$, $z = 0$

27. $\begin{cases} 2x + 3y + 4z = -6 \\ -x + 4y - 6z = 6 \\ 3x - 2y + 2z = 2 \end{cases}$

Multiplying the second equation by 2 and adding it to the first equation:
$$11y - 8z = 6$$
Multiplying the second equation by 3 and adding it to the third equation:
$$10y - 16z = 20$$
Multiplying the first of these new equations by -2:
$$-22y + 16z = -12$$
$$10y - 16z = 20$$
Adding these last two equations:
$$-12y = 8$$
$$y = \frac{8}{-12} = -\frac{2}{3}$$
Substituting $y = -\frac{2}{3}$ in $11y - 8z = 6$:
$$11\left(-\frac{2}{3}\right) - 8z = 6$$
$$-8z = \frac{40}{3}$$
$$z = \frac{40}{3}\left(-\frac{1}{8}\right) = -\frac{5}{3}$$
Substituting $y = -\frac{2}{3}$ and $z = -\frac{5}{3}$ in second original equation:
$$-x + 4\left(-\frac{2}{3}\right) - 6\left(-\frac{5}{3}\right) = 6$$
$$-x = 6 + \frac{8}{3} - 10$$
$$-x = -\frac{4}{3}$$
$$x = \frac{4}{3}$$
Solution: $x = \frac{4}{3}$, $y = -\frac{2}{3}$, $z = -\frac{5}{3}$

29. $\begin{cases} 3x - 2y + z = -2 \\ 4x + 3y - 5z = 5 \\ 5x - 5y + 3z = -4 \end{cases}$

Multiplying the first equation by 5 and adding it to the second equation:
$$19x - 7y = -5$$
Multiplying the first equation by -3 and adding it to the third equation:
$$-4x + y = 2$$
Multiplying the second of these new equations by 7:
$$19x - 7y = -5$$
$$-28x + 7y = 14$$
Adding these last two equations:
$$-9x = 9$$
$$x = -1$$
Substituting $x = -1$ in $-4x + y = 2$:
$$-4(-1) + y = 2$$
$$y = 2 - 4 = -2$$
Substituting $x = -1$ and $y = -2$ in the first original equation:

$$3(-1) - 2(-2) + z = -2$$
$$z = -2 + 3 - 4$$
$$z = -3$$
Solution: $x = -1$, $y = -2$, $z = -3$

31. $\begin{cases} \frac{1}{3}r + \frac{2}{9}s = 24 \\ \frac{2}{9}r + \frac{1}{3}s = 26 \end{cases}$

Multiplying each equation by 9 to eliminate fractions:
$$3r + 2s = 216$$
$$2r + 3s = 234$$
Multiplying the first of these new equations by 3 and the second by 2:
$$9r + 6s = 648$$
$$4r + 6s = 468$$
Subtracting the second equation from the first:
$$5r = 180$$
$$r = 36$$
Substituting this in $3r + 2s = 216$:
$$108 + 2s = 216$$
$$2s = 108$$
$$s = 54$$
Solution: $r = 36$, $s = 54$

33. Let x represent the selling price of the first car and y represent the selling price of the second car.
$$\begin{cases} x + y = 13000 \\ y = x + 1400 \end{cases}$$
Substituting for y in the first equation:
$$x + (x + 1400) = 13000$$
$$2x + 1400 = 13000$$
$$2x = 11600$$
$$x = 5800$$
Substituting $x = 5800$ in the second equation:
$$y = 5800 + 1400 = 7200$$
The first car sold for $5800 and the second car sold for $7200.

35. Let x and y represent the two numbers.
$$\begin{cases} x + y = \frac{1}{3} \\ x - y = 3 \end{cases}$$
Adding the two equations:
$$2x = \frac{10}{3}$$
$$x = \left(\frac{10}{3}\right)\left(\frac{1}{2}\right) = \frac{5}{3}$$
Substituting $x = \frac{5}{3}$ in the first equation:
$$\frac{5}{3} + y = \frac{1}{3}$$
$$y = -\frac{4}{3}$$
The two numbers are $\frac{5}{3}$ and $-\frac{4}{3}$.

37. Let $x =$ numerator and $y =$ denominator of the original fraction:

$$\begin{cases} \dfrac{x+1}{y+1} = \dfrac{3}{5} \\[2mm] \dfrac{x-1}{y-1} = \dfrac{5}{9} \end{cases}$$

Cross multiplying in each equation:

$$\begin{cases} 5x + 5 = 3y + 3 \\ 9x - 9 = 5y - 5 \end{cases}$$

$$\begin{cases} 5x - 3y = -2 \\ 9x - 5y = 4 \end{cases}$$

Multiplying the first equation by 5 and the second by 3:

$$25x - 15y = -10$$
$$27x - 15y = 12$$

Subtracting the first equation from the second:

$$2x = 22$$
$$x = 11$$

Substituting $x = 11$ in $5x - 3y = -2$:

$$55 - 3y = -2$$
$$3y = 57$$
$$y = 19$$

Hence the original fraction is $\frac{11}{19}$.

39. Let x represent the length of the piece of wire that will be bent into a square.

Then $\frac{x}{4}$ represents the length of a side of the square.

Let y represent the width of the rectangle.

Then $2y$ represents the length of the rectangle, and $2y + 2(2y) = 6y$ represents the perimeter of the rectangle, or the length of the piece of wire that forms the rectangle.

$$\begin{cases} x + 6y = 39 \\[1mm] \dfrac{x}{4} = y + 1 \end{cases}$$

Multiplying this second equation by 4:

$$x = 4y + 4$$

Substituting into the first equation:

$$(4y + 4) + 6y = 39$$
$$10y = 35$$
$$y = \frac{35}{10} = \frac{7}{2}$$

Substituting $y = \frac{7}{2}$ into $x = 4y + 4$:

$$x = 4\left(\frac{7}{2}\right) + 4 = 18$$

The wire is cut 18 inches from one end.

41. Let x represent the number of $10 tickets sold.

Then $10x$ represents the amount of money received for all the $10 tickets sold.

Let y represent the number of $15 tickets sold.

Then $15y$ represents the amount of money received for all the $15 tickets sold.

$$\begin{cases} x + y = 45000 \\ 10x + 15y = 495000 \end{cases}$$

Multiplying the first equation by -10:

$$-10x - 10y = -450000$$
$$10x + 15y = 495000$$

Adding these two equations:

$$5y = 45000$$
$$y = 9000$$

Substituting $y = 9000$ in the first original equation:

$$x + 9000 = 45000$$
$$x = 36000$$

36,000 $10 tickets and 9000 $15 tickets were sold.

43. Let t represent the number of hours it would take Huck Finn to float downstream in his raft.

Let x represent the speed of the current, which is also the speed of the raft, in miles per hour.

Let y represent the speed of the motor boat in still water.

Then $y + x$ represents the speed of the boat going downstream, and $y - x$ represents the speed of the boat going upstream.

Using the formula $D = RT$, the distance from A to B can be represented in 3 different ways: tx, $5(y + x)$, and $7(y - x)$. We can then form several equations:

$$\begin{cases} tx = 5(y + x) \\ 5(y + x) = 7(y - x) \end{cases}$$

Solving the second equation for y in terms of x:

$$5y + 5x = 7y - 7x$$
$$2y = 12x$$
$$y = 6x$$

Substituting $6x$ for y in the first equation and solving:

$$tx = 5(6x + x)$$
$$tx = 35x$$

Since the motorboat takes longer to go upstream than downstream, the speed of the current, x, cannot be 0. Therefore we can divide both sides of the preceding equation by x to get

$$t = 35$$

It would take Huck Finn 35 hours to float downstream on his raft.

45. Let x represent the number of pounds of $3.60 coffee the grocer must include in her blend.

Then $3.60x$ represents the monetary value of the x pounds of coffee, in dollars.

Let y represent the number of pounds of $2.90 coffee the grocer must include in her blend.

Then $2.90y$ represents the monetary value of the y pounds of coffee, in dollars. $3.30(x + y)$ represents the dollar value of the entire mixture of coffee.

$$\begin{cases} x + y = 100 \\ 3.60x + 2.90y = 3.30(x + y) \end{cases}$$
$$x + y = 100$$
$$36x + 29y = 33x + 33y$$
Expanding, collecting, and rearranging:
$$x = 100 - y$$
$$3x - 4y = 0$$
Substituting the first equation into the second:
$$3(100 - y) - 4y = 0$$
$$300 = 7y$$
$$y = \frac{300}{7} = 42\frac{6}{7}$$
$$x = 100 - \frac{300}{7} = \frac{400}{7} = 57\frac{1}{7}$$

The grocer should mix $42\frac{6}{7}$ lbs of \$2.90 coffee with $57\frac{1}{7}$ lbs of \$3.60 coffee.

47. Let x represent the selling price of a dress. Then $.2x$ represents the profit Susan makes on one dress. The selling price of the dress must equal the cost of the dress plus the profit Susan wants to earn on this dress:
$$x = 40 + .2x$$
$$.8x = 40$$
$$x = \frac{40}{.8} = 50$$

The selling price of one dress must be \$50.
Let y represent the selling price of a coat. Then $.5y$ represents the profit Susan makes on one coat.
$$y = 100 + .5y$$
$$.5y = 100$$
$$y = \frac{100}{.5} = 200$$

The selling price of a coat must be \$200.
Now, let c represent the number of coats Susan bought (and sold). Let d represent the number of dresses Susan bought (and sold).
$$\begin{cases} 100c + 40d = 4800 \\ 200c + 50d = 4800 + 1800 \end{cases}$$
Multiplying the first equation by -2:
$$-200c - 80d = -9600$$
$$200c + 50d = 6600$$
Adding these last two equations:
$$-30d = -3000$$
$$d = 100$$
Substituting $d = 100$ in the first original equation:
$$100c + 40(100) = 4800$$
$$100c = 800$$
$$c = 8$$
Susan bought 100 dresses and 8 coats.

49. $$\begin{cases} 3x - 4y + 2z = -5 \\ 4x - 5y + 5z = 2 \end{cases}$$
Multiply the first equation by 4 and the second equation by -3:
$$12x - 16y + 8z = -20$$
$$-12x + 15y - 15z = -6$$
Adding these last two equations:
$$-y - 7z = -26$$
$$y = 26 - 7z$$
Substituting $y = 26 - 7z$ in the first original equation:
$$3x - 4(26 - 7z) + 2z = -5$$
$$3x - 104 + 28z + 2z = -5$$
$$3x = 99 - 30z$$
$$x = 33 - 10z$$
Solution: $x = 33 - 10z$, $y = 26 - 7z$

Problem Set 3.4

1. $\sqrt{50} = \sqrt{25 \cdot 2} = 5\sqrt{2}$

3. $\sqrt{\frac{1}{4}} = \frac{1}{2}$

5. $\frac{\sqrt{45}}{\sqrt{20}} = \sqrt{\frac{45}{20}} = \sqrt{\frac{9}{4}} = \frac{3}{2}$

7. $\sqrt{11^2 \cdot 4} = 11 \cdot 2 = 22$

9. $\frac{5 + \sqrt{72}}{5} = \frac{5 + \sqrt{36 \cdot 2}}{5} = \frac{5 + 6\sqrt{2}}{5}$

11. $\frac{18 + \sqrt{-9}}{6} = \frac{18 + \sqrt{9}i}{6} = \frac{18 + 3i}{6} = \frac{3(6 + i)}{3(2)} = \frac{6 + i}{2}$

13. $$x^2 = 25$$
$$x = \pm 5$$

15. $$(x - 3)^2 = 16$$
$$x - 3 = \pm 4$$
$$x = 3 \pm 4$$
$$x = 7 \text{ or } x = -1$$

17. $$(2x + 5)^2 = 100$$
$$2x + 5 = \pm 10$$
$$2x = -5 \pm 10$$
$$x = \frac{-5 \pm 10}{2}$$
$$x = \frac{5}{2} \text{ or } x = -\frac{15}{2}$$

19. $$m^2 = -9$$
$$m = \pm\sqrt{-9} = \pm 3i$$

21. $$x^2 = 3x$$
$$x^2 - 3x = 0$$
$$x(x - 3) = 0$$
$$x = 0 \text{ or } x - 3 = 0, \; x = 3$$

23.
$$x^2 - 9 = 0$$
$$(x+3)(x-3) = 0$$
$$x+3 = 0 \quad \text{or} \quad x-3 = 0$$
$$x = -3 \quad \text{or} \quad x = 3$$

25.
$$m^2 - .0144 = 0$$
$$(m+.12)(m-.12) = 0$$
$$m+.12 = 0 \quad \text{or} \quad m-.12 = 0$$
$$m = -.12 \quad \text{or} \quad m = .12$$

27.
$$x^2 - 3x - 10 = 0$$
$$(x-5)(x+2) = 0$$
$$x-5 = 0 \quad \text{or} \quad x+2 = 0$$
$$x = 5 \quad \text{or} \quad x = -2$$

29.
$$3x^2 + 5x - 2 = 0$$
$$(3x-1)(x+2) = 0$$
$$3x-1 = 0 \quad \text{or} \quad x+2 = 0$$
$$3x = 1$$
$$x = \tfrac{1}{3} \quad \text{or} \quad x = -2$$

31.
$$6x^2 - 13x - 28 = 0$$
$$(2x-7)(3x+4) = 0$$
$$2x-7 = 0 \quad \text{or} \quad 3x+4 = 0$$
$$x = \tfrac{7}{2} \quad \text{or} \quad x = -\tfrac{4}{3}$$

33.
$$x^2 + 8x = 9$$
$$x^2 + 8x + 16 = 9 + 16$$
$$(x+4)^2 = 25$$
$$x+4 = \pm\sqrt{25}$$
$$x = -4 \pm 5$$
$$x = 1 \text{ or } x = -9$$

35.
$$z^2 - z = \tfrac{3}{4}$$
$$z^2 - z + \tfrac{1}{4} = \tfrac{3}{4} + \tfrac{1}{4}$$
$$\left(z - \tfrac{1}{2}\right)^2 = 1$$
$$z - \tfrac{1}{2} = \pm\sqrt{1}$$
$$z = \tfrac{1}{2} \pm 1$$
$$z = \tfrac{3}{2} \text{ or } z = -\tfrac{1}{2}$$

37.
$$x^2 + 4x = -9$$
$$x^2 + 4x + 4 = -9 + 4$$
$$(x+2)^2 = -5$$
$$x+2 = \pm i\sqrt{5}$$
$$x = -2 \pm i\sqrt{5}$$

39. $x^2 + 8x + 12 = 0$
$$a = 1,\ b = 8,\ c = 12$$
$$x = \frac{-8 \pm \sqrt{(8)^2 - 4(1)(12)}}{2(1)}$$
$$= \frac{-8 \pm \sqrt{64 - 48}}{2}$$

$$= \frac{-8 \pm \sqrt{16}}{2}$$
$$= \frac{-8 \pm 4}{2}$$
$$x = -\tfrac{4}{2} = -2 \text{ or } x = \tfrac{-12}{2} = -6$$

41. $x^2 + 5x + 3 = 0$
$$a = 1,\ b = 5,\ c = 3$$
$$x = \frac{-5 \pm \sqrt{(5)^2 - 4(1)(3)}}{2(1)}$$
$$= \frac{-5 \pm \sqrt{25 - 12}}{2}$$
$$= \frac{-5 \pm \sqrt{13}}{2}$$

43. $3x^2 - 6x - 11 = 0$
$$a = 3,\ b = -6,\ c = -11$$
$$x = \frac{-(-6) \pm \sqrt{(-6)^2 - 4(3)(-11)}}{2(3)}$$
$$= \frac{6 \pm \sqrt{36 + 132}}{6}$$
$$= \frac{6 \pm \sqrt{168}}{6}$$
$$= \frac{6 \pm 2\sqrt{42}}{6}$$
$$= \frac{3 \pm \sqrt{42}}{3}$$

45. $x^2 + 5x + 5 = 0$
$$a = 1,\ b = 5,\ c = 5$$
$$x = \frac{-5 \pm \sqrt{(5)^2 - 4(1)(5)}}{2(1)}$$
$$= \frac{-5 \pm \sqrt{25 - 20}}{2}$$
$$= \frac{-5 \pm \sqrt{5}}{2}$$

47. $2z^2 - 6z + 11 = 0$
$$a = 2,\ b = -6,\ c = 11$$
$$z = \frac{-(-6) \pm \sqrt{(-6)^2 - 4(2)(11)}}{2(2)}$$
$$= \frac{6 \pm \sqrt{36 - 88}}{4}$$
$$= \frac{6 \pm \sqrt{-52}}{4}$$
$$= \frac{6 \pm 2i\sqrt{13}}{4}$$
$$= \frac{3 \pm i\sqrt{13}}{2}$$

49. $2x^2 - \pi x - 1 = 0$

$a = 2, \ b = -\pi, \ c = -1$

$$x = \frac{-(-\pi) \pm \sqrt{(-\pi)^2 - 4(2)(-1)}}{2(2)}$$

$$= \frac{\pi \pm \sqrt{\pi^2 + 8}}{4}$$

$$\approx \frac{3.14159 \pm 4.22725}{4}$$

$$\approx 1.8422 \text{ or } -.2714$$

51. $x^2 + .8235x - 1.3728 = 0$

$a = 1, \ b = .8235, \ c = -1.3728$

$$x = \frac{-.8235 \pm \sqrt{(.8235)^2 - 4(1)(-1.3728)}}{2(1)}$$

$$\approx \frac{-.8235 \pm \sqrt{6.16935}}{2}$$

$$\approx \frac{-.8235 \pm 2.4838}{2}$$

$$\approx .8302 \text{ or } -1.6537$$

53. $(y - 2)^2 = 4x^2$

$y - 2 = \pm 2x$

$y = 2 \pm 2x$

$y = 2 + 2x \text{ or } y = 2 - 2x$

55. $(y + 3x)^2 = 9x^2$

$y + 3x = \pm 3x$

$y = -3x \pm 3x$

$y = 0 \text{ or } y = -6x$

57. $(y + 2x)^2 - 8(y + 2x) + 15 = 0$

$[(y + 2x) - 5][(y + 2x) - 3] = 0$

$(y + 2x) - 5 = 0 \qquad \text{or} \quad (y + 2x) - 3 = 0$

$\qquad y = 5 - 2x \quad \text{or} \qquad\qquad y = 3 - 2x$

59. $(2x - 1)^2 = \frac{9}{4}$

$2x - 1 = \pm \frac{3}{2}$

$2x = 1 \pm \frac{3}{2}$

$x = \frac{1}{2}\left(1 \pm \frac{3}{2}\right)$

$x = \frac{5}{4} \text{ or } x = -\frac{1}{4}$

61. $\qquad 2x^2 = 4x - 2$

$2x^2 - 4x + 2 = 0$

$2(x^2 - 2x + 1) = 0$

$2(x - 1)^2 = 0$

$x = 1 \text{ or } 1 \text{ [i.e., 1 is a double root]}$

63. $y^2 + 2y + 4 = 0$

$a = 1, \ b = 2, \ c = -4$

$$y = \frac{-2 \pm \sqrt{(2)^2 - 4(1)(-4)}}{2(1)}$$

$$= \frac{-2 \pm \sqrt{4 + 16}}{2}$$

$$= \frac{-2 \pm \sqrt{20}}{2}$$

$$= \frac{-2 \pm 2\sqrt{5}}{2}$$

$$= -1 \pm \sqrt{5}$$

65. $2m^2 + 2m + 1 = 0$

$a = 2, \ b = 2, \ c = 1$

$$m = \frac{-2 \pm \sqrt{(2)^2 - 4(2)(1)}}{2(2)}$$

$$= \frac{-2 \pm \sqrt{4 - 8}}{4}$$

$$= \frac{-2 \pm \sqrt{-4}}{4}$$

$$= \frac{-2 \pm 2i}{4}$$

$$= \frac{-1 \pm i}{2}$$

67. $\qquad x^4 - 5x^2 - 6 = 0$

$(x^2 - 6)(x^2 + 1) = 0$

$x^2 - 6 = 0 \qquad \text{or} \quad x^2 + 1 = 0$

$x^2 = 6 \qquad \text{or} \qquad x^2 = -1$

$x = \pm\sqrt{6} \quad \text{or} \qquad x = \pm i$

[or one could let $u = x^2$ and solve $u^2 - 5u - 6 = 0$]

69. $\left(x - \frac{4}{x}\right)^2 - 7\left(x - \frac{4}{x}\right) + 12 = 0$

Let $u = x - \frac{4}{x}$:

$u^2 - 7u + 12 = 0$

$(u - 3)(u - 4) = 0$

$u - 3 = 0 \quad \text{or} \quad u - 4 = 0$

$u = 3 \quad \text{or} \qquad u = 4$

If $u = 3, \ x - \frac{4}{x} = 3$

$$x\left[x - \frac{4}{x}\right] = x[3]$$

$x^2 - 4 = 3x$

$x^2 - 3x - 4 = 0$

$(x - 4)(x + 1) = 0$

$x - 4 = 0 \quad \text{or} \quad x + 1 = 0$

$x = 4 \quad \text{or} \qquad x = -1$

If $u = 4$, $x - \frac{4}{x} = 4$

$$x\left[x - \frac{4}{x}\right] = x[4]$$
$$x^2 - 4 = 4x$$
$$x^2 - 4x - 4 = 0$$
$a = 1$, $b = -4$, $c = -4$
$$x = \frac{-(-4) \pm \sqrt{(-4)^2 - 4(1)(-4)}}{2(1)}$$
$$= \frac{4 \pm \sqrt{16 + 16}}{2}$$
$$= \frac{4 \pm \sqrt{32}}{2}$$
$$= \frac{4 \pm 4\sqrt{2}}{2}$$
$$= 2 \pm 2\sqrt{2}$$

Since none of these values of x makes a denominator 0, the solutions of the equation are:
$x = 4$, $x = -1$, $x = 2 \pm 2\sqrt{2}$

71.
$$\frac{x^2 + 2}{x^2 - 1} = \frac{x + 4}{x + 1}$$
$$(x+1)(x-1)\left[\frac{x^2+2}{(x+1)(x-1)}\right] = (x+1)(x-1)\left[\frac{x+4}{x+1}\right]$$
$$x^2 + 2 = (x-1)(x+4)$$
$$x^2 + 2 = x^2 + 3x - 4$$
$$3x = 6$$
$$x = 2$$

Since $x = 2$ does not make a denominator 0, $x = 2$ is the solution.

73.
$$\frac{1}{x} + \frac{1}{x-1} = \frac{8}{3}$$
Multiply by $3x(x-1)$ to clear fractions:
$$3(x-1) + 3x = 8x(x-1)$$
$$3x - 3 + 3x = 8x^2 - 8x$$
$$8x^2 - 14x + 3 = 0$$
$$(4x-1)(2x-3) = 0$$
$$4x - 1 = 0 \text{ or } 2x - 3 = 0$$
$$x = \tfrac{1}{4} \text{ or } \qquad x = \tfrac{3}{2}$$
Neither solution makes a denominator 0, so the two solutions are $x = \frac{1}{4}$ and $x = \frac{3}{2}$.

75.
$$\sqrt{2x+1} = \sqrt{x} + 1$$
Square both sides:
$$(\sqrt{2x+1})^2 = (\sqrt{x}+1)^2$$
$$2x + 1 = (\sqrt{x})^2 + 2\sqrt{x} + 1$$
$$2x + 1 = x + 2\sqrt{x} + 1$$
Isolate the radical term:
$$x = 2\sqrt{x}$$
Square both sides again:

$$(x)^2 = (2\sqrt{x})^2$$
$$x^2 = 4x$$
$$x^2 - 4x = 0$$
$$x(x-4) = 0$$
$$x = 0 \text{ or } x - 4 = 0$$
$$x = 4$$
We must check these solutions in the original radical equation:
Check $x = 0$:
$$\sqrt{2(0)+1} \overset{?}{=} \sqrt{0} + 1$$
$$\sqrt{1} \overset{?}{=} 1$$
$$1 = 1$$
Check $x = 4$:
$$\sqrt{2(4)+1} \overset{?}{=} \sqrt{4} + 1$$
$$\sqrt{9} \overset{?}{=} 2 + 1$$
$$3 = 3$$
The solutions are $x = 0$ and $x = 4$.

77. $\begin{cases} xy = 20 \\ -3 = 2x - y \end{cases}$
Solving the second equation for y:
$$y = 2x + 3$$
Substituting into the first equation:
$$x(2x+3) = 20$$
$$2x^2 + 3x = 20$$
$$2x^2 + 3x - 20 = 0$$
$$(2x-5)(x+4) = 0$$
$$2x - 5 = 0 \text{ or } x + 4 = 0$$
$$x = \tfrac{5}{2} \text{ or } \qquad x = -4$$
Substituting in $y = 2x + 3$:
When $x = \frac{5}{2}$, $y = 2\left(\frac{5}{2}\right) + 3 = 8$
When $x = -4$, $y = 2(-4) + 3 = -5$
The solutions are $x = \frac{5}{2}$, $y = 8$ and $x = -4$, $y = -5$.

79. $\begin{cases} 2x^2 - xy + y^2 = 14 \\ x - 2y = 0 \end{cases}$
Substituting $x = 2y$ from the second equation into the first:
$$8y^2 - 2y^2 + y^2 = 14$$
$$7y^2 = 14$$
$$y^2 = 2$$
$$y = \pm\sqrt{2}$$
Substituting in $x = 2y$:
When $y = \sqrt{2}$, $x = 2\sqrt{2}$
When $y = -\sqrt{2}$, $x = -2\sqrt{2}$
The solutions are $x = 2\sqrt{2}$, $y = \sqrt{2}$ and $x = -2\sqrt{2}$, $y = -\sqrt{2}$.

81. Let x represent the length of the rectangle.
Let y represent the width of the rectangle.
$$\begin{cases} 2x + 2y = 26 \\ xy = 30 \end{cases}$$

Solving the first equation for y:
$$2y = 26 - 2x$$
$$y = 13 - x$$
Substituting in the second equation:
$$x(13 - x) = 30$$
$$13x - x^2 = 30$$
$$x^2 - 13x + 30 = 0$$
$$(x - 10)(x - 3) = 0$$
$$x - 10 = 0 \quad \text{or} \quad x - 3 = 0$$
$$x = 10 \quad \text{or} \quad x = 3$$
Substituting in $y = 13 - x$:
When $x = 10$, $y = 13 - 10 = 3$
When $x = 3$, $y = 13 - 3 = 10$
The dimensions of the rectangle are 3 by 10.

83. Let x represent the length of a side of the original square, in inches. Then $x - 4$ represents the length of the base of the tray, and $x - 4$ represents the width of the base of the tray.
The height of the tray is 2 inches.
Using the formula $V = lwh$,
$$2(x - 4)(x - 4) = 128$$
$$x^2 - 8x + 16 = 64$$
$$x^2 - 8x - 48 = 0$$
$$(x - 12)(x + 4) = 0$$
$$x - 12 = 0 \quad \text{or} \quad x + 4 = 0$$
$$x = 12 \quad \text{or} \quad x = -4$$
Since a negative length makes no sense, the original square was 12 inches by 12 inches.

85. Let T be Tom's final position, and D be Dick's final position. Let $t =$ time in hours after noon when they lose contact.
We use the formula $d = rt$.
$AT = 4t$ and
$AD = 4(t - 1)$
Using the Pythagorean Theorem:
$$TD = \sqrt{(4t)^2 + [4(t - 1)]^2}$$
But $TD = 8$
$$8 = \sqrt{4^2(t^2 + t^2 - 2t + 1)}$$
$$8 = 4\sqrt{2t^2 - 2t + 1}$$
$$\sqrt{2t^2 - 2t + 1} = 2$$
Squaring each side:
$$2t^2 - 2t + 1 = 4$$
$$2t^2 - 2t - 3 = 0$$
Using the quadratic formula:
$$t = \frac{2 \pm \sqrt{4 + 24}}{4}$$
$$= \frac{2 \pm \sqrt{28}}{4} = \frac{2 \pm 2\sqrt{7}}{4} = \frac{1 \pm \sqrt{7}}{2}$$

The negative sign leads to a negative distance, so radio contact is lost at time $t = \frac{1 + \sqrt{7}}{2}$ hours after noon, or about 1.8229 hours after noon. This is $1.8229 \times \frac{60 \text{ minutes}}{\text{hour}}$ or 109.37 minutes after noon.
Rounded off, they lost contact at about 1:49 P.M.

87. Let x represent Samantha's speed over the last 12 miles, in mph. Then $x + 3$ represents Samantha's speed over the first 18 miles, in mph.
2 hours and 12 minutes $= 2\frac{12}{60}$ hours $= 2\frac{1}{5} = \frac{11}{5}$ hours
Using the formula $T = \frac{D}{R}$,
$$\frac{11}{5} = \frac{18}{x + 3} + \frac{12}{x}$$
Multiply each side of the equation by $5x(x + 3)$ to clear fractions:
$$11x(x + 3) = 18(5x) + 12(5)(x + 3)$$
$$11x^2 + 33x = 90x + 60x + 180$$
$$11x^2 - 117x - 180 = 0$$
$$(11x + 15)(x - 12) = 0$$
$$11x + 15 = 0 \quad \text{or} \quad x - 12 = 0$$
$$x = \frac{-15}{11} \quad \text{or} \quad x = 12$$
Since a negative speed is impossible, Samantha rode at a rate of 12 mph for the last 12 miles.

89.
Let x represent the length of the shorter segment and y represent the length of the longer segment.
$x + y$ represents the length of the entire segment.
We want the value of $\phi = \frac{y}{x}$.
We know that
$$\frac{x + y}{y} = \frac{y}{x}$$
$$\frac{x}{y} + 1 = \frac{y}{x}$$
If $\phi = \frac{y}{x}$, then $\frac{x}{y} = \frac{1}{\phi}$.
Substituting in the last equation:
$$\frac{1}{\phi} + 1 = \phi$$
Multiply by ϕ to clear fractions:
$$1 + \phi = \phi^2$$
$$\phi^2 - \phi - 1 = 0$$
$$a = 1, b = -1, c = -1$$
$$\phi = \frac{-(-1) \pm \sqrt{(-1)^2 - 4(1)(-1)}}{2(1)}$$
$$\phi = \frac{1 \pm \sqrt{5}}{2}$$
Since line segments must have positive lengths, choose the positive value of ϕ; $\phi = \frac{1 + \sqrt{5}}{2}$

1. $x \geq 0$
Conditional inequality—not true when x is replaced with a negative number.

3. $x^2 + 1 > 0$
Unconditional inequality—x^2 will always be non-negative.

5. $x - 2 < -5$
Conditional inequality—not true when x is replaced with any number greater than or equal to -3.

7. $x(x + 4) \leq 0$
Conditional inequality—notice it is not true, for example, when x is replaced with 3.

9. $(x + 1)^2 > x^2$
Conditional inequality—notice it is not true, for example, when x is replaced with -1.

11. We are given: $(x + 1)^2 > x^2 + 2x$
This is equivalent to $x^2 + 2x + 1 > x^2 + 2x$,
which is equivalent to $1 > 0$,
which is unconditionally true.

13. $3x + 7 < x - 5$
$2x < -12$
$x < -6$

Solution: $\{x : x < -6\}$

15. $\frac{2}{3}x + 1 > \frac{1}{2}x - 3$
$\frac{1}{6}x > -4$
$x > -24$

Solution: $\{x : x > -24\}$

17. $\frac{3}{4}x - \frac{1}{2} < \frac{1}{6}x + 2$
Multiply by 12 to clear fractions:
$9x - 6 < 2x + 24$
$7x < 30$
$x < \frac{30}{7}$

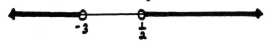

Solution: $\{x : x < \frac{30}{7}\}$

19. $(x - 2)(x + 5) \leq 0$
Find the split points:
$x - 2 = 0 \quad x + 5 = 0$
$x = 2 \quad\quad x = -5$

The split points divide the number line into 3 regions. Choose a test point in each region.

Region	Test point	Value of $(x-2)(x+5)$
$x < -5$	$x = -6$	$+$
$-5 < x < 2$	$x = 0$	$-$
$x > 2$	$x = 3$	$+$

$(x - 2)(x + 5) < 0$ when $-5 < x < 2$ and equal to 0 at the split points $x = 2$ and $x = -5$.
Solution: $\{x : -5 \leq x \leq 2\}$

21. $(2x - 1)(x + 3) > 0$
Find the split points:
$2x - 1 = 0 \quad x + 3 = 0$
$x = \frac{1}{2} \quad\quad x = -3$
The split points divide the number line into 3 regions. Choose a test point in each region.

Region	Test point	Value of $(2x-1)(x+3)$
$x < -3$	$x = -4$	$+$
$-3 < x < \frac{1}{2}$	$x = 0$	$-$
$x > \frac{1}{2}$	$x = 1$	$+$

$(2x - 1)(x + 3) > 0$ when $x < -3$ or $x > \frac{1}{2}$.
Solution: $\{x : x < -3 \text{ or } x > \frac{1}{2}\}$

23. $x^2 - 5x + 4 \geq 0$
$(x - 1)(x - 4) \geq 0$
Find the split points:
$x - 4 = 0 \quad x - 1 = 0$
$x = 4 \quad\quad x = 1$
The split points divide the number line into 3 regions. Choose a test point in each region.

Region	Test point	Value of $(2x-1)(x+3)$
$x < 1$	$x = 0$	$+$
$1 < x < 4$	$x = 2$	$-$
$x > 4$	$x = 5$	$+$

$x^2 - 5x + 4 \geq 0$ when $x \leq 1$ or $x \geq 4$.
Solution: $\{x : x \leq 1 \text{ or } x \geq 4\}$

25. $2x^2 - 7x + 3 < 0$
$(2x - 1)(x - 3) < 0$
Find the split points:
$2x - 1 = 0 \quad x - 3 = 0$
$x = \frac{1}{2} \quad\quad x = 3$
The split points divide the number line into 3 regions. Choose a test point in each region.

Region	Test point	Value of $(2x-1)(x-3)$
$x < \frac{1}{2}$	$x = 0$	$+$
$\frac{1}{2} < x < 3$	$x = 1$	$-$
$x > 3$	$x = 4$	$+$

$(2x-1)(x-3) < 0$ when $\frac{1}{2} < x < 3$.
Solution: $\{x : \frac{1}{2} < x < 3\}$

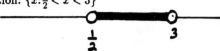

27. $(x+4)(x)(x-3) \geq 0$
Find the split points:

$$x+4=0 \qquad x=0 \qquad x-3=0$$
$$x=-4 \qquad\qquad x=3$$

The split points divide the number line into 4 regions. Choose a test point in each region.

Region	Test point	Value of $(x+4)(x)(x-3)$
$x < -4$	$x = -5$	$-$
$-4 < x < 0$	$x = -1$	$+$
$0 < x < 3$	$x = 1$	$-$
$x > 3$	$x = 4$	$+$

$(x+4)(x)(x-3) > 0$ when $-4 < x < 0$ or when $x > 3$; $(x+4)(x)(x-3) = 0$ at the split points $x = -4$, $x = 0$, and $x = 3$.
Solution: $\{x : -4 \leq x \leq 0 \text{ or } x \geq 3\}$

29. $(x-2)^2(x-5) < 0$
Noticing that $(x-2)^2$ is always nonnegative, $(x-2)^2(x-5)$ will be less than 0 only when

$$x-5 < 0$$
$$x < 5$$

Notice also, however, that in the region $x < 5$, the point $x = 2$ will make $(x-2)^2(x-5) = 0$, which we cannot allow.
Solution: $\{: x < 5 \text{ and } x \neq 2\}$

31. $\frac{x-5}{x+2} \leq 0$
Find the split points:

$$x-5=0 \qquad x+2=0$$
$$x=5 \qquad x=-2$$

The split points divide the number line into 3 regions. Choose a test point in each region.

Region	Test point	Value of $\frac{x-5}{x+2}$
$x < -2$	$x = -3$	$+$
$-2 < x < 5$	$x = 0$	$-$
$x > 5$	$x = 6$	$+$

$\frac{x-5}{x+2} < 0$ when $-2 < x < 5$;

$\frac{x-5}{x+2} = 0$ at the split point $x = -5$.
(Note that at $x = -2$, $\frac{x-5}{x+2}$ is undefined.)

Solution: $\{x : -2 < x \leq 5\}$

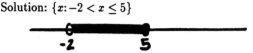

33. $\frac{x(x+2)}{x-5} > 0$
Find the split points:

$$x=0 \qquad x+2=0 \qquad x-5=0$$
$$x=-2 \qquad x=5$$

The split points divide the number line into 4 regions. Choose a test point in each region.

Region	Test point	Value of $\frac{x(x+2)}{x-5}$
$x < -2$	$x = -3$	$-$
$-2 < x < 0$	$x = -1$	$+$
$0 < x < 5$	$x = 1$	$-$
$x > 5$	$x = 6$	$+$

Solution: $\{x : -2 < x < 0 \text{ or } x > 5\}$

35.
$$\frac{5}{x-3} > \frac{4}{x-2}$$

$$\frac{5}{x-3} - \frac{4}{x-2} > 0$$

$$\frac{5(x-2) - 4(x-3)}{(x-3)(x-2)} > 0$$

$$\frac{x+2}{(x-3)(x-2)} > 0$$

Find the split points:

$$x+2=0 \qquad x-3=0 \qquad x-2=0$$
$$x=-2 \qquad x=3 \qquad x=2$$

The split points divide the number line into 4 regions. Choose a test point in each region.

Region	Test point	Value of $\frac{x+2}{(x-3)(x-2)}$
$x < -2$	$x = -3$	$-$
$-2 < x < 2$	$x = 0$	$+$
$2 < x < 3$	$x = 2.5$	$-$
$x > 3$	$x = 4$	$+$

Solution: $\{x : -2 < x < 2 \text{ or } x > 3\}$

37.
$$|2x+3| < 2$$
$$-2 < 2x+3 < 2$$
$$-5 < 2x < -1$$
$$-\frac{5}{2} < x < -\frac{1}{2}$$

39.
$$|-2x-1| \le 1$$
$$-1 \le -2x-1 \le 1$$
$$0 \le -2x \le 2$$
$$0 \ge x \ge -1$$
or
$$-1 \le x \le 0$$

41. $|5x-1| \ge 9$
$$5x-1 \ge 9 \quad \text{or} \quad 5x-1 \le -9$$
$$5x \ge 10 \quad \text{or} \quad 5x \le -8$$
$$x \ge 2 \quad \text{or} \quad x \le -\frac{8}{5}$$

43. $|2x-3| > 6$
$$2x-3 > 6 \quad \text{or} \quad 2x-3 < -6$$
$$2x > 9 \quad \text{or} \quad 2x < -3$$
$$x > \frac{9}{2} \quad \text{or} \quad x < -\frac{3}{2}$$

45. $0 < x < 6$

Midpoint: $\dfrac{0+6}{2} = 3$

Radius: $\dfrac{6-(0)}{2} = 3$

x must be within 3 units of the midpoint 3, or
$$|x-3| < 3$$

47. $-1 \le x \le 7$

Midpoint: $\dfrac{-1+7}{2} = 3$

Radius: $\dfrac{7-(-1)}{2} = 4$

x must therefore be within 4 units of the midpoint 3, or
$$|x-3| \le 4$$

49. $2 < x < 11$

Midpoint: $\dfrac{11+2}{2} = \dfrac{13}{2}$

Radius: $\dfrac{11-2}{2} = \dfrac{9}{2}$

x must be within $\frac{9}{2}$ units of $\frac{13}{2}$ or,
$$\left|x-\frac{13}{2}\right| < \frac{9}{2}$$

51. $x^2 - 7 < 0$

Find the split points:
$$(x-\sqrt{7})(x+\sqrt{7}) = 0$$
$$x - \sqrt{7} = 0 \qquad x + \sqrt{7} = 0$$
$$x = \sqrt{7} \qquad x = -\sqrt{7}$$

The split points divide the number line into 3 regions. Choose a test point in each region.

Region	Test point	Value of $x^2 - 7$
$x < -\sqrt{7}$	$x = -3$	$+$
$-\sqrt{7} < x < \sqrt{7}$	$x = 0$	$-$
$x > \sqrt{7}$	$x = 3$	$+$

Solution: $\{x : -\sqrt{7} < x < \sqrt{7}\}$

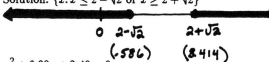

53. $x^2 - 4x + 2 \ge 0$

Use the quadratic formula to find the split points:
$$x = \frac{-(-4) \pm \sqrt{(-4)^2 - 4(1)(2)}}{2(1)}$$
$$= \frac{4 \pm \sqrt{16-8}}{2}$$
$$= \frac{4 \pm \sqrt{8}}{2}$$
$$= \frac{4 \pm 2\sqrt{2}}{2}$$
$$= 2 \pm \sqrt{2}$$

The split points divide the number line into 3 regions. Choose a test point in each region.

Region	Test point	Value of $x^2 - 4x + 2$
$x < 2 - \sqrt{2}$	$x = 0$	$+$
$2 - \sqrt{2} < x < 2 + \sqrt{2}$	$x = 1$	$-$
$x > 2 + \sqrt{2}$	$x = 5$	$+$

$x^2 - 4x + 2 > 0$ when $x < 2 - \sqrt{2}$ or when $x > 2 + \sqrt{2}$; $x^2 - 4x + 2 = 0$ at the split-points.
Solution: $\{x : x \le 2 - \sqrt{2} \text{ or } x \ge 2 + \sqrt{2}\}$

0   2-√2   2+√2
    (.586)   (2.414)

55. $x^2 + 6.32x + 3.49 > 0$

Use the quadratic formula to find the split points:
$$x = \frac{-6.32 \pm \sqrt{(6.32)^2 - 4(1)(3.49)}}{2(1)}$$
$$= \frac{-6.32 \pm \sqrt{25.9824}}{2}$$

$x \approx -.611$ or $x \approx -5.709$

The split points divide the number line into 3 regions. Choose a test point in each region.

Region	Test point	Value of $x^2 + 6.32x + 3.49$
$x < -5.709$	$x = -6$	$+$
$-5.709 < x < -.611$	$x = -1$	$-$
$x > -.611$	$x = 0$	$+$

Solution: $\{x : x < -5.709 \text{ or } x > -.611\}$

-5.709   -.611 0

57. $x^2 + 8x + 20$
Completing the square,
$$x^2 + 8x + 20 = x^2 + 8x + 16 - 16 + 20$$
$$= (x+4)^2 + 4$$
Since the smallest value $(x+4)^2$ can take on is 0 (when $x = -4$), the smallest value $(x+4)^2 + 4$ can assume is 4.

59. $x^2 - 2x + 101$
Completing the square,
$$x^2 - 2x + 101 = x^2 - 2x + 1 - 1 + 101$$
$$= (x-1)^2 + 100$$
Since the smallest value $(x-1)^2$ can take on is 0 (when $x = 1$), the smallest value $(x-1)^2 + 100$ can assume is 100.

61. $\frac{1}{2}x + \frac{3}{4} > \frac{2}{3}x - \frac{4}{3}$
Multiply by 12 to clear fractions:
$$6x + 9 > 8x - 16$$
$$-2x > -25$$
$$x < \frac{25}{2}$$
Solution: $\{x : x < \frac{25}{2}\}$

63. $2x^2 + 5x - 3 < 0$
$(2x - 1)(x + 3) < 0$
Find the split points:
$$2x - 1 = 0 \qquad x + 3 = 0$$
$$x = \frac{1}{2} \qquad x = -3$$
The split points divide the number line into 3 regions. Choose a test point in each region.

Region	Test point	Value of $(2x-1)(x+3)$
$x < -3$	$x = -4$	$+$
$-3 < x < \frac{1}{2}$	$x = 0$	$-$
$x > \frac{1}{2}$	$x = 1$	$+$

Solution: $\{x : -3 < x < \frac{1}{2}\}$

65. $(x+1)^2(x-1)(x-4)(x-8) < 0$
Find the split points:
$$(x+1)^2 = 0 \quad x - 1 = 0 \quad x - 4 = 0 \quad x - 8 = 0$$
$$x = -1 \qquad x = 1 \qquad x = 4 \qquad x = 8$$
The split points divide the number line into 5 regions. Choose a test point in each region.

Region	Test point	Value of $(x+1)^2(x-1)(x-4)(x-8)$
$x < -1$	$x = -2$	$-$
$-1 < x < 1$	$x = 0$	$-$
$1 < x < 4$	$x = 2$	$+$
$4 < x < 8$	$x = 5$	$-$
$x > 8$	$x = 9$	$+$

Solution: $\{x : x < -1 \text{ or } -1 < x < 1 \text{ or } 4 < x < 8\}$

67.
$$\frac{1}{x-2} + 1 < \frac{2}{x+2}$$
$$\frac{1}{x-2} + 1 - \frac{2}{x+2} < 0$$
$$\frac{(x+2) + (x-2)(x+2) - 2(x-2)}{(x-2)(x+2)} < 0$$
$$\frac{x + 2 + x^2 - 4 - 2x + 4}{(x-2)(x+2)} < 0$$
$$\frac{x^2 - x + 2}{(x-2)(x+2)} < 0$$
Find the split-points:
$$x - 2 = 0 \quad x + 2 = 0 \quad x^2 - x + 2 = 0$$
$$x = 2 \qquad x = -2 \qquad x = \frac{1 \pm \sqrt{1 - 4(1)(2)}}{2}$$
$$x = \frac{1 \pm i\sqrt{7}}{2}$$
The only real-valued split points are 2 and -2. They divide the number line into 3 regions. Choose a test point in each region.

Region	Test point	Value of $\frac{x^2-x+2}{(x-2)(x+2)}$
$x < -2$	$x = -3$	$+$
$-2 < x < 2$	$x = 0$	$-$
$x > 2$	$x = 3$	$+$

Solution: $\{x : -2 < x < 2\}$

69. $|4x - 3| \geq 2$
$$4x - 3 \geq 2 \quad \text{or} \quad 4x - 3 \leq -2$$
$$4x \geq 5 \qquad\qquad 4x \leq 1$$
$$x \geq \frac{5}{4} \qquad\qquad x \leq \frac{1}{4}$$
Solution: $\{x : x \leq \frac{1}{4} \text{ or } x \geq \frac{5}{4}\}$

71.
$$|3 - 4x| < 7$$
$$-7 < 3 - 4x < 7$$
$$-10 < \quad -4x < 4$$
$$\frac{-10}{-4} > \qquad x > \frac{4}{-4}$$
$$\frac{5}{2} > \qquad x > -1$$
Solution: $\{x : -1 < x < \frac{5}{2}\}$

73.
$$|x - 2| < |x + 3|$$
Since both sides are positive, squaring both sides will preserve the inequality.
$$|x - 2|^2 < |x + 3|^2$$
Since $|a|^2 = a^2$, this can be written:
$$(x - 2)^2 < (x + 3)^2$$
$$x^2 - 4x + 4 < x^2 + 6x + 9$$
$$-10x < 5$$
$$x > \frac{5}{-10}$$
$$x > -\frac{1}{2}$$
Solution: $\{x : x > -\frac{1}{2}\}$

48

75. $|x^2 - 2x - 4| > 4$

$$x^2 - 2x - 4 > 4 \quad \text{or} \quad x^2 - 2x - 4 < -4$$
$$x^2 - 2x - 8 > 0 \quad \text{or} \quad x^2 - 2x < 0$$
$$(x-4)(x+2) > 0 \quad \text{or} \quad x(x-2) < 0$$

Case I. $(x-4)(x+2) > 0$

Split points:
$$x - 4 = 0 \quad x + 2 = 0$$
$$x = 4 \quad x = -2$$

Region	Test point	Value of $(x-4)(x+2)$
$x < -2$	$x = -3$	$+$
$-2 < x < 4$	$x = 0$	$-$
$x > 4$	$x = 5$	$+$

Solution: $\{x : x < -2 \text{ or } x > 4\}$

Case II. $x(x-2) < 0$

Split points:
$$x = 0 \quad x - 2 = 0$$
$$x = 2$$

Region	Test point	Value of $(x-4)(x+2)$
$x < 0$	$x = -1$	$+$
$0 < x < 2$	$x = 1$	$-$
$x > 2$	$x = 3$	$+$

Solution: $\{x : 0 < x < 2\}$

Uniting the solution sets of case I and case II, $|x^2 - 2x - 4| > 4$ for
$$\{x : x < -2 \text{ or } 0 < x < 2 \text{ or } x > 4\}$$

77. The quadratics will have real solutions if, in the quadratic formula, the quantity $b^2 - 4ac$ under the radical sign is ≥ 0.

a) $x^2 + 4x + k = 0$
$$16 - 4k \geq 0$$
$$4k \leq 16$$
$$k \leq 4$$

b) $x^2 - kx + 9 = 0$
$$k^2 - 36 \geq 0$$
$$k^2 \geq 36$$
$$k \geq 6 \text{ or } k \leq -6$$

c) $x^2 + kx + k = 0$
$$k^2 - 4k \geq 0$$
$$k(k-4) \geq 0$$
$$k \leq 0 \text{ or } k \geq 4$$

d) $x^2 + kx + k^2 = 0$
$$k^2 - 4k^2 \geq 0$$
$$-3k^2 \geq 0$$
$$k^2 \leq 0$$

$k = 0$ is the only real value of k leading to real solutions for this quadratic.

79. Let S represent the salary of the new professor. Then $\dfrac{6(32000) + S}{7}$ represents the new average salary of the 7 professors.

$$31000 < \frac{192000 + S}{7} < 35000$$
$$217000 < 192000 + S < 245000$$
$$25000 < S < 53000$$

The professor's salary must fall between \$25,000 and \$53,000.

81. a) Complete the square:
$$-16t^2 + 64t + 80 = -16(t^2 - 4t) + 80$$
$$= -16(t^2 - 4t + 4) + 64 + 80$$
$$= -16(t-2)^2 + 144$$

Since the largest value $-16(t-2)^2$ can take on is 0 (when $t = 2$), the largest value $-16(t-2)^2 + 144$ can assume is 144. Thus, the maximum height is 144 feet above the ground.

b) We want to know when
$$-16t^2 + 64t + 80 > 96$$
$$t^2 - 4t - 5 < -6$$
$$t^2 - 4t + 1 < 0$$

Find the split points:
$$t = \frac{4 \pm \sqrt{(4)^2 - 4(1)(1)}}{2(1)}$$
$$= \frac{4 \pm \sqrt{12}}{2}$$
$$= \frac{4 \pm 2\sqrt{3}}{2}$$
$$= 2 \pm \sqrt{3}$$

The split points are $2 - \sqrt{3}$ and $2 + \sqrt{3}$.

Region	Test point	Value of $t^2 - 4t + 1$
$t < 2 - \sqrt{3}$	$t = 0$	$+$
$2 - \sqrt{3} < t < 2 + \sqrt{3}$	$t = 2$	$-$
$t > 2 + \sqrt{3}$	$t = 5$	$+$

The ball will be higher than 96 feet for t between $2 - \sqrt{3}$ seconds and $2 + \sqrt{3}$ seconds.

c) The ball will hit the ground when the height $= 0$.
$$-16t^2 + 64t + 80 = 0$$
$$t^2 - 4t - 5 = 0$$
$$(t-5)(t+1) = 0$$
$$t - 5 = 0 \quad \text{or} \quad t + 1 = 0$$
$$t = 5 \quad \text{or} \quad t = -1$$

Since a negative time makes no sense, the ball will hit the ground after 5 seconds.

83.
$$2 < 2\pi\sqrt{\frac{l}{980}} < 3$$

Since all the values in the inequalities are non-negative, squaring will preserve the inequality.

$$4 < 4\pi^2\left(\frac{l}{980}\right) < 9$$

$$\frac{4(980)}{4\pi^2} < l < \frac{9(980)}{4\pi^2}$$

$$\frac{980}{\pi^2} < l < \frac{2205}{\pi^2}$$

85. If a and b are the legs and c is the hypotenuse of a right triangle, the Pythagorean theorem says $a^2 + b^2 = c^2$. Hence $c^2 > a^2$ and $c^2 > b^2$. Since a, b, and c are positive, $c > a$ and $c > b$.

Let n be an integer greater than 2.
Then $c^{n-2} > a^{n-2}$ and $c^{n-2} > b^{n-2}$.

$$\begin{aligned}
c^n &= c^2 c^{n-2} \\
&= (a^2 + b^2)c^{n-2} \\
&= a^2 c^{n-2} + b^2 c^{n-2} \\
&> a^2 a^{n-2} + b^2 b^{n-2} = a^n + b^n
\end{aligned}$$

Therefore $a^n + b^n < c^n$.

Problem Set 3.6

1. Let d = distance from ship to monitor, in feet.
From $d = rt$, $t = \frac{d}{r}$.
Time by water + 13 seconds = time by air.

$$\frac{d}{5000} + 13 = \frac{d}{1100}$$

$$\frac{d}{1100} - \frac{d}{5000} = 13$$

$$\frac{d}{11} - \frac{d}{50} = 1300$$

$$50d - 11d = 1300 \cdot 11 \cdot 50$$

$$39d = 1300 \cdot 11 \cdot 50$$

$$d = \frac{1300 \cdot 11 \cdot 50}{39} = 18{,}333 \text{ feet}$$

3. Let d represent the distance between the two cities, in miles. Using $t = \frac{d}{r}$,

$\frac{2d}{120}$ represents the length of time, in hours, a round trip takes on a windless day.

$\frac{d}{140} + \frac{d}{100}$ represents the length of time, in hours, the round trip took on the windy day.

Time on the windy day = Time on the windless day $+ \frac{1}{4}$ hour

$$\frac{d}{140} + \frac{d}{100} = \frac{2d}{120} + \frac{1}{4}$$

Multiply by 20:

$$\frac{d}{7} + \frac{d}{5} = \frac{d}{3} + 5$$

Multiply by $(7)(5)(3) = 105$:
$$15d + 21d = 35d + 525$$
$$d = 525 \text{ miles}$$

5. The minute hand moves at the rate of $\frac{360 \text{ deg}}{\text{hour}}$.

The hour hand moves at the rate of $\frac{1}{12} \cdot \frac{360 \text{ deg}}{\text{hour}}$.

Let t = the time (in hours) after 4:00 P.M. when the minute hand will overtake the hour hand.
The hour hand starts out 120 degrees ahead of the minute hand. The distance (in degrees) the minute hand travels = the distance (in degrees) the hour hand travels + 120°. Using $D = RT$,

$$360t = 30t + 120$$
$$330t = 120$$
$$t = \frac{120}{330} \text{ hours} = \frac{4}{11} \text{ hours}$$
$$t = \frac{4}{11} \text{ hours} \times \frac{60 \text{ min}}{\text{hour}} \approx 21.82 \text{ minutes}$$

7. Let x = no. of hours for new machine to do the job

$\frac{1}{x}$ = portion of work done by new machine in one hour

$$\frac{1}{x} + \frac{1}{8} = \frac{1}{3}$$

$$\frac{1}{x} = \frac{1}{3} - \frac{1}{8} = \frac{8-3}{24} = \frac{5}{24}$$

$$x = \frac{24}{5} \text{ hours} = 4\frac{4}{5} \text{ hours} = 4 \text{ hrs } 48 \text{ min}$$

9. Let t represent the number of hours between the airplane's takeoff and return to the carrier.
Then $t - 2$ represents the number of hours the plane spends flying east.
If the plane flew west for 2 hours at 600 mph, it flew $2(600) = 1200$ miles from the original position of the carrier. Using $D = RT$:
The carrier will have moved $30t$ miles west before the plane returns.
The distance the plane must fly east = $1200 - 30t$.
$$500(t - 2) = 1200 - 30t$$
$$500t - 1000 = 1200 - 30t$$
$$530t = 2200$$
$$t \approx 4.15 \text{ hours}$$

11. Let x represent the number of standard valves that should be made.
Let y represent the number of deluxe valves that should be made.
$5x$ represent the time (in minutes) the lathe is in use to make x standard valves.
$9y$ represents the time (in minutes) the lathe is in use to make y deluxe valves.

$10x$ represents the time (in minutes) the drill press is in use to make x standard valves.

$15y$ represents the time (in minutes) the drill press is in use to make y deluxe valves.

Converting the hours that the drill press and lathe are available into minutes, we can say

$$\begin{cases} 5x + 9y = 240 \\ 10x + 15y = 420 \end{cases}$$

Multiplying the first equation by -2:

$$-10x - 18y = -480$$
$$10x + 15y = 420$$

Adding these last two equations:

$$-3y = -60$$
$$y = 20$$

Substituting $y = 20$ in the first original equation:

$$5x + 9(20) = 240$$
$$5x = 60$$
$$x = 12$$

12 standard valves and 20 deluxe valves should be made.

13. Let $x =$ rate of car in miles per hour.

Time for car to go 450 miles at x mph is

$$t = \frac{d}{r} = \frac{450}{x}$$

Time for car to go 450 miles at $x + 15$ mph is $\frac{450}{x+15}$

$$\frac{450}{x} - \frac{450}{x + 15} = \frac{3}{2}$$

$$\frac{300}{x} - \frac{300}{x + 15} = 1$$

$$300(x + 15) - 300x = x(x + 15)$$
$$300x + 4500 - 300x = x^2 + 15x$$
$$x^2 + 15x - 4500 = 0$$
$$(x - 60)(x + 75) = 0$$
$$x = 60 \text{ mph}$$

($x = -75$ is not a meaningful solution to the question.)

15. Let d represent the length of the street, in meters.

Let x represent Jack's walking speed, in meters per hour.

Let y represent Jill's walking speed, in meters per hour.

When Jack and Jill met for the first time, Jack traveled 300 meters, and Jill traveled $d - 300$ meters. They must have traveled for the same length of time. Using $T = \frac{D}{R}$,

$$\frac{300}{x} = \frac{d - 300}{y}$$

Before they met for the second time, Jack had traveled an additional $(d - 300) + 400 = d + 100$ meters, and Jill traveled an additional $300 + (d - 400) = d - 100$ meters. Again, they must have traveled for the same length of time.

$$\frac{d + 100}{x} = \frac{d - 100}{y}$$

From the first equation, we get

$$\frac{y}{x} = \frac{d - 300}{300}$$

From the second equation, we get

$$\frac{y}{x} = \frac{d - 100}{d + 100}$$

Thus,

$$\frac{d - 300}{300} = \frac{d - 100}{d + 100}$$
$$(d - 300)(d + 100) = 300(d - 100)$$
$$d^2 - 200d - 30,000 = 300d - 30,000$$
$$d^2 - 500d = 0$$
$$d(d - 500) = 0$$
$$d = 0 \text{ or } d = 500$$

Since a street 0 meters long makes no sense, the street must be 500 meters long.

17. Let $x =$ distance (in feet) between the 80-candlepower light and the surface to be illuminated.

Using the formula $l = \frac{d}{c^2}$, a 20-candlepower light at 10 feet will give illumination

$$l = \frac{20}{(10)^2} \text{ foot-candles. We want}$$

$$\frac{80}{x^2} = \frac{20}{(10)^2}$$
$$20x^2 = 100(80)$$
$$x^2 = \frac{100(80)}{20} = 400$$
$$x = 20 \text{ feet}$$

19. Let $x =$ the distance in feet that the center of the bridge drops.

Each half of the "V" of the bridge would be 100 feet 4 inches, or $100\frac{1}{3}$ feet, or $\frac{301}{3}$ feet. Then by the Pythagorean Theorem,

$$x = \sqrt{\left(\frac{301}{3}\right)^2 - 100^2} = 8.17 \text{ feet}$$

21. Let x represent the number of kiloliters of the 20% solution the chemist should drain off.

Then $5 - x$ represents the number of kiloliters of 20% solution in the final mixture, and x represents the number of kiloliters of 80% solution in the final mixture.

$.20(5 - x) + .80x$ represents the amount of sulfuric acid in the final mixture.

$$.20(5 - x) + .80x = .30(5)$$
$$2(5 - x) + 8x = 3(5)$$
$$10 - 2x + 8x = 15$$
$$6x = 5$$
$$x = \frac{5}{6} \text{ kiloliters}$$

23. Let x represent the number of moles of oxygen required to produce 4.52 milligrams of carbon dioxide. Then $\frac{x}{2}$ represents the number of moles of carbon required.

$16x$ represents the weight of oxygen required, in grams. $12 \cdot \frac{x}{2}$ represents the weight of carbon required, in grams.

$$16x + 12 \cdot \frac{x}{2} = .00452$$
$$22x = .00452$$
$$x = .0002054$$

Milligrams of oxygen required
$$= .0002054 \text{ moles} \times \frac{16 \text{ grams}}{\text{mole}}$$
$$= .00329 \text{ grams}$$
$$= 3.29 \text{ milligrams}$$

25. Let $x =$ number of grams of sodium chloride in original sample. Then $.5 - x =$ number of grams of sodium bromide in original sample.

Number of grams of chlorine in original sample $= .606x$. Number of grams of bromine in original sample $= .776(.5 - x)$.

Let $y =$ number of grams of silver chloride in precipitate. $1.1 - y =$ number of grams of silver bromide in the precipitate.

Number of grams of chlorine in precipitate $= .247y$.
Number of grams of bromine in precipitate $= .425(1.1 - y)$.

Assuming that the amount of chlorine and bromine is constant,
$$.606x = .247y$$
$$.776(.5 - x) = .425(1.1 - y)$$

From the first equation,
$$y = \frac{606}{427}x$$

Multiplying the second equation by 1000 and by 10:
$$776(5 - 10x) = 425(11 - 10y)$$

Substituting:
$$776(5 - 10x) = 425\left(11 - 10 \cdot \frac{606}{427}\right)x$$

Expanding and multiplying by 247:
$$776 \cdot 5 \cdot 247 - 776 \cdot 10 \cdot 247x$$
$$= 425 \cdot 11 \cdot 247 - 425 \cdot 10 \cdot 606x$$
$$x = \frac{425 \cdot 11 \cdot 247 - 776 \cdot 5 \cdot 247}{425 \cdot 10 \cdot 606 - 776 \cdot 10 \cdot 247}$$
$$= \frac{1,154,725 - 958,360}{2,575,500 - 1,916,720} = \frac{196,365}{658,780} \approx .2980737$$
$$= \text{number of grams of sodium chloride in original sample.}$$

$.5 - x \approx .2019263$ grams of sodium bromide in original sample.

Rounding, .298 g NaCl and .202 g NaBr.

27. Let x represent the maximum number of stocks Jane can buy.

Then $59x$ represents the cost of the stock and $4182 - 59x$ represents the amount of money left in the account.

$$4182 - 59x \geq 2000$$
$$-59x \geq -2182$$
$$x \leq 36.98$$

Assuming Jane cannot buy a fraction of a share of stock, Jane can buy at most 36 shares.

29. Let x represent the number of pounds of walnuts Susan can use. Then $25 - x$ represents the number of pounds of cashews Susan can use.

$1.30x$ represents the monetary value of x pounds of walnuts. $2.30(25 - x)$ represents the monetary value of $(25 - x)$ pounds of cashews.

Since the monetary value of the total mixture is $1.74(25)$,
$$1.30x + 2.30(25 - x) = 1.74(25)$$
$$130x + 230(25) - 230x = 174(25)$$
$$-100x = 25(174 - 230)$$
$$-100x = -1400$$
$$x = 14$$

Susan should use 14 pounds of walnuts and $(25 - 14) = 11$ pounds of cashews.

31. Let $x =$ the number of units to be produced. Then $4x =$ the total variable cost of producing x units and $6x =$ the total revenues received from selling x units. Using $P = TR - (FC + VC)$,
$$15,000 = 6x - (32,000 + 4x)$$
$$15,000 = 6x - 32,000 - 4x$$
$$47,000 = 2x$$
$$x = \frac{47,000}{2} = 23,500 \text{ units}$$

33. Let P represent the profit before taxes, in dollars.
Then $P - .30P = 40000$
$$.7P = 40000$$
$$P = \frac{40000}{.7} = \$57,142.86$$

Let x represent the total sales ABC Company must have to yield a before-tax-profit of $57,142.86.
Using $P = TR - (FC + VC)$,
$$57,142.86 = x - (100,000 + .8x)$$
$$57,142.86 = .2x - 100,000$$
$$.2x = 157,142.86$$
$$x = \frac{157,142.86}{.2} = 785,714.30$$

The ABC Company must have total sales amounting to \$785,714.30.

35. Let g represent the number of graduate students the university should admit.
Let u represent the number of undergraduate students the university should admit.
$$\begin{cases} 900g + 600u = 2,181,600 \\ \frac{g}{6} + \frac{u}{15} = 300 \end{cases}$$

Multiply the second equation by $(-600)(15)$:
$$900g + 600u = 2,181,600$$
$$-1500g - 600u = -2,700,000$$
Adding these last two equations:
$$-600g = -518,400$$
$$g = \frac{-518,400}{-600} = 864$$

Substituting $g = 864$ in the second original equation:
$$\frac{864}{6} + \frac{u}{15} = 300$$
$$\frac{u}{15} = 156$$
$$u = 2340$$

The university should admit 2340 undergraduates and 864 graduate students.

37.

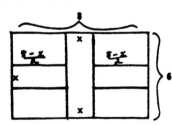

Let x = width of the cross.

Area of cross
$$= 6x + 2(x)\left(\frac{8-x}{2}\right) = 6x + x(8 - x) = 14x - x^2$$
Area of white background $= \frac{6 \cdot 8}{2} = 24$
$$24 = 14x - x^2$$
$$x^2 - 14x + 24 = 0$$
$$(x - 2)(x - 12) = 0$$
$x = 2$ or $x = 12$. But $x = 12$ fails to meet the physical conditions of the problem, so $x = 2$ feet = width of the cross.

39. Let x represent the length of one leg of the right triangle. Let y represent the length of the other leg of the right triangle.
By the Pythagorean Theorem, $\sqrt{x^2 + y^2}$ represents the length of the hypotenuse.
Since the perimeter equals 30,
$$x + y + \sqrt{x^2 + y^2} = 30$$
Since the area equals 30,
$$\tfrac{1}{2}xy = 30$$
$$y = \frac{60}{x}$$

Square the first equation to eliminate the radical:
$$\left[\sqrt{x^2 + y^2}\right]^2 = [30 - (x + y)]^2$$
$$x^2 + y^2 = (30)^2 - 60(x + y) + (x + y)^2$$
$$x^2 + y^2 = 900 - 60x - 60y + x^2 + 2xy + y^2$$
$$0 = 900 - 60x - 60y + 2xy$$

Substituting $y = \frac{60}{x}$:
$$0 = 900 - 60x - 60\left(\frac{60}{x}\right) + 2x\left(\frac{60}{x}\right)$$

Multiply by $\frac{x}{60}$ to clear fractions and reduce the equation:
$$0 = 15x - x^2 - 60 + 2x$$
$$x^2 - 17x + 60 = 0$$
$$(x - 5)(x - 12) = 0$$
$$x - 5 = 0 \text{ or } x - 12 = 0$$
$$x = 5 \text{ or } \qquad x = 12$$

Substituting in $y = \frac{60}{x}$, if $x = 5$, $y = \frac{60}{5} = 12$, and if $x = 12$, $y = \frac{60}{12} = 5$.

Thus, the legs of the triangle are of lengths 5 and 12 and the hypotenuse is $\sqrt{(5)^2 + (12)^2} = \sqrt{169} = 13$.

41.

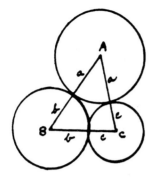

The diagram makes the following relationships clear:

$$\begin{cases} a+b = 13 \\ b+c = 15 \\ a+c = 18 \end{cases}$$

Subtracting the second equation from the first:

$$a - c = -2$$

Adding this last equation to the third original equation:

$$2a = 16$$
$$a = 8$$

Substituting $a = 8$ in the first original equation:

$$8 + b = 13$$
$$b = 5$$

Substituting $b = 5$ in the second original equation:

$$5 + c = 15$$
$$c = 10$$

The radii are $a = 8$, $b = 5$, and $c = 10$.

43.

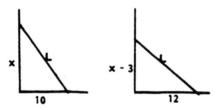

Let $x =$ the original distance from the upper end of the ladder to the ground.

Let $L =$ the length of the ladder.

When the ladder slips, the length of the ladder is unchanged, so we apply the Pythagorean Theorem to each triangle, and equate the two values of L^2:

$$10^2 + x^2 = 12^2 + (x-3)^2$$
$$100 + x^2 = 144 + x^2 - 6x + 9$$
$$6x = 53$$
$$x = \frac{53}{6}$$

Length of ladder $= \sqrt{\left(\frac{53}{6}\right)^2 + 100} \approx 13.34$ feet

45. Let x represent the length of wire used to form the square, in inches.

Then $40 - x$ represents the length of wire used to form the rectangle and $\frac{x}{4}$ represents the length of each side of the square.

Let y represent the width of the rectangle.

Then $3y$ represents the length of the rectangle.

Since the perimeter of the rectangle must be $40 - x$,

$$2y + 6y = 40 - x$$
$$y = \frac{40 - x}{8}$$

Since the combined area of the square and the rectangle is $55\frac{3}{4}$,

$$\left(\frac{x}{4}\right)^2 + 3y^2 = \frac{223}{4}$$

Substituting $y = \frac{40 - x}{8}$:

$$\left(\frac{x}{4}\right)^2 + 3\left(\frac{40 - x}{8}\right)^2 = \frac{223}{4}$$

$$\frac{x^2}{16} + \frac{3}{64}(1600 - 80x + x^2) = \frac{223}{4}$$

Multiplying by 64 and removing parentheses:

$$4x^2 + 4800 - 240x + 3x^2 = 3568$$
$$7x^2 - 240x + 1232 = 0$$
$$(7x - 44)(x - 28) = 0$$
$$7x - 44 = 0 \quad \text{or} \quad x - 28 = 0$$
$$x = \frac{44}{7} \qquad\qquad x = 28$$

There are two solutions: In one case, the wire is cut into two pieces of lengths 28 inches and 12 inches. In the second case, the wire is cut into two pieces of lengths $\frac{44}{7}$ inches and $\left(40 - \frac{44}{7}\right) = \frac{236}{7}$ inches.

Chapter 3 Review Problem Set

1. False; you may be multiplying by 0. Notice that $x = 2$ and $x^2 = 2x$ are not equivalent. $x^2 = 2x$ has $x = 0$ as a solution as well as $x = 2$.

2. True.

3. False; $x^2 = 0$ has $x = 0$ as its only solution.

4. False. $(2, 7)$ is also a solution. There are infinitely many.

5. True.

6. False. The solutions are complex (not real).

7. False; by definition $\sqrt{9}$ is the principal root of 9: $\sqrt{9} = 3$.

8. True

9. False; $|x| > 5$ means $x > 5$ or $x < -5$.

10. False. We don't know if $x + 1$ is positive.

11. a) Identity:
$$5(1-x) = 5 - 5x$$
$$5 - 5x = 5 - 5x$$

b) Conditional: $(x+2)^2 - x^2 = 4$
$$x^2 + 4x + 4 - x^2 = 4$$
$$4x + 4 = 4$$
$$4x = 0$$
$$x = 0$$

c) Identity:
$$x^2 - a^2 = (x-a)(x+a)$$
$$x^2 - a^2 = x^2 - a^2$$

d) Identity: $(x+2)(x-3) - x^2 + x = -6$
$$x^2 - x - 6 - x^2 + x = -6$$
$$-6 = -6$$

12. a) $2x + 3 = 2(x-1)$
$$2x + 3 = 2x - 2$$
$$3 = -2$$

b) $\dfrac{x+1}{x+2} = 1$
$$x + 1 = x + 2$$
$$1 = 2$$

13. a) $2\left(x + \dfrac{2}{3}\right) = x + \dfrac{1}{2}$
$$2x + \dfrac{4}{3} = x + \dfrac{1}{2}$$
$$2x - x = \dfrac{1}{2} - \dfrac{4}{3}$$
$$x = -\dfrac{5}{6}$$

b)
$$\dfrac{9}{x-4} = \dfrac{15}{x-2}$$
$$(x-2)(x-4)\left(\dfrac{9}{x-4}\right) = (x-2)(x-4)\left(\dfrac{15}{x-2}\right)$$
$$9(x-2) = 15(x-4)$$
$$9x - 18 = 15x - 60$$
$$-18 + 60 = 15x - 9x$$
$$42 = 6x$$
$$7 = x$$

c)
$$(x+3)(3x-2) = (3x+1)(x-4)$$
$$3x^2 + 7x - 6 = 3x^2 - 11x - 4$$
$$7x + 11x = -4 + 6$$
$$18x = 2$$
$$x = \dfrac{2}{18} = \dfrac{1}{9}$$

d)
$$\dfrac{x}{2x+2} - 2 = \dfrac{4-9x}{6x+6}$$
$$6(x+1)\left[\dfrac{x}{2(x+1)} - 2\right] = \left[\dfrac{4-9x}{6(x+1)}\right]6(x+1)$$
$$3x - 12(x+1) = 4 - 9x$$
$$3x - 12x - 12 = 4 - 9x$$
$$-9x + 9x = 4 + 12$$
$$0 = 16 \qquad \text{No solution}$$

14. Solver for s: $1 + rt = \dfrac{1-2s}{s}$
$$1 + rt = \dfrac{1}{s} - 2$$
$$3 + rt = \dfrac{1}{s}$$
$$s = \dfrac{1}{3+rt}$$

15. a) Let x represent the desired Fahrenheit temperature. Then $\frac{x}{3}$ represents the Celsius reading. Substituting in the formula $F = \frac{9}{5}C + 32$:
$$x = \dfrac{9}{5}\left(\dfrac{x}{3}\right) + 32$$
$$x = \dfrac{3x}{5} + 32$$
$$5x = 3x + 160$$
$$2x = 160$$
$$x = 80 \text{ degrees F}$$

b) Since $\qquad F = \frac{9}{5}C + 32,$
$$C = \tfrac{5}{9}(F - 32)$$
Let F represent the Fahrenheit temperature at which the Fahrenheit reading is higher than the Celsius reading. Then,
$$F > \tfrac{5}{9}(F - 32)$$
$$9F > 5(F - 32)$$
$$9F > 5F - 160$$
$$4F > -160$$
$$F > -40 \text{ degrees Fahrenheit}$$

16. a) $\begin{cases} 2x + y = 3 \\ 3x + 2y = 6 \end{cases}$
$\begin{cases} 4x + 2y = 6 \\ 3x + 2y = 6 \end{cases}$
$x = 0;\ 2(0) + y = 3,\ y = 3$

b) $\begin{cases} \frac{1}{2}x - \frac{1}{4}y = 1 \\ x + 2y = 5 \end{cases}$
$\begin{cases} 4x - 2y = 8 \\ x + 2y = 5 \end{cases}$
$$\cdot\ 5x = 13$$
$$x = \dfrac{13}{5}$$
Substituting in the second original equation:
$$2y = 5 - \dfrac{13}{5} = \dfrac{12}{5},\ y = \dfrac{6}{5}$$

17. a)
$$\dfrac{6 + \sqrt{18}}{12} = \dfrac{6 + \sqrt{9\cdot2}}{12}$$
$$= \dfrac{6 + 3\sqrt{2}}{12}$$
$$= \dfrac{3(2 + \sqrt{2})}{3(4)}$$
$$= \dfrac{2 + \sqrt{2}}{4}$$

b)
$$\frac{\sqrt{28}}{\sqrt{63}} = \sqrt{\frac{28}{63}}$$
$$= \sqrt{\frac{4}{9}}$$
$$= \frac{2}{3}$$

c)
$$\sqrt{13^2 \cdot 2^4} = \sqrt{(13 \cdot 2^2)^2}$$
$$= 13 \cdot 2^2$$
$$= 52$$

d)
$$\frac{3 - \sqrt{-9}}{3} = \frac{3 - \sqrt{9}i}{3}$$
$$= \frac{3 - 3i}{3}$$
$$= \frac{3(1-i)}{3}$$
$$= 1 - i$$

18. There are no real values for which $x^2 < -1$.

19. $3x^2 = 192$
$x^2 = 64$
$x = \pm 8$

20. $(x-1)^2 = 4$
$x - 1 = \pm 2$
$x = 3$ or $x = -1$

21. $x^2 + 4x + 4 = 0$
$(x+2)^2 = 0$
$x + 2 = 0$
$x = -2$

22. $x^2 - 3x = 0$
$x(x-3) = 0$
$x = 0$ or $x = 3$

23. $x^2 - 2x - 35 = 0$
$(x-7)(x+5) = 0$
$x - 7 = 0$ or $x + 5 = 0$
$x = 7$ or $x = -5$

24. $3x^2 - x - 2 = 0$
$(3x+2)(x-1) = 0$
$3x + 2 = 0$ or $x - 1 = 0$
$x = -\frac{2}{3}$ or $x = 1$

25. $(x-2)(x-3) = 12$
$x^2 - 5x + 6 = 12$
$x^2 - 5x - 6 = 0$
$(x-6)(x+1) = 0$
$x - 6 = 0$ or $x + 1 = 0$
$x = 6$ or $x = -1$

26. $3x^2 + x = 1 \Rightarrow 3x^2 + x - 1 = 0$
$a = 3,\ b = 1,\ c = -1$
$$x = \frac{-1 \pm \sqrt{(1)^2 - 4(3)(-1)}}{2(3)} = \frac{-1 \pm \sqrt{13}}{6}$$

27. $x^2 + 2x - 2 = 0$
$a = 1,\ b = 2,\ c = -2$
$$x = \frac{-2 \pm \sqrt{(2)^2 - 4(1)(-2)}}{2(1)}$$
$$= \frac{-2 \pm \sqrt{12}}{2}$$
$$= \frac{-2 \pm 2\sqrt{3}}{2}$$
$$= -1 \pm \sqrt{3}$$

28. $(x+1)(x-1) \cdot \frac{1}{x+1} = \frac{x}{x-1}(x+1)(x-1)$
$x - 1 = x(x+1) = x^2 + x$
$-1 = x^2$
$x = \pm\sqrt{-1} = \pm i$

29. $(x-1)(x-2) = 1$
$x^2 - 3x + 2 = 1$
$x^2 - 3x + 1 = 0$
$a = 1,\ b = -3,\ c = 1$
$$x = \frac{-(-3) \pm \sqrt{(-3)^2 - 4(1)(1)}}{2(1)}$$
$$= \frac{3 \pm \sqrt{5}}{2}$$

30. $x - 2\sqrt{x} - 8 = 0$. For convenience, let $y = \sqrt{x}$:
$y^2 - 2y - 8 = 0$
$(y-4)(y+2) = 0$
$y = 4$ or $y = -2$
$\sqrt{x} = 4$ or $\sqrt{x} = -2$
$x = 16,\ \sqrt{x}$ cannot be negative

31.
$(x-1)^2 + 2(x-1) - 8 = 0$
$[(x-1) + 4][(x-1) - 2] = 0$
$(x+3)(x-3) = 0$
$x + 3 = 0$ or $x - 3 = 0$
$x = -3$ or $x = 3$

32. $\frac{6}{(x+2)^2} + \frac{1}{x+2} - 1 = 0$
$$\frac{6(x+2)^2}{(x+2)^2} + \frac{1(x+2)^2}{x+2} - 1(x+2)^2 = 0$$
$6 + x + 2 - x^2 - 4x - 4 = 0$
$-x^2 - 3x + 4 = 0$
$x^2 + 3x - 4 = 0$
$(x+4)(x-1) = 0$
$x = -4$ or $x = 1$

33.
$$(3x - 2y)^2 = 9$$
$$3x - 2y = \pm 3$$
$$-2y = -3x \pm 3$$
$$y = \frac{-3x \pm 3}{-2}$$
$$y = \frac{3x \mp 3}{2}$$
$$y = \frac{3x - 3}{2} \text{ or } y = \frac{3x + 3}{2}$$

34. $(y + 2x^2)^2 - 5(y + 2x^2) + 4 = 0$
If we let $w = y + 2x^2$, we get
$$w^2 - 5w + 4 = 0$$
$$(w - 1)(w - 4) = 0$$

$w = 1$	or	$w = 4$
$y + 2x^2 = 1$	or	$y + 2x^2 = 4$
$y = 1 - 2x^2$	or	$y = 4 - 2x^2$

35.
$$2x - 3 < 4 - 3x$$
$$5x < 7$$
$$x < \frac{7}{5}$$

36.
$$5 - 2x \geq 2(x + 4)$$
$$5 - 2x \geq 2x + 8$$
$$-3 \geq 4x$$
$$x \leq -\frac{3}{4}$$

37. $x^2 - 2x - 24 > 0$
$(x - 6)(x + 4) > 0$
Find the split points:

$x - 6 = 0$	$x + 4 = 0$
$x = 6$	$x = -4$

The split points divide the number line into 3 regions. Choose a test point in each region.

Region	Test point	Value of $(x-6)(x+4)$
$x < -4$	$x = -5$	$+$
$-4 < x < 6$	$x = 0$	$-$
$x > 6$	$x = 7$	$+$

Solution: $\{x : x < -4 \text{ or } x > 6\}$

38. $x^2 + 4x + 4 > 0$
$(x + 2)^2 > 0$
$\{x : x \neq -2\}$

39. $\frac{2x + 1}{x - 3} \geq 0$. Find the split points:

$2x + 1 = 0$	$x - 3 = 0$
$x = -\frac{1}{2}$	$x = 3$

The split points divide the number line into 3 regions. Choose a test point in each region.

Region	Test point	Value of $\frac{2x+1}{x-3}$
$x < -\frac{1}{2}$	$x = -1$	$+$
$-\frac{1}{2} < x < 3$	$x = 0$	$-$
$x > 3$	$x = 4$	$+$

Since $\frac{2x + 1}{x - 3} = 0$ when $x = -\frac{1}{2}$, we want to include $x = -\frac{1}{2}$ in the solution set.
Since $\frac{2x + 1}{x - 3}$ is undefined when $x = 3$, the split point $x = 3$ could never be included in the solution set.
Solution: $\{x : x \leq -\frac{1}{2} \text{ or } x > 3\}$

40. $1 - \frac{4}{x^2} < 0$
$\frac{x^2 - 4}{x^2} < 0$ Since the denominator is always positive (never 0), this inequality is equivalent to
$$x^2 - 4 < 0$$
$$(x - 2)(x + 2) < 0$$
$$-2 < x < 2$$
Solution: $\{x : -2 < x < 2, x \neq 0\}$

41.
$$x^2 + 2x < 2$$
$$x^2 + 2x - 2 < 0$$
Find the split points using the quadratic formula:
$$x = \frac{-2 \pm \sqrt{(2)^2 - 4(1)(-2)}}{2(1)}$$
$$= \frac{-2 \pm \sqrt{12}}{2}$$
$$= \frac{-2 \pm 2\sqrt{3}}{2}$$
$$= -1 \pm \sqrt{3}$$
The split points divide the number line into 3 regions. Choose a test point in each region.

Region	Test point	Value of $x^2 + 2x - 2$
$x < -1 - \sqrt{3}$	$x = -4$	$+$
$-1 - \sqrt{3} < x < -1 + \sqrt{3}$	$x = 0$	$-$
$x > -1 + \sqrt{3}$	$x = 4$	$+$

Solution: $\{x : -1 - \sqrt{3} < x < -1 + \sqrt{3}\}$

42. $x^2 + 6x + 20 > 0$. The split points are $x = \frac{-6 \pm \sqrt{36 - 80}}{2}$. Since the split points are imaginary, this quantity is either always positive or always negative. Since letting $x = 0$ we get a positive value, the solution set is the set of all real numbers.

43.
$$|x + 5| < 5$$
$$-5 < x + 5 < 5$$
$$-10 < \quad x < 0$$

44. $|3x + 7| \geq 4$

$3x + 7 \geq 4$	or	$3x + 7 \leq -4$
$3x \geq -3$	or	$3x \leq -11$
$x \geq -1$	or	$x \leq \frac{-11}{3}$

45. $|2x+1| = 7$
$$2x + 1 = 7 \quad \text{or} \quad 2x + 1 = -7$$
$$2x = 6 \qquad\qquad 2x = -8$$
$$x = 3 \qquad\qquad x = -4$$
$$x = 3 \text{ or } x = -4$$

46. $|5 - x| = 2$
$$5 - x = 2 \quad \text{or} \quad 5 - x = -2$$
$$x = 3 \quad \text{or} \qquad x = 7$$

47. $|x| = 3 - x$
$$x = 3 - x \quad \text{or} \quad x = -(3 - x)$$
$$2x = 3 \qquad\qquad x = -3 + x$$
$$x = \tfrac{3}{2} \qquad\qquad 0 = -3$$
$x = \tfrac{3}{2}$ is the only solution.

48. $|x + 1| = |x - 2|$
$$x + 1 = x - 2 \quad \text{or} \quad x + 1 = -(x - 2) = -x + 2$$
$$1 = -2 \qquad \text{or} \qquad 2x = 1$$
$$\qquad\qquad\qquad\qquad x = \tfrac{1}{2}$$
Solution: $x = \tfrac{1}{2}$

49. You may be dividing by 0; in fact, the solution $x = 0$ is lost.

50. a) The ball strikes the ground when $s = 0$.
$$-16t^2 + 96t = 0$$
$$16t(-t + 6) = 0$$
$$t = 0 \text{ or } t = 6$$
$t = 0$ is when the ball first leaves the ground, so the ball strikes the ground 6 seconds later.

b) We need to solve
$$-16t^2 + 96t > 80$$
$$-16t^2 + 96t - 80 > 0$$
$$-t^2 + 6t - 5 > 0$$
$$t^2 - 6t + 5 < 0$$
$$(t - 1)(t - 5) < 0$$
$$1 < t < 5$$
So the ball's height is greater than 80 feet between 1 and 5 seconds after it leaves the ground.

51. Let x be the amount invested in the 7% fund. Then $25,000 - x$ represents the amount invested in the 8.5% fund.
$.07x$ represents the interest earned in one year from the 7% fund.
$.085(25,000 - x)$ represents the interest earned in one year from the 8.5% fund.
Since the total yearly interest is \$1981,
$$.07x + .085(25,000 - x) = 1981$$
$$.07x + 2125 - .085x = 1981$$
$$-0.015x = -144$$
$$x = \frac{-144}{-0.015} = 9600$$

Mr. Ito had \$9600 invested in the 7% fund and $(25,000 - 9600) = \$15,400$ invested in the 8.5% fund.

52. Let l = original length and w = original width. The original area is then lw. We are given:
$$\begin{cases} l = w + 20 \\ (w - 2)(l + 4) = lw - 16 \end{cases}$$
Substituting for l in the second equation:
$$(w - 2)(w + 20 + 4) = (w + 20)w - 16$$
$$(w - 2)(w + 24) = w^2 + 20w - 16$$
$$w^2 + 22w - 48 = w^2 + 20w - 16$$
$$2w = 32$$
$$w = 16$$
The original dimensions were 16 cm and 36 cm.

53. Let x represent the rate, in miles per hour, of the second train. Using $D = RT$,
$3x$ represents the distance the second train traveled by 11 A.M. The first train traveled $5(60) = 300$ miles by 11 A.M.
$$3x + 300 = 450$$
$$3x = 150$$
$$x = 50$$
The speed of the second train was 50 mph.

54. In the first three-fifths of the season they have played $\tfrac{3}{5}(150) = 90$ games. If they have won 60% of these games, they have won $.60 \cdot 90 = 54$ games. To win $66.\overline{66}\%$ of their 150 games they must win a total of 100 games; they must win 46 of their remaining 60 games, a percentage of $\tfrac{46}{60} \cdot 100\% = 76.\overline{66}\%$.

55. Let x represent the number of reserved seats sold. Let y represent the number of general admission tickets sold.
Then $7x$ represents the amount of money received for reserved seats and $4.50y$ represents the amount of money received for general admission tickets.
$$\begin{cases} x + y = 625 \\ 7x + 4.50y = 3375 \end{cases}$$
Multiplying the first equation by -7:
$$-7x - 7y = -4375$$
$$7x + 4.50y = 3375$$
Adding these last two equations:
$$-2.5y = -1000$$
$$y = \frac{-1000}{-2.5} = 400$$
Substituting $y = 400$ in the first original equation:
$$x + 400 = 625$$
$$x = 225$$
225 reserved seats were sold and 400 general admission tickets were sold.

56. Let x be the number of liters of water that must evaporate. We have 16 liters of salt to begin, none of which evaporates. Our final solution contains $100 - x$ liters; we want $.25(100 - x) = 16 \Rightarrow$ $100 - x = 64 \Rightarrow x = 36$. Therefore, 36 liters of water must evaporate.

Chapter 4: Coordinates and Curves

1. $d(A,B) = |-3-2| = |-5| = 5$

3. $d(A,B) = \left|\frac{11}{4} - \left(-\frac{5}{4}\right)\right| = \left|\frac{16}{4}\right| = 4$

5. $d(A,B) = |3.26 - 4.96| = |-1.70| = 1.70$

7. $d(A,B) = |2-\pi - (\pi-3)| = |5-2\pi| = 2\pi - 5$

9. Since $d(A,B) = 2$,

$\qquad |5-b| = 2$

Therefore,

$\qquad 5-b = 2 \ \ \text{or} \ \ 5-b = -2$

$\qquad\quad b = 3 \ \ \text{or} \qquad b = 7$

11. Since $d(A,B) = 4$,

$\qquad |-2-b| = 4$

Therefore,

$\qquad -2-b = 4 \quad \text{or} \quad -2-b = -4$

$\qquad\quad b = -6 \ \ \text{or} \qquad\quad b = 2$

13. Since $d(A,B) = \frac{3}{4}$,

$\qquad \left|\frac{5}{2} - b\right| = \frac{3}{4}$

Therefore,

$\qquad \frac{5}{2} - b = \frac{3}{4} \quad \text{or} \quad \frac{5}{2} - b = -\frac{3}{4}$

$\qquad\quad b = \frac{7}{4} \quad \text{or} \qquad b = \frac{13}{4}$

15.

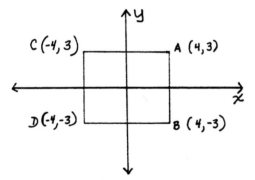

$d(A,C) = |4-(-4)| = 8$
$d(A,B) = |3-(-3)| = 6$
Since sides AC and BD are horizontal and sides AB and CD are vertical, all the angles are right angles, and so quadrilateral $ABDC$ is a rectangle. Since consecutive sides are not of equal length, the rectangle is not a square.

17. See the figure in the next column.

$d(A,B) = \sqrt{(2-1)^2 + (6-3)^2} = \sqrt{(1)^2 + (3)^2} = \sqrt{10}$

$d(C,D) = \sqrt{(4-3)^2 + (7-4)^2} = \sqrt{(1)^2 + (3)^2} = \sqrt{10}$

Therefore, $AB = CD$.

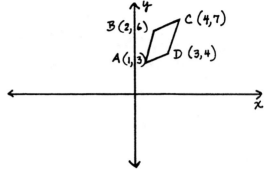

$d(B,C) = \sqrt{(4-2)^2 + (7-6)^2} = \sqrt{(2)^2 + (1)^2} = \sqrt{5}$

$d(A,D) = \sqrt{(3-1)^2 + (4-3)^2} = \sqrt{(2)^2 + (1)^2} = \sqrt{5}$

Therefore, $BC = AD$.

$d(A,C) = \sqrt{(4-1)^2 + (7-3)^2} = \sqrt{(3)^2 + (4)^2} = \sqrt{25}$
$\qquad\qquad\qquad\qquad\qquad\qquad\qquad\quad = 5$

Notice that $(AB)^2 + (BC)^2 = 10 + 5 = 15$ and $(AC)^2 = (5)^2 = 25$. Since $(AB)^2 + (BC)^2 \neq (AC)^2$, angle B is not a right angle.

Since the opposite sides of the quadrilateral are of equal lengths, the figure is parallelogram.

19. $d(P_1, P_2) = \sqrt{(2-5)^2 + (-1-3)^2}$

$\qquad\qquad = \sqrt{(-3)^2 + (-4)^2}$

$\qquad\qquad = \sqrt{9+16}$

$\qquad\qquad = \sqrt{25}$

$\qquad\qquad = 5$

The midpoint of $P_1 P_2$ is

$\left(\frac{2+5}{2}, \frac{-1+3}{2}\right) = \left(\frac{7}{2}, 1\right)$

21. $d(P_1 P_2) = \sqrt{(4-2)^2 + (2-4)^2}$

$\qquad\qquad = \sqrt{(2)^2 + (-2)^2}$

$\qquad\qquad = \sqrt{4+4}$

$\qquad\qquad = \sqrt{8}$

$\qquad\qquad = 2\sqrt{2}$

The midpoint of $P_1 P_2$ is

$\left(\frac{4+2}{2}, \frac{2+4}{2}\right) = (3,3)$

23. $d(P_1 P_2) = \sqrt{(\sqrt{3}-0)^2 + (0-\sqrt{6})^2}$

$\qquad\qquad = \sqrt{(\sqrt{3})^2 + (-\sqrt{6})^2}$

$\qquad\qquad = \sqrt{3+6}$

$\qquad\qquad = \sqrt{9}$

$\qquad\qquad = 3$

The midpoint of $P_1 P_2$ is

$\left(\frac{\sqrt{3}+0}{2}, \frac{0+\sqrt{6}}{2}\right) = \left(\frac{\sqrt{3}}{2}, \frac{\sqrt{6}}{2}\right)$

25. $d(P_1 P_2) = \sqrt{(6.714 - 1.234)^2 + [8.341 - (-5.132)]^2}$
$$= \sqrt{5.480^2 + 13.473^2}$$
$$= \sqrt{211.552129}$$
$$\approx 14.545$$

Midpoint of segment $P_1 P_2$ is
$$\left(\frac{6.714 + 1.234}{2}, \frac{8.341 - 5.132}{2}\right) = \left(\frac{7.948}{2}, \frac{3.209}{2}\right)$$
$$= (3.974, 1.605)$$

27. a) $d(A, B) = \sqrt{(2-1)^2 + (6-3)^2}$
$$= \sqrt{(1)^2 + (3)^2} = \sqrt{10}$$
$$d(B, C) = \sqrt{(4-2)^2 + (7-6)^2}$$
$$= \sqrt{(2)^2 + (1)^2} = \sqrt{5}$$
$$d(C, D) = \sqrt{(4-3)^2 + (7-4)^2}$$
$$= \sqrt{(1)^2 + (3)^2} = \sqrt{10}$$
$$d(D, A) = \sqrt{(3-1)^2 + (4-3)^2}$$
$$= \sqrt{(2)^2 + (1)^2} = \sqrt{5}$$

b) The midpoint of AC is
$$\left(\frac{1+4}{2}, \frac{3+7}{2}\right) = \left(\frac{5}{2}, 5\right)$$
The midpoint of BD is
$$\left(\frac{2+3}{2}, \frac{6+4}{2}\right) = \left(\frac{5}{2}, 5\right)$$

c) Part (a) agrees with the fact that opposite sides of a parallelogram have the same length; part (b) agrees with the fact that the diagonals of a parallelogram bisect each other.

29. a)

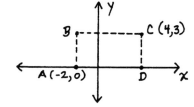

B has coordinates $(-2, 3)$; D has coordinates $(4, 0)$.

b)

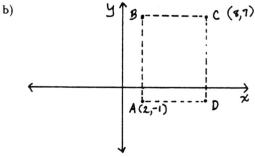

B has coordinates $(2, 7)$; D has coordinates $(8, -1)$.

31. Let the vertices be $A(2, -4)$, $B(4, 0)$, and $C(8, -2)$
$$d(A, B) = \sqrt{(4-2)^2 + [0 - (-4)]^2}$$
$$= \sqrt{2^2 + 4^2} = \sqrt{20}$$
$$d(B, C) = \sqrt{(8-4)^2 + (-2-0)^2}$$
$$= \sqrt{4^2 + 2^2} = \sqrt{20}$$
$$d(C, A) = \sqrt{(8-2)^2 + [-2-(-4)]^2}$$
$$= \sqrt{6^2 + (-2+4)^2}$$
$$= \sqrt{36+4} = \sqrt{40}$$

Since
$$[d(A, B)]^2 + [d(B, C)]^2 = (\sqrt{20})^2 + (\sqrt{20})^2 = 20 + 20$$
$$= 40 = [d(A, C)]^2,$$
the converse of the Pythagorean Theorem tells us that the triangle is a right triangle.

33. a) Label the points $A(0, 0)$, $B(3, 4)$, and $C(-6, -8)$.
$$d(A, B) = \sqrt{(3-0)^2 + (4-0)^2}$$
$$= \sqrt{(3)^2 + (4)^2} = \sqrt{25} = 5$$
$$d(B, C) = \sqrt{(-6-3)^2 + (-8-4)^2}$$
$$= \sqrt{(-9)^2 + (-12)^2}$$
$$= \sqrt{225} = 15$$
$$d(A, C) = \sqrt{(-6-0)^2 + (-8-0)^2}$$
$$= \sqrt{(-6)^2 + (-8)^2}$$
$$= \sqrt{100} = 10$$

Since
$$d(A, B) + d(A, C) = 5 + 10 = 15 = d(B, C),$$
the three points must lie on the same line.

b) Label the points $A(-4, 1)$, $B(-1, 5)$, and $C(5, 13)$.
$$d(A, B) = \sqrt{[-4 - (-1)]^2 + (1-5)^2}$$
$$= \sqrt{(-3)^2 + (-4)^2}$$
$$= \sqrt{25} = 5$$
$$d(B, C) = \sqrt{(-1-5)^2 + (5-13)^2}$$
$$= \sqrt{(-6)^2 + (-8)^2}$$
$$= \sqrt{100} = 10$$
$$d(A, C) = \sqrt{(-4-5)^2 + (1-13)^2}$$
$$= \sqrt{(-9)^2 + (-12)^2}$$
$$= \sqrt{225} = 15$$

Since $d(A, B) + d(B, C) = 5 + 10 = 15 = d(A, C)$, the three points lie on the same straight line.

35.

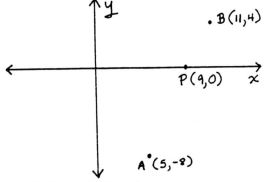

Substituting in the formulas
$$x = (1-t)x_1 + tx_2$$
$$y = (1-t)y_1 + ty_2,$$
$$x = (1 - \tfrac{2}{3})(5) + \tfrac{2}{3}(11)$$
$$= \tfrac{1}{3}(5) + \tfrac{2}{3}(11)$$
$$= \tfrac{5}{3} + \tfrac{22}{3} = \tfrac{27}{3} = 9$$
$$y = (1 - \tfrac{2}{3})(-8) + \tfrac{2}{3}(4)$$
$$= \tfrac{1}{3}(-8) + \tfrac{2}{3}(4)$$
$$= -\tfrac{8}{3} + \tfrac{8}{3} = 0$$
The point P has coordinates $(9,0)$.

37.

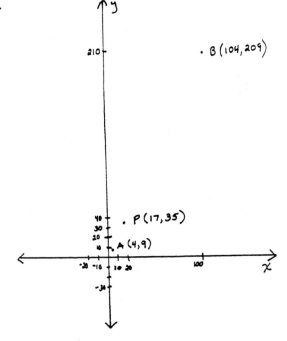

Substituting in the formulas
$$x = (1-t)x_1 + tx_2$$
$$y = (1-t)y_1 + ty_2,$$
$$x = \left(1 - \tfrac{13}{100}\right)4 + \tfrac{13}{100} \cdot 104$$
$$= \tfrac{87}{100} \cdot 4 + \tfrac{1352}{100} = \tfrac{348}{100} + \tfrac{1352}{100} = \tfrac{1700}{100} = 17$$
$$y = \left(1 - \tfrac{13}{100}\right) \cdot 9 + \tfrac{13}{100} \cdot 209$$
$$= \tfrac{87}{100} \cdot 9 + \tfrac{2717}{100} = \tfrac{783}{100} + \tfrac{2717}{100} = \tfrac{3500}{100} = 35$$
The point P has coordinates $(17,35)$.

39.
$$d(A,B) = \sqrt{(214-0)^2 + (17-0)^2}$$
$$= \sqrt{(214)^2 + (17)^2}$$
$$= \sqrt{46085} \approx 214.674 \text{ miles}$$
$$d(B,C) = \sqrt{(230-214)^2 + (179-17)^2}$$
$$= \sqrt{(16)^2 + (162)^2}$$
$$= \sqrt{26500} \approx 162.788 \text{ miles}$$
$$d(A,C) = \sqrt{(230-0)^2 + (179-0)^2}$$
$$= \sqrt{(230)^2 + (179)^2}$$
$$= \sqrt{84941} \approx 291.446 \text{ miles}$$
The cost of shipping by truck equals
$$3.71[d(A,B) + d(B,C)] \approx 3.71(214.674 + 162.788)$$
$$= (3.71)(377.462) \approx \$1400.38$$
The cost of shipping by air equals
$$4.81[d(A,C)] \approx 4.81(291.446) \approx \$1401.86$$
It is $(\$1401.86 - \$1400.38) = \$1.48$ cheaper to ship by truck.

41. Label the points $A(0,0)$, $B(-3,4)$, and $C(x,y)$.
$$d(A,B) = \sqrt{(-3-0)^2 + (4-0)^2}$$
$$= \sqrt{(-3)^2 + (4)^2}$$
$$= \sqrt{25} = 5$$
$$d(A,C) = \sqrt{(x-0)^2 + (y-0)^2}$$
$$= \sqrt{x^2 + y^2}$$
$$d(B,C) = \sqrt{[x-(-3)]^2 + (y-4)^2}$$
$$= \sqrt{(x+3)^2 + (y-4)^2}$$
Since A, B, and C are to be the vertices of an equilateral triangle, we must have
$$d(A,B) = d(A,C) = d(B,C)$$
Translating into separate equations, we get
$$\begin{cases} \sqrt{x^2 + y^2} = 5 \\ \sqrt{(x+3)^2 + (y-4)^2} = 5 \end{cases}$$

Squaring each side of each equation we get
$$x^2 + y^2 = 25$$
$$(x+3)^2 + (y-4)^2 = 25$$
Expanding the second equation:
$$x^2 + 6x + 9 + y^2 - 8y + 16 = 25$$
We can replace $x^2 + y^2$ with 25, to get
$$6x + 9 + 25 - 8y + 16 = 25$$
$$6x + 25 = 8y$$
$$y = \frac{6x + 25}{8}$$
Substituting into $x^2 + y^2 = 25$:
$$x^2 + \left(\frac{6x+25}{8}\right)^2 = 25$$
$$x^2 + \frac{36x^2 + 300x + 625}{64} = 25$$
$$64x^2 + 36x^2 + 300x + 625 = 1600$$
$$100x^2 + 300x - 975 = 0$$
$$4x^2 + 12x - 39 = 0$$
Using the quadratic formula:
$$x = \frac{-12 \pm \sqrt{(12)^2 - 4(4)(-39)}}{2(4)}$$
$$= \frac{-12 \pm \sqrt{768}}{8} = \frac{-12 \pm 16\sqrt{3}}{8} = \frac{-3 \pm 4\sqrt{3}}{2}$$
Since x must be in the first quadrant,
$$x = \frac{-3 + 4\sqrt{3}}{2}$$
$$y = \frac{6x+25}{8} = \frac{-9 + 12\sqrt{3} + 25}{8} = \frac{16 + 2\sqrt{3}}{8}$$
$$= \frac{4 + 3\sqrt{3}}{2}$$

43.

$e^2 = \left(\frac{a}{2}\right)^2 + \left(\frac{b}{2}\right)^2, f^2 = \left(\frac{a}{2}\right)^2 + \left(\frac{b}{2}\right)^2,$

$d^2 = \left(\frac{a}{2}\right)^2 + \left(\frac{b}{2}\right)^2$

The midpoint, M, of the hypotenuse has coordinates $\left(\frac{a+0}{2}, \frac{b+0}{2}\right) = \left(\frac{a}{2}, \frac{b}{2}\right)$

$$AM = \sqrt{\left(\frac{a}{2} - 0\right)^2 + \left(\frac{b}{2} - 0\right)^2}$$
$$= \sqrt{\frac{a^2}{4} + \frac{b^2}{4}} = \frac{1}{2}\sqrt{a^2 + b^2}$$

$$BM = \sqrt{\left(a - \frac{a}{2}\right)^2 + \left(0 - \frac{b}{2}\right)^2}$$
$$= \sqrt{\left(\frac{a}{2}\right)^2 + \left(\frac{-b}{2}\right)^2}$$
$$= \sqrt{\frac{a^2}{4} + \frac{b^2}{4}}$$
$$= \frac{1}{2}\sqrt{a^2 + b^2}$$

$$CM = \sqrt{\left(a - \frac{a}{2}\right)^2 + \left(b - \frac{b}{2}\right)^2}$$
$$= \sqrt{\left(\frac{a}{2}\right)^2 + \left(\frac{b}{2}\right)^2}$$
$$= \sqrt{\frac{a^2}{4} + \frac{b^2}{4}}$$
$$= \frac{1}{2}\sqrt{a^2 + b^2}$$

Since the triangle was perfectly general, the midpoint M of the hypotenuse is equidistant from the three vertices A, B, and C.

45. Using the formulas
$$x = (1-t)x_1 + tx_2$$
$$y = (1-t)y_1 + ty_2,$$
let $t = \frac{2}{3}$, $x_1 = 2$, $y_1 = 1$, $x = 4$, $y = 5$.
Find x_2 and y_2:
$$4 = \left(1 - \frac{2}{3}\right)2 + \frac{2}{3}x_2$$
$$4 = \frac{2}{3} + \frac{2}{3}x_2$$
$$\frac{10}{3} = \frac{2}{3}x_2$$
$$5 = x_2$$
$$5 = \left(1 - \frac{2}{3}\right)1 + \frac{2}{3}y_2$$
$$5 = \frac{1}{3} + \frac{2}{3}y_2$$
$$\frac{14}{3} = \frac{2}{3}y_2$$
$$7 = y_2$$
The point we seek has coordinates $(5, 7)$.

47. Introduce coordinates with Jane's house at the origin. Then Susan's house has coordinates $S(3,5)$ and Tammy's house has coordinates $T(-4,7)$.
$$d(S,T) = \sqrt{(-4-3)^2 + (7-5)^2}$$
$$= \sqrt{(-7)^2 + (2)^2}$$
$$= \sqrt{53}$$
$$\approx 7.28 \text{ miles}$$

49. Without loss of generality, we can position the triangle on a coordinate system as shown below, with $a^2 + b^2 = c^2$.

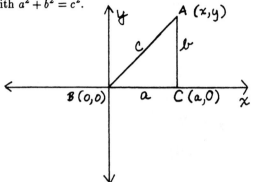

If we can show that $x = a$ then we will have shown that AC is a vertical line, perpendicular to BC, and therefore angle C is a right angle.
$$b^2 = (x-a)^2 + y^2$$
$$c^2 = x^2 + y^2$$
$$a^2 = a^2$$
Since $a^2 + b^2 = c^2$,
$$a^2 + (x-a)^2 + y^2 = x^2 + y^2$$
$$a^2 + x^2 - 2ax + a^2 + y^2 = x^2 + y^2$$
$$2a^2 - 2ax = 0$$
$$2ax = 2a^2$$
$$x = a$$
So, ABC is a right triangle. Notice that substituting a for x in $b^2 = (x-a)^2 + y^2$, we get
$$b^2 = y^2$$
$$y = \pm b$$
which is what we expect if ABC is a right triangle.

51. Dividing the horizontal distance between the points into 5 equal parts, we see that there should be $\frac{21-(-4)}{5} = \frac{25}{5} = 5$ horizontal units between each equally spaced point. Similarly, there should be $\frac{38-3}{5} = \frac{35}{5} = 7$ vertical units between each equally spaced point.
The four points we seek, then, are
$(-4+5, 3+7) = (1, 10)$
$(1+5, 10+7) = (6, 17)$
$(6+5, 17+7) = (11, 24)$
$(11+5, 24+7) = (16, 31)$

53.

$C (x_3, y_3)$
$A (x_1, y_1)$
$B (x_2, y_2)$
N
M
P

The midpoints of sides BC, CA, and AB are M, N, and P respectively. Using the midpoint formula, the coordinates of these points are
$M\left(\frac{x_2 + x_3}{2}, \frac{y_2 + y_3}{2}\right)$, $N\left(\frac{x_1 + x_3}{2}, \frac{y_1 + y_3}{2}\right)$, and
$P\left(\frac{x_1 + x_2}{2}, \frac{y_1 + y_2}{2}\right)$.

The coordinates of the point $\frac{2}{3}$ of the way from A to M:
$$x_A = \left(1 - \frac{2}{3}\right)x_1 + \frac{2}{3}\left(\frac{x_2 + x_3}{2}\right) = \frac{x_1}{3} + \frac{x_2}{3} + \frac{x_3}{3}$$
$$y_A = \left(1 - \frac{2}{3}\right)y_1 + \frac{2}{3}\left(\frac{y_2 + y_3}{2}\right) = \frac{y_1}{3} + \frac{y_2}{3} + \frac{y_3}{3}$$
The coordinates of the point $\frac{2}{3}$ of the way from B to N:
$$x_B = \left(1 - \frac{2}{3}\right)x_2 + \frac{2}{3}\left(\frac{x_1 + x_3}{2}\right) = \frac{x_2}{3} + \frac{x_1}{3} + \frac{x_3}{3}$$
$$y_B = \left(1 - \frac{2}{3}\right)y_2 + \frac{2}{3}\left(\frac{y_1 + y_3}{2}\right) = \frac{y_2}{3} + \frac{y_1}{3} + \frac{y_3}{3}$$
The coordinates of the point $\frac{2}{3}$ of the way from C to P:
$$x_C = \left(1 - \frac{2}{3}\right)x_3 + \frac{2}{3}\left(\frac{x_1 + x_2}{2}\right) = \frac{x_3}{3} + \frac{x_1}{3} + \frac{x_2}{3}$$
$$y_C = \left(1 - \frac{2}{3}\right)y_3 + \frac{2}{3}\left(\frac{y_1 + y_2}{2}\right) = \frac{y_3}{3} + \frac{y_1}{3} + \frac{y_2}{3}$$
Since the coordinates of all three points are the same, the 3 medians of a triangle intersect at one point.

55. Position the rectangle on a coordinate system as shown below.

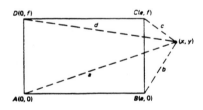

Using the distance formula,
$$a^2 = (x-0)^2 + (y-0)^2$$
$$= x^2 + y^2$$
$$c^2 = (x-e)^2 + (y-f)^2$$
$$b^2 = (x-e)^2 + (y-0)^2$$
$$= (x-e)^2 + y^2$$
$$d^2 = (x-0)^2 + (y-f)^2$$
$$= x^2 + (y-f)^2$$
Thus, $a^2 + c^2 = x^2 + y^2 + (x-e)^2 + (y-f)^2$ and $b^2 + d^2 = (x-e)^2 + y^2 + x^2 + (y-f)^2$.
Hence, $a^2 + c^2 = b^2 + d^2$.

64

x	y
-2	0
-1	3
0	4
1	3
2	0

1. $y = 3x - 2$

Check for symmetries.

Replace x with $-x$: $y = -3x - 2$

Not symmetric with respect to y-axis.

Replace y with $-y$: $-y = 3x - 2$
$$y = -3x + 2$$

Not symmetric with respect to x-axis.

Replace x with $-x$ and y with $-y$:
$$-y = -3x - 2$$
$$y = 3x + 2$$

Not symmetric with respect to the origin.

x	y
-2	-8
-1	-5
0	-2
1	1
2	4

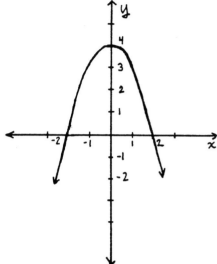

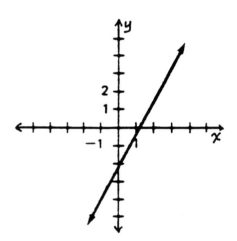

3. $y = -x^2 + 4$

Check for symmetries:

Replace x with $-x$: $y = -(-x)^2 + 4 = -x^2 + 4$

Symmetric with respect to y-axis.

Replace y with $-y$: $-y = -x^2 + 4$
$$y = x^2 - 4$$

Not symmetric with respect to x-axis.

Replace x with $-x$ and y with $-y$:
$$-y = -(-x)^2 + 4$$
$$y = x^2 - 4$$

Not symmetric with respect to the origin.

5. $y = x^2 - 4x$

Check for symmetries.

Replace x with $-x$: $y = (-x)^2 - 4(-x) = x^2 + 4x$

Not symmetric with respect to y-axis.

Replace y with $-y$: $-y = x^2 - 4x$
$$y = -x^2 + 4x$$

Not symmetric with respect to x-axis.

Replace x with $-x$ and y with $-y$:
$$-y = (-x)^2 - 4(-x)$$
$$y = -x^2 - 4x$$

Not symmetric with respect to the origin.

x	y
-2	12
-1	5
0	0
1	-3
2	-4
3	-3
4	0

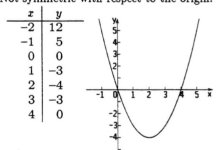

7. $y = -x^3$

Check for symmetries.

Replace x with $-x$: $y = -(-x)^3 = x^3$

Not symmetric with respect to y-axis.

Replace y with $-y$: $-y = -x^3$
$$y = x^3$$

Not symmetric with respect to x-axis.

65

Replace x with $-x$ and y with $-y$:
$$-y = -(-x)^3$$
$$y = -x^3$$
Symmetric with respect to the origin.

x	y
-3	27
-2	8
-1	1
0	0
1	-1
2	-8
3	-27

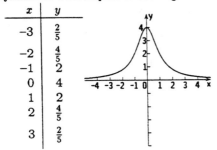

9. $y = \dfrac{4}{x^2 + 1}$

Check for symmetries.

Replace x with $-x$:
$$y = \frac{4}{(-x)^2 + 1}$$
$$y = \frac{4}{x^2 + 1}$$
Symmetric with respect to the y-axis.

Replace y with $-y$:
$$-y = \frac{4}{x^2 + 1}$$
$$y = -\frac{4}{x^2 + 1}$$
Not symmetric with respect to the x-axis.

Replace x with $-x$ and y with $-y$:
$$-y = \frac{4}{(-x)^2 + 1}$$
$$y = -\frac{4}{x^2 + 1}$$

Not symmetric with respect to the origin.

x	y
-3	$\frac{2}{5}$
-2	$\frac{4}{5}$
-1	2
0	4
1	2
2	$\frac{4}{5}$
3	$\frac{2}{5}$

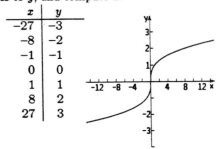

11. $x = 2y - 1$

Note that the graph will have none of the stated forms of symmetry. To construct a table of values, assign values to y, and compute x.

x	y
-5	-2
-3	-1
-1	0
1	1
3	2

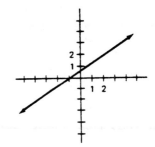

13. $x = -2y^2$

Note that the graph will be symmetric with respect to the x-axis. To construct a table of values, assign values to y, and compute x.

x	y
-8	-2
-2	-1
0	0
-2	1
-8	2

15. $x = y^3$

Note that the graph will be symmetric with respect to the origin. To construct a table of values, assign values to y, and compute x.

x	y
-27	-3
-8	-2
-1	-1
0	0
1	1
8	2
27	3

17. Circle with center $(0,0)$ and radius 6:
$$(x - 0)^2 + (y - 0)^2 = 6^2$$
$$x^2 + y^2 = 36$$

19. Circle with center $(4,1)$ and radius 5:
$$(x - 4)^2 + (y - 1)^2 = (5)^2$$
$$(x - 4)^2 + (y - 1)^2 = 25$$

21. Circle with center $(-2,1)$ and radius $\sqrt{3}$:
$$(x-[-2])^2 + (y-1)^2 = (\sqrt{3})^2$$
$$(x+2)^2 + (y-1)^2 = 3$$

23. $(x-2)^2 + y^2 = 16$
Circle with center $(2,0)$ and radius 4.

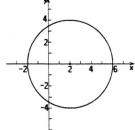

25. $(x+1)^2 + (y-3)^2 = 64$
Circle with center $(-1,3)$ and radius 8.

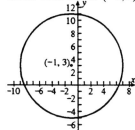

27.
$$x^2 + y^2 + 2x - 10y + 25 = 0$$
$$x^2 + 2x + y^2 - 10y = -25$$
$$x^2 + 2x + 1 + y^2 - 10y + 25 = -25 + 25 + 1$$
$$(x+1)^2 + (y-5)^2 = 1$$
Center: $(-1,5)$
Radius: 1

29.
$$x^2 + y^2 - 12x + 35 = 0$$
$$x^2 - 12x + y^2 = -35$$
$$x^2 - 12x + 36 + y^2 = -35 + 36$$
$$(x-6)^2 + y^2 = 1$$
Center: $(6,0)$
Radius: 1

31.
$$4x^2 + 4y^2 + 4x - 12y + 1 = 0$$
$$x^2 + y^2 + x - 3y + \tfrac{1}{4} = 0$$
$$x^2 + x + y^2 - 3y = -\tfrac{1}{4}$$
$$x^2 + x + \tfrac{1}{4} + y^2 - 3y + \tfrac{9}{4} = -\tfrac{1}{4} + \tfrac{1}{4} + \tfrac{9}{4}$$
$$(x+\tfrac{1}{2})^2 + (y-\tfrac{3}{2})^2 = \tfrac{9}{4}$$
Center: $(-\tfrac{1}{2}, \tfrac{3}{2})$
Radius: $\tfrac{3}{2}$

33. $y = \dfrac{12}{x}$

The graph will be symmetric with respect to the origin.

x	y
-4	-3
-3	-4
-2	-6
-1	-12
0	undefined
1	12
2	6
3	4
4	3

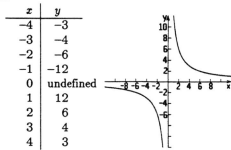

35. $y = 2(x-1)^2$
The graph will have none of the three forms of symmetry discussed.

x	y
-1	8
0	2
1	0
2	2
3	8

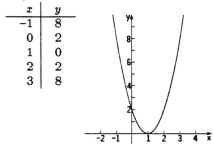

37. $x = 4y - y^2$
The graph will have none of the stated forms of symmetry. To construct a table of values, assign values to y, and compute x.

x	y
-21	-3
-12	-2
-5	-1
0	0
3	1
4	2
3	3
0	4

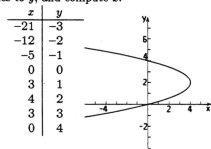

39. Since the circle is tangent to the x-axis, the length of the radius is the vertical distance from the center of the circle to the x-axis, or $|-7 - 0| = 7$.
The equation of the circle is
$$(x-5)^2 + (y-[-7])^2 = 7^2$$
$$(x-5)^2 + (y+7)^2 = 49$$

41. The length of the radius is the distance between the center and any point on the circle.
$$r^2 = [4 - (-3)]^2 + (3 - 2)^2$$
$$= (7)^2 + (1)^2$$
$$= 50$$
The equation of the circle is
$$(x - [-3])^2 + (y - 2)^2 = 50$$
$$(x + 3)^2 + (y - 2)^2 = 50$$

43. a)
$$x^2 + y^2 - 4x + 6y = -13$$
$$x^2 - 4x + 4 + y^2 + 6y + 9 = -13 + 4 + 9$$
$$(x - 2)^2 + (y - 3)^2 = 0$$
This is a circle with radius 0, called a "point circle."
The graph is the single point $(2, -3)$.

b)
$$2x^2 + 2y^2 - 2x + 6y = 3$$
$$x^2 - x + y^2 + 3y = \frac{3}{2}$$
$$x^2 - x + \frac{1}{4} + y^2 + 3y + \frac{9}{4} = \frac{3}{2} + \frac{1}{4} + \frac{9}{4}$$
$$\left(x - \frac{1}{2}\right)^2 + \left(y + \frac{3}{2}\right)^2 = \frac{16}{4} = 4$$
This circle has center $\left(\frac{1}{2}, -\frac{3}{2}\right)$ and radius 2.

c)
$$4x^2 + 4y^2 - 8x - 4y = -7$$
$$x^2 + y^2 - 2x - y = -\frac{7}{4}$$
$$x^2 - 2x + 1 + y^2 - y + \frac{1}{4} = -\frac{7}{4} + 1 + \frac{1}{4} = -\frac{1}{2}$$
$$(x - 1)^2 + \left(y - \frac{1}{2}\right)^2 = -\frac{1}{2}$$
This is a circle with imaginary radius, or an imaginary circle. It has no locus in the real plane.

d)
$$\sqrt{3}x^2 + \sqrt{3}y^2 - 6y = 2\sqrt{3}$$
$$x^2 + y^2 - \frac{6}{\sqrt{3}}y = 2$$
$$x^2 + y^2 - 2\sqrt{3}y + 3 = 2 + 3 = 5$$
$$x^2 + (y - \sqrt{3})^2 = 5$$
This is a circle with center $(0, \sqrt{3})$ and radius $\sqrt{5}$.

45. First find the center and radius of the circle.
$$x^2 + y^2 - 6x + 4y - 12 = 0$$
$$x^2 - 6x + y^2 + 4y = 12$$
$$x^2 - 6x + 9 + y^2 + 4y + 4 = 12 + 9 + 4$$
$$(x - 3)^2 + (y + 2)^2 = 25$$
Center: $(3, -2)$
Radius: 5
The distance from $\left(7, \frac{3}{2}\right)$ to the center of the circle is
$$\sqrt{(3 - 7)^2 + \left(-2 - \frac{3}{2}\right)^2} = \sqrt{(-4)^2 + (-3.5)^2}$$
$$= \sqrt{28.5}$$
Since $\sqrt{28.5} > 5$ (the length of the radius), $\left(7, \frac{3}{2}\right)$ must be outside the circle.

47. Label the points $A(1, 0)$, $B(8, 1)$, $C(7, 8)$, and $D(0, 7)$. The diagonals of the square intersect at their midpoints:

Midpoint of $\overline{AC} = \left(\frac{1 + 7}{2}, \frac{0 + 8}{2}\right)$
$$M = (4, 4)$$

a) The center of the circumscribed circle will be $M(4, 4)$ and the radius will be the length of any of the lines $\overline{MA}$, $\overline{MB}$, $\overline{MC}$, or $\overline{MD}$. Choose $\overline{MA}$.
$$r^2 = (4 - 1)^2 + (4 - 0)^2$$
$$= 3^2 + 4^2$$
$$= 25$$
The equation of the circumscribed circle is $(x - 4)^2 + (y - 4)^2 = 25$.

b) The center of the inscribed circle will be $M(4, 4)$ and the radius will be the length of the perpendicular line from M to any side of the square. The length of the perpendicular is $\frac{1}{2}$ the length of any side of the square. Choose $\overline{AB}$.
$$\tfrac{1}{2}(\overline{AB}) = \tfrac{1}{2}\sqrt{(8 - 1)^2 + (1 - 0)^2}$$
$$= \tfrac{1}{2}\sqrt{(7)^2 + (1)^2}$$
$$= \tfrac{1}{2}\sqrt{50}$$
Since $\quad r = \tfrac{1}{2}\sqrt{50}$,
$$r^2 = \tfrac{1}{4}(50)$$
$$= \tfrac{25}{2}$$
The equation of the inscribed circle is
$$(x - 4)^2 + (y - 4)^2 = \tfrac{25}{2}$$

49.

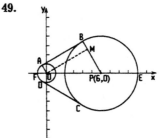

Radii OA and PB are both perpendicular to AB, which is tangent to both circles. Therefore OA is parallel to PB. Construct OM parallel to AB. Then $OABM$ is a rectangle, with $\overline{AB} = \overline{OM}$ and $\overline{OA} = \overline{BM} = 1$. $\overline{PB} = 4$. $PM = BP - BM = 3$.
Applying the Pythagorean Theorem to right triangle OMP,
$$\overline{OM} = \sqrt{(\overline{OP})^2 - (\overline{PM})^2}$$
$$= \sqrt{6^2 - 3^2}$$
$$= \sqrt{27} = 3\sqrt{3}$$
Since $\overline{AB} = \overline{DC} = \overline{OM}$, $\overline{AB} = \overline{DC} = 3\sqrt{3}$
To find the length of the arc CEB, note that the ratio $\dfrac{\overline{OM}}{\overline{PM}} = \dfrac{3\sqrt{3}}{3} = \sqrt{3}$.

For those who know trigonometry, that means that angle $OPM = \tan^{-1}\sqrt{3} = 60°$.

Arc CEB is subtended by a central angle of 240°.

The length of the arc must be $\frac{240}{360} = \frac{2}{3}$ the circumference of the large circle.

$$\text{Arc } CEB = \frac{2}{3} \cdot 2\pi \cdot 4 = \frac{16\pi}{3}$$

Angle FOA = angle OPM because OA and BP are parallel. Therefore arc DFA is subtended by a central angle of 120°.

The length of the arc must be $\frac{120}{360} = \frac{1}{3}$ the circumference of the small circle.

$$\text{Arc } FOA = \frac{1}{3} \cdot 2\pi \cdot 1 = \frac{2\pi}{3}$$

The length of the pulley
$$= AB + DC + \text{arc } CEB + \text{arc } FOA$$
$$= 3\sqrt{3} + 3\sqrt{3} + \frac{16\pi}{3} + \frac{2\pi}{3}$$
$$= 6\sqrt{3} + 6\pi$$

Problem Set 4.3

1. $m = \frac{8-3}{4-2} = \frac{5}{2}$

3. $m = \frac{2-0}{-4-3} = -\frac{2}{7}$

5. $m = \frac{0-5}{3-0} = -\frac{5}{3}$

7. $m = \frac{6.175 - 5.014}{4.315 - (-1.732)} = \frac{1.161}{4.315 + 1.732} = \frac{1.161}{6.047}$
$= .191996 \approx .1920$

9. The line through $(2,3)$ with slope 4:
$$y - 3 = 4(x - 2)$$
$$y - 3 = 4x - 8$$
$$4x - y - 5 = 0$$

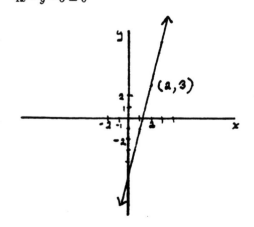

11. The line through $(3,-4)$ with slope -2:
$$y - (-4) = -2(x - 3)$$
$$y + 4 = -2x + 6$$
$$2x + y - 2 = 0$$

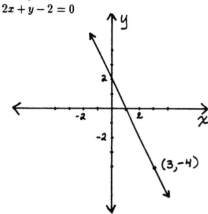

13. The line with y-intercept 4 and slope -2:
$$y = -2x + 4$$
$$2x + y - 4 = 0$$

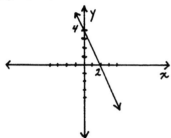

15. The line with y-intercept 5 and slope 0:
$$y = 0x + 5$$
$$y = 5$$
$$y - 5 = 0$$

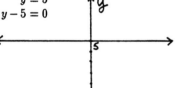

17. The line through $(2,3)$ and $(4,8)$:
First, compute the slope.
$$m = \frac{8-3}{4-2} = \frac{5}{2}$$
$$y - 3 = \frac{5}{2}(x - 2)$$
$$2y - 6 = 5x - 10$$
$$5x - 2y - 4 = 0$$
See graph on the next page

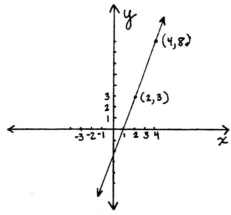

Since the slope is undefined, this is a vertical line with equation
$$x = 2$$
$$x - 2 = 0$$

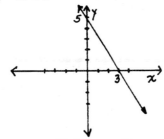

19. The line through $(3,0)$ and $(0,5)$:
First, compute the slope.
$$m = \frac{5-0}{0-3} = -\frac{5}{3}$$
$$y - 0 = -\frac{5}{3}(x-3)$$
$$3y = -5x + 15$$
$$5x + 3y - 15 = 0$$

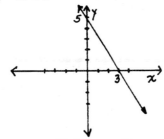

21. The line through $(\sqrt{3}, \sqrt{7})$ and $(\sqrt{2}, \pi)$:
First, compute the slope:
$$m = \frac{\pi - \sqrt{7}}{\sqrt{2} - \sqrt{3}} \approx -1.56$$
$$y - \sqrt{7} = -1.56(x - \sqrt{3})$$
$$y - 2.65 = -1.56(x - 1.73)$$
$$y - 2.65 = -1.56x + 2.70$$
$$1.56x + y - 5.35 = 0$$

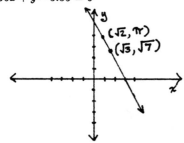

23. The line through $(2, -3)$ and $(2, 5)$:
First, compute the slope.
$$m = \frac{-3 - 5}{2 - 2} = \frac{-8}{0}$$

25. $y = 3x + 5$
$m = 3$, y-intercept $= 5$

27. $3y = 2x - 4$
$$y = \frac{2}{3}x - \frac{4}{3}$$
$m = \frac{2}{3}$, y-intercept $= -\frac{4}{3}$

29. $2x + 3y = 6$
$$3y = -2x + 6$$
$$y = -\frac{2}{3}x + 2$$
$m = -\frac{2}{3}$, y-intercept $= 2$

31. $y + 2 = -4(x - 1)$
$$y = -2 - 4x + 4$$
$$y = -4x + 2$$
$m = -4$, y-intercept $= 2$

33. a) The line through $(3, -3)$ parallel to $y = 2x + 5$.
Since the slope of $y = 2x + 5$ is 2, any line parallel to it will have $m = 2$.
The line through $(3, -3)$ has equation
$$y - (-3) = 2(x - 3)$$
$$y + 3 = 2x - 6$$
$$2x - y - 9 = 0$$

b) The line through $(3, -3)$ perpendicular to $y = 2x + 5$.
Since the slope of $y = 2x + 5$ is 2, any line perpendicular to it will have $m = -\frac{1}{2}$.
The line through $(3, -3)$ has equation
$$y - (-3) = -\frac{1}{2}(x - 3)$$
$$y + 3 = -\frac{1}{2}(x - 3)$$
$$2y + 6 = -x + 3$$
$$x + 2y + 3 = 0$$

c) The line through $(3, -3)$ parallel to $2x + 3y = 6$.
First, find the slope of
$$2x + 3y = 6$$
$$3y = -2x + 6$$
$$y = -\frac{2}{3}x + 2$$

70

The line we seek will have $m = -\frac{2}{3}$.
$$y - (-3) = -\frac{2}{3}(x - 3)$$
$$y + 3 = -\frac{2}{3}(x - 3)$$
$$3y + 9 = -2x + 6$$
$$2x + 3y + 3 = 0$$

d) The line through $(3, -3)$ perpendicular to $2x + 3y = 6$.
The line we seek will have $m = \frac{3}{2}$.
$$y - (-3) = \frac{3}{2}(x - 3)$$
$$y + 3 = \frac{3}{2}(x - 3)$$
$$2y + 6 = 3x - 9$$
$$3x - 2y - 15 = 0$$

e) The line through $(3, -3)$ parallel to the line through $(-1, 2)$ and $(3, -1)$.
$$m = \frac{-1 - 2}{3 - (-1)} = \frac{-3}{4}$$
$$y - (-3) = -\frac{3}{4}(x - 3)$$
$$y + 3 = -\frac{3}{4}(x - 3)$$
$$4y + 12 = -3x + 9$$
$$3x + 4y + 3 = 0$$

f) The line through $(3, -3)$ parallel to $x = 8$ will be a vertical line passing through $(3, -3)$. The equation of such a line is $x = 3$.

g) The line through $(3, -3)$ perpendicular to $x = 8$ will be a horizontal line passing through $(3, -3)$. The equation of such a line is $y = -3$.

35. The line $y + 2 = -\frac{1}{2}(x - 1)$ has slope $m = -\frac{1}{2}$. A line perpendicular to it will have $m = 2$.
$$y - (-4) = 2(x - 0)$$
$$y + 4 = 2x$$
$$y = 2x - 4$$

37. $\begin{cases} 2x + 3y = 4 \\ -3x + y = 5 \end{cases}$
$\begin{cases} 2x + 3y = 4 \\ 9x - 3y = -15 \end{cases}$
$11x = -11$ and $x = -1$
$y = 5 + 3x = 5 + 3(-1) = 5 - 3 = 2$
The point of intersection is $(-1, 2)$.
For the first line, $y = -\frac{2}{3}x + \frac{4}{3}$ and its slope is $-\frac{2}{3}$.

The line perpendicular to it has slope $\frac{3}{2}$.

The line through $(-1, 2)$ with slope $\frac{3}{2}$ is
$$y - 2 = \frac{3}{2}(x + 1)$$
$$2y - 4 = 3x + 3$$
The line is $2y - 3x - 7 = 0$.

39. $\begin{cases} 3x - 4y = 5 \\ 2x + 3y = 9 \end{cases}$
$\begin{cases} 9x - 12y = 15 \\ 8x + 12y = 36 \end{cases}$
$\overline{17x = 51}$
$$x = 3$$
$$3(3) - 4y = 5$$
$$-4y = -4$$
$$y = 1$$
The point of intersection is $(3, 1)$.
Find the slope of the first line.
$$3x - 4y = 5$$
$$-4y = -3x + 5$$
$$y = \frac{3}{4}x - \frac{5}{4}$$
Since the slope of the first line is $\frac{3}{4}$, the slope of a line perpendicular to it is $-\frac{4}{3}$.
The equation of the line we seek is
$$y - 1 = -\frac{4}{3}(x - 3)$$
$$3y - 3 = -4x + 12$$
$$4x + 3y - 15 = 0$$

41. Use the formula
$$D = \frac{|Ax_1 + By_1 + C|}{\sqrt{A^2 + B^2}}$$
with $x_1 = -3$, $y_1 = 2$, $A = 3$, $B = 4$, $C = -6$
$$D = \frac{|3(-3) + 4(2) - 6|}{\sqrt{3^2 + 4^2}}$$
$$= \frac{|-7|}{\sqrt{25}} = \frac{7}{5}$$

43. Use the formula
$$D = \frac{|Ax_1 + By_1 + C|}{\sqrt{A^2 + B^2}}$$
with $x_1 = -2$ and $y_1 = -1$
To find A, B, and C, write $5y = 12x + 1$ in standard form:
$$12x - 5y + 1 = 0$$
$A = 12$, $B = -5$, $C = 1$
$$D = \frac{|12(-2) + (-5)(-1) + 1|}{\sqrt{12^2 + (-5)^2}}$$
$$= \frac{|-18|}{\sqrt{169}} = \frac{18}{13}$$

45. First find a point on one of the lines, say $3x + 4y = 6$. $(2, 0)$ lies on this line.
Now find the distance from $(2, 0)$ to $3x + 4y = 12$. By definition, this will be the perpendicular distance.
$x_1 = 2$, $y_1 = 0$, $A = 3$, $B = 4$, $C = -12$
$$D = \frac{|3(2) + 4(0) - 12|}{\sqrt{3^2 + 4^2}} = \frac{|-6|}{\sqrt{25}} = \frac{6}{5}$$

47. a) $y = 3x - 2$, $m = 3$
$$6x - 2y = 0$$
$$-2y = -6x$$
$$y = 3x, \ m = 3$$
The two lines are parallel.

b) $x = -2$ is a vertical line.
$y = 4$ is a horizontal line.
The two lines are perpendicular.

c) $\quad x = 2(y - 2)$
$$y - 2 = \tfrac{1}{2}x, \ m = \tfrac{1}{2}$$
$$y = -\tfrac{1}{2}(x - 1), \ m = -\tfrac{1}{2}$$
The lines are neither parallel nor perpendicular.

d) $2x + 5y = 3$
$$5y = -2x + 3$$
$$y = -\tfrac{2}{5}x + \tfrac{3}{5}, \ m = -\tfrac{2}{5}$$
$$10x - 4y = 7$$
$$-4y = -10x + 7$$
$$y = \tfrac{5}{2}x - \tfrac{7}{4}, \ m = \tfrac{5}{2}$$
The two lines are perpendicular.

49. We find the intersection of $x + 2y = 1$ and $3x + 2y = 5$:
$$\begin{cases} 3x + 2y = 5 \\ x + 2y = 1 \end{cases}$$
Subtracting, $2x = 4$ and $x = 2$
$$2y = 1 - x = 1 - 2 = -1$$
$$y = -\tfrac{1}{2}$$
The point of intersection is $(2, -\tfrac{1}{2})$.
The line is to be parallel to $3x - 2y = 4$, which may be written $y = \tfrac{3}{2}x - 2$. From this, its slope is $\tfrac{3}{2}$.

The line through $(2, -\tfrac{1}{2})$ with slope $\tfrac{3}{2}$ is $y + \tfrac{1}{2} = \tfrac{3}{2}(x - 2)$. Multiplying each side by 2: $2y + 1 = 3x - 6$.
The desired line is $3x - 2y - 7 = 0$.

51. First find the slope of the given line segment.
$$m = \frac{6 - (-2)}{7 - 3} = \frac{8}{4} = 2$$
The perpendicular bisector of the segment will then have $m = -\tfrac{1}{2}$.
Find the midpoint of the given line segment:
$$\left(\frac{7 + 3}{2}, \frac{-2 + 6}{2} \right) = (5, 2)$$
The perpendicular bisector of the line segment will be a line through $(5, 2)$ with $m = -\tfrac{1}{2}$:
$$y - 2 = -\tfrac{1}{2}(x - 5)$$
$$2y - 4 = -x + 5$$
$$x + 2y - 9 = 0$$

53. The x-intercept has coordinates $(a, 0)$.
The y-intercept has coordinates $(0, b)$.
$$m = \frac{b - 0}{0 - a} = -\frac{b}{a}$$
The equation of the line is
$$y - b = -\frac{b}{a}(x - 0)$$
$$y - b = -\frac{b}{a}x$$
Divide through by b:
$$\frac{y}{b} - 1 = -\frac{x}{a}$$
$$\frac{x}{a} + \frac{y}{b} = 1$$

55. Let the intercepts be k and $2k$, so their coordinates are $(k, 0)$ and $(0, 2k)$. The slope of the line through these two points is $\dfrac{2k - 0}{0 - k} = \dfrac{2k}{-k} = -2$.
The line through $(3, 2)$ with slope -2 is
$$y - 2 = -2(x - 3)$$
$$2x + y - 8 = 0$$

57. Profit = Revenue − Expenses
Revenue = $20x$
Expenses = $16x + 8500$
$$P = 20x - (16x + 8500)$$
$$= 4x - 8500$$
If $x = 2000$,
$$P = 4(2000) - 8500 = -500$$
In other words, the company suffers a loss of $500.

59. a) After 20 years, the value of the house appreciated by $112,000 − $60,000 = $52,000.
Each year, it appreciated $\dfrac{\$52,000}{20} = \$2600$
After t years, $V = 60,000 + 2600t$

b) Let $V = 69,100$
$$69,100 = 60,000 + 2600t$$
$$2600t = 9100$$
$$t = 3.5 \text{ years}$$
The house was worth $69,100 3.5 years after January 1, 1970, or, on July 1, 1973.

61. Find a point on one of the lines, say $12x - 5y = 2$. $(1, 2)$ lies on the line.
Find the distance between $(1, 2)$ and the line $12x - 5y = 7$.
$x_1 = 1$, $y_1 = 2$, $A = 12$, $B = -5$, $C = -7$
$$D = \frac{\left| 12(1) - 5(2) - 7 \right|}{\sqrt{12^2 + (-5)^2}}$$
$$= \frac{|-5|}{\sqrt{169}} = \frac{5}{13}$$

63. Without loss of generality, let the vertices of the triangle be $A(0,0)$, $B(a,0)$, and $C(b,c)$.

Midpoint of AC is $\left(\frac{b}{2},\frac{c}{2}\right)$

Midpoint of BC is $\left(\frac{a+b}{2},\frac{c}{2}\right)$

The slope of the line joining these two midpoints is 0 since the y-coordinates are equal. But this is also the slope of the side AB, so the lines are parallel. Incidentally, notice that the length of this segment is $\sqrt{\left(\frac{a+b}{2}-\frac{b}{2}\right)^2+\left(\frac{c}{2}-\frac{c}{2}\right)^2}=\sqrt{\frac{a^2}{4}}=\frac{a}{2}$

65.

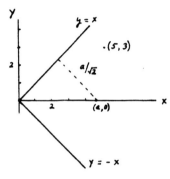

Because of the location of $(5,3)$ relative to lines $y=\pm x$, the point we seek will lie on the positive half of the x-axis. Call the point $(a,0)$.
The distance between $(a,0)$ and $(5,3)$ is

$$\sqrt{(a-5)^2+(-3)^2}=\sqrt{a^2-10a+34}$$

With $x_1=a$, $y_1=0$, $A=-1$, $B=1$, $C=0$, the distance between $(a,0)$ and the line $y=x$ is

$$\frac{|a(-1)+1(0)+0|}{\sqrt{(-1)^2+1^2}}=\frac{a}{\sqrt{2}}$$

But we must have

$$\frac{a}{\sqrt{2}}=\sqrt{a^2-10a+34}$$
$$\frac{a^2}{2}=a^2-10a+34$$
$$0=a^2-20a+68$$
$$a=\frac{20\pm\sqrt{(20)^2-4(1)(68)}}{2(1)}=\frac{20\pm\sqrt{128}}{2}$$
$$=\frac{20\pm8\sqrt{2}}{2}=10\pm4\sqrt{2}$$

The two points we seek are $(10+4\sqrt{2},0)$ and $(10-4\sqrt{2},0)$.

67.

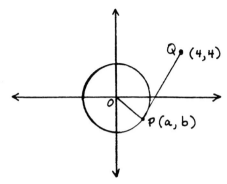

Let the coordinates of P be (a,b).

The slope of OP is $\frac{b-0}{a-0}=\frac{b}{a}$.

Since QP is perpendicular to OP, the slope of QP is $-\frac{a}{b}$. But the slope of QP is also $\frac{b-4}{a-4}$. So,

$$\frac{b-4}{a-4}=-\frac{a}{b}$$
$$b(b-4)=-a(a-4)$$
$$b^2-4b=-a^2+4a$$
$$a^2+b^2=4a+4b$$

Since $P(a,b)$ satisfies the equation of the circle, we know that $a^2+b^2=4$. So,

$$4a+4b=4$$
$$b=1-a$$

Substituting $b=1-a$ in $a^2+b^2=4$,

$$a^2+(1-a)^2=4$$
$$a^2+1-2a+a^2=4$$
$$2a^2-2a-3=0$$
$$a=\frac{-(-2)\pm\sqrt{(-2)^2-4(2)(3)}}{2(2)}$$
$$=\frac{2\pm\sqrt{28}}{4}$$
$$=\frac{2\pm2\sqrt{7}}{4}$$
$$=\frac{1\pm\sqrt{7}}{2}$$

Since P is in the fourth quadrant, reject $a=\frac{1-\sqrt{7}}{2}$.

$$a=\frac{1+\sqrt{7}}{2}$$
$$b=1-\left(\frac{1+\sqrt{7}}{2}\right)$$
$$=1-\frac{1}{2}-\frac{\sqrt{7}}{2}$$

The coordinates of P are $\left(\frac{1+\sqrt{7}}{2},\frac{1-\sqrt{7}}{2}\right)$.

For Problems 1–9, recall that the graph of $y = a(x-h)^2 + k$ is a parabola with vertex (h, k).

1. $y = 3x^2$
Vertex: $(0, 0)$

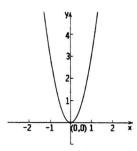

3. $y = x^2 + 5$
Vertex: $(0, 5)$

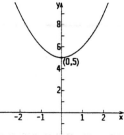

5. $y = (x-4)^2$
Vertex: $(4, 0)$

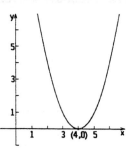

7. $y = 2(x-1)^2 + 5$
Vertex: $(1, 5)$

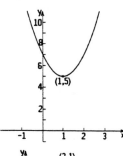

9. $y = -4(x-2)^2 + 1$
Vertex: $(2, 1)$

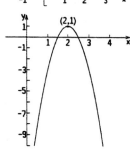

11.
$$y = 2(x-1)^2 + 7$$
$$= 2(x^2 - 2x + 1) + 7$$
$$= 2x^2 - 4x + 2 + 7$$
$$= 2x^2 - 4x + 9$$

13.
$$-2y + 5 = (x-5)^2$$
$$-2y + 5 = x^2 - 10x + 25$$
$$-2y = x^2 - 10x + 25 - 5$$
$$y = -\tfrac{1}{2}x^2 + 5x - 10$$

15. $y = x^2 + 2x$
Find vertex:
$$x = \frac{-2}{2} = -1$$
$$y = (-1)^2 + 2(-1) = -1$$
Vertex: $(-1, -1)$
Additional points: $(-3, 3)$,
   $(-2, 0)$, $(0, 0)$, $(1, 3)$

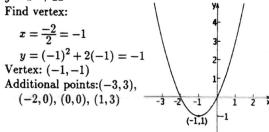

17. $y = -2x^2 + 8x + 1$
Find vertex:
$$x = \frac{-8}{2(-2)} = 2$$
$$y = -2(2)^2 + 8(2) + 1 = 9$$
Vertex: $(2, 9)$
Additional points: $(0, 1)$,
   $(1, 7)$, $(3, 7)$, $(4, 1)$

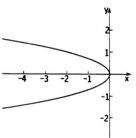

19. $x = -2y^2$
Vertex: $(0, 0)$

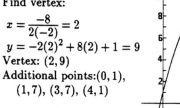

21. $x = -2(y+2)^2 + 8$
Vertex: $(8, -2)$

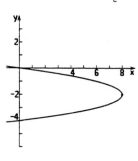

23. $x = y^2 + 4y + 2$
Find the vertex:

y-coordinate: $\frac{-4}{2} = -2$

x-coordinate:
$(-2)^2 + 4(-2) + 2 = -2$
Additional points:
(Assign values to y)
$(2, -4), (-1, -3),$
$(-1, -1), (2, 0)$

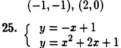

25. $\begin{cases} y = -x + 1 \\ y = x^2 + 2x + 1 \end{cases}$
Equating values of y:
$-x + 1 = x^2 + 2x + 1$
$x^2 + 3x = 0$
$x(x + 3) = 0$
$\quad x = 0$ or -3
$\quad y = 1$ or 4 respectively
The points of intersection of the parabola and the line are $(0, 1)$ and $(-3, 4)$

27. $\begin{cases} y = -2x + 1 \\ y = -x^2 - x + 3 \end{cases}$
Equate the values of y:
$-2x + 1 = -x^2 - x + 3$
$x^2 - x - 2 = 0$
$(x - 2)(x + 1) = 0$
$\quad x = 2$ or $x = -1$
$\quad y = -3$ or $y = 3$ respectively
The points of intersection are $(2, -3)$ and $(-1, 3)$.

29. $\begin{cases} y = 1.5x + 3.2 \\ y = x^2 - 2.9x \end{cases}$
Equate the values of y:
$x^2 - 2.9x = 1.5x + 3.2$
$x^2 - 4.4x - 3.2 = 0$

$x = \dfrac{-(4.4) \pm \sqrt{(-4.4)^2 - 4(1)(-3.2)}}{2(1)}$

$x = \dfrac{4.4 \pm 5.67}{2}$

$x = 5.04$ or $x = -.64$
$y = 10.76$ or $y = 2.24$ respectively
The points of intersection are $(5.04, 10.76)$ and $(-.64, 2.24)$

31. Since the vertex is $(0, 0)$, the equation will have the form $y = kx^2$, where k is to be determined. Since $(-6, 3)$ is on the graph, its coordinates satisfy the equation.
$3 = k(-6)^2$
$3 = 36k$
$k = \frac{3}{36} = \frac{1}{12}$
The equation is $y = \frac{1}{12}x^2$.

33. Since the vertex is $(-2, 0)$, the equation will have the form $y = k(x + 2)^2$, where k is to be determined. Since $(6, -8)$ is on the graph, its coordinates satisfy the equation.
$-8 = k(6 + 2)^2$
$-8 = 64k$
$k = \frac{-8}{64} = -\frac{1}{8}$
The equation is $y = -\frac{1}{8}(x + 2)^2$.

35. The equation is of the form $y = ax^2 + c$, with a and c to be determined. Since $(1, 1)$ and $(2, 7)$ are on the graph, each point satisfies the equation.
$\begin{cases} 1 = a(1)^2 + c \\ 7 = a(2)^2 + c \end{cases}$
$1 = a + c$
$7 = 4a + c$
Subtract the upper equation from the lower:
$6 = 3a$
$2 = a$
Let $a = 2$ in $1 = a + c$:
$1 = 2 + c$
$-1 = c$
The equation is $y = 2x^2 - 1$

37. a) $y = 2x^2 - 8x + 4$
Find the vertex.
$x = \dfrac{-(-8)}{2(2)} = 2$
$y = 2(2)^2 - 8(2) + 4 = -4$
Vertex: $(2, -4)$

b) $y = 2x^2 - 8x + 8$
Find the vertex.
$x = \dfrac{-(-8)}{2(2)} = 2$
$y = 2(2)^2 - 8(2) + 8 = 0$
Vertex: $(2, 0)$

c) $y = 2x^2 - 8x + 11$
Find the vertex.
$x = 2$
$y = 2(2)^2 - 8(2) + 11 = 3$
Vertex: $(2, 3)$

39. a) $y = \frac{1}{2}x^2 - 2x$
Find the vertex.
$x = \dfrac{-(-2)}{2(\frac{1}{2})} = 2$
$y = \frac{1}{2}(2)^2 - 2(2) = -2$
Vertex: $(2, -2)$
Additional points:
$(-2, 6), (0, 0), (4, 0), (6, 6)$

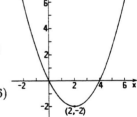

b) $x = \frac{1}{2}y^2 - 2y$

This will be a horizontal parabola. We can generate points on the parabola by exchanging the x and y coordinates obtained in part (a).

Vertex: $(-2, 2)$

Additional points:

$(6, -2), (0, 0), (0, 4), (6, 6)$

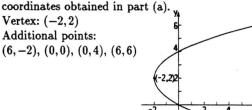

41. a) $\begin{cases} y = x^2 - 2x + 6 \\ y = -3x + 8 \end{cases}$

$x^2 - 2x + 6 = -3x + 8$

$x^2 + x - 2 = 0$

$(x + 2)(x - 1) = 0$

$x = -2$ or $x = 1$

$y = 14$ or $y = 5$, respectively.

The points of intersection are $(-2, 14)$ and $(1, 5)$.

b) $\begin{cases} y = x^2 - 4x + 6 \\ y = 2x - 3 \end{cases}$

$x^2 - 4x + 6 = 2x - 3$

$x^2 - 6x + 9 = 0$

$(x - 3)^2 = 0$

$x = 3$

$y = 3$

The only point of intersection is $(3, 3)$.

c) $\begin{cases} y = -x^2 + 2x + 4 \\ y = -2x + 9 \end{cases}$

$-2x + 9 = -x^2 + 2x + 4$

$x^2 - 4x + 5 = 0$

$x = \dfrac{-(-4) \pm \sqrt{(-4)^2 - (4)(1)(5)}}{2(1)}$

$= \dfrac{4 \pm \sqrt{-4}}{2}$

Since x is not a real number, there are no points of intersection in the real plane.

d) $\begin{cases} y = x^2 - 2x + 7 \\ y = 11 - x^2 \end{cases}$

$x^2 - 2x + 7 = 11 - x^2$

$2x^2 - 2x - 4 = 0$

$x^2 - x - 2 = 0$

$(x - 2)(x + 1) = 0$

$x = 2$ or $x = -1$

$y = 7$ or $y = 10$, respectively

The points of intersection are $(2, 7)$ and $(-1, 10)$.

43. Since point $(10, 40)$ lies on the curve of $y = a(x - 2)(x - 8)$, its coordinates satisfy the equation.

$40 = a(10 - 2)(10 - 8)$

$40 = a(8)(2) = 16a$

$a = \dfrac{40}{16} = \dfrac{5}{2}$

The parabola is $y = \frac{5}{2}(x - 2)(x - 8)$

$y = \frac{5}{2}(x^2 - 10x + 16)$

$= \frac{5}{2}x^2 - 25x + 40$

$x\text{-coordinate of vertex} = \dfrac{-(-25)}{2\left(\frac{5}{2}\right)} = \dfrac{25}{5} = 5$

$y\text{-coordinate of vertex} = \frac{5}{2}(5)^2 - 25(5) + 40 = -\dfrac{45}{2}$

Vertex: $\left(5, -\frac{45}{2}\right)$

45. Positioning the parabola on a coordinate system so that its vertex is at $(0, 10)$, it must have an equation of the form $y = kx^2 + 10$, where k is determined. $(10, 15)$ is a point on the parabola and therefore must satisfy the equation.

$15 = k(10)^2 + 10$

$5 = 100k$

$k = \dfrac{5}{100} = \dfrac{1}{20}$

The equation of the parabola is $y = \frac{1}{20}x^2 + 10$.

The distance $\overline{PQ}$ is the value of y when $x = 5$:

$y = \frac{1}{20}(5)^2 + 10$

$= \dfrac{25}{20} + 10$

$= \dfrac{45}{4}$ feet

47. Profit = Revenue − Expenses

Revenue = $x(300 - 100x)$

Expenses = $2(300 - 100x)$

$P = x(300 - 100x) - 2(300 - 100x)$

$= 300x - 100x^2 - 600 + 200x$

$= -100x^2 + 500x - 600$

The graph of the profit equation will be a parabola opening downwards. The maximum value of P will occur at the vertex of the parabola when

$x = \dfrac{-b}{2a} = \dfrac{-500}{2(-100)} = \dfrac{-500}{-200} = 2.5$

To maximize profits, the retailer should charge $2.50 for each truck.

49. Label the vertices A, B, and C as shown in the diagram. Perpendiculars are dropped to the x-axis from A, B, and C, and are labeled D, E, and F, respectively. The x-coordinates of A and D are each a, and of B and E are each b. Since F is the midpoint of D and E, the x-coordinate of F and of C is $\dfrac{a+b}{2}$.

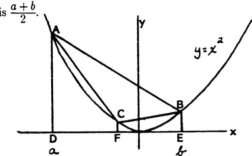

Points A, B, and C lie on the parabola $y = x^2$; their y-coordinates are each the square of their x-coordinates.

$$A(a, a^2), \; B(b, b^2), \; C\left(\frac{a+b}{2}, \frac{(a+b)^2}{4}\right)$$

We find the areas of three trapezoids:

Area $ADEB$

$$= \left(\frac{AD + BE}{2}\right)(b - a) = \frac{(a^2 + b^2)}{2} \cdot (b - a)$$

Area $ADFC$

$$= \frac{a^2 + \dfrac{(a+b)^2}{4}}{2}\left(\frac{a+b}{2} - a\right)$$

$$= \left(\frac{4a^2 + a^2 + 2ab + b^2}{8}\right)\left(\frac{b-a}{2}\right)$$

$$= \frac{(5a^2 + 2ab + b^2)(b-a)}{16}$$

Area $CFEB = \dfrac{\dfrac{(a+b)^2}{4} + b^2}{2}\left(b - \frac{a+b}{2}\right)$

$$= \left(\frac{a^2 + 2ab + b^2 + 4b^2}{8}\right)\left(\frac{b-a}{2}\right)$$

$$= \frac{(a^2 + 2ab + 5b^2)(b-a)}{16}$$

The area of the desired triangle is the first of these two areas minus the other two:

Area $\triangle ABC = \dfrac{(a^2 + b^2)(b-a)}{2}$

$$- \frac{(5a^2 + 2ab + b^2)(b-a)}{16} - \frac{(a^2 + 2ab + 5b^2)(b-a)}{16}$$

$$= \frac{b-a}{16}(8a^2 + 8b^2 - 5a^2 - 2ab - b^2 - a^2 - 2ab - 5b^2)$$

$$= \frac{b-a}{16}(2a^2 - 4ab + 2b^2) = \frac{b-a}{8}(a^2 - 2ab + b^2)$$

$$= \frac{(b-a)(a-b)^2}{8} = \frac{(b-a)^3}{8}$$

51.

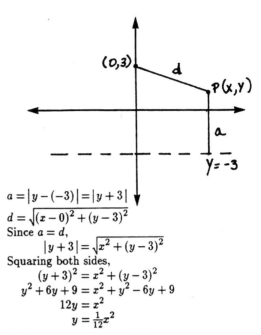

$a = |y - (-3)| = |y + 3|$
$d = \sqrt{(x-0)^2 + (y-3)^2}$
Since $a = d$,
$\quad |y + 3| = \sqrt{x^2 + (y-3)^2}$
Squaring both sides,
$\quad (y+3)^2 = x^2 + (y-3)^2$
$\quad y^2 + 6y + 9 = x^2 + y^2 - 6y + 9$
$\quad\quad\quad 12y = x^2$
$\quad\quad\quad\; y = \frac{1}{12}x^2$

53. If R has coordinates (a, b) then

$$\overline{FR} = \sqrt{(a-0)^2 + (b-p)^2}$$
$$= \sqrt{a^2 + (b-p)^2}$$

Since (a, b) satisfies the equation of the parabola,
$$a^2 = 4pb$$
Substituting,
$$\overline{FR} = \sqrt{4pb + (b-p)^2}$$
$$= \sqrt{4pb + b^2 - 2pb + p^2}$$
$$= \sqrt{b^2 + 2pb + p^2}$$
$$= \sqrt{(b+p)^2}$$
$$= b + p$$
$$\overline{RG} = p - b$$
$$\overline{FR} + \overline{RG} = b + p + p - b = 2p$$

Problem Set 4.5

1. $\dfrac{x^2}{25} + \dfrac{y^2}{9} = 1$

$a = 5$, $b = 3$, center $(0, 0)$
Major diameter endpoints $(-5, 0)$, $(5, 0)$
Minor diameter endpoints $(0, 3)$, $(0, -3)$

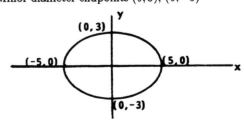

3. $\frac{x^2}{9} + \frac{y^2}{25} = 1$

$a = 5$, $b = 3$, center $(0,0)$
Major diameter endpoints $(0,5)$, $(0,-5)$
Minor diameter endpoints $(3,0)$, $(-3,0)$

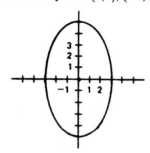

5. $\frac{(x-2)^2}{25} + \frac{(y+1)^2}{9} = 1$

$a = 5$, $b = 3$, center $(2,-1)$
Major diameter endpoints $(7,-1)$, $(-3,-1)$
Minor diameter endpoints $(2,-4)$, $(2,2)$

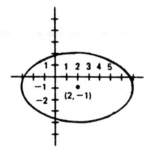

7. $\frac{(x+3)^2}{9} + \frac{y^2}{16} = 1$

$a = 4$, $b = 3$, center $(-3,0)$
Axis vertical
Major diameter endpoints $(-3,4)$, $(-3,-4)$
Minor diameter endpoints $(-6,0)$, $(0,0)$

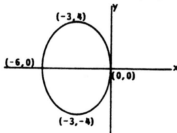

9. The midpoint of the major and minor axes will be the center of the ellipse: $(0,0)$.
a is $\frac{1}{2}AB$, or 6.

b is $\frac{1}{2}CD$, or 3.

Since the major axis is horizontal, a^2 goes with x^2.
$$\frac{x^2}{36} + \frac{y^2}{9} = 1$$

11. The midpoint of the major and minor axes will be the center of the ellipse: $(0,0)$.
a is $\frac{1}{2}AB$, or 6.

b is $\frac{1}{2}CD$, or 4.
Since the major axis is vertical, a^2 goes with y^2.
$$\frac{x^2}{16} + \frac{y^2}{36} = 1$$

13. The midpoint of the major and minor axes will be the center of the ellipse:
$$\left(\frac{-4+8}{2}, \frac{3+3}{2}\right) = (2,3)$$
a is $\frac{1}{2}AB - \frac{1}{2}|8-(-4)| - 6$

b is $\frac{1}{2}CD = \frac{1}{2}|5-1| = 2$
Since the major axis is horizontal, a^2 goes with x^2.
$$\frac{(x-2)^2}{36} + \frac{(y-3)^2}{4} = 1$$

15. $\frac{x^2}{16} - \frac{y^2}{9} = 1$

$a = 4$, $b = 3$, center $(0,0)$
Let $y = 0$ to find
the vertices:
$$\frac{x^2}{16} = 1$$
$$x^2 = 16$$
$$x = \pm 4$$
Vertices: $(4,0)$, $(-4,0)$

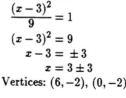

17. $\frac{y^2}{9} - \frac{x^2}{16} = 1$

$a = 3$, $b = 4$, center $(0,0)$
Let $x = 0$ to find
the vertices:
$$\frac{y^2}{9} = 1$$
$$y^2 = 9$$
$$y = \pm 3$$
Vertices: $(0,3)$, $(0,-3)$

19. $\frac{(x-3)^2}{9} - \frac{(y+2)^2}{16} = 1$ $a = 3$, $b = 4$, center $(3,-2)$

Set $y = -2$ to find
the vertices:
$$\frac{(x-3)^2}{9} = 1$$
$$(x-3)^2 = 9$$
$$x - 3 = \pm 3$$
$$x = 3 \pm 3$$
Vertices: $(6,-2)$, $(0,-2)$

21. $\dfrac{(y+3)^2}{4} - \dfrac{x^2}{25} = 1$

$a = 2$, $b = 5$, center $(0, -3)$

Set $x = 0$ to find the vertices:

$\dfrac{(y+3)^2}{4} = 1$

$(y+3)^2 = 4$

$y + 3 = \pm 2$

$y = -3 \pm 2$

Vertices: $(0, -1)$, $(0, -5)$

23. Since the vertices lie on the x-axis, this is a horizontal hyperbola, with an equation of the form
$\dfrac{x^2}{a^2} - \dfrac{y^2}{b^2} = 1$. $a = 4$

The slope of an asymptote $= \dfrac{b}{a} = \dfrac{5}{4}$

$\dfrac{b}{4} = \dfrac{5}{4}$

$b = 5$

The equation of the hyperbola is
$\dfrac{x^2}{16} - \dfrac{y^2}{25} = 1$

25. Since the vertex lies on a horizontal line from the center, this is a horizontal hyperbola with an equation of the form $\dfrac{(x-h)^2}{a^2} - \dfrac{(y-k)^2}{b^2} = 1$.

$a = 8 - 6 = 2$

The slope of an asymptote $= \dfrac{b}{a} = 1$

$\dfrac{b}{2} = 1$

$b = 2$

The equation of the hyperbola is
$\dfrac{(x-6)^2}{4} - \dfrac{(y-3)^2}{4} = 1$

27. Since the vertices lie on a line parallel to the y-axis, this is a vertical hyperbola with an equation of the form $\dfrac{(y-k)^2}{a^2} - \dfrac{(x-h)^2}{b^2} = 1$.

$a = \dfrac{15-3}{2} = 6$

To find b, find the slope of an asymptote.

$3x - 2y + 6 = 0$

$-2y = -3x - 6$

$y = \dfrac{3}{2}x + 3$

The slope of an asymptote $= \dfrac{a}{b} = \dfrac{3}{2}$

$\dfrac{6}{b} = \dfrac{3}{2}$

$b = 4$

The center is the point midway between vertices:

$\left(\dfrac{4+4}{2}, \dfrac{3+15}{2}\right) = (4, 9)$

The equation of the hyperbola is
$\dfrac{(y-9)^2}{36} - \dfrac{(x-4)^2}{16} = 1$

29. $9x^2 + 16y^2 + 36x - 96y = -36$

$9(x^2 + 4x) + 16(y^2 - 6y) = -36$

$9(x^2 + 4x + 4) + 16(y^2 - 6y + 9)$
$\qquad = -36 + 9(4) + 16(9)$

$9(x+2)^2 + 16(y-3)^2 = 144$

$\dfrac{9(x+2)^2}{144} + \dfrac{16(y-3)^2}{144} = 1$

$\dfrac{(x+2)^2}{16} + \dfrac{(y-3)^2}{9} = 1$

This is the equation of an ellipse with the major axis parallel to the x-axis.

Center: $(-2, 3)$; vertices: $(2, 3)$, $(-6, 3)$

31. $4x^2 - 9y^2 - 16x - 18y - 29 = 0$

$4(x^2 - 4x) - 9(y^2 + 2y) = 29$

$4(x^2 - 4x + 4) - 9(y^2 + 2y + 1) = 29 + 4(4) - 9(1)$

$4(x-2)^2 - 9(y+1)^2 = 36$

$\dfrac{4(x-2)^2}{36} - \dfrac{9(y+1)^2}{36} = 1$

$\dfrac{(x-2)^2}{9} - \dfrac{(y+1)^2}{4} = 1$

This is the equation of a horizontal hyperbola.

Center: $(2, -1)$; vertices: $(5, -1)$, $(-1, -1)$

33. $25x^2 + y^2 - 4y = 96$

$25x^2 + y^2 - 4y + 4 = 96 + 4$

$25x^2 + (y-2)^2 = 100$

$\dfrac{25x^2}{100} + \dfrac{(y-2)^2}{100} = 1$

$\dfrac{x^2}{4} + \dfrac{(y-2)^2}{100} = 1$

This is the equation of an ellipse with major axis on the y-axis.

Center: $(0, 2)$; vertices: $(0, -8)$, $(0, 12)$

35. $4x^2 - y^2 - 32x - 4y + 69 = 0$

$4(x^2 - 8x) - (y^2 + 4y) = -69$

$4(x^2 - 8x + 16) - (y^2 + 4y + 4) = -69 + 4(16) - 4$

$4(x-4)^2 - (y+2)^2 = -9$

$(y+2)^2 - 4(x-4)^2 = 9$

$\dfrac{(y+2)^2}{9} - \dfrac{4(x-4)^2}{9} = 1$

$\dfrac{(y+2)^2}{9} - \dfrac{(x-4)^2}{9/4} = 1$

This is the equation of a vertical hyperbola.

Center: $(4, -2)$

Vertices: $(4, -5)$, $(4, 1)$

37. $\dfrac{x^2}{64} + \dfrac{y^2}{16} = 1$

This is an ellipse with its major axis horizontal. Center is $(0,0)$, and the vertices are $(8,0)$ and $(-8,0)$. The ends of the minor diameter are $(0,4)$ and $(0,-4)$.

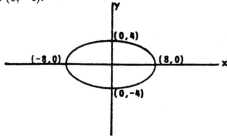

39. $\dfrac{x^2}{64} - \dfrac{y^2}{16} = 1$

Since the coefficients of x^2 and y^2 are of opposite sign, this is a hyperbola. Because the x^2 term is positive, it is a horizontal hyperbola, with $a = 8$. The center is $(0,0)$ and the vertices are $(0,-8)$ and $(0,8)$.

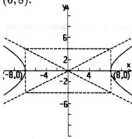

41. $\dfrac{(x-2)^2}{25} + \dfrac{(y-1)^2}{4} = 1$

This is an ellipse with major axis parallel to the x-axis. The center is $(2,1)$ and the vertices are $(7,1)$ and $(-3,1)$.

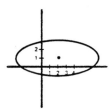

43.
$$x^2 - 2y^2 - 6x + 8y = 1$$
$$x^2 - 6x - 2(y^2 - 4y) = 1$$
$$x^2 - 6x + 9 - 2(y^2 - 4y + 4) = 1 + 9 - 2(4)$$
$$(x-3)^2 - 2(y-2)^2 = 2$$
$$\dfrac{(x-3)^2}{2} - (y-2)^2 = 1$$

Since the coefficients of the squared expressions are of opposite signs, this is a hyperbola. The hyperbola opens horizontally. (Set $y = 2$ to find the x-coefficients of the vertices.) Center is at $(3,2)$. Vertices are at $(3 + \sqrt{2}, 2)$ and $(3 - \sqrt{2}, 2)$. $a = \sqrt{2}$ and $b = 1$.

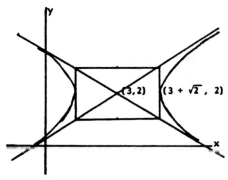

45. If the minor diameter is of length 10, then $b = 5$. The point midway between the vertices is the center of the ellipse: $\left(\dfrac{10-4}{2}, \dfrac{2+2}{2}\right) = (3,2)$.

Since the vertices lie on a line parallel to the x-axis, a^2 will go with x^2, and $a = \frac{1}{2}\left|10 - (-4)\right| = 7$.

The equation of the ellipse is $\dfrac{(x-3)^2}{49} + \dfrac{(y-2)^2}{25} = 1$

47. This is a horizontal hyperbola with center $(0,0)$ and $a = 6$. The asymptote must be a line through the origin, so the slope of the asymptote can be found using the slope formula.
$$\dfrac{b}{a} = \dfrac{2-0}{3-0}$$
$$\dfrac{b}{6} = \dfrac{2}{3}$$
$$b = 4$$

The equation of the hyperbola is $\dfrac{x^2}{36} - \dfrac{y^2}{16} = 1$.

49. Find the y-coordinates for which $x = \pm 2.5$ on the ellipse $\dfrac{x^2}{24} + \dfrac{y^2}{19} = 1$.

$$\dfrac{(\pm 2.5)^2}{24} + \dfrac{y^2}{19} = 1$$
$$y^2 = \left(1 - \dfrac{6.25}{24}\right)19 = \dfrac{19(24 - 6.25)}{24}$$
$$= \dfrac{19(17.75)}{24} \approx 14.052083$$
$$y \approx \pm 3.7486$$

51. a) $\frac{x^2}{7} + \frac{y^2}{11} = 1$

$a = \sqrt{7},\ b = \sqrt{11}$

Area $= \pi\sqrt{7}\sqrt{11}$

$\quad = \pi\sqrt{77}$

$\quad \approx 27.5674$

b) $\frac{x^2}{111} + y^2 = 1$

$a = \sqrt{111},\ b = 1$

Area $= \pi\sqrt{111}\sqrt{1}$

$\quad = \pi\sqrt{111}$

$\quad \approx 33.0987$

53. Place the mug on a coordinate system with the center of the hyperbola on the origin. Then the vertex of the hyperbola is at $(0,1)$ and $a = 1$. The hyperbola has an equation of the form $\frac{y^2}{1} - \frac{x^2}{b^2} = 1$.

We can find b^2 by replacing x and y with a point on the hyperbola, namely $(4,9)$.

$$\frac{81}{1} - \frac{16}{b^2} = 1$$

$$-\frac{16}{b^2} = -80$$

$$b^2 = \frac{16}{80} = \frac{1}{5}$$

The equation of the hyperbola is $y^2 - 5x^2 = 1$.
When $x = 2$,

$$y^2 - 5(4) = 1$$
$$y^2 = 21$$
$$y = \pm\sqrt{21}$$

Since we are interested in the depth of the mug at this point on the hyperbola,

Depth $= 9 - \sqrt{21}$ cm ≈ 4.42 cm

55.

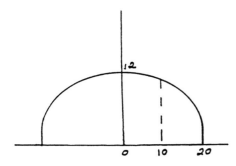

The equation of the ellipse has the form

$$\frac{x^2}{a^2} + \frac{y^2}{b^2} = 1$$

$a = 20,\ b = 12$, so

$$\frac{x^2}{400} + \frac{y^2}{144} = 1$$

We want to find y when $x = 10$:

$$\frac{(10)^2}{400} + \frac{y^2}{144} = 1$$

$$\frac{y^2}{144} = 1 - \frac{100}{400}$$

$$\frac{y^2}{144} = \frac{3}{4}$$

$$y^2 = 108$$

$$y = \pm\sqrt{108} = \pm 6\sqrt{3}$$

The height of the arch is $6\sqrt{3} \approx 10.39$ feet.

57. Let (x,y) be the coordinates of Arnold's nose. Let X and Y be the [varying] x- and y-intercepts of the ends of the ladder with the wall and the ground. Then $X^2 + Y^2 = 25^2$. The three triangles in the figure are all similar. From these similarities,

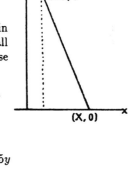

$\frac{X}{x} = \frac{25}{4}$ and $X = \frac{25x}{4}$.

In similar fashion,

$$\frac{Y}{Y - y} = \frac{25}{4}$$

$$4Y = 25Y - 25y$$

$$21Y = 25y$$

$$Y = \frac{25y}{21}.$$

Substituting these equivalents of X and Y:

$$\left(\frac{25x}{4}\right)^2 + \left(\frac{25y}{21}\right)^2 = 25^2$$

$$\frac{x^2}{16} + \frac{y^2}{144} = 1$$

The path of Arnold's nose is part of an ellipse.

59. $d(P,F_1) = \sqrt{[x - (-4)]^2 + (y - 0)^2}$

$\quad = \sqrt{(x + 4)^2 + y^2}$

$d(P,F_2) = \sqrt{(x - 4)^2 + (y - 0)^2}$

$\quad = \sqrt{(x - 4)^2 + y^2}$

Since $d(P,F_1) + d(P,F_2) = 10$,

$$\sqrt{(x + 4)^2 + y^2} + \sqrt{(x - 4)^2 + y^2} = 10$$

$$\sqrt{(x + 4)^2 + y^2} = 10 - \sqrt{(x - 4)^2 + y^2}$$

Squaring both sides,

$$(x + 4)^2 + y^2$$
$$= 100 - 20\sqrt{(x - 4)^2 + y^2} + (x - 4)^2 + y^2$$

$$x^2 + 8x + 16 + y^2$$
$$= 100 - 20\sqrt{(x-4)^2 + y^2} + x^2 - 8x + 16 + y^2$$
$$16x - 100 = -20\sqrt{(x-4)^2 + y^2}$$
$$4x - 25 = -5\sqrt{(x-4)^2 + y^2}$$
Squaring both sides,
$$16x^2 - 200x + 625 = 25(x^2 - 8x + 16 + y^2)$$
$$16x^2 - 200x + 625 = 25x^2 - 200x + 400 + 25y^2$$
$$225 = 9x^2 + 25y^2$$
$$\frac{9x^2}{225} + \frac{25y^2}{225} = 1$$
$$\frac{x^2}{25} + \frac{y^2}{9} = 1$$

61. $d(P, F_1) = \sqrt{(x+5)^2 + y^2}$
$d(P, F_2) = \sqrt{(x-5)^2 + y^2}$

Since $\left| d(P, F_1) - d(P, F_2) \right| = 6$,
$$\left| \sqrt{(x+5)^2 + y^2} - \sqrt{(x-5)^2 + y^2} \right| = 6$$
$$\sqrt{(x+5)^2 + y^2} - \sqrt{(x-5)^2 + y^2} = \pm 6$$
$$\sqrt{(x+5)^2 + y^2} = \sqrt{(x-5)^2 + y^2} \pm 6$$
Squaring both sides,
$$(x+5)^2 + y^2 = (x-5)^2 + y^2 \pm 12\sqrt{(x-5)^2 + y^2} + 36$$
$$x^2 + 10x + 25 + y^2$$
$$= x^2 - 10x + 25 + y^2 \pm 12\sqrt{(x-5)^2 + y^2} + 36$$
$$20x - 36 = \pm 12\sqrt{(x-5)^2 + y^2}$$
$$5x - 9 = \pm 3\sqrt{(x-5)^2 + y^2}$$
Squaring both sides,
$$25x^2 - 90x + 81 = 9(x^2 - 10x + 25 + y^2)$$
$$25x^2 - 90x + 81 = 9x^2 - 90x + 225 + 9y^2$$
$$16x^2 - 9y^2 = 144$$
$$\frac{x^2}{9} - \frac{y^2}{16} = 1$$

63. If the stakes are thought of as points F_1 and F_2 of Problem 58, and the loop of the rope is thought of as the distance $PF_1 + PF_2 +$ the constant distance F_1F_2, the situation is exactly that of Problem 60. The 80-foot loop minus the 30-foot distance F_1F_2 is the sum $PF_1 + PF_2$. That is, $2a = 50$, and $a = 25$. The distance between the two stakes is $2c$, so $c = 15$. Then $b^2 = a^2 - c^2$, and $b^2 = 25^2 - 15^2 = 625 - 225 = 400$, so $b = 20$. The equation of the ellipse is therefore
$$\frac{x^2}{25^2} + \frac{y^2}{20^2} = 1$$
$$\frac{x^2}{625} + \frac{y^2}{400} = 1$$

Chapter 4 Review Problem Set

1. True

2. True

3. False:
$$d = \sqrt{[\pi - (-\pi)]^2 + (3-2)^2}$$
$$= \sqrt{(2\pi)^2 + (1)^2}$$
$$= \sqrt{4\pi^2 + 1}$$

4. True

5. False: Complete the square.
$$x^2 + y^2 + 4x + 4 = 0 + 4$$
$$x^2 + (y+2)^2 = 4$$
$$r^2 = 4$$
$$r = 2$$

6. False. This parabola opens to the left.

7. False. It will not be a parabola if $a = 0$.

8. False. For example, the graph of $x^2 + y^2 = -4$ is empty.

9. True

10. True

11. Horizontal parabola

12. Circle

13. Vertical ellipse

14. Straight line

15. Horizontal parabola

16. Hyperbola

17. $4x^2 - 25y^2 = 0$
$$4x^2 = 25y^2$$
$$2x = \pm 5y$$
Two intersecting straight lines

18. The single point $(-2, 3)$

19. $3x + y = 4$

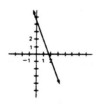

20. $x^2 + y^2 = 16$

This is a circle with center $(0,0)$ and radius $\sqrt{16} = 4$

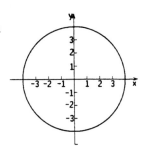

21. $y = x^2 + \frac{1}{x}$

First notice that the graph is not symmetrical about the x-axis, the y-axis, or the origin. Also notice there will be no point on the graph for $x = 0$.

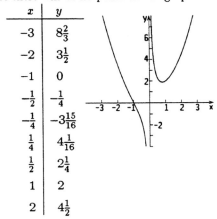

x	y
-3	$8\frac{2}{3}$
-2	$3\frac{1}{2}$
-1	0
$-\frac{1}{2}$	$-\frac{1}{4}$
$-\frac{1}{4}$	$-3\frac{15}{16}$
$\frac{1}{4}$	$4\frac{1}{16}$
$\frac{1}{2}$	$2\frac{1}{4}$
1	2
2	$4\frac{1}{2}$

22. $y = x^3 - 4x$

Notice that the graph is symmetrical about the origin.

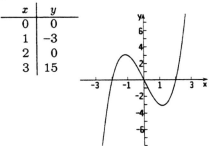

x	y
0	0
1	-3
2	0
3	15

23. $y = \sqrt{x} + 2$

First notice that the graph is not symmetrical about the x-axis, the y-axis, or the origin. Also notice that x can only be assigned non-negative values.

x	y
0	2
1	3
4	4

See graph in the next column.

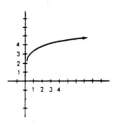

24. $\frac{x^2}{25} + \frac{y^2}{9} = 1$

This is an ellipse with its major axis horizontal. Center is $(0,0)$, and the vertices are $(5,0)$ and $(-5,0)$. The ends of the minor diameter are $(0,3)$ and $(0,-3)$.

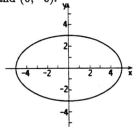

25. a)

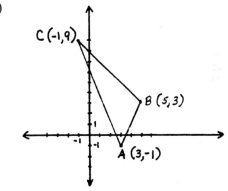

b) $\overline{AC} = \sqrt{[3 - (-1)]^2 + (-1 - 9)^2}$

$\qquad = \sqrt{(4)^2 + (-10)^2}$

$\qquad = \sqrt{116} = 2\sqrt{29}$

$\overline{AB} = \sqrt{(5 - 3)^2 + [3 - (-1)]^2}$

$\qquad = \sqrt{(2)^2 + (4)^2}$

$\qquad = \sqrt{20} = 2\sqrt{5}$

$\overline{BC} = \sqrt{[5 - (-1)]^2 + (3 - 9)^2}$

$\qquad = \sqrt{(6)^2 + (-6)^2} = \sqrt{72} = 6\sqrt{2}$

26. a) $AB: m = \dfrac{3+1}{5-3} = 2;$ $AC: m = \dfrac{9+1}{-1-3} = -\dfrac{5}{2};$

$BC: m = \dfrac{9-3}{-1-5} = -1$

b) $AB:$ $y - 3 = 2(x - 5)$
$y - 3 = 2x - 10$
$2x - y - 7 = 0$

$AC:$ $y - 9 = -\dfrac{5}{2}[x - (-1)]$
$2y - 18 = -5(x + 1)$
$2y - 18 = -5x - 5$
$5x - 2y + 13 = 0$

$BC:$ $y - 3 = -1(x - 5)$
$y - 3 = -x + 5$
$x + y - 8 = 0$

27. a) The midpoint of $AB = \left(\dfrac{5+3}{2}, \dfrac{-1+3}{2}\right) = (4, 1)$

The midpoint of $AC = \left(\dfrac{-1+3}{2}, \dfrac{9-1}{2}\right) = (1, 4)$

b) The slope of the line joining the midpoints is

$\dfrac{4-1}{1-4} = \dfrac{3}{-3} = -1$

Since -1 is also the slope of BC (Problem 26), the two lines are parallel.

The length of the line joining the midpoints is
$\sqrt{(4-1)^2 + (1-4)^2} = \sqrt{(3)^2 + (-3)^2} = \sqrt{18} = 3\sqrt{2}$

The length of BC is $6\sqrt{2}$ (Problem 25), so the line joining the midpoints is one-half the length of BC.

28. a) Since the slope of $BC = -1, (y + 1) = -1(x - 3),$
$y = -x + 2$

b) The slope of the perpendicular line $= 1$. The equation is $y + 1 = x - 3 \Rightarrow y = x - 4$

29. $x = 4 \Rightarrow x - 4 = 0 \Rightarrow x + 0y - 4 = 0$

30. Slope $= \dfrac{-1-3}{5-2} = -\dfrac{4}{3}.$ $y - 3 = -\dfrac{4}{3}(x - 2)$

$3y - 9 = -4x + 8$
$4x + 3y - 17 = 0$

31. Find the slope of
$5x + 7y = 2$
$7y = -5x + 2$
$y = -\dfrac{5}{7}x + \dfrac{2}{7}$
$m = -\dfrac{5}{7}$

The line we seek will have slope $-\dfrac{5}{7}$. Since the line passes through $(0, 0)$ its equation is
$y - 0 = -\dfrac{5}{7}(x - 0)$
$7y = -5x$
$5x + 7y = 0$

32. The slope of the given line is $-\dfrac{2}{3}$. The slope of the desired line $= \dfrac{3}{2}$.
$y = \dfrac{3}{2}x + 6$
$2y = 3x + 12$
$3x - 2y + 12 = 0$

33. The line passes through the two points $(0, -1)$ and $(2, 0)$. $m = \dfrac{0 - (-1)}{2 - 0} = \dfrac{1}{2}$

$y - (-1) = \dfrac{1}{2}(x - 0)$
$y + 1 = \dfrac{1}{2}x$
$2y + 2 = x$
$x - 2y - 2 = 0$

34. The slope of the tangent line to the circle $x^2 + y^2 + 25$ at the point (x, y) is given by $-\dfrac{x}{y}$, $(y \neq 0)$. The slope at the point $(3, -4)$ is $\dfrac{3}{4}$. The equation of the tangent line is given by
$y + 4 = \dfrac{3}{4}(x - 3)$
$4y + 16 = 3x - 9$
$3x - 4y - 25 = 0$

35. The parabola is of the form $y = ax^2 + bx + c$, so it is a vertical parabola. It opens up since $a > 0$.
Vertex:

x-coordinate: $\dfrac{-(-8)}{2(2)} = 2$

y-coordinate: $2(2)^2 - 8(2) + 1 = -7$
The vertex is $(2, -7)$.

36. $x + 2 = \dfrac{1}{12}(y + 3)^2$
The parabola is of the form $x - h = a(y - k)^2$, so it is a horizontal parabola. Since $a > 0$, it opens to the right. The vertex is (h, k) or $(-2, -3)$.

37. $y - 3 = -\dfrac{2}{3}(x + 1)^2$
The parabola is of the form $y - k = a(x - h)^2$, so it is a vertical parabola. Since $a < 0$, it opens down. The vertex is (h, k) or $(-1, 3)$.

38. $x = -2(y^2 - 2y)$. Completing the square:
$x = -2(y^2 - 2y + 1) + 2 = -(y - 1)^2 + 2.$
The vertex is $(2, 1)$; the parabola opens to the left.

39. The equation will have the form $y = kx^2$.
To find k, let $x = 2$ and $y = -12$.
$-12 = k(2)^2$
$-12 = 4k$
$-3 = k$
The equation of the parabola is $y = -3x^2$.

40. The general form of the equation would be $x + 2 = a(y - 3)^2$. Substituting the x- and y-values,
$-4 + 2 = a(1 - 3)^2$
$-2 = 4a$
$a = -\dfrac{1}{2}$. The equation is $x + 2 = -\dfrac{1}{2}(y - 3)^2$.

41. This is a vertical ellipse centered on the origin.
$a = 3$, $b = \dfrac{4}{2} = 2$.

The equation is $\dfrac{x^2}{4} + \dfrac{y^2}{9} = 1$.

42. The center of the ellipse is $(3,1)$, $a = 4$, and $b = 3$.

The equation is $\dfrac{(x-3)^2}{9} + \dfrac{(y-1)^2}{16} = 1$.

43. The center of the circle is $\left(\dfrac{3+3}{2}, \dfrac{-3+5}{2}\right) = (3,1)$.

The equation of the circle is
$$(x-3)^2 + (y-1)^2 = (5)^2$$
$$(x-3)^2 + (y-1)^2 = 25$$

44. The symmetry of these four points implies that the center of the ellipse will be the point $(-7,3)$, $a = 7$, and $b = 2$. The equation of this ellipse is given by

$$\dfrac{(x+7)^2}{49} + \dfrac{(y-3)^2}{4} = 1$$

45. Since the vertices lie on a line parallel to the x-axis, this is a horizontal hyperbola with an equation of

the form $\dfrac{(x-h)^2}{a^2} - \dfrac{(y-k)^2}{b^2} = 1$.

$a = \frac{1}{2}\left|5 - (-1)\right| = 3$

The center of the hyperbola is at

$\left(\dfrac{5-1}{2}, \dfrac{-3-3}{2}\right) = (2,-3)$

The asymptote passes through the center $(2,3)$ and the point $(8,5)$. The slope of the asymptote is

$\dfrac{b}{a} = \dfrac{5-(-3)}{8-2} = \dfrac{4}{3}$

$\dfrac{b}{3} = \dfrac{4}{3}$

$b = 4$

The equation of the hyperbola is

$$\dfrac{(x-2)^2}{9} - \dfrac{(y+3)^2}{16} = 1$$

46. The center is the midpoint of these given points, and the radius is $\frac{1}{2}$ the distance between them. Therefore, center $= (2.5, 3)$ and

radius $= \frac{1}{2}\sqrt{(4-1)^2 + [7-(-1)]^2} = \frac{1}{2}\sqrt{73}$.

The equation is given by $(x-2.5)^2 + (y-3)^2 = \dfrac{73}{4}$

47.
$$x^2 - 6x + y^2 + 2y = -1$$
$$x^2 - 6x + 9 + y^2 + 2y + 1 = -1 + 9 + 1$$
$$(x-3)^2 + (y+1)^2 = 9$$

This is the equation of a circle with center $(3,-1)$ and radius of length $\sqrt{9} = 3$.

48.
$$y^2 + 6y + 4x = 7$$
$$y^2 + 6y + 9 + 4x = 7 + 9$$
$$(y+3)^2 + 4x = 16$$
$$(y+3)^2 = -4(x-4)$$

We have a parabola with vertex $(4,-3)$ that opens to the left.

49.
$$4x^2 - 8x - 9y^2 = 32$$
$$4(x^2 - 2x + 1) - 9y^2 = 32 + 4(1)$$
$$4(x-1)^2 - 9y^2 = 36$$

$$\dfrac{(x-1)^2}{9} - \dfrac{y^2}{4} = 1$$

This is the equation of a horizontal hyperbola with center $(1,0)$ and vertices $(4,0)$ and $(-2,0)$. The asymptotes pass through the center $(1,0)$ and have

slope $\dfrac{b}{a} = \dfrac{2}{3}$. The equations of the asymptotes are

$y = \pm\frac{2}{3}(x-1)$

50.
$$x^2 + 6x + 9 + 4(y^2 - 10y + 25) = -9 + 9 + 100$$
$$(x+3)^2 + 4(x-5)^2 = 100$$
$$\dfrac{(x+3)^2}{100} + \dfrac{(x-5)^2}{25} = 1$$

Horizontal ellipse with center $(-3,5)$, length of major axis $= 20$, and length of minor axis $= 10$.

51. $\begin{cases} y = 2x^2 + x - 14 \\ y = -x - 2 \end{cases}$

Equate the y-values.
$$2x^2 + x - 14 = -x - 2$$
$$2x^2 + 2x - 12 = 0$$
$$x^2 + x - 6 = 0$$
$$(x+3)(x-2) = 0$$
$$x = -3 \text{ or } x = 2$$
$$y = 1 \text{ or } y = -4 \text{ respectively}$$

The points of intersection are $(-3,1)$ and $(2,-4)$.

52. Solving the hyperbolic equation $y^2 - x^2 = \frac{5}{2}$ for y^2, we get $y^2 = x^2 + \frac{5}{2}$. Substituting into the elliptic equation $9x^2 + 4y^2 = 36$,
$$9x^2 + 4\left(x^2 + \frac{5}{2}\right) = 36$$
$$13x^2 + 10 = 36$$
$$13x^2 = 26$$
$$x^2 = 2$$

$y^2 = 2 + \frac{5}{2} = \frac{9}{2}$, $y = \pm\dfrac{3\sqrt{2}}{2}$. Therefore, the points of

intersection are $\left(\pm\sqrt{2}, \pm\dfrac{3\sqrt{2}}{2}\right)$.

53. Place the arch on a coordinate system with the origin at the midpoint of the base.

a) The equation of the parabola is of the form
$$y - 9 = ax^2$$
Since we know that the point $(3, 0)$ satisfies the equation, we can find the value of a.
$$0 - 9 = a(3)^2$$
$$-9 = 9a$$
$$-1 = a$$
The equation of the parabola is
$$y - 9 = -x^2$$
When $x = 2$,
$$y - 9 = -4$$
$$y = 5$$
The arch is 5 feet high at the point of interest.

b) The equation of the ellipse is of the form
$$\frac{x^2}{b^2} + \frac{y^2}{a^2} = 1$$
$a = 9$, $b = \frac{6}{2} = 3$
So,
$$\frac{x^2}{9} + \frac{y^2}{81} = 1$$
When $x = 2$,
$$\frac{4}{9} + \frac{y^2}{81} = 1$$
$$\frac{y^2}{81} = \frac{5}{9}$$
$$y^2 = 45$$
$$y = \pm\sqrt{45}$$
$$= \pm 3\sqrt{5}$$
The arch is $3\sqrt{5} \approx 6.71$ feet high at the point of interest.

54. The length of the belt $= 2 \cdot \sqrt{6^2 + 8^2} + 2 \cdot \pi \cdot 3$
$$= 20 + 6\pi$$

55. $M = x^2 - 100x + 2486$
The minimum value of M will occur at the vertex of the parabola.
x-coordinate of the vertex $= \dfrac{-(-100)}{2} = 50$

M-coordinate of the vertex
$$= (50)^2 - 100(50) + 2486 = -14$$
The minimum value of M is -14.

56. a) The maximum height occurs at the vertex of this parabola, when $x = \dfrac{1}{.0064}$. Substituting this into the height equation, we get that the maximum height $= 78.125$ feet.

b) The horizontal distance is determined by setting $y = 0$: $-0.0032x^2 + x = 0$. Since $x \neq 0$, divide by x:
$-0.032x + 1 = 0$. Range $= \dfrac{1}{.0032} = 312.5$ feet.

Chapter 5: Functions and their Graphs

Problem Set 5.1

1. $f(x) = x^2 - 4$

   a) $f(-2) = (-2)^2 - 4 = 4 - 4 = 0$

   b) $f(0) = 0^2 - 4 = -4$

   c) $f\left(\frac{1}{2}\right) = \left(\frac{1}{2}\right)^2 - 4 = \frac{1}{4} - 4 = -3\frac{3}{4}$

   d) $f(.1) = (.1)^2 - 4 = .01 - 4 = -3.99$

   e) $f(\sqrt{2}) = (\sqrt{2})^2 - 4 = 2 - 4 = -2$

   f) $f(a) = a^2 - 4$

   g) $f\left(\frac{1}{x}\right) = \left(\frac{1}{x}\right)^2 - 4 = \frac{1}{x^2} - 4 = \frac{1 - 4x^2}{x^2}$

   h) $f(x+1) = (x+1)^2 - 4 = x^2 + 2x + 1 - 4$
   $$= x^2 + 2x - 3$$

3. $f(x) = \frac{1}{x-4}$

   a) $f(8) = \frac{1}{8-4} = \frac{1}{4}$

   b) $f(2) = \frac{1}{2-4} = \frac{1}{-2} = -\frac{1}{2}$

   c) $f(\frac{9}{2}) = \frac{1}{\frac{9}{2}-4} = \frac{1}{\frac{1}{2}} = 2$

   d) $f(\frac{31}{8}) = \frac{1}{\frac{31}{8}-4} = \frac{1}{\frac{-1}{8}} = -8$

   e) $f(4) = \frac{1}{4-4} = \frac{1}{0}$ Undefined

   f) $f(4.01) = \frac{1}{4.01-4} = \frac{1}{.01} = 100$

   g) $f\left(\frac{1}{x}\right) = \frac{1}{\frac{1}{x}-4} = \frac{x}{1-4x}$

   h) $f(x^2) = \frac{1}{x^2-4}$

   i) $f(2+h) = \frac{1}{2+h-4} = \frac{1}{h-2}$

   j) $f(2-h) = \frac{1}{2-h-4} = \frac{1}{-h-2} = -\frac{1}{h+2}$

5. The natural domain of $f(x) = x^2 - 4$ is the set of all real numbers.

7. Natural domain of $\frac{1}{x^2-4}$: all real numbers except $x = 2$ and $x = -2$, which would make the denominator zero.

9. Natural domain of
   $$h(x) = \frac{2}{x^2 - x - 6} = \frac{2}{(x-3)(x+2)}:$$
   all real numbers except $x = 3$ and $x = -2$, which would make the denominator zero.

11. Natural domain of $F(x) = \frac{1}{x^2+4}$: all real numbers

13. $G(x) = \sqrt{x-2}$ has a natural domain consisting of all real numbers which make $x - 2 \geq 0$. That is, $x \geq 2$.

15. Natural domain of $H(x) = \frac{1}{5 - \sqrt{x}}$ consists of all real x such that $x \geq 0$ (to make the radical a real number) *and* such that $\sqrt{x} \neq 5$ (to prevent the denominator from being 0.) Thus, the domain is $\{x : x \geq 0 \text{ and } x \neq 25\}$.

17. a) $f(6) = 9$
    b) $f(9) = 5$
    c) $f(16) = 3$
    The natural domain for this function is the set of counting numbers, or the positive integers.

19. $f(x) = (x+2)^2$
    $$(\,[\#]\,[+]\,[2]\,)\,[x^2]\,[=]$$
    $f(2.9) = 24.01$

21. $f(x) = 3(x+2)^2 - 4$
    $$(\,[\#]\,[+]\,[2]\,)\,[x^2]\,[\times]\,[3]\,[-]\,[4]\,[=]$$
    $f(2.9) = 68.03$

23. $f(x) = \left(3x + \frac{2}{\sqrt{x}}\right)^3$
    $$(\,[3]\,[\times]\,[\#]\,[+]\,[2]\,[\div]\,[\#]\,[\sqrt{\ }]\,)\,[y^x]\,[3]\,[=]$$
    $f(2.9) = 962.80311$

25. $f(x) = \frac{\sqrt{x^5 - 4}}{2 + \frac{1}{x}}$
    $$(\,[\#]\,[y^x]\,[5]\,[-]\,[4]\,)\,[\sqrt{\ }]\,[\div]\,(\,[2]\,[+]\,(\,[\#]\,[x^{-1}]\,)\,)\,[=]$$
    $f(2.9) = 6.047940738$

27. $x + \sqrt{5}$

29. $2x^2 + 3x$

31. $\sqrt{3(x-2)^2 + 9}$

33. $g(x,y) = 3xy - 5x$
    $g(2,5) = 3(2)(5) - 5(2) = 30 - 10 = 20$

35. $g(5,2) = 3(5)(2) - 5(5) = 30 - 25 = 5$

37. $G(x,y) = \frac{5x + 3y}{2x - y}$
    $$G(1,1) = \frac{5(1) + 3(1)}{2(1) - 1} = \frac{5+3}{2-1} = 8$$

39. $G(\frac{1}{2}, 1) = \frac{5(\frac{1}{2}) + 3(1)}{2(\frac{1}{2}) - 1} = \frac{\frac{11}{2}}{0}$ Undefined

41. $g(2x, 3y) = 3(2x)(3y) - 5(2x) = 18xy - 10x$

43. $g\left(x,\frac{1}{x}\right) = 3(x)\left(\frac{1}{x}\right) - 5(x) = 3 - 5x$

45. $y = kx$
To find k, let $x = 3$ and $y = 12$.
$12 = 3k$
$4 = k$
Therefore,
$y = 4x$

47. $y = \frac{k}{x}$
To find k, let $x = \frac{1}{5}$ and $y = 5$.
$5 = \frac{k}{\frac{1}{5}}$
$k = 5\left(\frac{1}{5}\right) = 1$
Therefore,
$y = \frac{1}{x}$

49. $I = \frac{ks}{d^2}$ and $I = 9$ when $s = 4$ and $d = 12$
$9 = \frac{k \cdot 4}{12^2}$
$k = \frac{9 \cdot 144}{4} = 9 \cdot 36 = 324$
The formula is $I = \frac{324s}{d^2}$.

51. a) $R = kv^2$
To find k, let $v = 600$ and $R = 16,000$
$16000 = k(600)^2$
$k = \frac{16000}{360000} = \frac{2}{45}$
The formula is $R = \frac{2v^2}{45}$

b) Let $v = 800$:
$R = \frac{2(800)^2}{45} \approx 28{,}444$ feet

53. $f(x) = x^2 - \frac{2}{x}$

a) $f(2) = (2)^2 - \frac{2}{2} = 4 - 1 = 3$

b) $f(-1) = (-1)^2 - \frac{2}{-1} = 1 + 2 = 3$

c) $f\left(\frac{1}{2}\right) = \left(\frac{1}{2}\right)^2 - \frac{2}{\frac{1}{2}} = \frac{1}{4} - 4 = -\frac{15}{4}$

d) $f(\sqrt{2}) = (\sqrt{2})^2 - \frac{2}{\sqrt{2}} = 2 - \frac{2}{\sqrt{2}} = 2 - \sqrt{2}$

e) $f(2 - \sqrt{2}) = (2 - \sqrt{2})^2 - \frac{2}{2 - \sqrt{2}}$
$= 4 - 4\sqrt{2} + 2 - \frac{2(2 + \sqrt{2})}{(2 - \sqrt{2})(2 + \sqrt{2})}$
$= 6 - 4\sqrt{2} - \frac{4 + 2\sqrt{2}}{4 - 2}$
$= 6 - 4\sqrt{2} - \frac{4 + 2\sqrt{2}}{2}$
$= 6 - 4\sqrt{2} - 2 - \sqrt{2}$
$= 4 - 5\sqrt{2}$

f) $f(.01) = (.01)^2 - \frac{2}{.01} = .0001 - 200 = -199.9999$

g) $f\left(\frac{1}{x}\right) = \left(\frac{1}{x}\right)^2 - \frac{2}{\left(\frac{1}{x}\right)} = \frac{1}{x^2} - 2x = \frac{1 - 2x^3}{x^2}$

h) $f(a^2) = (a^2)^2 - \frac{2}{(a)^2} = a^4 - \frac{2}{a^2} = \frac{a^6 - 2}{a^2}$

i) $f(a + b) = (a + b)^2 - \frac{2}{a + b}$
$= \frac{(a + b)^3 - 2}{a + b}$
$= \frac{a^3 + 3a^2b + 3ab^2 + b^3 - 2}{a + b}$

55. a) Natural domain of $f(t) = \frac{t + 2}{t^2(t + 3)}$ is all real t except those that make the denominator zero; i.e., when $t = 0$ or $t = -3$.

b) Natural domain of $g(t) = \sqrt{t^2 - 4}$ is all real numbers such that
$t^2 - 4 \geq 0$
$t^2 \geq 4$
$t \geq 2$ or $t \leq -2$

c) The natural domain of $h(t) = \frac{3t + 1}{1 - \sqrt{2t}}$ is the set of all real $t \geq 0$ (to keep $\sqrt{2t}$ a real number) except $t = \frac{1}{2}$ (which would make the denominator zero.)

d) $k(s, t) = \frac{3st\sqrt{9 - s^2}}{t^2 - 1}$
The natural domain will be all those values of s for which $9 - s^2 \geq 0$, and all those values of t for which the denominator is not zero. In other words, $|s| \leq 3$ [i.e., $-3 \leq s \leq 3$], and all real t except $t = \pm 1$.

57.

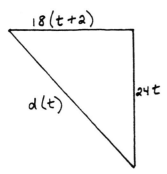

Using the formula $d = rt$,

$24t$ represents the distance the southbound ship has covered t hours after noon.

$18(t+2)$ represents the distance the westbound ship has covered t hours after noon.

Using the Pythagorean Theorem,

$$d(t) = \sqrt{(24t)^2 + [18(t+2)]^2}$$
$$= \sqrt{(6)^2 16t^2 + (6)^2 9[t^2 + 4t + 4]}$$
$$= 6\sqrt{16t^2 + 9t^2 + 36t + 36}$$
$$= 6\sqrt{25t^2 + 36t + 36}$$

59.

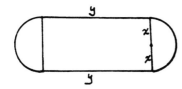

The perimeter of the track is 2 miles $= 2y + 2\pi x$

Therefore,

$$y = \frac{2 - 2\pi x}{2}$$
$$y = 1 - \pi x$$

The area of the track is the area of the rectangle plus the area of each semicircle.

$$A(x) = 2x(1 - \pi x) + \pi x^2$$

Since $A(x)$ represents a physical area, we cannot let it be a negative number. We must keep

$$1 - \pi x > 0$$
$$1 > \pi x$$
$$\frac{1}{\pi} > x$$

Since x must also be greater than 0, $0 < x < \frac{1}{\pi}$.

If $0 < x < \frac{1}{\pi}$, $A(x) > 0$ and

$$A(x) < 2\left(\frac{1}{\pi}\right)\left[1 - \pi\left(\frac{1}{\pi}\right)\right] + \pi\left(\frac{1}{\pi}\right)^2$$
$$A(x) < \frac{1}{\pi}$$

In other words, the range of $A(x)$ is the set of numbers between 0 and $\frac{1}{\pi}$.

61. a) If the perimeter of the triangle is x, each side is of length $\frac{x}{3}$.

Heron's formula for the area of a triangle is
$$A = \sqrt{s(s-a)(s-b)(s-c)}$$
where s is half the perimeter, and a, b, and c are the three sides of the triangle.

For our triangle, $a = b = c = \frac{x}{3}$, and $s = \frac{x}{2}$

$$F(x) = \sqrt{\frac{x}{2}\left(\frac{x}{2} - \frac{x}{3}\right)^3}$$
$$= \sqrt{\frac{x}{2}\left(\frac{x}{6}\right)^3}$$
$$= \sqrt{\frac{x^4}{6^2 \cdot 12}}$$
$$= \frac{x^2}{6 \cdot 2\sqrt{3}}$$
$$= \frac{x^2\sqrt{3}}{6 \cdot 2 \cdot 3}$$
$$= \frac{x^2\sqrt{3}}{36}$$

b)

The area of the hexagon equals the sum of the areas of 6 equilateral triangles of perimeter $3x$. Again using Heron's formula, with $a = b = c = x$ and $s = \frac{3x}{2}$,

$$F(x) = 6\sqrt{\frac{3x}{2}\left(\frac{3x}{2} - x\right)^3}$$
$$= 6\sqrt{\frac{3x}{2}\left(\frac{x}{2}\right)^3}$$
$$= 6\sqrt{\frac{3x^4}{2^4}}$$
$$= \frac{6x^2}{2^2}\sqrt{3}$$
$$= \frac{3x^2\sqrt{3}}{2}$$

c)

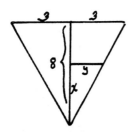

Let y be the radius of the cone at a height of x feet from the vertex. Noting the similar triangles in the above figure, we see that

$$\frac{y}{3} = \frac{x}{8}$$

$$y = \frac{3x}{8}$$

Using the formula $V = \frac{1}{3}\pi r^2 h$ with $r = \frac{3x}{8}$ and $h = x$,

$$F(x) = \frac{1}{3}\pi\left(\frac{3x}{8}\right)^2 x$$

$$= \frac{\pi \cdot 9x^3}{3 \cdot 64}$$

$$= \frac{3\pi x^3}{64}$$

d) The total cost of producing x refrigerators is $1300 + 240x$. The cost per refrigerator is

$$F(x) = \frac{1300 + 240x}{x}$$

e) If you drive 100 miles or less, the cost of the car is $18(10) = \$180$. If you drive more than 100 miles, the cost is $180 + .22(x - 100)$.

$$F(x) = \begin{cases} 180, & 0 \le x \le 100 \\ 180 + .22(x - 100), & x > 100 \end{cases}$$

63. $S(x, y, z) = \frac{kxy^2}{z}$

Expressing all units in inches, $x = 2$, $y = 6$, and $z = 10 \cdot 12 = 120$

$$1000 = \frac{k \cdot 2 \cdot 6^2}{120}$$

$$k = \frac{120(1000)}{2 \cdot 36} = \frac{5000}{3}$$

So the formula is $s = \frac{5000xy^2}{3z}$ where all linear measurements are in inches.

If placed flatwise,

$$s = \frac{5000}{3} \cdot \frac{6 \cdot 2^2}{120} = \frac{1000}{3} = 333\frac{1}{3} \text{ pounds.}$$

65. $f(x)f(y) - f(xy) = x + y$

Since the relationship is true for all real numbers, we can replace x and y with any values we choose.

Let $y = 0$.

$$f(x)f(0) - f(0) = x + 0$$
$$f(0)[(f(x) - 1] = x$$

We see from the above that $f(0) \ne 0$.

Let $x = y = 0$.

$$f(0)f(0) - f(0) = 0$$
$$f(0)[f(0) - 1] = 0$$

Since $f(0) \ne 0$,

$$f(0) - 1 = 0$$
$$f(0) = 1$$

Now let $x = t$ and $y = 0$:

$$f(t)f(0) - f(0) = t$$
$$f(t) - 1 = t$$
$$f(t) = 1 + t$$

Problem Set 5.2

1. $f(x) = 5$

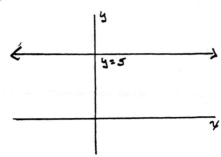

3. $f(x) = -3x + 5$

x	$f(x)$
0	5
1	2

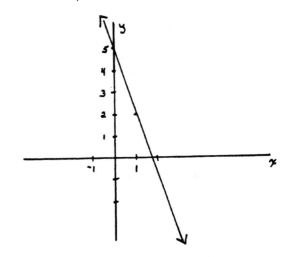

5. $f(x) = x^2 - 5x + 4 = (x-4)(x-1)$
This is a parabola with x-intercepts at $x = 4$ and $x = 1$, vertex at $(\frac{5}{2}, -\frac{9}{4})$, and y-intercept at $y = 4$.

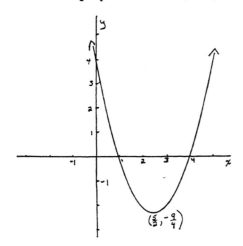

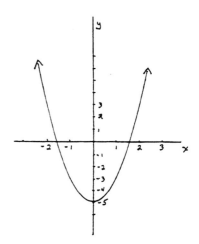

7. $f(x) = x^3 - 9x$
$\quad = x(x^2 - 9)$
$\quad = x(x+3)(x-3)$

x	$f(x)$
0	0
-3	0
3	0
1	-8
2	-10

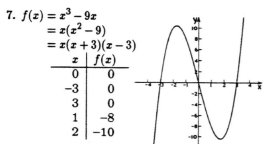

Notice that $f(x)$ is an odd function, so the graph will be symmetrical about the origin.

9. $f(x) = 2.12x^3 - 4.13x + 2$

x	$f(x)$
-3	-42.85
-2	-6.7
-1	4.01
0	2
1	-0.01
2	10.7

11. $f(x) = 2x^2 - 5$
$f(-x) = 2(-x)^2 - 5 = 2x^2 - 5 = f(x)$
Therefore, $f(x)$ is an even function, symmetric about the y-axis.

x	$f(x)$
0	-5
1	-3
2	3

13. $f(x) = x^2 - x - 1$
$f(-x) = (-x)^2 - (-x) - 1$
$\quad = x^2 + x - 1$
$\quad \neq f(x)$
$\quad \neq -f(x)$
Therefore, $f(x)$ is neither an odd nor an even function.

15. $f(x) = 4x^3 - x$
$f(-x) = 4(-x)^3 - (-x)$
$\quad = -4x^3 + x$
$\quad = -(4x^3 - x)$
$\quad = -f(x)$
Therefore, $f(x)$ is an odd function, symmetric about the origin.
$f(x) = x(4x^2 - 1)$
$\quad = x(2x+1)(2x-1)$

x	$f(x)$
$-\frac{1}{2}$	0
0	0
$\frac{1}{2}$	0
$\frac{1}{4}$	$-\frac{3}{16}$
1	3
2	30

x-intercepts (for rows $-\frac{1}{2}$, 0, $\frac{1}{2}$)

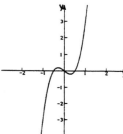

17. $f(x) = 2x^4 - 5x^2$
$f(-x) = 2(-x)^4 - 5(-x)^2$
$\quad = 2x^4 - 5x^2$
$\quad = f(x)$
Therefore, $f(x)$ is an even function, symmetric about the y-axis.

$f(x) = x^2(2x^2 - 5)$

x	y	
0	0	
$\sqrt{\frac{5}{2}}$	0	x-inter- cepts
$-\sqrt{\frac{5}{2}}$	0	
1	−3	
3	117	

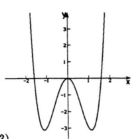

19. $f(x) = (x+1)(x-1)(x-3)$
$f(x) = 0$ at $x = -1$, $x = 1$, and $x = 3$.
Make a table of values including at least one value of x on each side of each zero.

x	y
2	−15
0	3
2	−3
4	15

21. $f(x) = x^2(x-4)$
$f(x) = 0$ at $x = 0$ and $x = 4$.
Make a table of values to include at least one value of x on each side of each zero.

x	y
−1	−5
1	−3
3	−9
5	25

23. $f(x) = (x+2)^2(x-2)^2$
$f(x) = 0$ at $x = -2$ and $x = 2$.
The graph will not cross the axis at these zeroes.
Make a table of values including at least one value of x on each side of each zero.

x	y
−3	25
0	16
3	25

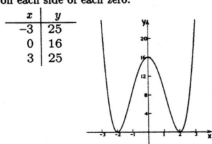

25. $f(x) = 2|x|$
This function is equivalent to the multipart function
$f(x) = \begin{cases} 2x & \text{if } x \geq 0 \\ -2x & \text{if } x < 0 \end{cases}$

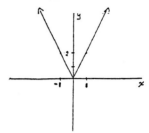

27. $f(x) = |x-2|$
This function is equivalent to the multipart function
$f(x) = \begin{cases} x-2 & \text{if } x \geq 2 \\ -(x-2) & \text{if } x < 2 \end{cases}$

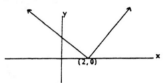

29. $f(x) = \begin{cases} x & \text{if } x < 0 \\ 2 & \text{if } x \geq 0 \end{cases}$

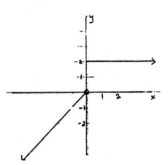

31. $f(x) = \begin{cases} -5 & \text{if } x \leq -3 \\ 4 - x^2 & \text{if } -3 < x \leq 3 \\ -5 & \text{if } x > 3 \end{cases}$

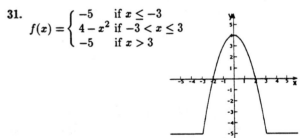

33. Any vertical line drawn through the graph must intersect the graph at most once.

35. $f(x) = \sqrt{x}$
The domain of the function is $\{x: x \geq 0\}$.

x	y
0	0
1	1
4	2
9	3

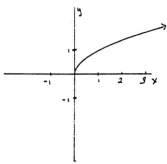

37. $f(x) = (x+2)(x-1)^2(x-3)$
$f(x) = 0$ at $x = -2$, $x = 1$, and $x = 3$.
The graph will not cross the x-axis at $x = 1$.
Make a table of values including at least one value
of x on each side of each zero.

x	y
-3	96
-1	-16
0	-6
2	-4
4	54

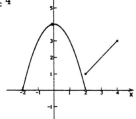

39. $f(x) = \begin{cases} 4 - x^2 & \text{if } -2 \leq x < 2 \\ x - 1 & \text{if } 2 \leq x \leq 4 \end{cases}$

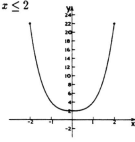

41. $f(x) = x^4 + x^2 + 2, -2 \leq x \leq 2$

x	y
-2	22
-1	4
0	2
1	4
2	22

43. As the exponent increases, the function becomes flatter and flatter between -1 and $+1$, and then rises very rapidly beyond these x values.

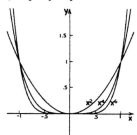

45. $f(x) = [x]$
$$f(x) = \begin{cases} -2, -2 \leq x < 1 \\ -1, -1 \leq x < 0 \\ 0, 0 \leq x < 1 \\ \vdots \end{cases}$$

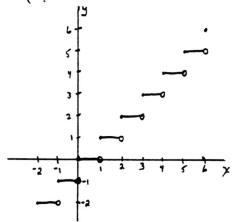

47. $C(x) = 15 + 10[x]$, $x > 0$

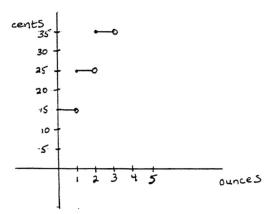

49. a) Let y represent the length of the fence.
   Then $xy = 400$

$$y = \frac{400}{x}$$

$$L = 4x + 3\left(\frac{400}{x}\right)$$

$$L = 4x + \frac{1200}{x}$$

b)

x	L
5	$4(5) + \dfrac{1200}{5} = 260$
10	$4(10) + \dfrac{1200}{10} = 160$
15	$4(15) + \dfrac{1200}{15} = 140$
20	$4(20) + \dfrac{1200}{20} = 140$
50	$4(50) + \dfrac{1200}{50} = 224$

c)

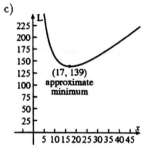

(17, 139)
approximate
minimum

d) The minimum value of L occurs when x is between 15 and 20 meters, at approximately 17 meters.

51. $f(3) = 792$
   $f(4.3) = 5022.32142$
   $f(-1.6) = -86.55584$

53.

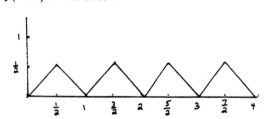

The area of each triangle is $\frac{1}{2}bh = \frac{1}{2}(1)\left(\frac{1}{2}\right) = \frac{1}{4}$.

The area of the region between the graph and the x-axis is $4\left(\frac{1}{4}\right) = 1$.

Problem Set 5.3

1. Set standard range.
   Press $\boxed{Y=}$
   Enter $X \wedge 3$
   Press $\boxed{\text{GRAPH}}$

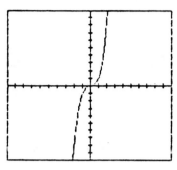

3. Set standard range.
   Press $\boxed{Y=}$
   Enter $X \wedge 3 - X \wedge 2 + 2$
   Press $\boxed{\text{GRAPH}}$

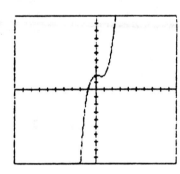

5. Set standard range.
   Press $\boxed{Y=}$
   Enter $-.2X \wedge 3 - X \wedge 2 + 2$
   Press $\boxed{\text{GRAPH}}$

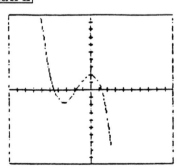

7. Set standard range.
Press $\boxed{Y=}$
Enter $X\,^\wedge\,4 - X\,^\wedge\,2 - 3$
Press $\boxed{\text{GRAPH}}$

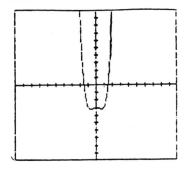

9. Set standard range.
Press $\boxed{Y=}$
Enter $.1X\,^\wedge\,4 - X\,^\wedge\,2 + 2X - 3$
Press $\boxed{\text{GRAPH}}$

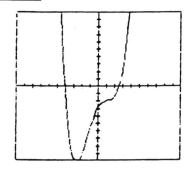

11. Set new range values:
Xmin $= -5$
Xmax $= 5$
Ymin $= -120$
Ymax $= 5$
Yscl $= .0001$
Press $\boxed{Y=}$
Enter $.05X\,^\wedge\,6 - 15X\,^\wedge\,2 + 5X - 3$
Press $\boxed{\text{GRAPH}}$

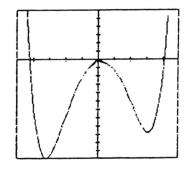

13. Set standard range.
Press $\boxed{Y=}$
Enter $X\,^\wedge\,3 - X\,^\wedge\,2 + 2$ as Y_1
Enter $\boxed{(-)}2X + 1$ as Y_2
Press $\boxed{\text{GRAPH}}$.
1) Press $\boxed{\text{ZOOM}}$.
2) Select Zoom in and press $\boxed{\text{ENTER}}$.
3) Move cursor to intersection point and press $\boxed{\text{ENTER}}$.
4) Repeat step 3 until desired accuracy is obtained.
$(-3.93, 1.785)$

15. Set standard range.
Press $\boxed{Y=}$
Enter $X\,^\wedge\,3 - X\,^\wedge\,2 + 2$ as Y_1
Enter $\boxed{\sqrt{\;}}\boxed{(}X + 8\boxed{)}$ as Y_2
Press $\boxed{\text{GRAPH}}$
Perform steps 1–4 as described in the solution to exercise 13. $(1.488, 3.080)$

17. Set standard range.
Press $\boxed{Y=}$
Enter $X\,^\wedge\,4 - X\,^\wedge\,2 - 3$ as Y_1
Enter $\boxed{\sqrt{\;}}X + 6$ as Y_2
Press $\boxed{\text{GRAPH}}$
Perform steps 1–4 as described in the solution to Exercise 13. $(1.940, 7.393)$

19. a) Set standard range.
Press $\boxed{Y=}$
Enter $-.2X\,^\wedge\,3 - X\,^\wedge\,2 + 2$
Press $\boxed{\text{GRAPH}}$
Move cursor to the turning point of the curve in Quadrant III. Perform the steps described in the solution to Exercise 13, zooming in on the minimum point repeatedly. $(-3.333, -1.703)$

b) It is easy to see from the graph that the global minimum on the interval $-10 \le x \le 10$ will occur at $x = 10$. $(10, -298)$

21. Set new range values:
Ymin $= -150$
Yscl $= .5$
Press $\boxed{Y=}$
Enter $.05X\,^\wedge\,6 - 15X\,^\wedge\,2 + 5X - 3$
Press $\boxed{\text{GRAPH}}$
The global minimum is the turning point of the curve in Quadrant III. Perform the steps described in the solution to Exercise 13, zooming in on the minimum point repeatedly. $(-3.203, -118.913)$

23. Set standard range.
Press $\boxed{Y=}$
Enter $\boxed{ABS}$ X
Press $\boxed{GRAPH}$

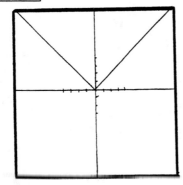

25. Set standard range.
Press $\boxed{Y=}$
Enter 2 $\boxed{ABS}$ $\boxed{(}$ X + 1 $\boxed{)}$
Press $\boxed{GRAPH}$

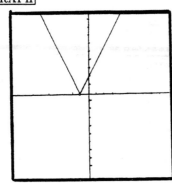

27. Set standard range.
Press $\boxed{Y=}$
Enter 8X2 $^\wedge$ $\boxed{(-)}$ X
Press $\boxed{GRAPH}$

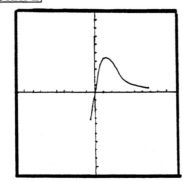

29. Set standard range.
Press $\boxed{Y=}$
Enter X $^\wedge$ 3 $-$ 3X $-$ 5

Press $\boxed{GRAPH}$
Follow the usual procedure for zooming in on a point, this time zooming in on the x-intercept.
$x = 2.279$

31. Set standard range.
Press $\boxed{Y=}$
Enter X $^\wedge$ 3 $-$ 3X $-$ 5
Press $\boxed{GRAPH}$
Zoom in on the local maximum in quadrant III, $(-1, -3)$. After locating the above point, press $\boxed{ZOOM}$ and select "Standard" to reset the range to standard values. Press $\boxed{ENTER}$ to redraw the graph. Now zoom in on the local minimum in quadrant IV, $(1, -7)$.

33. To graph a piecewise function, press $\boxed{MODE}$. Select Dot. (To place a $<$, $>$, $\leq$, $\geq$, or $\neq$ symbol into a formula, press $\boxed{2nd}$ $\boxed{TEST}$ to display a menu. Choose the correct symbol from the menu.)
Press $\boxed{Y=}$
Enter: $\boxed{(-)}$ 2 $\boxed{(}$ X $\boxed{2nd}$ $\boxed{TEST}$ $\boxed{<}$ $\boxed{(-)}$ 1 $\boxed{)}$ +
2X $\boxed{(}$ $\boxed{(-)}$ 1 $\boxed{2nd}$ $\boxed{TEST}$ $\boxed{\leq}$ X $\boxed{)}$ $\boxed{(}$ X $\boxed{2nd}$ $\boxed{TEST}$ $\boxed{\leq}$ 2 $\boxed{)}$
+ 4 $\boxed{(}$ X $\boxed{2nd}$ $\boxed{TEST}$ $\boxed{>}$ 2 $\boxed{)}$
Press $\boxed{TRACE}$ to display the graph.

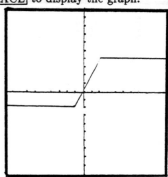

35. $y = 1(x < -1) + x^2(-1 \leq x)(x \leq 2) + 4(x > 2)$
Press $\boxed{MODE}$. Select Dot.
Press $\boxed{Y=}$
Enter: 1 $\boxed{(}$ X $\boxed{2nd}$ $\boxed{TEST}$ $\boxed{<}$ $\boxed{(-)}$ 1 $\boxed{)}$
+ X $\boxed{x^2}$ $\boxed{(}$ $\boxed{(-)}$ 1 $\boxed{2nd}$ $\boxed{TEST}$ $\boxed{\leq}$ X $\boxed{)}$
$\boxed{(}$ X $\boxed{2nd}$ $\boxed{TEST}$ $\boxed{\leq}$ 2 $\boxed{)}$
+ 4 $\boxed{(}$ X $\boxed{2nd}$ $\boxed{TEST}$ $\boxed{>}$ 2 $\boxed{)}$
Press $\boxed{TRACE}$ to display the graph.

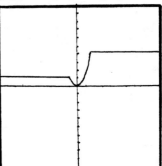

37. Set standard range values.

Press $\boxed{Y=}$

Enter INT x by pressing $\boxed{MATH}\boxed{\triangleright}$ and selecting INT (option 4). Then enter X.

Press $\boxed{GRAPH}$

39. Set standard range values.

Enter: X − INT X (Access INT as in Exercise 37.)

Press $\boxed{GRAPH}$

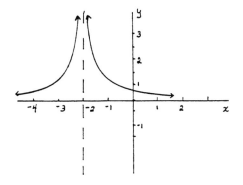

Problem Set 5.4

1. $f(x) = \dfrac{2}{x+2}$

Vertical asymptote: $x = -2$

Horizontal asymptote: $y = 0$

Plot several points on each side of the vertical asymptote.

x	$f(x)$
-4	-1
-3	-2
-1	2
0	1
1	$\frac{2}{3}$

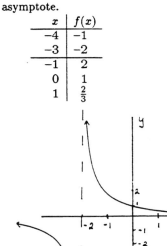

3. $f(x) = \dfrac{2}{(x+2)^2}$

Vertical asymptote: $x = -2$

Horizontal asymptote: $y = 0$

Plot several points on each side of the vertical asymptote.

x	$f(x)$
-4	$\frac{1}{2}$
-3	2
-1	2
0	$\frac{1}{2}$

(See graph, next column.)

5. $f(x) = \dfrac{2x}{x+2}$

Vertical asymptote: $x = -2$

Horizontal asymptote: $y = 2$

(Note that as x gets very large, $f(x) \to \dfrac{2x}{x} = 2$)

Plot several points on each side of the vertical asymptote.

x	$f(x)$
-4	4
-3	6
-1	-2
0	0
1	$\frac{2}{3}$

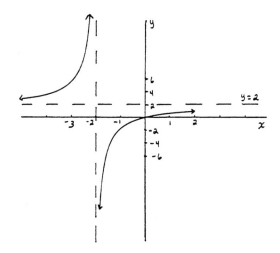

7. $f(x) = \dfrac{1}{(x+2)(x-1)}$

Vertical asymptotes: $x = 1$, $x = -2$

Horizontal asymptote: $y = 0$

Note that $f(x) > 0$ when $x < -2$ or $x > 1$;

$f(x) < 0$ when $-2 < x < 1$.

(See table and graph, next page.)

x	$f(x)$
-4	$\frac{1}{10}$
-3	$\frac{1}{4}$
-1	$-\frac{1}{2}$
0	$-\frac{1}{2}$
2	$\frac{1}{4}$
3	$\frac{1}{10}$

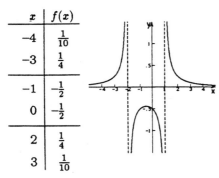

9. $f(x) = \dfrac{x+1}{(x+2)(x-1)}$

Vertical asymptotes: $x = -2$, $x = 1$
Horizontal asymptote: $y = 0$
(As x grows very large, $f(x) \to \dfrac{x}{x^2} = \dfrac{1}{x} \to 0$)

Plot several points on each side of the vertical asymptotes.

x	$f(x)$
-4	$-\frac{3}{10}$
-3	$-\frac{1}{2}$
-1.5	0.4
-1	0
0	$-\frac{1}{2}$
1.5	1.4
2	$\frac{3}{4}$

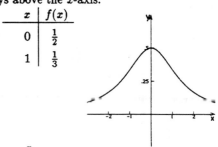

11. $f(x) = \dfrac{2x^2}{(x+2)(x-1)}$

Vertical asymptotes: $x = -2$, $x = 1$
Horizontal asymptote: $y = 2$
(As x grows very large, $f(x) \to \dfrac{2x^2}{x^2} = 2$)

Plot enough points so that you can get a good sense of what is really happening between the asymptotes.

x	$f(x)$
-4	3.2
-3	4.5
-1.5	-3.6
-1	-1
0	0
0.5	-0.4
1.5	2.6
2	2
3	1.8
4	1.77
5	1.78

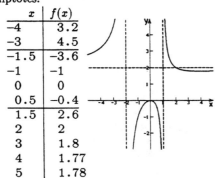

13. $f(x) = \dfrac{1}{x^2 + 2}$

Vertical asymptotes: None; $x^2 + 2$ is always positive.
Horizontal asymptote: $y = 0$
Substituting $-x$ for x, we see that $f(-x) = f(x)$, so it is an even function and symmetric about the y-axis. $f(x)$ has a maximum value when the denominator is as small as possible; that is, when $x = 0$. The function is always positive, and its graph is always above the x-axis.

x	$f(x)$
0	$\frac{1}{2}$
1	$\frac{1}{3}$

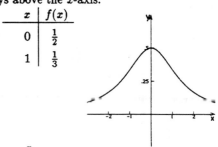

15. $f(x) = \dfrac{x}{x^2 + 2}$

Vertical asymptotes: None; $x^2 + 2$ is always positive.
Horizontal asymptote: $y = 0$
(As x grows very large, $f(x) \to \dfrac{x}{x^2} = \dfrac{1}{x} \to 0$)

Since $f(-x) = \dfrac{-x}{(-x)^2 + 2} = -\dfrac{x}{x^2 + 2} = -f(x)$,
the graph will be symmetric about the origin.
Since $x^2 + 2$ is always positive,
$f(x) > 0$ when $x > 0$ and $f(x) < 0$ when $x < 0$.

x	$f(x)$
0	0
1	$\frac{1}{3}$
2	$\frac{1}{3}$
3	$\frac{3}{11}$

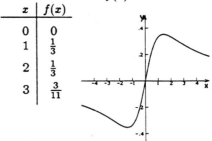

17. $f(x) = \dfrac{2x^2 + 1}{2x} = x + \dfrac{1}{2x}$

Vertical asymptote: $x = 0$
Oblique asymptote: $y = x$
(As x grows very large, $f(x) = x + \dfrac{1}{2x} \to x + 0 = x$)

Plot several points on each side of the vertical asymptote. Note that $f(x) > 0$ when $x > 0$ and $f(x) < 0$ when $x < 0$. Since $f(x) = -f(-x)$, the graph is symmetric about the origin.
(See table and graph, next page.)

98

x	$f(x)$
0.5	1.5
1	1.5
2	2.25
-0.5	1.5
-1	-1.5
-2	-2.25

19. $f(x) = \dfrac{x^2}{x-1} = x + 1 + \dfrac{1}{x-1}$ (By long division)

Vertical asymptote: $x = 1$
Oblique asymptote: $y = x + 1$
(As x grows very large,

$$f(x) = x + 1 + \frac{1}{x-1} \to x + 1 + 0 = x + 1)$$

Plot several points on each side of the vertical asymptote, noting that when $x > 1$, $f(x) > 0$, and when $x < 1$, $f(x) \le 0$.

x	$f(x)$
0.5	-0.5
0	0
-1	-0.5
1.5	4.5
2	4
3	4.5
4	5.3

21. $f(x) = \dfrac{(x+2)(x-4)}{x+2}$
$= x - 4$, when $x \ne -2$
The function is undefined at $x = -2$, but $x = -2$ is not an asymptote. The graph is the straight line $f(x) = x - 4$, with the single point $(-2, -6)$ deleted.

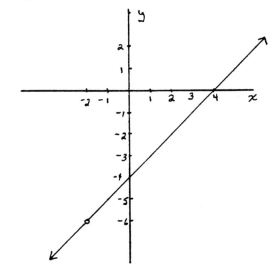

23. $f(x) = \dfrac{x^3 - x^2 - 12x}{x+3}$
$= \dfrac{x(x^2 - x - 12)}{x+3}$
$= \dfrac{x(x-4)(x+3)}{x+3}$
$= x(x-4)$, when $x \ne -3$.
The function is undefined when $x = -3$, but $x = -3$ is not an asymptote. The graph is the parabola $f(x) = x^2 - 4x$, with the single point $(-3, 21)$ deleted.

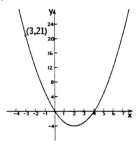

25. $f(x) = \dfrac{x}{x+5}$
Vertical asymptote: $x = -5$
Horizontal asymptote: $y = 1$
(As x grows very large, $f(x) \to \dfrac{x}{x} = 1$)

Plot several points on each side of the vertical asymptote.

x	$f(x)$
-7	$\frac{7}{2}$
-6	6
-4	-4
0	0
1	$\frac{1}{6}$

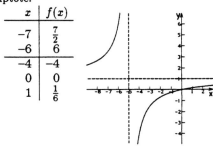

27. $f(x) = \dfrac{x^2 - 9}{x^2 - x - 2} = \dfrac{(x-3)(x+3)}{(x-2)(x+1)}$
Vertical asymptotes: $x = 2$, $x = -1$
Horizontal asymptote: $y = 1$
(As x grows very large, $f(x) \to \dfrac{x^2}{x^2} = 1$)

Plot several points on each side of each asymptote, being sure to include the x-intercepts: $x = 3$ and $x = -3$.
(See table and graph, next page.)

x	$f(x)$
-2	-1.25
-3	0
-4	0.34
-0.5	7
0	4.5
1	4
1.5	5.4
2.5	-1.6
3	0
4	0.7

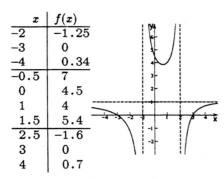

29. $f(x) = \dfrac{x^2 - 9}{x^2 - x - 6} = \dfrac{(x+3)(x-3)}{(x-3)(x+2)}$

$\qquad = \dfrac{x+3}{x+2}, \ x \neq 3$

Vertical asymptote: $x = -2$
Note that although the function is not defined at $x = 3$, $x = 3$ is not an asymptote. The graph will

be the graph of the rational function $f(x) = \dfrac{x+3}{x+2}$, with the point $(3, \frac{6}{5})$ deleted.
Horizontal asymptote: $y = 1$
(As x grows very large, $f(x) \to \frac{x}{x} = 1$)

Plot several points on each side of the vertical asymptote.

x	$f(x)$
-4	$\frac{1}{2}$
-3	0
-2.5	-1
-1	2
0	$\frac{3}{2}$
1	$\frac{4}{3}$

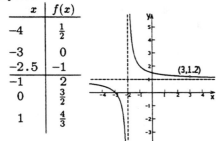

31. a) $f(x) = \dfrac{x}{x^2 + 1}$

Vertical asymptote: None; $x^2 + 1$ is always positive.
Horizontal asymptote: $y = 0$
(As x grows very large, $f(x) \to \frac{x}{x^2} = \frac{1}{x} \to 0$.)
Since $f(-x) = -f(x)$, the graph is symmetric about the origin. $f(x) > 0$ when $x > 0$, and $f(x) < 0$ when $x < 0$.

x	$f(x)$
-2	$-\frac{2}{5}$
-1	$-\frac{1}{2}$
0	0
1	$\frac{1}{2}$

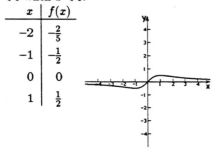

b) $f(x) = \dfrac{x^2}{x^2 + 1}$

Vertical asymptote: None
Horizontal asymptote: $y = 1$
(As x grows very large, $f(x) \to \frac{x^2}{x^2} = 1$)
Since $f(x) = f(-x)$, the graph is symmetric about the y-axis. Note that $f(x) \geq 0$ for all x.

x	$f(x)$
-2	$\frac{4}{5}$
-1	$\frac{1}{2}$
0	0
1	$\frac{1}{2}$

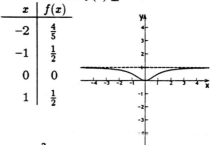

c) $f(x) = \dfrac{x^3}{x^2 + 1} = x - \dfrac{x}{x^2 + 1}$

Vertical asymptotes: None
Oblique asymptote: $y = x$
(As x grows very large,

$f(x) \to x - \dfrac{x}{x^2} = x - \dfrac{1}{x} \to x - 0 = x$)

Since $f(x) = -f(x)$, the graph is symmetric about the origin. $f(x) > 0$ when $x > 0$ and $f(x) < 0$ when $x < 0$.

x	$f(x)$
-2	$-\frac{8}{5}$
-1	$-\frac{1}{2}$
0	0
1	$\frac{1}{2}$

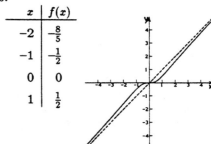

33. a) $f(x) = \dfrac{x^4}{x - 1}$

Vertical asymptote: $x = 1$
Horizontal asymptote: None (As x grows very large, so does $f(x)$.)
Oblique asymptote: None (By long division,

$f(x) = x^3 + x^2 + x + \dfrac{1}{x - 1}$. As x grows very large,

$f(x)$ does not approach the equation of a straight line.)

b) $f(x) = \dfrac{x^4}{x^2 - 1} = \dfrac{x^4}{(x+1)(x-1)}$
Vertical asymptotes: $x = -1$, $x = +1$
Horizontal asymptotes: None (As x grows very large, so does $f(x)$.)

Oblique asymptote: None (By long division,

$f(x) = x^2 + 1 + \dfrac{1}{x^2 - 1}$. As x grows very large, $f(x)$

does not approach the equation of a straight line.)

c) $f(x) = \dfrac{x^4}{x^3 - 1}$

Vertical asymptote: $x = 1$

Horizontal asymptote: None (As x grows very large, so does $f(x)$.)

Oblique asymptote: $y = x$ (By long division,

$f(x) = x + \dfrac{x}{x^3 - 1}$. As x grows very large,

$f(x) \to x + 0 = x$)

d) $f(x) = \dfrac{x^4}{x^4 - 1} = \dfrac{x^4}{(x^2 + 1)(x + 1)(x - 1)}$

Vertical asymptotes: $x = 1$ and $x = -1$

Horizontal asymptote: $y = 1$ (As x grows very large, $f(x) \to \dfrac{x^4}{x^4} = 1$)

Oblique asymptote: None

e) $f(x) = \dfrac{x^4}{x^5 - 1}$

Vertical asymptote: $x = 1$

Horizontal asymptote: $y = 0$ (As x grows very large, $f(x) \to \dfrac{x^4}{x^5} = \dfrac{1}{x} \to 0$)

Oblique asymptote: None

f) $f(x) = \dfrac{x^4}{x^6 - 1} = \dfrac{x^4}{(x^3 - 1)(x^3 + 1)}$

Vertical asymptotes: $x = 1$, $x = -1$

Horizontal asymptotes: $y = 0$ (As x grows very large, $f(x) \to \dfrac{x^4}{x^6} = \dfrac{1}{x^2} \to 0$)

Oblique asymptote: None

35. The cost of manufacturing x gizmos is $20{,}000 + 50x$
The cost per gizmo

$$= U(x) = \dfrac{20{,}000 + 50x}{x} = 50 + \dfrac{20{,}000}{x}$$

$U(x)$ has a horizontal asymptote at $U = 50$ and a vertical asymptote at $x = 0$.

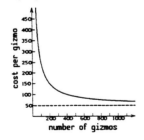

37. a) Since 20 liters of solution flow into the tank each minute, and each liter contains 50 grams of salt, $50(20) = 1000$ grams of salt enter the tank each minute.

$A(t) = 1000t$

b) After t minutes, there are $400 + 20t$ liters of fluid in the tank and $1000t$ grams of salt.

$$c(t) = \dfrac{1000t}{400 + 20t} = \dfrac{50t}{20 + t}$$

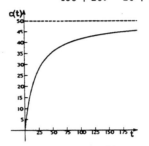

c) As $t \to \infty$, $c(t) \to \dfrac{50t}{t} = 50$.

39. If $x = 3$ is an asymptote, some term will have to include a factor of $x - 3$. If $y = 2x + 3$ is to be an asymptote for a large value of x, the terms $2x + 3$ will have to be present. Hence initially we know that $f(x) = \dfrac{A}{x - 3} + 2x + 3$, where A is to be determined. Substituting coordinates $(2, 5)$ in this equation,

$$5 = \dfrac{A}{2 - 3} + 2 \cdot 2 + 3$$
$$5 = -A + 7$$
$$A = 2$$

The function is $f(x) = \dfrac{2}{x - 3} + 2x + 3$.

Combining, $y = f(x) = \dfrac{2 + (x - 3)(2x + 3)}{x - 3}$

$$y = \dfrac{2 + 2x^2 - 3x - 9}{x - 3} = \dfrac{2x^2 - 3x - 7}{x - 3}$$

41. Set standard range values.
Press $\boxed{Y=}$
Enter $\boxed{(}x + 1\boxed{)}\,\boxed{\div}\,\boxed{(}x - 2\boxed{)}$ as Y_1
Enter X^2 + 2 as Y_2
Press $\boxed{\text{GRAPH}}$
Zoom in on the point of intersection, $(2.433, 7.922)$.

Problem Set 5.5

1. Let $f(x) = x^2 - 2x + 2$ and $g(x) = \dfrac{2}{x}$.

a) $(f + g)(2) = 2^2 - 2 \cdot 2 + 2 + \dfrac{2}{2} = 4 - 4 + 2 + 1 = 3$

b) $(f+g)(0) = 0^2 - 2 \cdot 0 + 2 + \frac{2}{0}$ = undefined or no number

c) $(f-g)(1) = 1^2 - 2 \cdot 1 + 2 - \frac{2}{1} = 1 - 2 + 2 - 2 = -1$

d) $(f \cdot g)(-1) = [(-1)^2 - 2(-1) + 2] \cdot \frac{2}{-1}$
$= (1 + 2 + 2)(-2) = -10$

e) $\left(\frac{f}{g}\right)(2) = \frac{2^2 - 2 \cdot 2 + 2}{\frac{2}{2}} = \frac{4 - 4 + 2}{1} = \frac{2}{1} = 2$

f) $\left(\frac{g}{f}\right)(2) = \frac{\frac{2}{2}}{2^2 - 2 \cdot 2 + 2} = \frac{1}{4 - 4 + 2} = \frac{1}{2}$

g) $(f \circ g)(-1) = f[g(-1)] = f(\frac{2}{-1}) = f(-2)$
$= (-2)^2 - 2(-2) + 2 = 4 + 4 + 2 = 10$

h) $(g \circ f)(-1) = g[f(-1)]$
$= g[(-1)^2 - 2(-1) + 2] = g(5) = \frac{2}{5}$

i) $(g \circ g)(3) = g[g(3)] = g\left(\frac{2}{3}\right) = \frac{2}{\frac{2}{3}} = 3$

3. Let $f(x) = x^2$ and $g(x) = x - 2$.
$(f+g)(x) = x^2 + x - 2$; Domain: all real x
$(f-g)(x) = x^2 - (x-2) = x^2 - x + 2$; D: all real x
$(f \cdot g)(x) = x^2(x-2) = x^3 - 2x$; Domain: all real x
$\left(\frac{f}{g}\right)(x) = \frac{x^2}{x-2}$; Domain: all real x except $x = 2$

5. Let $f(x) = x^2$ and $g(x) = \sqrt{x}$
$(f+g)(x) = x^2 + \sqrt{x}$; Domain: all real $x \geq 0$
$(f-g)(x) = x^2 - \sqrt{x}$; Domain: all real $x \geq 0$
$(f \cdot g)(x) = x^2\sqrt{x}$; Domain: all real $x \geq 0$
$\left(\frac{f}{g}\right)(x) = \frac{x^2}{\sqrt{x}}$; Domain: all real $x > 0$

7. Let $f(x) = \frac{1}{x-2}$ and $g(x) = \frac{x}{x-3}$
Domain of $f(x)$: all real x except $x = 2$
Domain of $g(x)$: all real x except $x = 3$

a) $(f+g)(x) = \frac{1}{x-2} + \frac{x}{x-3} = \frac{x^2 - x - 3}{(x-2)(x-3)}$
Domain: all real x except $x = 2, 3$

b) $(f-g)(x) = \frac{1}{x-2} - \frac{x}{x-3} = \frac{-x^2 + 3x - 3}{(x-2)(x-3)}$
Domain: all real x except $x = 2, 3$

c) $(f \cdot g)(x) = \frac{1}{x-2} \cdot \frac{x}{x-3} = \frac{x}{(x-2)(x-3)}$
Domain: all real x except $x = 2, 3$

d) $\left(\frac{f}{g}\right)(x) = \frac{\frac{1}{x-2}}{\frac{x}{x-3}} = \frac{x-3}{x(x-2)}$

Domain: all real x except $x = 0, 2, 3$. [0 is excluded because $g(0) = 0$, and division by 0 is not defined. x also appears in the denominator of the final result, which would also exclude $x = 0$. Similarly, 2 is excluded since $(x-2)$ is a factor in the final denominator. $x = 3$ is excluded even though in the final result it leads to the perfectly proper number 0; $g(3)$ is itself undefined, so any function involving $g(x)$ is undefined at $x = 3$.

9. $f(x) = x^2$ and $g(x) = x - 2$

x	$f(x)$	$g(x)$	$(f+g)(x)$
-2	4	-4	0
-1	1	-3	-2
0	0	-2	-2
1	1	-1	0
2	4	0	4

11. $f(x) = \frac{1}{x}$ and $g(x) = x$

x	$f(x)$	$g(x)$	$(f-g)(x)$
-3	$-\frac{1}{3}$	-3	$2\frac{2}{3}$
-2	$-\frac{1}{2}$	-2	$1\frac{1}{2}$
-1	-1	-1	0
$-\frac{1}{2}$	-2	$-\frac{1}{2}$	$-1\frac{1}{2}$
0	undefined	0	undefined

Since the graph will be symmetric about the origin, we don't need any more points to get a good sense of the graph.

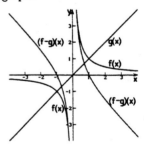

102

13. Let $f(x) = x^2$ and $g(x) = x - 2$.
a) $(g \circ f)(x) = g(f(x)) = g(x^2) = x^2 - 2$
Domain: all real x

b) $(f \circ g)(x) = f(g(x)) = f(x - 2) = (x - 2)^2$
Domain: all real x

15. Let $f(x) = \frac{1}{x}$ and $g(x) = x + 3$
a) $(g \circ f)(x) = g(f(x)) = g\left(\frac{1}{x}\right) = \frac{1}{x} + 3 = \frac{3x + 1}{x}$
Domain: all real x, except $x = 0$

b) $(f \circ g)(x) = f(g(x)) = f(x + 3) = \frac{1}{x + 3}$
Domain: all real x, except $x = -3$

17. Let $f(x) = \sqrt{x - 2}$ and $g(x) = x^2 - 2$
a) $(g \circ f)(x) = g(f(x))$
$= g(\sqrt{x - 2})$
$= (\sqrt{x - 2})^2 - 2$
$= x - 4$
Domain: all real $x \geq 2$, because of the restriction on $f(x)$.

b) $(f \circ g)(x) = f(g(x))$
$= f(x^2 - 2)$
$= \sqrt{x^2 - 2 - 2}$
$= \sqrt{x^2 - 4}$
Domain: all real x such that $x^2 - 4 \geq 0$; that is $|x| \geq 2$

19. Let $f(x) = 2x - 3$ and $g(x) = \frac{1}{2}(x + 3)$
a) $(g \circ f)(x) = g(f(x))$
$= g(2x - 3)$
$= \frac{1}{2}(2x - 3 + 3)$
$= x$
Domain: all real x

b) $(f \circ g)(x) = f(g(x))$
$= f(\frac{1}{2}(x + 3))$
$= 2[\frac{1}{2}(x + 3)] - 3$
$= x$
Domain: all real x

21. $H(x) = (x + 4)^3$
First you add 4 to x, so let $f(x) = x + 4$. Then you cube the result, so let $g(x) = x^3$.

23. $H(x) = \sqrt{x + 2}$
First you add 2 to x, so let $f(x) = x + 2$. Then you take the square root of the result, so let $g(x) = \sqrt{x}$.

25. $H(x) = \frac{1}{(2x + 5)^3}$. First you double x and add 5, so $f(x) = 2x + 5$. Then you raise the result to the power -3, so $g(x) = x^{-3} = \frac{1}{x^3}$.

27. $H(x) = |x^3 - 4|$
First you cube x and subtract 4, so $f(x) = x^3 - 4$. Then you take the absolute value of the result, so $g(x) = |x|$.

29. a) $f(x) = x^2$

b) $g(x) = (x - 2)^2$
Translate $f(x)$ 2 units to the right.

c) $h(x) = x^2 - 4$
Translate $f(x)$ 4 units down.

d) $j(x) = (x - 2)^2 + 1$
Translate $f(x)$ 2 units right and 1 unit up.

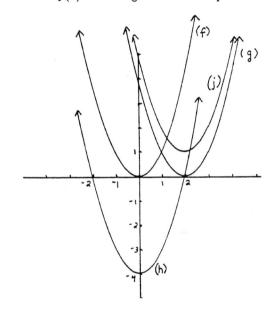

31. a) $f(x) = \sqrt{x}$

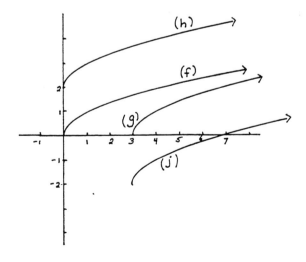

b) $g(x) = \sqrt{x-3}$
Translate $f(x)$ 3 units to the right.

c) $h(x) = \sqrt{x} + 2$
Translate $f(x)$ 2 units up.

d) $j(x) = \sqrt{x-3} - 2$
Translate $f(x)$ 3 units right and 2 units down.

33. Let $f(x) = 2x + 3$ and $g(x) = x^3$
a) $(f+g)(x) = x^3 + 2x + 3$

b) $(g-f)(x) = x^3 - (2x+3) = x^3 - 2x - 3$

c) $(f \cdot g)(x) = (2x+3)(x^3) = 2x^4 + 3x^3$

d) $\left(\dfrac{f}{g}\right) = \dfrac{2x+3}{x^3}$

e) $(f \circ g)(x) = f(g(x)) = f(x^3) = 2x^3 + 3$

f) $(g \circ f)(x) = g(f(x)) = g(2x+3) = (2x+3)^3$

g) $(f \circ f)(x) = f(f(x))$
$ = f(2x+3)$
$ = 2(2x+3) + 3$
$ = 4x + 9$

h) $(g \circ g \circ g)(x) = g(g(g(x)))$
$ = g(g(x^3))$
$ = g([x^3]^3)$
$ = g(x^9)$
$ = (x^9)^3$
$ = x^{27}$

35. Let $f(x) = x^2 - 4$, $g(x) = |x|$, and $h(x) = \dfrac{1}{x}$.

$(h \circ g \circ f)(x) = h(g(f(x)))$
$ = h(g(x^2 - 4))$
$ = h(|x^2 - 4|)$

$ = \dfrac{1}{|x^2 - 4|}$

Domain: all real x except $x = 2$ and $x = -2$

37. Define difference quotient $D(x)$:
$D(x) = \dfrac{f(x+h) - f(x)}{h}$ where h is a constant.

a) If $f(x) = x^2$, then
$D(x) = \dfrac{(x+h)^2 - x^2}{h}$

$ = \dfrac{x^2 + 2hx + h^2 - x^2}{h}$

$ = \dfrac{2hx + h^2}{h}$

$ = 2x + h$

b) If $f(x) = 2x + 3$, then
$D(x) = \dfrac{2(x+h) + 3 - (2x+3)}{h}$

$ = \dfrac{2x + 2h + 3 - 2x - 3}{h}$

$ = \dfrac{2h}{h} = 2$

c) If $f(x) = \dfrac{1}{x}$, then

$D(x) = \dfrac{\dfrac{1}{x+h} - \dfrac{1}{x}}{h}$

$ = \dfrac{\dfrac{x - x - h}{x(x+h)}}{h}$

$ = \dfrac{-h}{x(x+h)} \cdot \dfrac{1}{h}$

$ = \dfrac{-1}{x(x+h)}$

d) If $f(x) = \dfrac{2}{x-2}$,

$D(x) = \dfrac{\dfrac{2}{x+h-2} - \dfrac{2}{x-2}}{h}$

$ = \dfrac{\dfrac{2x - 4 - 2x - 2h + 4}{(x+h-2)(x-2)}}{h}$

$ = \dfrac{-2h}{h(x+h-2)(x-2)}$

$ = \dfrac{-2}{(x+h-2)(x-2)}$

39. a) $f(x) = x^2$ and $g(x) = \sqrt{x}$
$(f \circ g)(x) = f(g(x)) = f(\sqrt{x}) = (\sqrt{x})^2 = x$
$(g \circ f)(x) = g(f(x)) = g(x^2) = \sqrt{x^2} = |x|$

b) $f(x) = x^3$ and $g(x) = \sqrt[3]{x}$
$(f \circ g)(x) = f(g(x)) = f(\sqrt[3]{x}) = (\sqrt[3]{x})^3 = x$
$(g \circ f)(x) = g(f(x)) = g(x^3) = \sqrt[3]{x^3} = x$

c) $f(x) = x^2$ and $g(x) = x^3$
$(f \circ g)(x) = f(g(x)) = f(x^3) = (x^3)^2 = x^6$
$(g \circ f)(x) = g(f(x)) = g(x^2) = (x^2)^3 = x^6$

d) $f(x) = x^2$ and $g(x) = \dfrac{1}{x^3}$

$(f \circ g)(x) = f(g(x)) = f\left(\dfrac{1}{x^3}\right) = \left(\dfrac{1}{x^3}\right)^2 = \dfrac{1}{x^6}$

$(g \circ f)(x) = g(f(x)) = g(x^2) = \dfrac{1}{(x^2)^3} = \dfrac{1}{x^6}$

41. Let $f(x) = \left[\dfrac{1-\sqrt{x}}{1+\sqrt{x}}\right]^2$. Then

$$f(f(x)) = f\left(\left(\dfrac{1-\sqrt{x}}{1+\sqrt{x}}\right)^2\right)$$

$$= \left[\dfrac{1 - \sqrt{\left(\dfrac{1-\sqrt{x}}{1+\sqrt{x}}\right)^2}}{1 + \sqrt{\left(\dfrac{1-\sqrt{x}}{1+\sqrt{x}}\right)^2}}\right]^2$$

$$= \left[\dfrac{1 - \left(\dfrac{1-\sqrt{x}}{1+\sqrt{x}}\right)}{1 + \left(\dfrac{1-\sqrt{x}}{1+\sqrt{x}}\right)}\right]^2$$

$$= \left[\dfrac{1 + \sqrt{x} - 1 + \sqrt{x}}{1 + \sqrt{x} + 1 - \sqrt{x}}\right]^2$$

$$= \left[\dfrac{2\sqrt{x}}{2}\right]^2$$

$$= (\sqrt{x})^2$$

$$= x$$

So,
$$x^2 + \tfrac{1}{4} = x$$
$$x^2 - x + \tfrac{1}{4} = 0$$
$$(x - \tfrac{1}{2})^2 = 0$$
$$x - \tfrac{1}{2} = 0$$
$$x = \tfrac{1}{2}$$

43. a) Since Steven walked for t hours, $x = 3.5t$
Since Carol walked for $t - 1$ hours, $y = 4(t-1)$

b) $d = \sqrt{x^2 + y^2}$

c) $d = \sqrt{(3.5t)^2 + [4(t-1)]^2}$
$ = \sqrt{12.25t^2 + 16(t-1)^2}$

d) At 4:30 P.M., $t = 4.5$
$$d = \sqrt{12.25(4.5)^2 + 16(3.5)^2}$$
$$= \sqrt{444.0625}$$
$$= 21.07 \text{ miles}$$

45. $f(x) = |x+1| - |x| + |x-1|$

x	$f(x)$
-2	$1 - 2 + 3 = 2$
-1	$0 - 1 + 2 = 1$
0	$1 - 0 + 1 = 2$
$+1$	$2 - 1 + 0 = 1$
$+2$	$3 - 2 + 1 = 2$

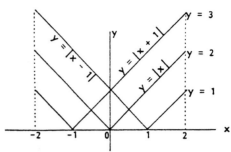

In each unit interval $[-2, -1]$, $[-1, 0]$, etc., the three functions are each linear. Hence the sum of these linear functions is also linear and each segment is a straight line. The points at the ends of these segments are $(-2, 2)$, $(-1, 1)$, $(0, 2)$, $(1, 1)$, and $(2, 2)$.

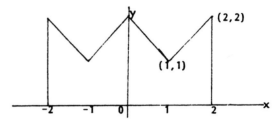

The area of each trapezoid is $\tfrac{3}{2}$, and the total area from $x = -2$ to $x = 2$ is $4 \cdot \tfrac{3}{2} = 6$.

47. a) Let $h(x) = f(x)g(x)$.
Then $\begin{aligned}[t] h(-x) &= f(-x)g(-x) \\ &= f(x)[-g(x)] \\ &= -f(x)g(x) \\ &= -h(x) \end{aligned}$
$h(x)$ is an odd function.

b) Let $h(x) = \dfrac{f(x)}{g(x)}$
Then $\begin{aligned}[t] h(-x) &= \dfrac{f(-x)}{g(-x)} \\[4pt] &= \dfrac{f(x)}{-g(x)} \\[4pt] &= -\dfrac{f(x)}{g(x)} \\[4pt] &= -h(x) \end{aligned}$
$h(x)$ is an odd function.

c) Let $h(x) = [g(x)]^2$
Then $\begin{aligned}[t] h(-x) &= [g(-x)]^2 \\ &= [-g(x)]^2 \\ &= [g(x)]^2 \\ &= h(x) \end{aligned}$
$h(x)$ is an even function.

d) Let $h(x) = [g(x)]^3$
Then $h(-x) = [g(-x)]^3 = [-g(x)]^3$
$\qquad = -[g(x)]^3$
$\qquad = -h(x)$
$h(x)$ is an odd function.

e) Let $h(x) = f(x) + g(x)$
Then $h(-x) = f(-x) + g(-x)$
$\qquad = f(x) - g(x)$
$h(-x)$ equals neither $h(x)$ nor $-h(x)$, and is therefore neither an even nor an odd function.

f) Let $h(x) = g(g(x))$
Then $h(-x) = g(g(-x)) = g(-g(x))$
$\qquad = -g(g(x))$
$\qquad = -h(x)$
$h(x)$ is an odd function.

g) Let $h(x) = f(f(x))$
Then $h(-x) = f(f(-x))$
$\qquad = f(f(x))$
$\qquad = h(x)$
$h(x)$ is an even function.

h) Let $h(x) = 3f(x) + [g(x)]^2$
Then $h(-x) = 3f(-x) + [g(-x)]^2$
$\qquad = 3f(x) + [-g(x)]^2$
$\qquad = 3f(x) + [g(x)]^2$
$\qquad = h(x)$
$h(x)$ is an even function.

i) Let $h(x) = g(x) + g(-x)$
Then $h(-x) = g(-x) + g[-(-x)]$
$\qquad = -g(x) - g(-x)$
$\qquad = -[g(x) + g(-x)]$
$\qquad = -h(x)$
$h(x)$ is an even function.

49. a)

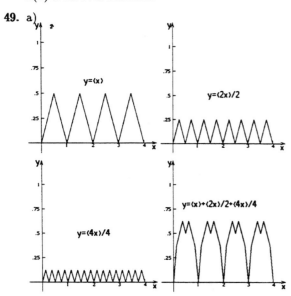

b) The area under $y = f(x)$ is $4(\frac{1}{2})(1)(\frac{1}{2}) = 1$
$\qquad$ (4 triangles with base 1 and height $\frac{1}{2}$)

The area under $y = g(x)$ is $8(\frac{1}{2})(\frac{1}{2})(\frac{1}{4}) = \frac{1}{2}$
$\qquad$ 8 triangles with base $\frac{1}{2}$ and height $\frac{1}{4}$)

The area under $y = h(x)$ is $16(\frac{1}{2})(\frac{1}{4})(\frac{1}{8}) = \frac{1}{4}$

The area under $y = F(x)$ is $1 + \frac{1}{2} + \frac{1}{4} = 1\frac{3}{4}$

51. Set standard range values.
Press [Y=]
Enter [√][ABS]x as Y$_1$
Enter [(][X$^\wedge$3$-$4[)][÷][(][1$+$X$^\wedge$2[)] as Y$_2$
Enter [√][ABS]X$+$[(][X$^\wedge$3$-$4[)][÷][(]1$+$X$^\wedge$2[)]as Y$_3$
Press [GRAPH]
Zoom in on the intersection of the first two graphs, $(2.481, 1.575)$.

Problem Set 5.6

1. a) The graphs which are graphs of functions with x as domain variable are i, ii, iii, iv, vii, and viii.

b) The functions which are one-to-one are i, ii, and viii.

c) The functions which have inverses are i, ii, and viii.

3. Let $f(x) = 3x - 2$.
a) $f^{-1}(1) = a$ if $f(a) = 1$, that is,
$\qquad 3a - 2 = 1$
$\qquad\quad 3a = 3$
$\qquad\quad\ a = 1$
$f^{-1}(1) = 1$

b) $f^{-1}(-3) = a$ if $f(a) = -3$, that is,
$\qquad 3a - 2 = -3$
$\qquad\quad 3a = -1$
$\qquad\quad\ a = -\frac{1}{3}$
$f^{-1}(-3) = -\frac{1}{3}$

c) $f^{-1}(14) = a$ if $f(a) = 14$, that is,
$\qquad 3a - 2 = 14$
$\qquad\quad 3a = 16$
$\qquad\quad\ a = \frac{16}{3}$
$f^{-1}(14) = \frac{16}{3}$

5. Let $y = 5x$
Solve for x:
$$x = \frac{y}{5}$$
$$f^{-1}(y) = \frac{y}{5}$$
Replace y with x:
$$f^{-1}(x) = \frac{x}{5}$$
Check:
$$f(f^{-1}(x)) = f\left(\frac{x}{5}\right)$$
$$= 5\left(\frac{x}{5}\right)$$
$$= x$$

7. Let $y = 2x - 7$
Solve for x:
$$2x = y + 7$$
$$x = \frac{y+7}{2}$$
$$f^{-1}(y) = \frac{y+7}{2}$$
Replace y with x:
$$f^{-1}(x) = \frac{x+7}{2}$$
Check:
$$f(f^{-1}(x)) = f\left(\frac{x+7}{2}\right)$$
$$= 2\left(\frac{x+7}{2}\right) - 7$$
$$= x$$

9. Let $y = \sqrt{x} + 2$
Solve for x:
$$\sqrt{x} = y - 2$$
$$x = (y-2)^2$$
$$f^{-1}(y) = (y-2)^2$$
Replace y with x:
$$f^{-1}(x) = (x-2)^2$$
Check:
$$f(f^{-1}(x)) = f[(x-2)^2]$$
$$= \sqrt{(x-2)^2} + 2$$
$$= x - 2 + 2$$
$$= x$$
Notice that in the original function, $y \geq 2$; therefore, in $f^{-1}(x)$, $x \geq 2$.

11. Let $y = \frac{x}{x-3}$.
Solve for x:
$$yx - 3y = x$$
$$yx - x = 3y$$
$$x(y-1) = 3y$$

$$x = \frac{3y}{y-1}$$
$$f^{-1}(y) = \frac{3y}{y-1}$$
Replace y with x:
$$f^{-1}(x) = \frac{3x}{x-1}$$
Check:
$$f(f^{-1}(x)) = f\left(\frac{3x}{x-1}\right)$$
$$= \frac{\frac{3x}{x-1}}{\frac{3x}{x-1} - 3}$$
$$= \frac{3x}{3x - 3(x-1)}$$
$$= \frac{3x}{3}$$
$$= x$$

13. Let $y = (x-2)^3 + 2$
Solve for x:
$$(x-2)^3 = y - 2$$
$$x - 2 = \sqrt[3]{y-2}$$
$$x = 2 + \sqrt[3]{y-2}$$
$$f^{-1}(y) = 2 + \sqrt[3]{y-2}$$
Replace y with x:
$$f^{-1}(x) = 2 + \sqrt[3]{x-2}$$
Check:
$$f(f^{-1}(x)) = f(2 + \sqrt[3]{x-2})$$
$$= (2 + \sqrt[3]{x-2} - 2)^3 + 2$$
$$= \sqrt[3]{x-2})^3 + 2$$
$$= x - 2 + 2$$
$$= x$$

15. $f(x) = \sqrt{x} + 2,\ x \geq 0$
$f^{-1}(x) = (x-2)^2,\ x \geq 2$

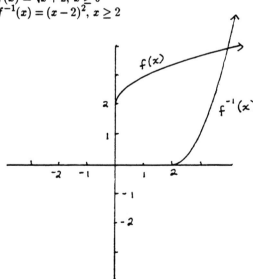

17. Draw the line $y = x$.
Reflect the given graph through that line.

a)

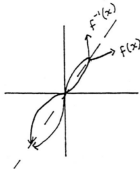

b)

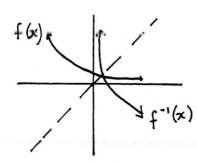

c)

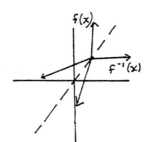

19. Show that $f(x) = \dfrac{3x}{x+2}$ and $g(x) = \dfrac{2x}{3-x}$ are inverses of one another. To do this, we show that $f[g(x)] = x$ and $g[f(x)] = x$.

$$f[g(x)] = \frac{3\dfrac{2x}{3-x}}{\dfrac{2x}{3-x}+2}$$

$$= \frac{\dfrac{3\cdot 2x}{3-x}}{\dfrac{2x}{3-x}+2}\cdot\frac{3-x}{3-x}$$

$$= \frac{6x}{2x+2(3-x)}$$

$$= \frac{6x}{2x+6-2x} = \frac{6x}{6} = x$$

$$g[f(x)] = \frac{2\left(\dfrac{3x}{x+2}\right)}{3-\dfrac{3x}{x+2}}$$

$$= \frac{\dfrac{6x}{x+2}}{3-\dfrac{3x}{x+2}}\cdot\frac{x+2}{x+2}$$

$$= \frac{6x}{3(x+2)-3x}$$

$$= \frac{6x}{3x+6-3x}$$

$$= \frac{6x}{6} = x$$

21. $f(x) = (x-1)^2$
This is a parabola opening upward with vertex at $(1,0)$.
Restrict the domain to $x \geq 1$; the range is $y \geq 0$.
To find $f^{-1}(x)$, let
$$y = (x-1)^2$$
Solve for x:
$$x-1 = \pm\sqrt{y}$$
$$x = 1 \pm \sqrt{y}$$
Since we restricted the domain to $x \geq 1$, we choose
$$x = 1 + \sqrt{y}$$
$$f^{-1}(y) = 1 + \sqrt{y}$$
Replace y with x:
$$f^{-1}(x) = 1 + \sqrt{x}$$
Note: The domain of $f^{-1}(x)$ is the range of $f(x)$; that is, $x \geq 0$. The range of $f^{-1}(x)$ is the domain of $f(x)$; that is, $y \geq 1$.

23. $f(x) = (x+1)^2 - 4$
This is a parabola opening upward with vertex at $(-1,-4)$.
Restrict the domain to $x \geq -1$; the range is $y \geq -4$.
To find $f^{-1}(x)$, let
$$y = (x+1)^2 - 4$$
Solve for x:
$$(x+1)^2 = y+4$$
$$x+1 = \pm\sqrt{y+4}$$
$$x = -1 \pm \sqrt{y+4}$$
Since we restricted the domain to $x \geq -1$, we choose
$$x = -1 + \sqrt{y+4}$$
$$f^{-1}(y) = -1 + \sqrt{y+4}$$
Replace y with x:
$$f^{-1}(x) = -1 + \sqrt{x+4}$$
Note: The domain of $f^{-1}(x)$ is the range of $f(x)$; that is, $x \geq -4$. The range of $f^{-1}(x)$ is the domain of $f(x)$; that is, $y \geq -1$.

25. $f(x) = x^2 + 6x + 7$

This is a parabola opening upward with vertex at
$\left(\frac{-b}{2a}, f\left(\frac{-b}{2a}\right)\right) = (-3, -2)$.

Restrict the domain to $x \geq -3$; the range is $y \geq -2$.
To find $f^{-1}(x)$, let
$$y = x^2 + 6x + 7$$
Solve for x by completing the square:
$$y = x^2 + 6x + 9 - 9 + 7$$
$$y = (x + 3)^2 - 2$$
$$x + 3 = \pm\sqrt{y + 2}$$
$$x = -3 \pm \sqrt{y + 2}$$
Since we restricted the domain to $x \geq -3$, we choose
$$x = -3 + \sqrt{y + 2}$$
$$f^{-1}(y) = -3 + \sqrt{y + 2}$$
Replace y with x:
$$f^{-1}(x) = -3 + \sqrt{x + 2}$$
Note: The domain of $f^{-1}(x)$ is the range of $f(x)$;
that is, $x \geq -2$. The range of $f^{-1}(x)$ is the domain
of $f(x)$; that is, $y \geq -3$.

27. $f(x) = |x + 2|$

This is a V-shaped absolute value function with
vertex at $(-2, 0)$.
Restrict the domain to $x \geq -2$; the range is $y \geq 0$.
To find $f^{-1}(x)$, let $y = |x + 2|$.
With the restriction on x, $|x + 2| = x + 2$, so
$$y = x + 2$$
Solve for x:
$$x = y - 2$$
$$f^{-1}(y) = y - 2$$
Replace y with x:
$$f^{-1}(x) = x - 2$$
Note: The domain of $f^{-1}(x)$ is the range of $f(x)$;
that is, $x \geq 0$. The range of $f^{-1}(x)$ is the domain of
$f(x)$; that is, $y \geq -2$.

29. $f(x) = \frac{(x - 1)^2}{1 + 2x - x^2} = \frac{x^2 - 2x + 1}{-x^2 + 2x + 1}$

This is a rational function, with vertical
asymptotes at $x = 1 \pm \sqrt{2}$ (solve $-x^2 + 2x + 1 = 0$)
and horizontal asymptote at $y = -1$.

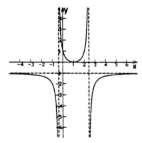

A sketch of the graph reveals that if we restrict the
domain to $x \geq 1$, the function will be one-to-one.
Notice that the range of the function will be
$\{y : y \geq 0 \text{ or } y < -1\}$.
To find the inverse, let
$$y = \frac{(x - 1)^2}{-x^2 + 2x + 1}$$
Complete the square of the denominator:
$$y = \frac{(x - 1)^2}{-(x^2 - 2x + 1) + 1 + 1}$$
$$y = \frac{(x - 1)^2}{-(x - 1)^2 + 2}$$
$$-(x - 1)^2 y + 2y = (x - 1)^2$$
$$2y = (x - 1)^2 + (x - 1)^2 y$$
$$2y = (x - 1)^2 (1 + y)$$
$$\frac{2y}{1 + y} = (x - 1)^2$$
$$(x - 1) = \pm\sqrt{\frac{2y}{1 + y}}$$
$$x = 1 \pm \sqrt{\frac{2y}{1 + y}}$$
Since we restricted the domain to $x \geq 1$, choose
$$x = 1 + \sqrt{\frac{2y}{1 + y}}$$
$$f^{-1}(y) = 1 + \sqrt{\frac{2y}{1 + y}}$$
Replace y with x:
$$f^{-1}(x) = 1 + \sqrt{\frac{2x}{1 + x}}$$
Note: The domain of $f^{-1}(x)$ is the range of $f(x)$;
that is, $x \geq 0$ or $x < -1$. The range of $f^{-1}(x)$ is the
domain of $f(x)$; that is, $y \geq 1$.

31. $f(x) = \frac{1}{x - 1}$

This is rational function with vertical asymptote at
$x = 1$ and horizontal asymptote at $y = 0$.

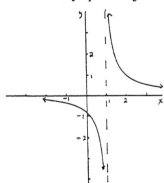

$f(x)$ is a one-to-one function.

a) $f(3) = \frac{1}{3-1} = \frac{1}{2}$

b) $f^{-1}(\frac{1}{2}) = 3$, by part (a).

c) $f(0) = \frac{1}{0-1} = -1$

d) $f^{-1}(-1) = 0$, by part (c).

e) Let $f^{-1}(3) = a$.
Then $f(a) = 3$.

But $f(a) = \frac{1}{a-1}$.

Therefore,

$$\frac{1}{a-1} = 3$$
$$1 = 3a - 3$$
$$4 = 3a$$
$$\frac{4}{3} = a$$

$$f^{-1}(3) = \frac{4}{3}$$

f) Let $f^{-1}(-2) = a$.
Then $f(a) = -2$.

But $f(a) = \frac{1}{a-1}$.

Therefore,

$$\frac{1}{a-1} = -2$$
$$1 = -2a + 2$$
$$2a = 1$$
$$a = \frac{1}{2}$$

$$f^{-1}(-2) = \frac{1}{2}$$

33. Let $y = \frac{1}{x-1}$

Solve for x:
$$y(x-1) = 1$$
$$x - 1 = \frac{1}{y}$$

$$x = 1 + \frac{1}{y}$$

$$f^{-1}(y) = 1 + \frac{1}{y}$$

Replace y with x.

$$f^{-1}(x) = 1 + \frac{1}{x} = \frac{x+1}{x}$$

$f^{-1}(x)$ is a rational function with vertical asymptote at $x = 0$ and a horizontal asymptote at $y = 1$.
(See graph, next column.)

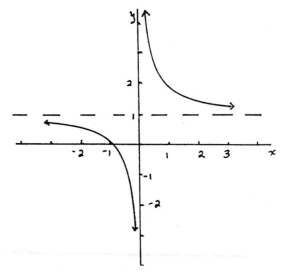

35. $f(x) = x^2 - 2x - 3$
$f(x)$ is a parabola facing upward with vertex at
$\left(-\frac{b}{2a}, f\left(\frac{-b}{2a}\right)\right) = (1, -4)$.

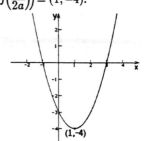

If we restrict the domain to $x \geq 1$, the resulting function will be one-to-one. The range will be $y \geq -4$.
To find the inverse, let
$$y = x^2 - 2x - 3$$
Solve for x by completing the square.
$$y = x^2 - 2x + 1 - 1 - 3$$
$$y = (x-1)^2 - 4$$
$$(x-1)^2 = y + 4$$
$$x - 1 = \pm\sqrt{y+4}$$
$$x = 1 \pm \sqrt{y+4}$$
Since we restricted the domain to $x \geq 1$, choose
$$x = 1 + \sqrt{y+4}$$
$$f^{-1}(y) = 1 + \sqrt{y+4}$$
Replace y with x:
$$f^{-1}(x) = 1 + \sqrt{x+4}$$
Note that the domain of $f^{-1}(x)$ is the range of $f(x)$; that is, $x \geq -4$. The range of $f^{-1}(x)$ is the domain of $f(x)$; that is, $y \geq 1$.

37. Let $y = \dfrac{(2x^2 - 4x - 1)}{(x-1)^2}$, $x > 1$

Solve for x:
$$y(x-1)^2 = 2x^2 - 4x - 1$$
Complete the square of the trinomial on the right:
$$y(x-1)^2 = 2(x^2 - 2x + 1) - 2 - 1$$
$$y(x-1)^2 = 2(x-1)^2 - 3$$
$$3 = 2(x-1)^2 - y(x-1)^2$$
$$3 = (x-1)^2(2-y)$$
$$(x-1)^2 = \frac{3}{2-y}$$
$$x - 1 = \pm\sqrt{\frac{3}{2-y}}$$
$$x = 1 \pm\sqrt{\frac{3}{2-y}}$$
Since $x > 1$, choose
$$x = 1 + \sqrt{\frac{3}{2-y}}$$
$$f^{-1}(y) = 1 + \sqrt{\frac{3}{2-y}}$$
Replace y with x:
$$f^{-1}(x) = 1 + \sqrt{\frac{3}{2-x}}$$

39. $s = -16t^2 + 96$, $0 \le t \le \sqrt{6}$

a) Solve for t:
$$16t^2 = 96 - s$$
$$t^2 = \frac{96 - s}{16}$$
$$t = \pm\sqrt{\frac{96 - s}{16}}$$
$$t = \pm\frac{\sqrt{96 - s}}{4}$$
Since $t \ge 0$, choose
$$t = \frac{\sqrt{96 - s}}{4}$$
$$f^{-1}(s) = \frac{\sqrt{96 - s}}{4}$$

b) Let $s = 60$.
$$t = f^{-1}(60) = \frac{\sqrt{96 - 60}}{4} = \frac{\sqrt{36}}{4} = \frac{6}{4} = 1.5$$
$t = 1.5$ seconds.

41. For f to be its own inverse, the graph of f must be symmetric about the line $y = x$. This means that the xy-equation determining f will be unchanged if x and y are interchanged.

43. a) Let $y = \dfrac{(ax + b)}{(cx + d)}$

Solve for x:
$$cyx + dy = ax + b$$
$$cyx - ax = b - dy$$
$$x(cy - a) = b - dy$$
$$x = \frac{b - dy}{cy - a}$$
$$f^{-1}(y) = \frac{b - dy}{cy - a}$$
Replace y with x:
$$f^{-1}(x) = \frac{b - dx}{cx - a}$$

b) $f(f^{-1}(x))$ must equal x.
$$f(f^{-1}(x)) = f\left(\frac{b - dx}{cx - a}\right)$$
$$= \frac{a\left(\dfrac{b - dx}{cx - a}\right) + b}{c\left(\dfrac{b - dx}{cx - a}\right) + d}$$
Multiply numerator and denominator by $(cx - a)$:
$$f(f^{-1}(x)) = \frac{a(b - dx) + b(cx - a)}{c(b - dx) + d(cx - a)}$$
$$= \frac{ab - adx + bcx - ba}{cb - cdx + cdx - ad}$$
$$= \frac{x(bc - ad)}{bc - ad}$$
$$= x, \text{ if } bc - ad \ne 0$$

c) For $f(x) = f^{-1}(x)$, we must have
$$\frac{ax + b}{cx + d} = \frac{b - dx}{cx - a}$$
$$(ax + b)(cx - a) = (b - dx)(cx + d)$$
$$acx^2 + bcx - a^2x - ab = bcx - cdx^2 + bd - d^2x$$
$$acx^2 + (bc - a^2)x - ab = -cdx^2 + (bc - d^2)x + bd$$
For this to be true for all values of x, we must have $a = -d$.

45. a) $f(x) = x^5 + 2x - 1$ is one-to-one.

b) $f(x) = x^5 - .02x - 1$ is not one-to-one.
To see this, zoom in on the graph near the y-intercept. Use TRACE to move the cursor along the curve, and notice that the y-values first decrease, then increase, doubling back over their previous values. Two different values of x must have the same value of y.

c) $f(x) = x^5 + .06x - 1$ is one-to-one.
When you zoom in on the graph near the y-intercept, and use the TRACE function to move along the curve, you see that the y-values are constantly increasing.

d) $f(x) = .2x^4 - x^3 + 2x - 3$, $x \geq 3.53$ is not one-to-one. Move the cursor to the point on the graph below $x \approx 3.5$. Zoom in on that part of the curve and use the $\boxed{\text{TRACE}}$ feature to see that values of y are repeated to the right of $x = 3.53$.

Chapter 5 Review Problem Set

1. True: $f(-6.5) = |-6.5 + 3| = |-3.5| = 3.5$
$f(.5) = |.5 + 3| = |3.5| = 3.5$

2. False, since the graph of $y = f(x)$ does not exist (has a "hole") above $x = 2$, while the graph of $y = g(x)$ is continuous.

3. False: We want $x^2 - x - 12 \geq 0$
Solve $(x + 3)(x - 4) = 0$
$x + 3 = 0$ $x - 4 = 0$
$x = -3$ $x = 4$
$x^2 - x - 12 \geq 0$ when $x \leq -3$ or $x \geq 4$.

4. True.

5. True: The graph of $f(x)$ is a parabola opening upward with vertex at $(-3, 0)$. Restricting the domain to $x \leq -3$ will result in a function that is one-to-one.

6. False. If $f(x) = x^3$, then $f(f(x)) = (x^3)^3 = x^9 \neq x^6$

7. False: As x grows very large, $f(x) \to \frac{x}{x^2} = \frac{1}{x} \to 0$. The horizontal asymptote is $y = 0$.

8. True. The only possible zero is $x = 2$.

9. True.

10. False. $(f \circ g)(x) = f(g(x)) = \sqrt[3]{x^3 + 2} - 2 \neq x$

11. Not a function of x.

12. Yes, this represents a function of x.

13. Yes, this represents a function of x.

14. Not a function of x.

15. Let $f(x) = x^2 - 4$ and $g(x) = \frac{1}{\sqrt{x}}$.

a) $g(.81) = \frac{1}{\sqrt{.81}} = \frac{1}{.9} = \frac{1}{\frac{9}{10}} = \frac{10}{9}$

b) $f(\sqrt{7}) = (\sqrt{7})^2 - 4 = 7 - 4 = 3$

c) $f(g(.01)) = f\left(\frac{1}{\sqrt{.01}}\right)$
$= \left(\frac{1}{\sqrt{.01}}\right)^2 - 4$
$= \frac{1}{.01} - 4 = \frac{1}{\frac{1}{100}} - 4 = 100 - 4 = 96$

d) $f(2) \cdot g(0) = [2^2 - 4] \cdot [\frac{1}{0}]$ Undefined

e) $g(g(16)) = g\left(\frac{1}{\sqrt{16}}\right) = g\left(\frac{1}{4}\right) = \frac{1}{\sqrt{\frac{1}{4}}} = \frac{1}{\frac{1}{2}} = 2$

f) $\dfrac{f(0)}{g(4)} = \dfrac{0^2 - 4}{\frac{1}{\sqrt{4}}} = \dfrac{-4}{\frac{1}{2}} = -8$

16. Domain of $f(x) = \frac{\sqrt{x}}{x - 1}$ is $\{x : x \geq 0, \, x \neq 1\}$

17. $\dfrac{f(x + h) - f(x)}{h} = \dfrac{\frac{1}{(x+h)^2} - \frac{1}{x^2}}{h}$
$= \dfrac{x^2 - (x + h)^2}{(x + h)^2 x^2} \cdot \dfrac{1}{h}$
$= \dfrac{x^2 - (x^2 + 2xh + h^2)}{hx^2(x + h)^2}$
$= \dfrac{-2xh - h^2}{hx^2(x + h)^2}$
$= \dfrac{-2x - h}{x^2(x + h)^2}$
$= -\dfrac{2x + h}{x^2(x + h)^2}$

18. $z = k\dfrac{x^2}{\sqrt[3]{y}}$ and $-8 = k\dfrac{(-2)^2}{\sqrt[3]{-1}} = -4k \Rightarrow k = 2$; $z = \dfrac{2x^2}{\sqrt[3]{y}}$.
When $x = 3$ and $y = .001$ we get
$z = \dfrac{2 \cdot 3^2}{\sqrt[3]{.001}} = \dfrac{18}{.1} = 180$

19. $f(x) = (x - 2)^2$
The graph will be the parabola $y = x^2$ translated 2 units to the right.

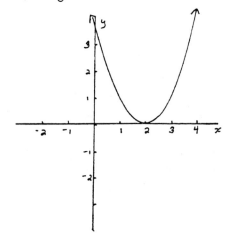

20. $f(x) = x^3 + 2x$
$= x(x^2 + 2)$

x	$f(x)$
-2	-12
-1	-3
0	0
1	3
2	12

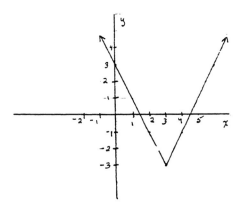

Notice that $f(x)$ is an odd function, so the graph will be symmetrical about the origin.

21. $f(x) = 2|x - 3| - 3$

This will be the typical V-shaped absolute value graph with vertex at $(3, -3)$. It will be narrower, however, than the graph of $y = |x|$.

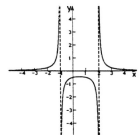

22. $f(x) = \dfrac{1}{x^2 - x - 2} = \dfrac{1}{(x - 2)(x + 1)}$

This is a rational function, with vertical asymptotes at $x = 2$ and $x = -1$ and horizontal asymptote $y = 0$.

23. $f(x) = [2x]$

x	$2x$	$[2x]$
-1.5	-3	-3
-1.25	-2.5	-3
-1	-2	-2
-0.75	-1.5	-2
-0.5	-1	-1
-0.25	-0.5	-1
0	0	0
0.25	0.5	0
0.5	1	1
0.75	1.5	1
1	2	2

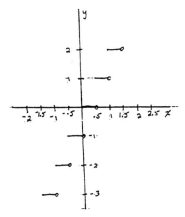

24.
$$f(x) = \begin{cases} 0 & \text{if } x \le 0 \\ x^2 & \text{if } 0 < x < 1 \\ 1 & \text{if } x \ge 1 \end{cases}$$

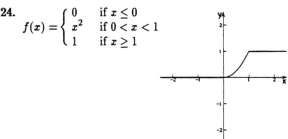

25. Since $g(x) = g(-x)$, we know that $g(x)$ is symmetric about the y-axis. If we sketch the graph of $g(x)$ for $x \ge 0$, we can sketch $g(x)$ for $x < 0$ by symmetry.

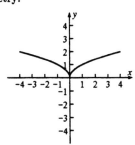

113

26. Since $g(-x) = -g(x)$, we know that $g(x)$ is symmetric about the origin. If we sketch the graph of $g(x)$ for $x \geq 0$, we can sketch $g(x)$ for $x < 0$ by symmetry.

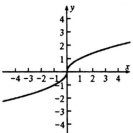

27. a) Reflect the graph through the line $y = x$.

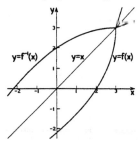

b) Reflect the graph through the line $y = x$.

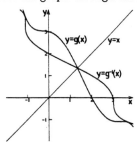

28. a) $y = f(x+4)$
Translate the graph of $f(x)$ 4 units to the left.

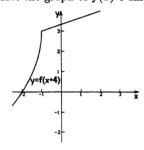

b) $y = g(x) - 3$
Translate the graph of $g(x)$ 3 units down.

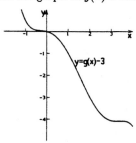

29. The graph of $y = f(x+3) - 4$ is the graph of $y = f(x)$ translated 3 units left and 4 units down.

30. $y = x^2 + 1/x$

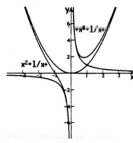

31. Let $f(x) = \sqrt[3]{x-2} + 3$ and $g(x) = (x-3)^3$.
a) $f(x+2) = \sqrt[3]{x+2-2} + 3 = \sqrt[3]{x} + 3$

b) $g(f(x)) = g(\sqrt[3]{x-2} + 3)$
$= (\sqrt[3]{x-2} + 3 - 3)^3$
$= (\sqrt[3]{x-2})^3$
$= x - 2$

c) Let $y = (x-3)^3$
$x - 3 = \sqrt[3]{y}$
$x = \sqrt[3]{y} + 3$
$g^{-1}(y) = \sqrt[3]{y} + 3$
Replace y with x:
$g^{-1}(x) = \sqrt[3]{x} + 3$

d) Let $y = \sqrt[3]{x-2} + 3$
$\sqrt[3]{x-2} = y - 3$
$x - 2 = (y-3)^3$
$x = 2 + (y-3)^3$
$f^{-1}(y) = 2 + (y-3)^3$
Replace y with x:
$f^{-1}(x) = 2 + (x-3)^3$

32. $f(x) = \dfrac{x+3}{2}$. Logically, f^{-1} should first multiply by 2 and then subtract 3. $f^{-1}(x) = 2x - 3$.

114

33. Let $g^{-1}(2) = a$.
Then $g(a) = 2$
But $g(a) = \dfrac{a}{a+2}$.
Therefore,
$$\dfrac{a}{a+2} = 2$$
$$a = 2a + 4$$
$$-4 = a$$
$$g^{-1}(2) = -4$$

34. $f(x) = y = \dfrac{x}{x-5}$. Solve for x in terms of y:
$$xy - 5y = x$$
$$xy - x = 5y$$
$$x(y-1) = 5y$$
$$f^{-1}(y) = x = \dfrac{5y}{y-1}$$
$$f^{-1}(x) = \dfrac{5x}{x-1}$$

35. First you cube x and subtract 7. So, $g(x) = x^3 - 7$.
Then you take the square root of the result. So,
$f(x) = \sqrt{x}$.

36. Suppose $f(x) = 2$. $\quad \dfrac{2x+1}{x+2} = 2$
$$2x + 1 = 2x + 4$$
$$1 = 4, \text{ which is not true}$$

37. Let $f(x) = \dfrac{2x+1}{x+2}$.
Domain: all real x except $x = -2$.

38. $f(x) = y = (2x-5)^2$. Restrict the domain so that
$2x - 5 \geq 0$. [These are the points to the right of
the vertex of this parabola.] $x \geq \frac{5}{2}$
Solve for x in terms of y:
$$2x - 5 = \sqrt{y}$$
$$2x = \sqrt{y} + 5$$
$$f^{-1}(y) = x = \tfrac{1}{2}(\sqrt{y} + 5)$$
$$f^{-1}(x) = \tfrac{1}{2}(\sqrt{x} + 5)$$

39. Let $f(x) = 2|x|$. $\quad f(-x) = 2|-x| = 2|x| = f(x)$.
$f(x)$ is an even function.

40. Let $f(x) = \dfrac{x^2}{x^2+1}$. $f(-x) = \dfrac{(-x)^2}{(-x)^2+1} = \dfrac{x^2}{x^2+1}$
$f(x)$ is an even function.

41. Let $f(x) = \dfrac{2x}{x^2+1}$. $f(-x) = \dfrac{2(-x)}{(-x)^2+1} = -\dfrac{2x}{x^2+1}$
$\qquad\qquad = -f(x)$. $f(x)$ is an odd function.

42. Let $f(x) = \sqrt[3]{x^3+8}$. $f(-x) = \sqrt[3]{-x^3+8}$
$f(x)$ is neither even nor odd since $f(-x) \neq f(x)$
and $f(-x) \neq -f(x)$.

43. Let $f(x) = x^5 + x^3 + 1$.
$f(-x) = (-x)^5 + (-x)^3 + 1 = -x^5 - x^3 + 1$
$f(x)$ is neither even nor odd.

44. Let $f(x) = (x^3 + x)^3(x^2 - 1)$.
$f(-x) = (-x^3 - x)^3(x^2 - 1) = -(x^3 + x)^3(x^2 - 1)$
$\quad = -f(x)$. $f(x)$ is an odd function.

45. a) Let $h(x) = f(x) \cdot g(x)$.
$\quad h(-x) = f(-x) \cdot g(-x)$
$\qquad\quad = f(x) \cdot [-g(x)]$
$\qquad\quad = -f(x) \cdot g(x)$
$\qquad\quad = -h(x)$
$h(x)$ is an odd function.

b) Let $h(x) = (f \circ g)(x) = f(g(x))$.
$\quad h(-x) = f(g(-x))$
$\qquad\quad = f(-g(x))$
$\qquad\quad = f(g(x))$
$\qquad\quad = h(x)$
$h(x)$ is an even function.

46. This would be the bottom half of the given circle:
$$g(x) = -\sqrt{25 - x^2}$$

47. a) $S = \overline{AC} + \overline{CD}$.
$\overline{CD} = 12 - x$. Using the Pythagorean Theorem,
$$\overline{AC} = \sqrt{\overline{BC}^2 + \overline{AB}^2} = \sqrt{x^2 + 9}$$
Therefore,
$$S = \sqrt{x^2 + 9} + 12 - x$$

b) $T = $ time required to row from A to C plus time
required to walk from C to D.
Using the formula $t = \dfrac{d}{r}$,
$$T = \dfrac{\sqrt{x^2+9}}{2.5} + \dfrac{12-x}{3.5}$$

48. a) $V = $ volume of cylinder + volume of hemisphere
$V = \pi x^2 x + \frac{1}{2} \cdot \frac{4}{3}\pi x^3 = \pi x^3 + \frac{2}{3}\pi x^3 = \frac{5}{3}\pi x^3$

b) $S = $ surface area of topless cylinder
$\qquad\qquad$ + surface area of hemisphere
$= (\pi x^2 + 2\pi x \cdot x) + \frac{1}{2} \cdot 4\pi x^2 = 3\pi x^2 + 2\pi x^2 = 5\pi x^2$

Chapter 6:
Exponential and Logarithmic Functions

Problem Set 6.1

1. $\sqrt{9} = 3$

3. $\sqrt[5]{32} = 2$

5. $(\sqrt[3]{7})^3 = 7$

7. $\sqrt[3]{\left(\frac{3}{2}\right)^3} = \frac{3}{2}$

9. $(\sqrt{5})^4 = \left[(\sqrt{5})^2\right]^2 = [5]^2 = 25$

11. $\sqrt{3}\sqrt{27} = \sqrt{81} = 9$

13. $\dfrac{\sqrt[3]{16}}{\sqrt[3]{2}} = \sqrt[3]{\dfrac{16}{2}} = \sqrt[3]{8} = 2$

15. $\sqrt[3]{10^{-6}} = \sqrt[3]{(10^{-2})^3} = 10^{-2} = \dfrac{1}{100}$

17. $\dfrac{1}{\sqrt{2}} = \dfrac{1}{\sqrt{2}} \cdot \dfrac{\sqrt{2}}{\sqrt{2}} = \dfrac{\sqrt{2}}{2}$

19. $\dfrac{\sqrt{10}}{\sqrt{2}} = \sqrt{\dfrac{10}{2}} = \sqrt{5}$

21. $\sqrt[3]{54x^4y^5} = \sqrt[3]{27x^3y^3 \cdot 2xy^2} = \sqrt[3]{(3xy)^3 \cdot 2xy^2}$
$= 3xy\sqrt[3]{2xy^2}$

23. $\sqrt[4]{(x+2)^4 y^7} = \sqrt[4]{(x+2)^4 y^4 \cdot y^3}$
$= \sqrt[4]{[(x+2)y]^4 \cdot y^3}$
$= (x+2)y\sqrt[4]{y^3}$

25. $\sqrt{x^2 + x^2 y^2} = \sqrt{x^2(1+y^2)} = x\sqrt{1+y^2}$

27. $\sqrt[3]{x^6 - 9x^3y} = \sqrt[3]{x^3(x^3 - 9y)} = x\sqrt[3]{x^3 - 9y}$

29. $\sqrt[3]{x^4 y^{-6} z^6} = \sqrt[3]{x^3 y^{-6} z^6 \cdot x}$
$= \sqrt[3]{(xy^{-2}z^2)^3 \cdot x}$
$= xy^{-2}z^2\sqrt[3]{x}$
$= \dfrac{xz^2\sqrt[3]{x}}{y^2}$

31. $\dfrac{2}{\sqrt{x}+3} = \dfrac{2}{(\sqrt{x}+3)} \cdot \dfrac{(\sqrt{x}-3)}{(\sqrt{x}-3)} = \dfrac{2(\sqrt{x}-3)}{x-9}$

33. $\dfrac{2}{\sqrt{x+3}} = \dfrac{2}{\sqrt{x+3}} \cdot \dfrac{\sqrt{x+3}}{\sqrt{x+3}} = \dfrac{2\sqrt{x+3}}{x+3}$

35. $\dfrac{1}{\sqrt[4]{8x^3}} = \dfrac{1}{\sqrt[4]{(2x)^3}} = \dfrac{1}{\sqrt[4]{(2x)^3}} \cdot \dfrac{\sqrt[4]{2x}}{\sqrt[4]{2x}} = \dfrac{\sqrt[4]{2x}}{\sqrt[4]{(2x)^4}} = \dfrac{\sqrt[4]{2x}}{2x}$

37. $\dfrac{2}{\sqrt{3}-\sqrt{2}} = \dfrac{2}{\sqrt{3}-\sqrt{2}} \cdot \dfrac{(\sqrt{3}+\sqrt{2})}{(\sqrt{3}+\sqrt{2})}$
$= \dfrac{2(\sqrt{3}+\sqrt{2})}{3-2}$
$= 2(\sqrt{3}+\sqrt{2})$

39. $\sqrt[3]{2x^{-2}y^4}\sqrt[3]{4xy^{-1}} = \sqrt[3]{2x^{-2}y^4 \cdot 4xy^{-1}}$
$= \sqrt[3]{8x^{-1}y^3}$
$= 2y\sqrt[3]{x^{-1}}$
$= 2y\sqrt[3]{\dfrac{x^{-1}x^3}{x^3}}$
$= \dfrac{2y}{x}\sqrt[3]{x^2}$

41. $\sqrt{50} - 2\sqrt{18} + \sqrt{8} = \sqrt{25 \cdot 2} - 2\sqrt{9 \cdot 2} + \sqrt{4 \cdot 2}$
$= 5\sqrt{2} - 6\sqrt{2} + 2\sqrt{2}$
$= \sqrt{2}$

43. $\sqrt[3]{192} + \sqrt[3]{-81} + \sqrt[3]{24} = \sqrt[3]{64 \cdot 3} + \sqrt[3]{-27 \cdot 3} + \sqrt[3]{8 \cdot 3}$
$= 4\sqrt[3]{3} - 3\sqrt[3]{3} + 2\sqrt[3]{3}$
$= 3\sqrt[3]{3}$

45. $\dfrac{2-\sqrt{5}}{2+\sqrt{5}} = \dfrac{(2-\sqrt{5})}{(2+\sqrt{5})} \cdot \dfrac{(2-\sqrt{5})}{(2-\sqrt{5})}$
$= \dfrac{4 - 4\sqrt{5} + 5}{4-5}$
$= \dfrac{9 - 4\sqrt{5}}{-1}$
$= 4\sqrt{5} - 9$

47. $\dfrac{a}{\sqrt[5]{8a^4b^9}} = \dfrac{a}{\sqrt[5]{8a^4b^9}} \cdot \dfrac{\sqrt[5]{4ab}}{\sqrt[5]{4ab}}$
$= \dfrac{a\sqrt[5]{4ab}}{\sqrt[5]{32a^5b^{10}}}$
$= \dfrac{a\sqrt[5]{4ab}}{2ab^2}$
$= \dfrac{\sqrt[5]{4ab}}{2b^2}$

49. $\sqrt{x-1} = 5$
Square both sides:
$x - 1 = 25$
$x = 26$
Check:
$\sqrt{26-1} \overset{?}{=} 5$
$\sqrt{25} \overset{?}{=} 5$
$5 = 5$
So $x = 26$ *is a root.*

116

51. $\sqrt[3]{2x-1}=2$
Cube both sides:
$$2x-1=8$$
$$2x=9$$
$$x=\frac{9}{2}$$

53. $\sqrt{\frac{x}{x+2}}=4$
Square both sides:
$$\frac{x}{x+2}=16$$
$$x=16(x+2)$$
$$x=16x+32$$
$$-15x=32$$
$$x=-\frac{32}{15}$$
Check:
$$\sqrt{\frac{-\frac{32}{15}}{-\frac{32}{15}+2}}\overset{?}{=}4$$
$$\sqrt{\frac{-32}{-32+30}}\overset{?}{=}4$$
$$\sqrt{\frac{-32}{-2}}\overset{?}{=}4$$
$$\sqrt{16}\overset{?}{=}4$$
$$4=4$$
So $x=-\frac{32}{15}$ *is* a root.

55. $\sqrt{x^2+4}=x+2$
Square both sides:
$$x^2+4=x^2+4x+4$$
$$4=4x+4$$
$$0=4x$$
$$0=x$$
Check:
$$\sqrt{0^2+4}\overset{?}{=}0+2$$
$$\sqrt{4}\overset{?}{=}2$$
$$2\overset{?}{=}2$$
So $x=0$ *is* a root.

57. $\sqrt{2x+1}=x-1$
Square both sides:
$$2x+1=x^2-2x+1$$
$$x^2-4x=0$$
$$x(x-4)=0$$
$$x=0 \quad x-4=0$$
$$x=4$$
Check $x=0$:
$$\sqrt{2(0)+1}\overset{?}{=}0-1$$
$$\sqrt{1}\overset{?}{=}-1$$
$$1\neq-1$$
So $x=0$ *is not* a root.

Check $x=4$:
$$\sqrt{2(4)+1}\overset{?}{=}4-1$$
$$\sqrt{9}\overset{?}{=}3$$
$$3=3$$
So $x=4$ *is* a root.

59. $2\sqrt{x+1}=\sqrt{x}+2$
Square both sides:
$$4(x+1)=x+4\sqrt{x}+4$$
$$4x+4=x+4\sqrt{x}+4$$
$$3x=4\sqrt{x}$$
Square both sides again:
$$9x^2=16x$$
$$9x^2-16x=0$$
$$x(9x-16)=0$$
$$x=0 \quad\text{or}\quad 9x-16=0$$
$$x=\frac{16}{9}$$
Check $x=0$:
$$2\sqrt{0+1}\overset{?}{=}\sqrt{0}+2$$
$$2=2$$
So $x=0$ *is* a root.
Check $x=\frac{16}{9}$:
$$2\sqrt{\frac{16}{9}+1}\overset{?}{=}\sqrt{\frac{16}{9}}+2$$
$$2\sqrt{\frac{25}{9}}\overset{?}{=}\sqrt{\frac{16}{9}}+2$$
$$2\cdot\frac{5}{3}\overset{?}{=}\frac{4}{3}+2$$
$$\frac{10}{3}=\frac{10}{3}$$
So $x=\frac{16}{9}$ *is* a root.

61. $\dfrac{2}{\sqrt{x+h}}-\dfrac{2}{\sqrt{x}}=\dfrac{2}{\sqrt{x+h}}\cdot\dfrac{\sqrt{x}}{\sqrt{x}}-\dfrac{2}{\sqrt{x}}\cdot\dfrac{\sqrt{x+h}}{\sqrt{x+h}}$
$$=\frac{2\sqrt{x}-2\sqrt{x+h}}{\sqrt{x(x+h)}}$$

63. $\dfrac{1}{\sqrt{x+6}}+\sqrt{x+6}=\dfrac{1}{\sqrt{x+6}}+\sqrt{x+6}\cdot\dfrac{\sqrt{x+6}}{\sqrt{x+6}}$
$$=\frac{1+x+6}{\sqrt{x+6}}$$
$$=\frac{x+7}{\sqrt{x+6}}$$

65. $\dfrac{\sqrt[3]{(x+2)^2}}{2}-\dfrac{1}{\sqrt[3]{x+2}}=\dfrac{\sqrt[3]{(x+2)^2}}{2}\cdot\dfrac{\sqrt[3]{x+2}}{\sqrt[3]{x+2}}-\dfrac{1}{\sqrt[3]{x+2}}\cdot\dfrac{2}{2}$
$$=\frac{x+2-2}{2\sqrt[3]{x+2}}=\frac{x}{2\sqrt[3]{x+2}}$$

67. $\dfrac{1}{\sqrt{x^2+9}} - \dfrac{\sqrt{x^2+9}}{x^2} = \dfrac{1}{\sqrt{x^2+9}} \cdot \dfrac{x^2}{x^2} - \dfrac{\sqrt{x^2+9}}{x^2} \cdot \dfrac{\sqrt{x^2+9}}{\sqrt{x^2+9}}$

$$= \dfrac{x^2 - (x^2+9)}{x^2\sqrt{x^2+9}} = \dfrac{-9}{x^2\sqrt{x^2+9}}$$

69. a) $\sqrt[4]{16a^4b^8} = \sqrt[4]{(2ab^2)^4} = 2ab^2$

b) $\sqrt{27}\sqrt{3b^3} = \sqrt{81b^3} = \sqrt{81b^2 \cdot b} = 9b\sqrt{b}$

c) $\sqrt{12} + \sqrt{48} - \sqrt{27} = \sqrt{4\cdot3} + \sqrt{16\cdot3} - \sqrt{9\cdot3}$
$= 2\sqrt{3} + 4\sqrt{3} - 3\sqrt{3}$
$= 3\sqrt{3}$

d) $\sqrt{250a^4b^6} = \sqrt{25a^4b^6 \cdot 10} = 5a^2b^3\sqrt{10}$

e) $\sqrt[3]{\dfrac{-32x^2y^7}{4x^5y}} = \sqrt[3]{\dfrac{-8y^6}{x^3}} = \dfrac{-2y^2}{x}$

f) $\left(\sqrt[3]{\dfrac{y}{2x}}\right)^6 = \left[\left(\sqrt[3]{\dfrac{y}{2x}}\right)^3\right]^2 = \left(\dfrac{y}{2x}\right)^2 = \dfrac{y^2}{4x^2}$

g) $\sqrt{8a^5} + \sqrt{18a^3} = \sqrt{4a^4 \cdot 2a} + \sqrt{9a^2 \cdot 2a}$
$= 2a^2\sqrt{2a} + 3a\sqrt{2a}$
$= (2a^2 + 3a)\sqrt{2a}$

h) $\sqrt[4]{512} - \sqrt{50} + \sqrt[6]{128} = \sqrt[4]{256\cdot2} - \sqrt{25\cdot2} - \sqrt[6]{64\cdot2}$
$= 4\sqrt[4]{2} - 5\sqrt{2} - 2\sqrt[6]{2}$

i) $\sqrt[4]{a^4 + a^4b^4} = \sqrt[4]{a^4(1+b^4)} = a\sqrt[4]{1+b^4}$

j) $\dfrac{1}{\sqrt[3]{7bc^3}} = \dfrac{1}{c\sqrt[3]{7b}} \cdot \dfrac{\sqrt[3]{7^2b^2}}{\sqrt[3]{7^2b^2}} = \dfrac{\sqrt[3]{49b^2}}{c\sqrt[3]{7^3b^3}} = \dfrac{\sqrt[3]{49b^2}}{7bc}$

k) $\dfrac{2}{\sqrt{a}-b} = \dfrac{2}{(\sqrt{a}-b)} \cdot \dfrac{(\sqrt{a}+b)}{(\sqrt{a}+b)} = \dfrac{2(\sqrt{a}+b)}{a - b^2}$

l) $\sqrt{a}\left(\sqrt{a} + \dfrac{1}{\sqrt{a^3}}\right) = a + \dfrac{\sqrt{a}}{\sqrt{a^3}}$
$= a + \dfrac{1}{\sqrt{a^2}}$
$= a + \dfrac{1}{a}$
$= \dfrac{a^2+1}{a}$

71. a) $\sqrt[3]{1-5x} = -4$
Cube each side:
$1 - 5x = (-4)^3 = -64$
$5x = 65$
$x = 13$

b) $\sqrt{4x+1} = x+1$
Square each side:
$4x + 1 = x^2 + 2x + 1$
$x^2 - 2x = 0$
$x(x-2) = 0$
$x = 0 \text{ or } x = 2$
Check $x = 0$:
$\sqrt{4(0)+1} \overset{?}{=} 0+1$
$\sqrt{1} \overset{?}{=} 1$
$1 = 1$
So $x = 0$ *is* a root.
Check $x = 2$:
$\sqrt{4(2)+1} \overset{?}{=} 2+1$
$\sqrt{9} \overset{?}{=} 3$
$3 = 3$
So $x = 2$ is also a root.

c) $\sqrt{x+3} = 2 + \sqrt{x-5}$
Square each side:
$x + 3 = 4 + 4\sqrt{x-5} + x - 5$
$4 = 4\sqrt{x-5}$
$1 = \sqrt{x-5}$
Square each side again:
$1 = x - 5$
$6 = x$
Check:
$\sqrt{6+3} \overset{?}{=} 2 + \sqrt{6-5}$
$\sqrt{9} \overset{?}{=} 2 + \sqrt{1}$
$3 = 3$
So $x = 6$ *is* a root.

d) $\sqrt{12+x} = 4 + \sqrt{4+x}$
Square each side:
$12 + x = 16 + 8\sqrt{4+x} + 4 + x$
$-8 = 8\sqrt{4+x}$
We need go no further; the left side is negative, non-zero, and the right side is positive or 0. The "equation" has no solution.

e) $x - \sqrt{x} - 6 = 0$
$x - 6 = \sqrt{x}$
Square each side:
$x^2 - 12x + 36 = x$
$x^2 - 13x + 36 = 0$
$(x-4)(x-9) = 0$
$x = 4 \text{ or } 9$
Check $x = 9$:
$9 - \sqrt{9} - 6 \overset{?}{=} 0$
$9 - 3 - 6 \overset{?}{=} 0$
$0 = 0$
So $x = 9$ *is* a solution.

Check $x = 4$:
$$4 - \sqrt{4} - 6 \overset{?}{=} 0$$
$$4 - 2 - 6 \overset{?}{=} 0$$
$$-4 \neq 0$$
So $x = 4$ *is not* a solution.

f) $\quad \sqrt[3]{x^2} - 2\sqrt[3]{x} - 8 = 0$
$\quad (\sqrt[3]{x})^2 - 2\sqrt[3]{x} - 8 = 0$
We think of this as a quadratic equation with $\sqrt[3]{x}$ as the unknown, rather than x.
$$(\sqrt[3]{x} - 4)(\sqrt[3]{x} + 2) = 0$$
$\sqrt[3]{x} - 4 = 0 \qquad$ or $\qquad \sqrt[3]{x} + 2 = 0$
$\qquad \sqrt[3]{x} = 4 \qquad$ or $\qquad \sqrt[3]{x} = -2$
$\qquad x = 4^3 = 64 \quad$ or $\qquad x = (-2)^3 = -8$
Checking by substitution, we see that each solves the original equation.

73. The graphs are reflections of each other about the line $y = x$.

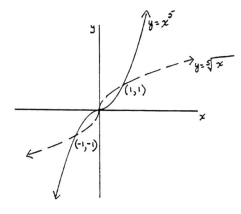

75. Let x represent the length of AC.
Let y represent the length of CB.
We want ACB to equal ADB:
$$x + y = 6 + 20$$
$$y = 26 - x$$
By the Pythagorean Theorem we know
$$(6 + x)^2 + (20)^2 = y^2$$
Let $y = 26 - x$ in the last equation:
$$(6 + x)^2 + 400 = (26 - x)^2$$
$$36 + 12x + x^2 + 400 = 676 - 52x + x^2$$
$$64x = 240$$
$$x = \frac{240}{64} = \frac{15}{4}$$

77. Label the hypotenuses of the right triangles L_1, L_2, L_3, L_4, L_5, and finally L, moving counterclockwise from the bottom of the figure.

$$L_1 = \sqrt{(1)^2 + (1)^2} = \sqrt{2}$$
$$L_2 = \sqrt{(\sqrt{2})^2 + (1)^2} = \sqrt{3}$$
$$L_3 = \sqrt{(\sqrt{3})^2 + (1)^2} = \sqrt{4}$$
$$L_4 = \sqrt{(\sqrt{4})^2 + (1)^2} = \sqrt{5}$$
$$L_5 = \sqrt{(\sqrt{5})^2 + (1)^2} = \sqrt{6}$$
$$L = \sqrt{(\sqrt{6})^2 + (1)^2} = \sqrt{7}$$

79. a) Show $\dfrac{\sqrt{6} + \sqrt{2}}{2} = \sqrt{2 + \sqrt{3}}$
$$\frac{\sqrt{6} + \sqrt{2}}{2} = \sqrt{\left(\frac{\sqrt{6} + \sqrt{2}}{2}\right)^2}$$
$$= \sqrt{\frac{6 + 2\sqrt{12} + 2}{4}}$$
$$= \sqrt{\frac{8 + 4\sqrt{3}}{4}}$$
$$= \sqrt{2 + \sqrt{3}}$$

b) Show $\sqrt{2 + \sqrt{3}} + \sqrt{2 - \sqrt{3}} = \sqrt{6}$
$$(\sqrt{2 + \sqrt{3}} + \sqrt{2 - \sqrt{3}})^2$$
$$= (2 + \sqrt{3}) + 2\sqrt{2 + \sqrt{3}}\sqrt{2 - \sqrt{3}} + (2 - \sqrt{3})$$
$$= 4 + 2\sqrt{(2 + \sqrt{3})(2 - \sqrt{3})}$$
$$= 4 + 2\sqrt{4 - 3}$$
$$= 4 + 2\sqrt{1}$$
$$= 6$$
Taking the positive square root of both sides:
$$\sqrt{2 + \sqrt{3}} + \sqrt{2 - \sqrt{3}} = \sqrt{6}$$

c) Show $\sqrt[3]{9\sqrt{3} - 11\sqrt{2}} = \sqrt{3} - \sqrt{2}$
$$(\sqrt{3} - \sqrt{2})^3 = (\sqrt{3})^3 - 3(\sqrt{3})^2\sqrt{2} + 3\sqrt{3}(\sqrt{2})^2 - (\sqrt{2})^3$$
$$= 3\sqrt{3} - 9\sqrt{2} + 6\sqrt{3} - 2\sqrt{2}$$
$$= 9\sqrt{3} - 11\sqrt{2}$$
Taking the cube root of both sides:
$$\sqrt{3} - \sqrt{2} = \sqrt[3]{9\sqrt{3} - 11\sqrt{2}}$$

81. Set standard range values.
Press $\boxed{Y=}$
Enter $\boxed{\sqrt{}}\boxed{(}X + 4\boxed{)} + \boxed{\sqrt{}}\boxed{(}5X\boxed{)}$ as Y_1
Enter $\boxed{\sqrt{}}\boxed{(}X - 1\boxed{)} + X\hat{} 2$ as Y_2
Press $\boxed{\text{GRAPH}}$
Zoom in on the point of intersection of the two curves to find that at that point $x = 2.167$.

Problem Set 6.2

1. $\sqrt[3]{7} = 7^{1/3}$

3. $\sqrt[3]{7^2} = 7^{2/3}$

5. $\dfrac{1}{\sqrt[3]{7}} = \dfrac{1}{7^{1/3}} = 7^{-1/3}$

7. $\dfrac{1}{\sqrt[3]{7^2}} = \dfrac{1}{7^{2/3}} = 7^{-2/3}$

9. $7\sqrt[3]{7} = 7^1 \cdot 7^{1/3} = 7^{4/3}$

11. $\sqrt[3]{x^2} = x^{2/3}$

13. $x^2\sqrt{x} = x^2 \cdot x^{1/2} = x^{5/2}$

15. $\sqrt{(x+y)^3} = (x+y)^{3/2}$

17. $\sqrt{x^2+y^2} = (x^2+y^2)^{1/2}$

19. $4^{3/2} = \sqrt[2]{4^3} = \sqrt[2]{16}$

21. $8^{-3/2} = \dfrac{1}{8^{3/2}} = \dfrac{1}{\sqrt{8^3}} = \dfrac{1}{8\sqrt{8}} = \dfrac{\sqrt{8}}{64} = \dfrac{2\sqrt{2}}{64} = \dfrac{\sqrt{2}}{32}$

23. $(x^4+y^4)^{1/4} = \sqrt[4]{x^4+y^4}$

25. $(x^2y^3)^{2/5} = \sqrt[5]{(x^2y^3)^2} = \sqrt[5]{x^4y^6} = y\sqrt[5]{x^4y}$

27. $(x^{1/2}+y^{1/2})^{1/2} = \sqrt{x^{1/2}+y^{1/2}} = \sqrt{\sqrt{x}+\sqrt{y}}$

29. $25^{1/2} = \sqrt{25} = 5$

31. $8^{2/3} = (\sqrt[3]{8})^2 = 2^2 = 4$

33. $9^{-3/2} = \dfrac{1}{9^{3/2}} = \dfrac{1}{(\sqrt{9})^3} = \dfrac{1}{3^3} = \dfrac{1}{27}$

35. $(-0.008)^{2/3} = (\sqrt[3]{-0.008})^2 = (-0.2)^2 = 0.04$

37. $(.0025)^{3/2} = (\sqrt{.0025})^3 = (.05)^3 = .000125$

39. $5^{2/3}5^{-5/3} = 5^{2/3-5/3} = 5^{-3/3} = 5^{-1} = \dfrac{1}{5}$

41. $16^{7/6}16^{-5/6}16^{-4/3} = 16^{7/6-5/6-4/3}$
$\qquad\qquad = 16^{-3/3} = 16^{-1} = \dfrac{1}{16}$

43. $(8^2)^{-2/3} = 8^{(2)(-2/3)}$
$\qquad\qquad = 8^{-4/3} = \dfrac{1}{8^{4/3}}$
$\qquad\qquad = \dfrac{1}{(\sqrt[3]{8})^4} = \dfrac{1}{2^4}$
$\qquad\qquad = \dfrac{1}{16}$

45. $(3a^{1/2})(-2a^{3/2}) = -6a^{1/2+3/2} = -6a^2$

47. $(2^{1/2}x^{-2/3})^6 = 2^{(1/2)(6)}x^{(-2/3)(6)} = 2^3x^{-4} = \dfrac{8}{x^4}$

49. $(xy^{-2/3})^3(x^{1/2}y)^2 = x^3y^{-2}xy^2 = x^4y^0 = x^4$

51. $\dfrac{(2x^{-1}y^{2/3})^2}{x^2y^{-2/3}} = \dfrac{2^2x^{-2}y^{4/3}}{x^2y^{-2/3}}$
$\qquad\qquad = \dfrac{4y^{4/3}y^{2/3}}{x^2x^2}$
$\qquad\qquad = \dfrac{4y^{6/3}}{x^4}$
$\qquad\qquad = \dfrac{4y^2}{x^4}$

53. $\left(\dfrac{x^{-2}y^{3/4}}{x^{1/2}}\right)^{12} = \dfrac{x^{-24}y^9}{x^6} = \dfrac{y^9}{x^{24}x^6} = \dfrac{y^9}{x^{30}}$

55. $y^{9/3}(2y^{1/2}-y^{-5/3}) = 2y^{2/3}y^{4/3} - y^{2/3}y^{-5/3}$
$\qquad\qquad = 2y^{6/3} - y^{-3/3}$
$\qquad\qquad = 2y^2 - y^{-1}$
$\qquad\qquad = 2y^2 - \dfrac{1}{y}$
$\qquad\qquad = \dfrac{2y^3-1}{y}$

57. $(x^{1/2}+y^{1/2})^2 = (x^{1/2})^2 + 2x^{1/2}y^{1/2} + (y^{1/2})^2$
$\qquad\qquad = x + 2\sqrt{xy} + y$

59. $\dfrac{(x+2)^{4/5}}{3} + \dfrac{2x}{(x+2)^{1/5}}$
$\qquad = \dfrac{(x+2)^{4/5}(x+2)^{1/5}}{3(x+2)^{1/5}} + \dfrac{3(2x)}{3(x+2)^{1/5}}$
$\qquad = \dfrac{(x+2)^1 + 6x}{3(x+2)^{1/5}}$
$\qquad = \dfrac{7x+2}{3(x+2)^{1/5}}$

61. $(x^2+1)^{1/3} - \dfrac{2x^2}{(x^2+1)^{2/3}}$
$\qquad = \dfrac{(x^2+1)^{1/3}(x^2+1)^{2/3}}{(x^2+1)^{2/3}} - \dfrac{2x^2}{(x^2+1)^{2/3}}$
$\qquad = \dfrac{x^2+1-2x^2}{(x^2+1)^{2/3}}$
$\qquad = \dfrac{1-x^2}{(x^2+1)^{2/3}}$

63. $\sqrt{2}\sqrt[3]{2} = 2^{1/2}2^{1/3} = 2^{3/6}2^{2/6} = 2^{5/6} = \sqrt[6]{2^5} = \sqrt[6]{32}$

65. $\sqrt[4]{2}\,\sqrt[6]{x} = 2^{1/4}x^{1/6}$
$= 2^{3/12}x^{2/12}$
$= (2^3)^{1/12}(x^2)^{1/12}$
$= (2^3x^2)^{1/12}$
$= \sqrt[12]{8x^2}$

67. $\sqrt[3]{x\sqrt{x}} = \sqrt[3]{x^1x^{1/2}} = \sqrt[3]{x^{3/2}} = (x^{3/2})^{1/3} = x^{1/2} = \sqrt{x}$

69. $\boxed{2}\,\boxed{y^x}\,\boxed{1.34}\,\boxed{=}\,2.53151$

71. $\boxed{\pi}\,\boxed{y^x}\,\boxed{1.34}\,\boxed{=}\,4.636399$

73. $\boxed{1.46}\,\boxed{y^x}\,2\,\boxed{\sqrt{}}\,\boxed{=}\,1.70777$

75. $\boxed{.9}\,\boxed{y^x}\,\boxed{50.2}\,\boxed{=}\,.0050463$

77. $f(x) = 4^x$

x	$f(x)$
-2	$\frac{1}{16}$
-1	$\frac{1}{4}$
1	4
2	16

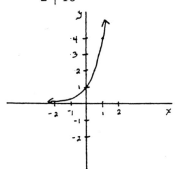

79. $f(x) = \left(\frac{2}{3}\right)^x$

x	$f(x)$
-2	$\frac{9}{4}$
-1	$\frac{3}{2}$
0	1
1	$\frac{2}{3}$
2	$\frac{4}{9}$

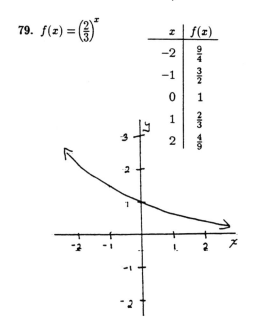

81. $f(x) = \pi^x$

x	$f(x)$
-2	0.101
-1	0.318
0	1
1	3.14
2	9.87

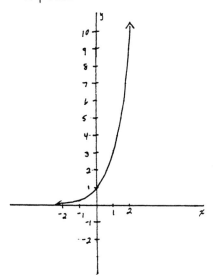

83. a) $\sqrt[5]{b^3} = b^{3/5}$

b) $\sqrt[8]{x^4} = x^{4/8} = x^{1/2}$

c) $\sqrt[3]{a^2 + 2ab + b^2} = \sqrt[3]{(a+b)^2} = (a+b)^{2/3}$

85. a) $27^{2/3}(.0625)^{-3/4} = (\sqrt[3]{27})^2((.5)^4)^{-3/4}$
$= 9(.5)^{-3}$
$= \dfrac{9}{(.5)^3} = \dfrac{9}{.125} = \dfrac{9}{\frac{1}{8}} = 72$

b) $\sqrt[3]{4}\,\sqrt{2} + \sqrt[6]{2} = 4^{1/3}2^{1/2} + 2^{1/6}$
$= 2^{2/3}2^{1/2} + 2^{1/6}$
$= 2^{4/6}2^{3/6} + 2^{1/6}$
$= 2^{7/6} + 2^{1/6}$
$= 2 \cdot 2^{1/6} + 2^{1/6}$
$= 2^{1/6}(2 + 1)$
$= 3 \cdot 2^{1/6}$

c) $\sqrt[3]{a^2}\,\sqrt[4]{a^3} = a^{2/3}a^{3/4} = a^{8/12}a^{9/12} = a^{17/12}$

d) $\sqrt{a\,\sqrt[3]{a^2}} = \sqrt{a \cdot a^{2/3}} = (a^{5/3})^{1/2} = a^{5/6}$

e) $\left[a^{3/2} + a^{-3/2}\right]^2 = (a^{3/2})^2 + 2a^{3/2}a^{-3/2} + (a^{-3/2})^2$
$= a^3 + 2a^0 + a^{-3}$
$= a^3 + 2 + \dfrac{1}{a^3} = \dfrac{a^6 + 2a^3 + 1}{a^3}$

f) $\left[a^{1/4}\left(a^{-5/4}+a^{3/4}\right)\right]^{-1} = \left[a^{1/4-5/4}+a^{1/4+3/4}\right]^{-1}$

$= \left[a^{-1}+a^{1}\right]^{-1}$

$= \left[\frac{1}{a}+a\right]^{-1}$

$= \left[\frac{1+a^2}{a}\right]^{-1}$

$= \frac{a}{1+a^2}$

g) $\left(\frac{\sqrt[3]{a^3 b^2}}{\sqrt[4]{a^6 b^3}}\right)^{-1} = \left(\frac{a^{3/3}b^{2/3}}{a^{6/4}b^{3/4}}\right)^{-1}$

$= \left(\frac{a^{2/2}b^{8/12}}{a^{3/2}b^{9/12}}\right)^{-1}$

$= \left(\frac{1}{a^{1/2}b^{1/12}}\right)^{-1}$

$= a^{1/2}b^{1/12}$

h) $\left(\frac{a^{-2}b^{2/3}}{b^{-1/2}}\right)^{-4} = \frac{a^8 b^{-8/3}}{b^2}$

$= \frac{a^8}{b^2 b^{8/3}}$

$= \frac{a^8}{b^{14/3}}$

$= \frac{a^8}{b^4 b^{2/3}}$

i) $\left[\frac{(27)^{4/3}-(27)^0}{(3^2+4^2)^{1/2}}\right]^{3/4} = \left[\frac{(\sqrt[3]{27})^4-1}{(9+16)^{1/2}}\right]^{3/4}$

$= \left[\frac{3^4-1}{25^{1/2}}\right]^{3/4}$

$= \left[\frac{81-1}{5}\right]^{3/4}$

$= \frac{80^{3/4}}{5^{3/4}}$

$= \frac{(16\cdot 5)^{3/4}}{5^{3/4}}$

$= \frac{16^{3/4}5^{3/4}}{5^{3/4}}$

$= 8$

j) $(16a^2 b^3)^{3/4} - 4ab^2(a^2 b)^{1/4}$

$= 16^{3/4}a^{6/4}b^{9/4} - 4ab^2 a^{2/4}b^{1/4}$

$= (\sqrt[4]{16})^3 a^{3/2}b^{9/4} - 4a^{3/2}b^{9/4}$

$= 8a^{3/2}b^{9/4} - 4a^{3/2}b^{9/4}$

$= 4a^{3/2}b^{9/4}$

k) $(\sqrt{3})^{3\sqrt{3}} - (3\sqrt{3})^{\sqrt{3}} + (\sqrt{3}^{\sqrt{3}})^{\sqrt{3}}$

$= (3^{1/2})^{3\sqrt{3}} - (3^{3/2})^{\sqrt{3}} + (3^{1/2})^3$

$= 3^{\frac{3\sqrt{3}}{2}} - 3^{\frac{3\sqrt{3}}{2}} + 3^{3/2}$

$= 3^{3/2}$

l) Notice that the multiplication is of the form $(x-y)(x^2+xy+y^2)$ which we know equals x^3-y^3. Therefore,

$(a^{1/3}-b^{1/3})(a^{2/3}+a^{1/3}b^{1/3}+b^{2/3})$

$= (a^{1/3})^3 - (b^{1/3})^3$

$= a - b$

87. a) $\qquad 4^{x+1} = (\frac{1}{2})^{2x}$

$(2^2)^{x+1} = (2^{-1})^{2x}$

$2^{2x+2} = 2^{-2x}$

Equating the exponents:

$2x+2 = -2x$

$4x = -2$

$x = -\frac{1}{2}$

b) $\qquad 5^{x^2-x} = 25$

$5^{x^2-x} = 5^2$

Equating the exponents:

$x^2 - x = 2$

$x^2 - x - 2 = 0$

$(x-2)(x+1) = 0$

$x-2 = 0 \text{ or } x+1 = 0$

$x = 2 \text{ or } \qquad x = -1$

c) $\qquad 2^{4x}4^{x-3} = (64)^{x-1}$

$2^{4x}(2^2)^{x-3} = (2^6)^{x-1}$

$2^{4x}2^{2x-6} = 2^{6x-6}$

$2^{6x-6} = 2^{6x-6}$

The equation is an identity, which will be true for all real values of x.

d) $\qquad (x^2+x+4)^{3/4} = 8$

Raise each side of the equation to the 4/3 power.

$[(x^2+x+4)^{3/4}]^{4/3} = 8^{4/3}$

$x^2 + x + 4 = (\sqrt[3]{8})^4$

$x^2 + x + 4 = 16$

$x^2 + x - 12 = 0$

$(x+4)(x-3) = 0$

$x = -4 \text{ or } \qquad x = 3$

e)
$$x^{2/3} - 3x^{1/3} = -2$$
$$(x^{1/3})^2 - 3(x^{1/3}) + 2 = 0$$
$$(x^{1/3} - 2)(x^{1/3} - 1) = 0$$
$$x^{1/3} - 2 = 0 \quad \text{or} \quad x^{1/3} - 1 = 0$$
$$x^{1/3} = 2 \quad \text{or} \quad x^{1/3} = 1$$
$$(x^{1/3})^3 = 2^3 \quad \text{or} \quad (x^{1/3})^3 = 1^3$$
$$x = 8 \quad \text{or} \quad x = 1$$

f)
$$2^{2x} - 2^{x+1} - 8 = 0$$
$$(2^x)^2 - 2(2^x) - 8 = 0$$
$$(2^x - 4)(2^x + 2) = 0$$
$$2^x - 4 = 0 \quad \text{or} \quad 2^x + 2 = 0$$
$$2^x = 4 \quad \text{or} \quad 2^x = -2$$

There is no value of x such that 2^x will equal -2, but $2^x = 4$ has the solution $x = 2$, since $2^2 = 4$.

89.
$$6^x > 4 \cdot 3^x - 27 \cdot 2^x + 108$$
$$6^x - 4 \cdot 3^x + 27 \cdot 2^x - 108 > 0$$
$$2^x 3^x - 4 \cdot 3^x + 27 \cdot 2^x - 4 \cdot 27 > 0$$

Factor by grouping:
$$3^x(2^x - 4) + 27(2^x - 4) > 0$$
$$(3^x + 27)(2^x - 4) > 0$$

Find the split points:
$$3^x + 27 = 0 \qquad 2^x - 4 = 0$$
$$3^x = -27 \qquad 2^x = 4$$
$$\text{No solution} \qquad x = 2$$

The number line is divided into only 2 regions by the split point $x = 2$.

Region	Test point	Value of $(3^x + 27)(2^x - 4)$
$x < 2$	$x = 0$	$-$
$x > 2$	$x = 3$	$+$

The solution set is $\{x : x > 2\}$.

91. $f(x) = 2^{-|x|}$

x	$f(x)$
-2	$2^{-2} = \frac{1}{4}$
-1	$2^{-1} = \frac{1}{2}$
0	$2^0 = 1$
1	$2^{-1} = \frac{1}{2}$
2	$2^{-2} = \frac{1}{4}$

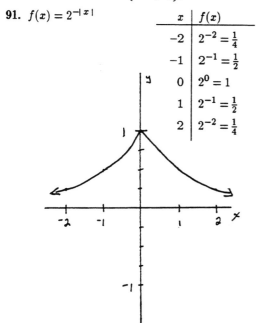

93. A polynomial function,
$$p(x) = k_n x^n + k_{n-1} x^{n-1} + \cdots + k_1 x + k \quad \text{with at}$$
least one of k_1 through $k_n \neq 0$, will become arbitrarily large (positive or negative) as x takes on arbitrarily large positive and negative values. The exponential function $f(x) = a^x$ with $a \neq 1$ and $a > 0$ must be asymptotic to the x-axis either for very large positive or very large negative x. Hence no polynomial can be equivalent to an exponential function.

95. Set new range values:
Xmin = -5
Xmax = 5
Ymin = -5
Ymax = 20
Press $\boxed{Y=}$
Enter π^X as Y_1
Enter X^π as Y_2
Press $\boxed{\text{GRAPH}}$
Zoom in on the apparent intersection point, to find that $x = 2.382$.

Problem Set 6.3

1. a) $y = 128(\frac{1}{2})^t$ decays, since $\frac{1}{2} < 1$.

b) $y = 5(\frac{5}{3})^t$ grows, since $\frac{5}{3} > 1$.

c) $y = 4(10)^9(1.03)^t$ grows, since $1.03 > 1$.

d) $1000(.99)^t$ decays, since $.99 < 1$.

3. a) $\boxed{1.08}\ \boxed{y^x}\ \boxed{20}\ \boxed{=}\ 4.66095714$

b) $\boxed{1.12}\ \boxed{y^x}\ \boxed{25}\ \boxed{=}\ 17.00006441$

c) $\boxed{1000}\ \boxed{\times}\ \boxed{1.04}\ \boxed{y^x}\ \boxed{40}\ \boxed{=}\ 4801.0206$

d) $\boxed{2000}\ \boxed{\times}\ \boxed{1.02}\ \boxed{y^x}\ \boxed{80}\ \boxed{=}\ 9750.87831$

5. After 10 years, the population will be $1000(1.04)^{10} \approx 1480$

7. a) 2020 is 45 years after 1975.
$$p(45) = 4(1.02)^{45} \approx 9.751 \text{ (billion)}$$

b) 2065 is 90 years after 1975.
$$p(90) = 4(1.02)^{90} \approx 23.773 \text{ (billion)}$$

9. $p(10) = 56000(1 - .02)^{10} = 56000(.98)^{10} \approx 45756$

11. After 15 years, the use of mercury will be
$$50(1 - .08)^{15} = 50(.92)^{15} \approx 14.31 \text{ kilograms}$$

13. Let y represent the amount of substance left after t years. Then

$$y = 30(\tfrac{1}{2})^{kt} \text{ with } k \text{ to be determined.}$$

Since the half-life of the element is 230 years, when $t = 230$, $y = \frac{30}{2} = 15$.

$$15 = 30(\tfrac{1}{2})^{230k}$$
$$\tfrac{1}{2} = (\tfrac{1}{2})^{230k}$$

Equating exponents:
$$230k = 1$$
$$k = \frac{1}{230}$$

The formula for decay becomes
$$y = 30(\tfrac{1}{2})^{t/230}$$

After 400 years,
$$y = 30(\tfrac{1}{2})^{400/230} \approx 8.99 \text{ kg}$$

15. Let y represent the amount of carbon-14 left after t years. Let a represent the initial amount of carbon-14. Then

$$y = a(\tfrac{1}{2})^{kt} \text{ with } k \text{ to be determined}$$

Since the half-life of carbon-14 is 5730 years, $y = \frac{a}{2}$

$$\frac{a}{2} = a(\tfrac{1}{2})^{5730k}$$
$$\tfrac{1}{2} = (\tfrac{1}{2})^{5730k}$$

Equating exponents:
$$5730k = 1$$
$$k = \frac{1}{5730}$$

The formula for decay becomes
$$y = a(\tfrac{1}{2})^{t/5730}$$

After 10,000 years,
$$y = a(\tfrac{1}{2})^{10000/5730}$$

Comparing this amount to the starting amount, the fraction of carbon-14 remaining is

$$\frac{a(\tfrac{1}{2})^{10000/5730}}{a} = (\tfrac{1}{2})^{10000/5730} \approx .298$$

17. $\frac{1}{7} = (\tfrac{1}{2})^{t/210}$

Since $\frac{1}{7} = .\overline{142857}$, we want to find a value of t such that $(\tfrac{1}{2})^{t/210}$ is as close to $.\overline{142857}$ as possible

t	$(\tfrac{1}{2})^{t/210}$
500	0.19198
600	0.13801
590	0.14264

$t \approx 590$ years

19. With $y = 800(2)^{t/9}$, we want to find a value of t such that

$$2700 = 800(2)^{t/9}$$
$$\frac{2700}{800} = (2)^{t/9}$$
$$3.375 = (2)^{t/9}$$

Try various values of t.

t	$(2)^{t/9}$
18	4
17	3.70
16	3.42
15	3.17
15.5	3.299
15.6	3.32
15.7	3.35
15.8	3.3765
15.9	3.402

$t \approx 15.8$ days

21. a) The amount after t years at 8% is
$$100(1.08)^8 = \$185.09$$

b) The amount after 8 years at 12% is
$$100(1.12)^8 = \$247.60$$

23. a) The amount after 40 years at 8% is
$$3500(1.08)^{40} = \$76,035.83$$

b) The amount after 40 years at 12% is
$$3500(1.12)^{40} = \$325,678.40$$

25. If r is the interest rate expressed as a percent, $\frac{r}{100}$ is the rate expressed as a fraction. Let A represent the amount of money accumulated after n years.

$$A = P\left(1 + \frac{r}{100}\right)^n$$

27. Principal is \$5000. 8% compounded semiannually is 4% per (6-month) period. 5 years is 10 periods.
$$\text{Amount} = 5000(1.04)^{10} = \$7401.22$$

29. Principal is \$3000. 9% compounded annually is 9% per (yearly) period. 10 years is 10 periods.
$$\text{Amount} = 3000(1.09)^{10} = \$7102.09$$

31. Principal is \$3000.
9% compounded quarterly is $\frac{9}{4}\% = 2.25\%$ per (quarterly) period. 10 years is 40 quarters (periods).
$$\text{Amount} = 3000(1.0225)^{40}$$
$$= 3000(2.435188965)$$
$$= \$7305.57$$

33. Principal is \$1000. 8% compounded monthly is
$$\frac{8\%}{12} = \frac{.08}{12} \text{ per (monthly) period.}$$
$$\text{Amount} = 1000\left(1 + \frac{.08}{12}\right)^{120} \approx \$2219.64$$

35. $t = -1$: $y = 5400(\frac{2}{3})^{-1} = 5400(\frac{3}{2}) = 8100$

$t = 0$: $\quad y = 5400(\frac{2}{3})^0 = 5400(1) = 5400$

$t = 1$: $\quad y = 5400(\frac{2}{3})^1 = 5400(\frac{2}{3}) = 3600$

$t = 2$: $\quad y = 5400(\frac{2}{3})^2 = 5400(\frac{4}{9}) = 2400$

$t = 3$: Another way to generate this sequence is to say that each value of y will be $\frac{2}{3}$ of the previous value of y. So here, $y = \frac{2}{3}(2400) = 1600$

37. If $(1.023)^T = 2$, $100(1.023)^{3T} = 100[(1.023)^T]^3$
$$= 100(2^3) = 100 \cdot 8 = 800$$

39. a) 2020 is 40 years after 1980, so $t = 40$.
$p(40) = 4600(1.016)^{40} \approx 8680$
2080 is 100 years after 1980, so $t = 100$.
$p(100) = 4600(1.016)^{100} \approx 22{,}497$

b) When will $p(t) = 9200$?
$9200 = 4600(1.016)^t$
$2 = (1.016)^t$

t	$(1.016)^t$
42	1.948
43	1.979
44	2.011

The doubling time is approximately 44 years.

41. a) Amount after 5 years compounded annually is
$100(1.08)^5 = \$146.93$

b) 8% compounded quarterly is $\frac{8\%}{4} = 2\%$

per quarter. 5 years is 20 quarters.
Amount $= 100(1.02)^{20} = \$148.59$

c) 8% compounded monthly is $\frac{8\%}{12} = \frac{.08}{12}$ per month. 5 years is 60 months.

Amount $= 100\left(1 + \frac{.08}{12}\right)^{60} = \$148.98$

d) 8% compounded daily is $\frac{8\%}{365} = \frac{.08}{365}$ per day. 5 years is 1825 days.

Amount $= 100\left(1 + \frac{.08}{365}\right)^{1825} = \$149.18$

43. a) We want to know the value of t at which
$(1.08)^t = 2$

t	$(1.08)^t$
5	1.469
7	1.713
9	1.999

$t \approx 9$ years

b) We want to know the value of t at which
$(1.065)^t = 2$

t	$(1.065)^t$
10	1.877
11	1.999

$t \approx 11$ years

45. a) Since the half-life of the element is 1690 years, when $t = 1690$, $q = \frac{30}{2} = 15$.

$$15 = 30(\tfrac{1}{2})^{1690k}$$

$$\tfrac{1}{2} = (\tfrac{1}{2})^{1690k}$$

Equating exponents:
$$1 = 1690k$$

$$\frac{1}{1690} = k$$

$$k = .0005917$$

b) The formula for decay becomes
$$q(t) = 30(\tfrac{1}{2})^{t/1690}$$
Then
$$q(2500) = 30(\tfrac{1}{2})^{2500/1690} \approx 10.76 \text{ mg}$$

47. Let A_0 represent the initial amount of carbon-14 and A represent the amount at any time t. Since the half-life of carbon-14 is 5730 years,
$$A = A_0(\tfrac{1}{2})^{t/5730}$$
At the time that we are interested in,
$$\frac{A}{A_0} = .76$$
Since $\frac{A}{A_0} = (\tfrac{1}{2})^{t/5730}$,
$$(\tfrac{1}{2})^{t/5730} = .76$$
To find the value of t that makes the above equation true, experiment with the calculator.
$t \approx 2270$ years

49. After 133 years, in 1987, the 10000 at 10% would have grown to $10000(1.1)^{133} = 10000(320056.8496)$
$= \$3{,}200{,}568{,}496$. A year later, in 1988, they could withdraw the 10% it had grown in that year, or $\$320{,}056{,}849.60$ The principal, reduced to the 1987 level, would continue to produce that same 10% interest forever. Such a fund is called a "perpetuity".

51.

t	$(1.03)^t$	$1 + .12t$
70	7.9178	9.4
90	14.300	11.8
*80	10.64	10.6
79	10.33	10.48

By experimenting with a calculator, we see that at $t = 80$ years, $(1.03)^t > 1 + .12t$ for the first time.

Problem Set 6.4

1. $4^3 = 64$ becomes $\log_4 64 = 3$

3. $27^{1/3} = 3$ becomes $\log_{27} 3 = \frac{1}{3}$

5. 4^0 becomes $\log_4 1 = 0$

7. $125^{-2/3} = \frac{1}{25}$ becomes $\log_{125}(\frac{1}{25}) = -\frac{2}{3}$

9. $10^{\sqrt{3}} = a$ becomes $\log_{10} a = \sqrt{3}$

11. 10^a becomes $\log_{10} \sqrt{3} = a$

13. $\log_5 625 = 4$ becomes $5^4 = 625$

15. $\log_4 8 = \frac{3}{2}$ becomes $4^{3/2} = 8$

17. $\log_{10}(.01) = -2$ becomes $10^{-2} = .01$

19. $\log_c c = 1$ becomes $c^1 = c$

21. $\log_c Q = y$ becomes $c^y = Q$

23. Let $\log_5 25 = x$.
    Then $5^x = 25$
    $\quad\ \ 5^x = 5^2$
    $\quad\ \ \ x = 2$

25. Let $\log_3 \frac{1}{3} = x$.
    Then $\ 3^x = \frac{1}{3}$
    $\quad\quad 3^x = 3^{-1}$
    $\quad\quad\ x = -1$

27. Let $\log_{27} 3 = x$.
    Then $27^x = 3$
    $\quad (3^3)^x = 3$
    $\quad\ \ 3^{3x} = 3^1$
    $\quad\quad\ \ x = \frac{1}{3}$

29. Let $\log_{10}(.0001) = x$.
    Then $10^x = .0001$
    $\quad\ \ 10^x = 10^{-4}$
    $\quad\quad\ x = -4$

31. Let $\log_3 1 = x$.
    Then $\ 3^x = 1$
    $\quad\quad 3^x = 3^0$
    $\quad\quad\ x = 0$

33. Let $\log_8 16 = x$.
    Then $\ 8^x = 16$
    $\quad (2^3)^x = 2^4$
    $\quad\ \ 2^{3x} = 2^4$
    $\quad\quad\ 3x = 4$
    $\quad\quad\quad x = \frac{4}{3}$

35. If $\log_c 8 = 3$,
    then $\quad c^3 = 8$
    $\quad\quad\ \ c = \sqrt[3]{8} = 2$

37. If $\log_9 c = -\frac{3}{2}$,
    then $9^{-3/2} = c$
    $$c = \frac{1}{(\sqrt{9})^3}$$
    $$c = \frac{1}{27}$$

39. If $\log_3(3^{-2.9}) = c$,
    then $\quad 3^c = 3^{-2.9}$
    $\quad\quad\quad c = -2.9$

41. $\quad c = 5^{2\log_5 7}$
    $\quad\quad = (5^{\log_5 7})^2$
    $\quad\quad = (7)^2$
    $\quad\quad = 49$

43. $\quad \log_{10} 6 = \log_{10}(3 \cdot 2)$
    $\quad\quad\quad\quad = \log_{10} 3 + \log_{10} 2$
    $\quad\quad\quad\quad = .477 + .301$
    $\quad\quad\quad\quad = .778$

45. $\quad \log_{10} 16 = \log_{10} 2^4$
    $\quad\quad\quad\quad = 4 \log_{10} 2$
    $\quad\quad\quad\quad = 4(.301)$
    $\quad\quad\quad\quad = 1.204$

47. $\quad \log_{10} \frac{1}{4} = \log_{10} 1 - \log_{10} 2^2$
    $\quad\quad\quad\quad = 0 - 2 \log_{10} 2$
    $\quad\quad\quad\quad = -2(.301)$
    $\quad\quad\quad\quad = -0.602$

49. $\quad \log_{10} 24 = \log_{10} 8 \cdot 3$
    $\quad\quad\quad\quad = \log_{10} 2^3 \cdot 3$
    $\quad\quad\quad\quad = 3 \log_{10} 2 + \log_{10} 3$
    $\quad\quad\quad\quad = 3(.301) + .477 = .903 + .477$
    $\quad\quad\quad\quad = 1.380$

51. $\quad \log_{10} \frac{8}{9} = \log_{10} 8 - \log_{10} 9$
    $\quad\quad\quad\quad = \log_{10} 2^3 - \log_{10} 3^2$
    $\quad\quad\quad\quad = 3 \log_{10} 2 - 2 \log_{10} 3$
    $\quad\quad\quad\quad = 3(.301) - 2(.477)$
    $\quad\quad\quad\quad = -0.051$

53. $\quad \log_{10} 5 = \log_{10} \frac{10}{2}$
    $\quad\quad\quad\quad = \log_{10} 10 - \log_{10} 2$
    $\quad\quad\quad\quad = 1 - .301$
    $\quad\quad\quad\quad = .699$

55. $\boxed{34}\ \boxed{\log} = 1.5314789$

57. $\boxed{.0123}\ \boxed{\log} = -1.910094889$

59. $\boxed{9723}\ \boxed{\log} = 3.987800286$

61. $3 \log_{10}(x + 1) + \log_{10}(4x + 7)$
    $\quad = \log_{10}(x + 1)^3 + \log_{10}(4x + 7)$
    $\quad = \log_{10}[(x + 1)^3(4x + 7)]$

63. $3\log_2(x+2) + \log_2 8x - 2\log_2(x+8)$
$= \log_2(x+2)^3 + \log_2 8x - \log_2(x+8)^2$
$= \log_2 \dfrac{8x(x+2)^3}{(x+8)^2}$

65. $\frac{1}{2}\log_6 x + \frac{1}{3}\log_6(x^3+3)$
$= \log_6 x^{1/2} + \log_6(x^3+3)^{1/3}$
$= \log_6 x^{1/2}(x^3+3)^{1/3}$
$= \log_6 \sqrt{x}\,\sqrt[3]{x^3+3}$

67. $\log_7(x+2) = 2$
$7^2 = x+2$
$x = 49 - 2$
$x = 47$

69. $\log_2(x+3) = -2$
$x+3 = 2^{-2}$
$x+3 = \dfrac{1}{2^2}$
$x = -3 + \dfrac{1}{4}$
$x = -\dfrac{11}{4}$

71. $\log_2 x - \log_2(x-2) = 3$
$\log_2 \dfrac{x}{x-2} = 3$
$\dfrac{x}{x-2} = 2^3$
$\dfrac{x}{x-2} = 8$
$x = 8(x-2)$
$x = 8x - 16$
$16 = 7x$
$\dfrac{16}{7} = x$

73. $\log_2(x-4) + \log_2(x-3) = 1$
$\log_2(x-4)(x-3) = 1$
$2^1 = (x-4)(x-3)$
$2^1 = x^2 - 7x + 12$
$x^2 - 7x + 12 - 2 = 0$
$x^2 - 7x + 10 = 0$
$(x-2)(x-5) = 0$
$x = 2 \text{ or } 5$
Checking, we find $x = 2$ leads to the logarithm of a negative number, so $x = 2$ is extraneous and the only solution is $x = 5$.

75. Use the change of base formula to get
$\log_2 128 = \dfrac{\log_{10} 128}{\log_{10} 2}$
$= \dfrac{2.10720997}{.301029995}$
$= 7$

77. Use the change of base formula to get
$\log_3 82 = \dfrac{\log_{10} 82}{\log_{10} 3}$
$= \dfrac{1.913813852}{.477121254}$
$= 4.01116872$

79. Use the change of base formula to get
$\log_6 39 = \dfrac{\log_{10} 39}{\log_{10} 6}$
$= \dfrac{1.591064607}{.77815125}$
$= 2.044672686$

81. a) If $x = \log_6 36$,
then $6^x = 36$
$6^x = 6^2$
$x = 2$

b) If $x = \log_4 2$,
then $4^x = 2$
$(2^2)^x = 2$
$2^{2x} = 2^1$
$2x = 1$
$x = \frac{1}{2}$

c) If $\log_{25} x = \frac{3}{2}$,
then $x = 25^{3/2} = (\sqrt{25})^3 = 5^3 = 125$

d) If $\log_4 x = \frac{5}{2}$,
then $x = 4^{5/2} = (\sqrt{4})^5 = 2^5 = 32$

e) If $\log_x 10\sqrt{10} = \frac{3}{2}$,
then $x^{3/2} = 10\sqrt{10}$
$x^{3/2} = 10 \cdot 10^{1/2}$
$x^{3/2} = 10^{3/2}$
$x = 10$

f) If $\log_x \frac{1}{8} = -\frac{3}{2}$, then $x^{-3/2} = \frac{1}{8}$
To isolate x, raise each side of the equation to the power $-\frac{2}{3}$.
$(x^{-3/2})^{-2/3} = (\frac{1}{8})^{-2/3}$
$x = 8^{2/3} = (\sqrt[3]{8})^2 = 2^2 = 4$

83. To evaluate $\dfrac{(\log_{27} 3)(\log_{27} 9)(3^{2\log_3 2})}{\log_3 27 - \log_3 9 + \log_3 1}$,
evaluate each term.
$\log_{27} 3 = \frac{1}{3}$ since $27^{1/3} = \sqrt[3]{27} = 3$
$\log_{27} 9 = \frac{2}{3}$ since $27^{2/3} = (\sqrt[3]{27})^2 = 9$
$3^{2\log_3 2} = 3^{\log_3(2^2)} = 3^{\log_3 4} = 4$
$\log_3 27 = 3$ since $3^3 = 27$
$\log_3 9 = 2$ since $3^2 = 9$
$\log_3 1 = 0$ since $3^0 = 1$

The expression is equal to

$$\frac{(\frac{1}{3})(\frac{2}{3})(4)}{3-2+0} = \frac{\frac{8}{9}}{1} = \frac{8}{9}$$

85. a) $\log_5(2x-1) = 2$
$$5^2 = 2x - 1$$
$$2x = 25 + 1 = 26$$
$$x = 13$$

b) $\log_4\left(\frac{x-2}{2x+3}\right) = 0$
$$4^0 = \frac{x-2}{2x+3}$$
$$1 = \frac{x-2}{2x+3}$$
$$2x + 3 = x - 2$$
$$2x - x = -2 - 3 = -5$$
$$x = -5$$

c) $\log_4(x-2) - \log_4(2x+3) = 0$
$$\log_4 \frac{x-2}{2x+3} = 0$$
$$\frac{x-2}{2x+3} = 4^0$$
$$\frac{x-2}{2x+3} = 1$$
$$x - 2 = 2x + 3$$
$$-5 = x$$

Checking, we see that each term becomes $\log_4(-7)$, which is undefined. Hence there is no solution. Note that (b) and (c) at first glance appear to be the same equation, but the slight change makes $x = -5$ impossible for part (c), emphasizing that one must check a solution in the *original* equation.

d) $\log_{10} x + \log_{10}(x-15) = 2$
$$\log_{10} x(x-15) = 2$$
$$10^2 = x(x-15)$$
$$100 = x^2 - 15x$$
$$x^2 - 15x - 100 = 0$$
$$(x-20)(x+5) = 0$$
$$x = 20 \text{ or } x = -5$$

$x = -5$ leads to the logarithms of negative numbers which is impossible, so $x = -5$ is extraneous. $x = 20$ checks, so it is the only solution.

e)
$$\frac{\log_2(x+1)}{\log_2(x-1)} = 2$$
$$\log_2(x+1) = 2\log_2(x-1)$$
$$\log_2(x+1) = \log_2(x-1)^2$$
$$\log_2(x-1)^2 - \log_2(x+1) = 0$$
$$\log_2 \frac{(x-1)^2}{x+1} = 0$$
$$2^0 = \frac{(x-1)^2}{x+1}$$

$$1 = \frac{(x-1)^2}{x+1}$$
$$x + 1 = (x-1)^2 = x^2 - 2x + 1$$
$$x^2 - 2x + 1 - x - 1 = 0$$
$$x^2 - 3x = 0$$
$$x(x-3) = 0$$
$$x = 0 \text{ or } x = 3$$

Checking in the original equation, $x = 0$ leads to the logarithm of a negative number, so 0 is an extraneous root. Substituting $x = 3$,
$$\frac{\log_2(3+1)}{\log_2(3-1)} = \frac{\log_4 4}{\log_2 2} = \frac{2}{1} = 2, \text{ so } x = 3 \text{ is a solution.}$$

f) $\log_8[\log_4(\log_2 x)] = 0$
$$8^0 = 1 = \log_4(\log_2 x)$$
$$4^1 = 4 = \log_2 x$$
$$2^4 = x$$
$$x = 16$$

87. a) $\log_a(x+y) = \log_a x + \log_a y$
$$\log_a(x+y) - \log_a x - \log_a y = 0$$
$$\log_a \frac{x+y}{xy} = 0$$
$$\frac{x+y}{xy} = a^0 = 1$$
$$x + y = xy$$
$$y - xy = -x$$
$$y(1-x) = -x$$
$$y = \frac{-x}{1-x} = \frac{x}{x-1}$$

b) $x = \log_a(y + \sqrt{y^2-1})$
$$y + \sqrt{y^2-1} = a^x$$
$$\sqrt{y^2-1} = a^x - y$$
Square both sides:
$$y^2 - 1 = (a^x)^2 - 2ya^x + y^2$$
$$2ya^x = a^{2x} + 1$$
$$y = \frac{a^{2x}+1}{2a^x}$$
$$= \frac{1}{2}\left(\frac{a^{2x}+1}{a^x}\right)$$
$$= \frac{1}{2}\left(a^x + \frac{1}{a^x}\right)$$
$$= \frac{1}{2}(a^x + a^{-x})$$

89. a) If $M = \log_{10}\left(\frac{A}{C}\right)$

then $\quad \frac{A}{C} = 10^M$
$$A = C \cdot 10^M$$

b) For the 1933 Japan earthquake, $M = 8.9$ and
$$A = C \cdot 10^{8.9}$$
For the 1989 San Francisco earthquake, $M = 7.1$ and $A = C \cdot 10^{7.1}$
The ratio of the two amplitudes is
$$\frac{C \cdot 10^{8.9}}{C \cdot 10^{7.1}} = 10^{1.8} \approx 63$$
The Japan earthquake was approximately 63 times as strong as the San Francisco earthquake.

91. Let $x = \log_a b$ and $y = \log_b a$. To show that $\log_a b = \dfrac{1}{\log_b a}$, we must show that $x = \dfrac{1}{y}$.

Since $x = \log_a b$, $a^x = b$. Since $y = \log_b a$, $b^y = a$.
Let $a = b^y$ in the equation $a^x = b$:
$$(b^y)^x = b$$
$$b^{yx} = b^1$$
Equating exponents:
$$yx = 1$$
$$x = \frac{1}{y}$$

93.

$y = 3^x$		$y = \log_3 x$	
x	3^x	x	$\log_3 x$
-2	$\frac{1}{9}$	$\frac{1}{9}$	-2
-1	$\frac{1}{3}$	$\frac{1}{3}$	-1
0	1	1	0
1	3	3	1
2	9	9	2

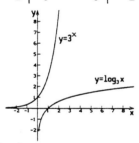

95. a) $f(x) = \log_2 x$

x	$\log_2 x$
$\frac{1}{4}$	-2
$\frac{1}{2}$	-1
1	0
2	1
4	2

b) $g(x) = \log_2(x + 1)$
The graph is the graph of $f(x)$ translated 1 unit left.

c) $h(x) = 3 + \log_2 x$
The graph is the graph of $f(x)$ translated 3 units up.

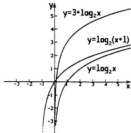

97. a) Set standard range values.
Press $\boxed{Y=}$
Enter $\boxed{(}\,\boxed{10^{\wedge}}\text{X} - \boxed{10^{\wedge}} - \text{X}\,\boxed{)}\,\boxed{\div}\,2$ as Y_1
Enter $\boxed{\log}\,\boxed{(}\,\text{X} + \boxed{\surd}\,\boxed{(}\,\text{X}^{\wedge}2 + 1\,\boxed{)}\,\boxed{)}$ as Y_2
Press $\boxed{\text{GRAPH}}$
The graphs appear to be reflections of each other in the line $y = x$; they appear to be inverse functions.

b) $f(g(x)) = f(\log_{10}(x + \sqrt{x^2 + 1})$

$$= \frac{10^{\log_{10}(x + \sqrt{x^2 + 1})} - 10^{-\log_{10}(x + \sqrt{x^2 + 1})}}{2}$$

$$= \frac{x + \sqrt{x^2 + 1} - (x + \sqrt{x^2 + 1})^{-1}}{2}$$

$$= \frac{x + \sqrt{x^2 + 1} - \dfrac{1}{x + \sqrt{x^2 + 1}}}{2}$$

$$= \frac{(x + \sqrt{x^2 + 1})(x + \sqrt{x^2 + 1}) - 1}{2(x + \sqrt{x^2 + 1})}$$

$$= \frac{x^2 + 2x\sqrt{x^2 + 1} + x^2 + 1 - 1}{2(x + \sqrt{x^2 + 1})}$$

$$= \frac{2x^2 + 2x\sqrt{x^2 + 1}}{2(x + \sqrt{x^2 + 1})}$$

$$= \frac{2x(x + \sqrt{x^2 + 1})}{2(x + \sqrt{x^2 + 1})}$$

$$= x$$

$f(x)$ and $g(x)$ are inverse functions.

Problem Set 6.5

1. $\ln e = 1$ since $e^1 = e$

3. $\ln 1 = 0$ since $e^0 = 1$

5. $\ln \sqrt{e} = \ln e^{1/2} = \frac{1}{2}\ln e = \frac{1}{2}(1) = \frac{1}{2}$

7. $\ln \dfrac{1}{e^3} = \ln(e^{-3}) = -3 \ln e = -3(1) = -3$

9. $e^{2 \ln 5} = e^{\ln 5^2} = e^{\ln 25} = 25$

11. $\ln(ae) = \ln a + \ln e = 2.5 + 1 = 3.5$

13. $\ln \sqrt{b} = \ln b^{1/2} = \frac{1}{2}\ln b = \frac{1}{2}(-4) = -2$

15. $\ln \frac{1}{a^3} = \ln a^{-3} = -3 \ln a = -3(2.5) = -7.5$

17. $\boxed{4.31}\,\boxed{\ln} = 1.460937904$

19. $\boxed{.127}\,\boxed{\ln} = -2.063568193$

21. $\boxed{(}\,\boxed{6.71}\,\boxed{\div}\,\boxed{42.3}\,\boxed{)}\,\boxed{\ln} = -1.841188135$

23. $\boxed{6.71}\,\boxed{\ln}\,\boxed{\div}\,\boxed{42.3}\,\boxed{\ln}\,\boxed{=} 0.508333025$

25. $\boxed{51.4}\,\boxed{y^x}\,\boxed{3}\,\boxed{=}\,\boxed{\ln} = 11.81891452$

27. $\boxed{51.4}\,\boxed{\ln}\,\boxed{y^x}\,\boxed{3}\,\boxed{=} 61.14613495$

29. $\boxed{2.12}\,\boxed{e^x} = 8.331137488$

31. $\boxed{.125}\,\boxed{+/-}\,\boxed{e^x} = 0.882496902$

33. If $\ln \sqrt{N} = 3.41$,
then $\sqrt{N} = e^{3.41}$
$N = (e^{3.41})^2$
$\boxed{3.41}\,\boxed{e^x}\,\boxed{x^2} = 915.9850101$

35. If $3^x = 20$,
$\ln 3^x = \ln 20$
$x \ln 3 = \ln 20$
$x = \dfrac{\ln 20}{\ln 3}$
$x = 2.7268$

37. If $2^{x-1} = .3$,
$\ln 2^{x-1} = \ln .3$
$(x-1)\ln 2 = \ln .3$
$x - 1 = \dfrac{\ln .3}{\ln 2}$
$x = 1 + \dfrac{\ln .3}{\ln 2}$
$x = -0.73697$

39. If $(1.4)^{x+2} = 19.6$,
$\ln(1.4)^{x+2} = \ln 19.6$
$(x+2)\ln 1.4 = \ln 19.6x$
$x \ln 1.4 + 2 \ln 1.4 = \ln 19.6$
$x \ln 1.4 = \ln 19.6 - 2 \ln 1.4$
$x = \dfrac{\ln 19.6 - 2 \ln 1.4}{\ln 1.4}$
$\quad = 6.843313779$

41. If P dollars are invested at 12% interest compounded annually, the amount of money, A, accumulated after t years is given by the formula $A = P(1.12)^t$. To find out when P doubles, let $A = 2P$ and solve the resulting equation for t.

$2P = P(1.12)^t$
$2 = (1.12)^t$
$\ln 2 = \ln(1.12)^t$
$\ln 2 = t \ln 1.12$
$t = \dfrac{\ln 2}{\ln 1.12} \approx 6.12$ years

43. If P dollars are invested at 15% interest compounded quarterly, the amount of money, A, accumulated after t years is given by the formula

$A = P\left(1 + \dfrac{.15}{4}\right)^{4t}$

To find out when P doubles, let $A = 2P$ and solve the resulting equation for t.
$2P = P(1.0375)^{4t}$
$2 = (1.0375)^{4t}$
$\ln 2 = \ln(1.0375)^{4t}$
$\ln 2 = 4t \ln(1.0375)$
$t = \dfrac{\ln 2}{4 \ln(1.0375)} \approx 4.71$ years

45. Since we start out with 50 grams of the element, to find the half-life, let $y = 25$ and solve the resulting equation for t.
$25 = 50e^{-.0125t}$
$.5 = e^{-.0125t}$
$\ln .5 = \ln e^{-.0125t}$
$\ln .5 = -0.125t \ln e$
$\ln .5 = -.0125t$
$t = \dfrac{\ln .5}{-.0125} \approx 55.45$ years

47. a) $\ln 10^5 = 5 \ln 10 = 11.5129$
$\ln 5^{10} = 10 \ln 5 = 16.0944$
5^{10} is larger.

b) $\ln 10^9 = 9 \ln 10 = 20.7233$
$\ln 9^{10} = 10 \ln 9 = 21.9722$
9^{10} is larger.

c) $\ln 10^{20} = 20 \ln 10 = 46.0517$
$\ln 20^{10} = 10 \ln 20 = 29.9573$
10^{20} is larger.

d) $\ln 10^{1000} = 1000 \ln 10 = 2302.5851$
$\ln 1000^{10} = 10 \ln 1000 = 69.0776$
10^{1000} is larger.

49.

$y = 3^x$		$y = x^3$	
x	3^x	x	x^3
0	1	0	0
1	3	1	1
2	9	2	8
3	27	3	27
4	81	4	64

(See graphs, next page.)

130

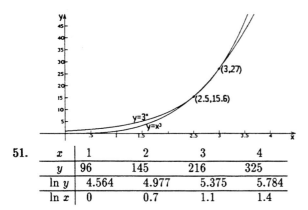

51.

x	1	2	3	4
y	96	145	216	325
$\ln y$	4.564	4.977	5.375	5.784
$\ln x$	0	0.7	1.1	1.4

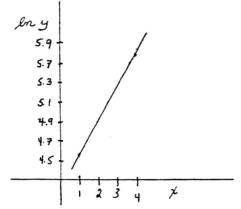

$\ln y$ is more closely a linear function of x than it is of $\ln x$, so we choose $y = ba^x$ as our model curve.

$\ln y = x \ln a + \ln b$

Let $Y = \ln y$ so that the equation of the line becomes $Y = x(\ln a) + \ln b$. An examination of the line shows that its approximate slope is

$$m = \frac{5.784 - 4.564}{4 - 1} \approx .40$$

Choosing $(x_1, y_1) = (1, 4.564)$, we can write the equation of the line as

$$Y - 4.564 = .40(x - 1)$$
$$Y = .40x + 4.564 - .40$$
$$Y = .4x + 4.164$$

Comparing this with $Y = \ln b + x \ln a$ gives

$\ln b = 4.164 \quad$ and $\quad \ln a = .4$
$b \approx 64 \qquad$ and $\qquad a \approx 1.5$

Thus the original data are described reasonably well by the equation $y = (64)1.5^x$.

53.

x	1	2	3	5
y	12	190	975	7490
$\ln y$	2.5	5.2	6.9	8.9
$\ln x$	0	0.7	1.1	1.6

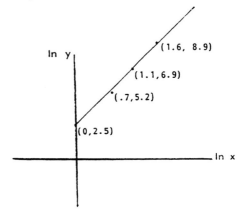

$\ln y$ is more closely a linear function of $\ln x$ than it is of x, so we choose $y = bx^a$ as our model curve.

$\ln y = \ln b + a \ln x$

Let $Y = \ln y$ and $X = \ln x$

$Y = aX + b, \ a \approx 4$

$\ln b = 2.5, \ b \approx 12$

The data are approximately described by $y = 12x^4$. A check shows that $f(1) = 12$, $f(2) = 192$, $f(3) = 972$, and $f(5) = 7500$, all of which are very close to the original data.

55. a) $\ln(e^{4.2}) = 4.2$

b) $e^{2 \ln 2} = e^{\ln(2^2)} = e^{\ln 4} = 4$

c) $\dfrac{\ln 3e}{2 + \ln 9} = \dfrac{\ln 3 + \ln e}{2 + \ln(3^2)}$

$\qquad = \dfrac{\ln 3 + 1}{2 + 2 \ln 3}$

$\qquad = \dfrac{\ln 3 + 1}{2(1 + \ln 3)}$

$\qquad = \dfrac{1}{2}$

57. a) $\ln[(e^{3.5})^2] = \ln(e^7) = 7$

b) $(\ln e^{3.5})^2 = (3.5)^2 = 12.25$

c) $\ln\left(\dfrac{1}{\sqrt{e}}\right) = \ln(e^{-1/2}) = -\dfrac{1}{2}$

d) $\dfrac{\ln 1}{\ln \sqrt{e}} = \dfrac{0}{\ln \sqrt{e}} = 0$

e) $e^{3 \ln 5} = e^{\ln(5^3)} = e^{\ln(125)} = 125$

f) $e^{\ln 1/2 + \ln 2/3} = (e^{\ln 1/2})(e^{\ln 2/3}) = \left(\dfrac{1}{2}\right)\left(\dfrac{2}{3}\right) = \dfrac{1}{3}$

59. a) If
$$10^{2x+3} = 200,$$
$$\log_{10} 10^{2x+3} = \log_{10} 200$$
$$(2x+3)\log_{10} 10 = \log_{10} 200$$
$$2x + 3 = \log_{10} 200$$
$$2x = \log_{10} 200 - 3$$
$$x = \frac{\log_{10} 200 - 3}{2}$$
$$x = -.349$$

b)
$$10^{2x} = 8^{x-1}$$
Taking $\log_{10}$ of each side,
$$2x \log_{10} 10 = (x-1)\log_{10} 8$$
$$2x = x \log_{10} 8 - \log_{10} 8$$
$$2x - x \log_{10} 8 = -\log_{10} 8$$
$$x(2 - \log_{10} 8) = -\log_{10} 8$$
$$x = \frac{-\log_{10} 8}{2 - \log_{10} 8}$$
$$x = -.823303622$$

c) If
$$10^{x^2+3x} = 200,$$
$$\log_{10} 10^{x^2+3x} = \log_{10} 200$$
$$(x^2 + 3x)\log_{10} 10 = \log_{10} 100$$
$$x^2 + 3x = \log_{10} 200$$
$$x^2 + 3x - \log_{10} 200 = 0$$
Using the quadratic formula
$$x = \frac{-3 \pm \sqrt{9 + 4\log_{10} 200}}{2} = \frac{-3 \pm 4.266}{2}$$
$$x = -3.633 \text{ or } x = .633$$

d) If
$$e^{-.32x} = \tfrac{1}{2},$$
$$\ln e^{-.32x} = \ln \tfrac{1}{2}$$
$$-.32x \ln e = \ln(.5)$$
$$-.32x = \ln(.5)$$
$$x = \frac{\ln(.5)}{-.32}$$
$$x = 2.166$$

e)
$$x^{\ln x} = 10$$
Taking ln of each side,
$$(\ln x)(\ln x) = \ln 10$$
$$(\ln x)^2 = \ln 10$$
$$\ln x = \pm \sqrt{\ln 10} = \pm 1.517427129$$
$$x = e^{1.517427129} = 4.56047657$$
$$\text{or } x = e^{-1.517427129} = .219275329$$

f)
$$(\ln x)^{\ln x} = x$$
Taking ln of each side,
$$(\ln x)[\ln(\ln x)] = \ln x$$
$$(\ln x)[\ln(\ln x)] - \ln x = 0$$
$$(\ln x)[\ln(\ln x) - 1] = 0$$
If $\ln x = 0$, then $x = 1$;
$x = 1$ does not satisfy the original equation

If $\ln(\ln x) - 1 = 0$, then
$$\ln(\ln x) = 1$$
$$\ln x = e^1 = e$$
$$x = e^e = 15.15426224$$

g)
$$x^{\ln x} = x$$
Taking ln of each side,
$$(\ln x)(\ln x) = \ln x$$
$$(\ln x)^2 - \ln x = 0$$
$$(\ln x)(\ln x - 1) = 0$$
If $\ln x = 0$, then $x = 1$
If $\ln x - 1 = 0$, then $\ln x = 1$ and so $x = e$.

h)
$$\ln x = (\ln x)^{\ln x}$$
Taking ln of each side,
$$\ln(\ln x) = (\ln x)[\ln(\ln x)]$$
$$\ln(\ln x) - (\ln x)[\ln(\ln x)] = 0$$
$$[\ln(\ln x)](1 - \ln x) = 0$$
If $\ln(\ln x) = 0$, then $\ln x = 1$ and so $x = e$.
If $1 - \ln x = 0$, then $\ln x = 1$ and so $x = e$.
Each factor leads to $x = e \approx 2.7182818285$.

i)
$$(x^2 - 5)^{\ln x} = x$$
Taking ln of each side,
$$(\ln x)[\ln(x^2 - 5)] = \ln x$$
$$(\ln x)[\ln(x^2 - 5)] - \ln x = 0$$
$$(\ln x)[\ln(x^2 - 5) - 1] = 0$$
If $\ln x = 0$, then $x = 1$.
If $\ln(x^2 - 5) - 1 = 0$, then
$$\ln(x^2 - 5) = 1$$
$$x^2 - 5 = e$$
$$x^2 = e + 5$$
$$x = \pm\sqrt{e + 5} = \pm 2.778179589$$
However, $x = -2.778179589$ does not check in the original equation.

61. If $y = 100e^{-3t}$, we know that half the initial amount of the substance is $\frac{100}{2} = 50$. To find the half-life, let $y = 50$ and solve for t.
$$50 = 100e^{-3t}$$
$$.5 = e^{-3t}$$
$$\ln(.5) = \ln(e^{-3t})$$
$$\ln(.5) = -3t \ln e$$
$$\ln(.5) = -3t$$
$$t = \frac{\ln(.5)}{-3} = 2.31 \text{ years}$$

63. We know that at the half-life, $A = \frac{A_0}{2}$. Let $A = \frac{A_0}{2}$ and $t = 240$ to find k.
$$\frac{A_0}{2} = A_0 e^{-240k}$$
$$.5 = e^{-240k}$$
$$\ln(.5) = \ln(e^{-240k}) = -240k(\ln e) = -240k$$
$$k = \frac{\ln(.5)}{-240} = .00289$$

65. Sketch the graph of $f(x) = y = \frac{1}{\sqrt{2\pi}} e^{-(1/2)x^2}$

Substituting $-x$ for x, we see $f(-x) = f(x)$; being an even function, it is symmetric about the y-axis.

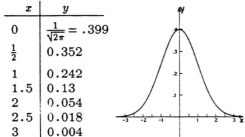

x	y
0	$\frac{1}{\sqrt{2\pi}} = .399$
$\frac{1}{2}$	0.352
1	0.242
1.5	0.13
2	0.054
2.5	0.018
3	0.004

67. Given that $e^x > 1 + x$ for $x > 0$, and that $\frac{\pi}{e} - 1 > 0$, let $x = \frac{\pi}{e} - 1$. Then

$$e^{(\pi/e - 1)} > 1 + \left(\frac{\pi}{e} - 1\right)$$
$$e^{\pi/e} e^{-1} > \frac{\pi}{e}$$
$$e^{\pi/e} > \pi$$

Raising both sides to the power e,
$$(e^{\pi/e})^e > \pi^e$$
$$e^\pi > \pi^e$$

69. a) Use the formula $A = P\left(1 + \frac{r}{m}\right)^{mt}$ with

$P = 100$, $t = 10$, $r = .12$, $m = 12$:
$$A = 100\left(1 + \frac{.12}{12}\right)^{120} = 100(1.01)^{120} = \$330.04$$

b) Use the formula $A = P\left(1 + \frac{r}{m}\right)^{mt}$ with

$P = 100$, $t = 10$, $r = .12$, $m = 365$:
$$A = 100\left(1 + \frac{.12}{365}\right)^{3650} = \$331.95$$

c) Use the formula $A = P\left(1 + \frac{r}{m}\right)^{mt}$ with

$P = 100$, $t = 10$, $r = .12$, $m = (365)(24) = 8760$:
$$A = 100\left(1 + \frac{.12}{8760}\right)^{87600} = \$332.01$$

d) Use the formula $A = Pe^{rt}$ with
$P = 100$, $r = .12$, $t = 10$:
$$A = 100e^{(1.2)} = \$332.01$$

71. Set new range values:
Xmin = 600   Xmax = 700
Xscal = .1   Yscal = .1
Press $\boxed{Y=}$
Enter .01X as Y_1 Enter ln x as Y_2
Press $\boxed{GRAPH}$
Zoom in on the area where the graphs appear to cross, eventually narrowing x down to approximately 647.278.

Chapter 6 Review Problem Set

1. True: $(^{15}\!\sqrt{3})^3 = 3^{3/15} = 3^{1/5} = \sqrt[5]{3}$

2. False, since $\sqrt[4]{16}$ denotes the positive 4th root only.

3. False: $\sqrt{5}\sqrt[3]{5} = 5^{1/2} \cdot 5^{1/3} = 5^{5/6} = \sqrt[6]{5^5}$

4. True

5. True: $\log_5 64 = \log_5 4^3 = 3\log_5 4$

6. True

7. True: $c^{17} = 2$
$(c^{17})^4 = (2)^4$
$c^{68} = 16$

8. False. $\log_2((-2)^2 - 2) = \log_2(4 - 2) = \log_2 2 = 1$

9. True

10. False. $f(x) = \ln(x^2)$ is defined for all x, $x \neq 0$, whereas $g(x) = 2\ln x$ is undefined for $x < 0$.

11. $\sqrt[3]{\frac{-27y^6}{z^{13}}} = \sqrt[3]{\frac{-27y^6 z^2}{z^{15}}} = \frac{-3y^2}{z^5} \sqrt[3]{z^2}$

12. $\sqrt[5]{64x^{10}y^6z^9} = \sqrt[5]{2 \cdot 2^5 \cdot (x^2)^5 \cdot y^5 y \cdot z^5 z^4}$
$$= \sqrt[5]{2^5(x^2)^5 y^5 z^5 \cdot 2yz^4} = 2x^2yz\sqrt[5]{2yz^4}$$

13. $\sqrt[3]{5\sqrt{5}} = \sqrt[3]{5 \cdot 5^{1/2}} = \sqrt[3]{5^{3/2}} = (\sqrt[3]{5^3})^{1/2} = (5)^{1/2} = \sqrt{5}$

14. $\frac{5}{2\sqrt{x} - 3} \cdot \frac{2\sqrt{x} + 3}{2\sqrt{x} + 3} = \frac{10\sqrt{x} + 15}{4x - 9}$

15. $\sqrt{81 + 9x^2} = \sqrt{9(9 + x^2)} = 3\sqrt{9 + x^2}$

16. $\sqrt{52} - \sqrt{117} + \frac{39}{\sqrt{13}} = \sqrt{4 \cdot 13} - \sqrt{9 \cdot 13} + \frac{39\sqrt{13}}{13}$
$$= 2\sqrt{13} - 3\sqrt{13} + 3\sqrt{13}$$
$$= 2\sqrt{13}$$

17. $\sqrt{2x - 5} = 9$
Square both sides:
$2x - 5 = 81$
$2x = 86$
$x = 43$
Check:
$\sqrt{2(43) - 5} \overset{?}{=} 9$
$\sqrt{81} \overset{?}{=} 9$
$9 = 9$
So $x = 43$ *is a root.*

18.
$$3\sqrt{x} = x - 4$$
$$(3\sqrt{x})^2 = (x-4)^2$$
$$9x = x^2 - 8x + 16$$
$$x^2 - 17x + 16 = 0$$
$$(x-16)(x-1) = 0$$
$$x = 16 \quad \text{or} \quad x = 1$$
Check:
$$3\sqrt{16} \stackrel{?}{=} 16 - 4 \qquad 3\sqrt{1} \stackrel{?}{=} 1 - 4$$
$$12 = 12 \qquad\qquad 3 \neq -3$$
So, $x = 12$ is the only solution.

19. $\sqrt[5]{x^3}x^{7/5} = x^{3/5}x^{7/5} = x^{10/5} = x^2$

20. $(\sqrt{a}\sqrt{ab})^2 = a \cdot ab = a^2 b$

21. $(x^2 y^{-4} z^6)^{-1/2} = x^{-1} y^2 z^{-3} = \dfrac{y^2}{xz^3}$

22. $\sqrt[3]{27\sqrt{64x^4}} = \sqrt[3]{27 \cdot 8x^2} = \sqrt[3]{27}\sqrt[3]{8x^2} = 3 \cdot 2x^{2/3} = 6x^{2/3}$

23. $(3a^{1/3}a^{-1/4})^3 = 3^3 a^1 a^{-3/4} = 27a^{1/4}$

24. $\sqrt[4]{\dfrac{32x^{-2}}{x^6 y^{-3}}} = \left(\dfrac{2^5 y^3}{x^8}\right)^{1/4} = \dfrac{2^{5/4}y^{3/4}}{x^2}$

25.

$$y = \left(\tfrac{5}{4}\right)^x \qquad\qquad y = \left(\tfrac{4}{5}\right)^x$$

x	y	x	y
-4	$256/625$	-4	$625/256$
-3	$64/125$	-3	$125/64$
-2	$16/25$	-2	$25/16$
-1	$4/5$	-1	$5/4$
0	1	0	1
1	$5/4$	1	$4/5$
2	$25/16$	2	$16/25$
3	$125/64$	3	$64/125$
4	$625/256$	4	$256/625$

The graphs are reflections of each other in the y-axis.

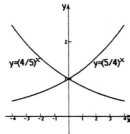

26. $P(t)$ is multiplied by 2 for every time t increases by 32. Therefore, $P(t) = 800,000 \cdot 2^{t/32}$. In 96 years, $P = 800,000 \cdot 2^{96/32} = 800,000 \cdot 2^3 = 6,400,000.$

27. Since the half-life of the substance is 1600 years, after 1600 years the ten grams will be reduced to 5 grams. After another 1600 years, th 5 grams will be reduced to 2.5 grams. That is, it will take 3200 years for 10 grams of the substance to decay to 2.5 grams.

28. a) $100 + 100 \times 0.12 \times 10 = \$220$

b) $100(1.12)^{10} \approx \$310.58$

c) $100\left(1 + \dfrac{.12}{12}\right)^{12\cdot 10} \approx \$330.04$

29. Let A represent the amount of money accumulated after t years when P dollars are invested at 8.5% compounded quarterly. Then
$$A = P\left(1 + \dfrac{.085}{4}\right)^{4t}$$
$$A = P(1.02125)^{4t}$$
To find the doubling time, let $A = 2P$ and solve for t:
$$2P = P(1.02125)^{4t}$$
$$2 = (1.02125)^{4t}$$

30. $100\left(1 + \dfrac{.09}{365}\right)^{365\cdot 2}$

31. If $\log_4 x = 3,$
then $x = 4^3$
$$x = 64$$

32. $\log_{25} x = \dfrac{-3}{2}$
$$x = 25^{-3/2} = \dfrac{1}{(\sqrt{25})^3} = \dfrac{1}{125}$$

33. If $2^{\log_2 3} = x,$
then $x = 3$

34. $\log_6 6^\pi = x \Rightarrow 6^x = 6^\pi \Rightarrow x = \pi$

35. $x = \log_{10}\sqrt{10,000}$
$$= \log_{10}(10,000)^{1/2}$$
$$= \log_{10}(10^4)^{1/2}$$
$$= \log_{10}(10)^2$$
$$= 2$$

36. $x = \log_8 32 = \dfrac{\log_2 32}{\log_2 8} = \dfrac{5}{3}$

37. If $\log_x 64 = \dfrac{3}{2},$
then $x^{3/2} = 64.$
Raise each side to the power $\tfrac{2}{3}$.
$$(x^{3/2})^{2/3} = 64^{2/3}$$
$$x = (\sqrt[3]{64})^2$$
$$= 16$$

38. $\ln x = 0 \Rightarrow x = e^0 = 1$

39. If $\log_2(x^2 - 1) = 3$,

then
$$x^2 - 1 = 2^3$$
$$x^2 = 9$$
$$x = \pm 3$$

40. $\log_2 x + \log_2(x + 2) = 3$
$$\log_2 x(x + 2) = 3$$
$$x^2 + 2x = 2^3$$
$$x^2 + 2x - 8 = 0$$
$$(x + 4)(x - 2) = 0$$
$$x = -4 \qquad x = 2$$

Since $\log_2(-4)$ is undefined, the only answer is $x = 2$.

41. $3 \log_4(x^2 + 1) - 2 \log_4 x + 2$
$$= \log_4(x^2 + 1)^3 - \log_4 x^2 + \log_4 16$$

$$= \log_4 \frac{16(x^2 + 1)^3}{x^2}$$

42. The graph of $y = \ln(x - 1) + 2$ is the graph of $y = \ln x$ translated 1 unit to the right and 2 units upwards.

43. $y = \log_3(x + 2)$
The vertical asymptote is the line $x = -2$.
The graph is the graph of $y = \log_3 x$ translated 2 units left.

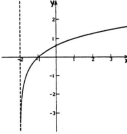

44. $\log_a b = \frac{2}{3}$

$$\frac{\log b}{\log a} = \frac{2}{3}$$

$$\frac{\log a}{\log b} = \frac{3}{2}$$

$$\log_b a = \frac{3}{2}$$

45.
$$3^{2x+1} = 20$$
$$\ln 3^{2x+1} = \ln 20$$
$$(2x + 1)\ln 3 = \ln 20$$
$$(2 \ln 3)x + \ln 3 = \ln 20$$
$$(2 \ln 3)x = \ln 20 - \ln 3$$
$$x = \frac{\ln 20 - \ln 3}{2 \ln 3}$$
$$x = .86342$$

46. After n quarters
$$160 = 100\left(1 + \frac{.085}{4}\right)^n$$
$$1.6 = 1.02125^n$$
$$\ln 1.6 = n \ln 1.02125$$
$$n = \frac{\ln 1.6}{\ln 1.02125} = 22.35$$
It will take 23 quarters or $5\frac{3}{4}$ years.

47. a) To find the half-life, let $y = \frac{y_0}{2}$ and solve for t.

$$\frac{y_0}{2} = y_0 e^{-.055t}$$

$$\frac{1}{2} = e^{-.055t}$$

$$\ln(\tfrac{1}{2}) = \ln(e^{-.055t})$$

$$\ln(\tfrac{1}{2}) = -.055t(\ln e)$$

$$t = \frac{\ln(\tfrac{1}{2})}{-.055}$$

$$t = 12.603 \text{ years}$$

b) Let $y = \frac{y_0}{10}$ and solve for t.

$$\frac{y_0}{10} = y_0 e^{-.055t}$$

$$\frac{1}{10} = e^{-.055t}$$

$$\ln(\tfrac{1}{10}) = -.055t(\ln e)$$

$$t = \frac{\ln(\tfrac{1}{10})}{-.055}$$

$$t = 41.865 \text{ years}$$

48.
$$10 = 40e^{-20k}$$
$$\frac{1}{4} = e^{-20k}$$
$$4 = e^{20k}$$
$$\ln 4 = 20k$$
$$k = \frac{\ln 4}{20} \approx 0.0693$$

49. $f(x) = \log_4 x$ and $g(x) = 4^x$ are inverse functions. Therefore, their graphs are reflections of each other in the line $y = x$.

50. $x^x = 500$

Take the logarithm of both sides:

$x \ln x = \ln 500 = 6.214608098$

Let $f(x) = x \ln x$.

$f(4) = 4 \ln 4 = 5.54518$; $f(5) = 5 \ln 5 = 8.04719$

Estimate the fraction of the distance from 4 to 5 at which answer lies:

$$\frac{\ln 500 - f(4)}{f(5) - f(4)} = \frac{6.21461 - 5.54518}{8.04719 - 5.54518} = .27$$

$f(4.27) = 6.198391$; $f(4.28) = 6.222919$

Estimate the fraction of the distance from 4.27 to 4.28 at which answer lies:

$$\frac{\ln 500 - f(4.27)}{f(4.28) - f(4.27)} = \frac{6.214608 - 6.198391}{6.222919 - 6.198391} = .6612$$

$f(4.276612) = 6.214606$; $f(4.276613) = 6.214609$

$x = 4.276613$ to six decimal places.

(The reason we first took the logarithm of both sides is that x^x grows too quickly for accurate estimates.)

Chapter 7: The Trigonometric Functions

Problem Set 7.1

1. $\sin 41.3° = .6600$

3. $\cos 49.2° = .6534$

5. $\tan 72.3° = 3.1334$

7. $\sin \theta = .2164$
   $\theta = 12.50°$

9. $\tan \theta = 2.311$
   $\theta = 66.60°$

11. $\cos \theta = .3535$
    $\theta = 69.30°$

13. $\sin 29° = \frac{x}{35}$
    $x = 35 \sin 29° = 16.97 \approx 17$

15. $\sin 14° = \frac{10}{x}$
    $x = \frac{10}{\sin 14°} = 41.34 \approx 41$

17.

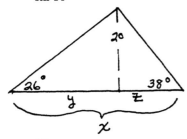

$\tan 26° = \frac{20}{y}$

$y = \frac{20}{\tan 26°} = 41.00$

$\tan 38° = \frac{20}{z}$

$z = \frac{20}{\tan 38°} = 25.60$

$x = y + z$

$\quad = 41.00 + 25.60 = 66.60 \approx 67$

19.

$\beta = 90° - 42° = 48°$
$\sin 42° = \frac{a}{35}$
$\quad a = 35 \sin 42° = 23.42 \approx 23$
$\cos 42° = \frac{b}{35}$
$\quad b = 35 \cos 42° = 26.01 \approx 26$

21.

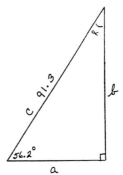

$\alpha = 90° - 56.2° = 33.8°$

$\sin 56.2° = \frac{b}{91.3}$

$\quad b = 91.3 \times \sin 56.2° = 75.9$

$\cos 56.2° = \frac{a}{91.3}$

$\quad a = 91.3 \times \cos 56.2° = 50.8$

23.

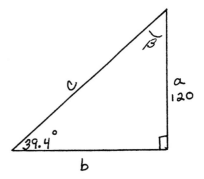

$\beta = 90° - 39.4° = 50.6°$

$\sin 39.4° = \frac{120}{c}$

$\quad c = \frac{120}{\sin 39.4°} = 189$

$\tan 39.4° = \frac{120}{b}$

$\quad b = \frac{120}{\tan 39.4°} = 146$

25.

$c = \sqrt{a^2 + b^2} = \sqrt{81 + 144} = \sqrt{225} = 15$

$\tan \beta = \frac{12}{9} = \frac{4}{3}$

$\beta = 53.1°$

$\tan \alpha = \frac{9}{12} = \frac{3}{4}$

$\alpha = 36.9°$

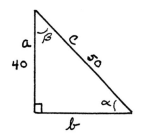

Problem Set 7.1

27.

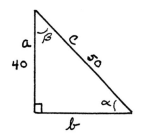

$$b = \sqrt{c^2 - a^2} = \sqrt{2500 - 1600} = \sqrt{900} = 30$$
$$\cos \alpha = \frac{30}{50} = .6$$
$$\alpha = 53.1°$$
$$\cos \beta = \frac{40}{50} = .8$$
$$\beta = 36.9°$$

29.

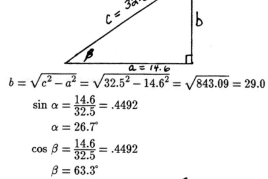

$$b = \sqrt{c^2 - a^2} = \sqrt{32.5^2 - 14.6^2} = \sqrt{843.09} = 29.0$$
$$\sin \alpha = \frac{14.6}{32.5} = .4492$$
$$\alpha = 26.7°$$
$$\cos \beta = \frac{14.6}{32.5} = .4492$$
$$\beta = 63.3°$$

31.

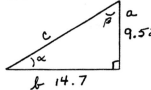

$$c = \sqrt{a^2 + b^2}$$
$$= \sqrt{9.52^2 + 14.7^2} = \sqrt{306.7204} = 17.5$$
$$\tan \alpha = \frac{9.52}{14.7} = .6476$$
$$\alpha = 32.9°$$
$$\tan \beta = \frac{14.7}{9.52} = 1.5441$$
$$\beta = 57.1°$$

33.

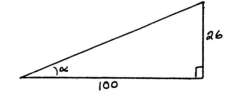

$$\tan \alpha = \frac{26}{100} = .26$$
$$\alpha = 14.6°$$

35.

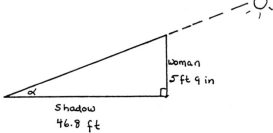

5 feet 9 inches = 5.75 feet
$$\tan \alpha = \frac{5.75}{46.8}$$
$$\alpha = 7°$$
The angle of elevation of the sun is 7.0°.

37.

$$\tan \alpha = \frac{5.75}{46.8} = \frac{3\frac{10}{12}}{x}$$
$$x = \frac{46}{12} \times \frac{46.8}{5.75} = 31.2$$
Sue's shadow is 31.2 feet long.

39. a) $\tan 14.5° = .25862$

b) $24.6 \cos 74.3° \rightarrow 24.6 \times 74.3 \boxed{\cos} = 6.6568$

c) $15.6 \, (\sin 14°)^2 / \cos 87° \rightarrow$

$15.6 \times 14 \boxed{\sin} \boxed{x^2} \div 87 \boxed{\cos} = 17.445$

41. Let x feet be the distance from the boat to the foot of the lighthouse.
By the properties of parallel lines, the given angle of depression equals the angle of elevation from the boat to the top of the lighthouse. Therefore,

$$\tan 9.4° = \frac{120}{x}$$
$$x = \frac{120}{\tan 9.4°} = 724.9 \text{ feet} \approx 725 \text{ feet}$$

43. Let y and z be the bases of the triangles on the left and right, respectively. Then

$$\tan 61° = \frac{31}{y}$$
$$y = \frac{31}{\tan 61°} = 17.18$$
$$\tan 34° = \frac{14}{z}$$
$$z = \frac{14}{\tan 34°} = 20.76$$
$$x = y + z = 17.18 + 20.76 = 37.94 \approx 37.9$$

45.

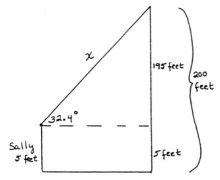

Let there be x feet of kite string out. Referring to the above diagram,

$$\sin 32.4° = \frac{195}{x}$$
$$x = \frac{195}{\sin 32.4°} = 363.9 \approx 364 \text{ feet}$$

47.

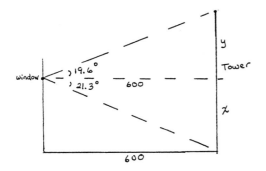

Let the height of the tower be $x + y$ meters, where x meters is the height of the office window, and y meters is the height the tower extends above the office window. Referring to the above diagram,

$$\tan 21.3° = \frac{x}{600}$$
$$x = 600 \tan 21.3° = 233.93$$
$$\tan 19.6° = \frac{y}{600}$$
$$y = 600 \tan 19.6° = 213.65$$
$$x + y = 233.93 + 213.65 = 447.58$$
The height of the tower is 447.6 meters

49.

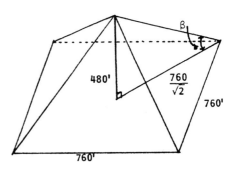

The diagonal of the square base is $760\sqrt{2}$ so the distance from a vertex to the center is $380\sqrt{2}$.

$$\tan \beta = \frac{480}{380\sqrt{2}} = .8932$$
$$\beta = 41.8°$$

51.

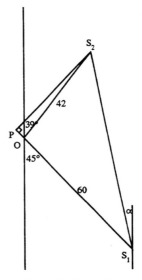

a) Using the formula $d = rt$,

$$\overline{OS_1} = 2.5 \times 24 = 60 \text{ miles}$$
$$\overline{OS_2} = 1.5 \times 28 = 42 \text{ miles}$$

Draw a perpendicular S_2P to the extension of S_1O.

Angle $S_2OP = 45° + 39° = 84°$
$$\sin 84° = \frac{\overline{S_2P}}{42}$$
$$\overline{S_2P} = 42 \sin 84° = 41.77$$
$$\cos 84° = \frac{\overline{PO}}{42}$$
$$\overline{PO} = 42 \cos 84° = 4.39$$
$$\overline{S_1P} = \overline{PO} + 60 = 64.39$$

$$\overline{S_1S_2} = \sqrt{\overline{S_1P}^2 + \overline{S_2P}^2}$$
$$= \sqrt{64.39^2 + 41.77^2} = 76.75 \text{ miles}$$
$$\tan(45° - \alpha) = \frac{\overline{S_2P}}{\overline{S_1P}} = \frac{41.77}{64.39} = .6487$$
$$45° - \alpha = 32.97°$$
$$\alpha = 45° - 32.97° = 12.03°$$

53.

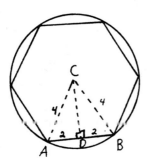

Equilateral triangle ABC is $\frac{1}{6}$ of the hexagon.

The altitude of ABC is $\overline{CD} = \sqrt{4^2 - 2^2} = 2\sqrt{3}$.

Therefore,
the area of the hexagon is $6(\frac{1}{2})(4)2\sqrt{3} = 24\sqrt{3}$

and its perimeter is $6 \cdot 4 = 24$

55.

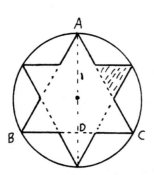

The required area
$$= \text{area of } \triangle ABC + 3(\text{area of shaded } \triangle).$$
In triangle ADC, $\overline{AD} = \frac{3}{2}$ and $\overline{DC} = \frac{1}{2}\overline{AC}$.
$$\overline{AC}^2 = (2\,\overline{DC})^2 = (\tfrac{3}{2})^2 + \overline{DC}^2$$
$$\Rightarrow 3\,\overline{DC}^2 = \frac{9}{4} \Rightarrow \overline{DC} = \frac{1}{2}\sqrt{3}$$
Area of $\triangle ABC = \frac{3}{2} \cdot \frac{1}{2}\sqrt{3} = \frac{3}{4}\sqrt{3}$

The sides of the shaded triangle are $\frac{1}{3}$ as long as those of the similar triangle ABC. Therefore

Shaded area $= \frac{1}{9} \cdot \frac{3}{4}\sqrt{3} = \frac{1}{12}\sqrt{3}$

Total area $= \frac{3}{4}\sqrt{3} + 3 \cdot \frac{1}{12}\sqrt{3} = \sqrt{3}$

Problem Set 7.2

1. $120° = 120° \times \frac{\pi \text{ radians}}{180°} = \frac{2\pi}{3}$ radians

3. $240° = 240° \times \frac{\pi \text{ radians}}{180°} = \frac{4\pi}{3}$ radians

5. $210° = 210° \times \frac{\pi \text{ radians}}{180°} = \frac{7\pi}{6}$ radians

7. $315° = 315° \times \frac{\pi \text{ radians}}{180°} = \frac{7\pi}{4}$ radians

9. $540° = 540° \times \frac{\pi \text{ radians}}{180°} = 3\pi$ radians

11. $-420° = -420° \times \frac{\pi \text{ radians}}{180°} = -\frac{7\pi}{3}$ radians

13. $160° = 160° \times \frac{\pi \text{ radians}}{180°} = \frac{8\pi}{9}$ radians

15. $\left(\frac{20}{\pi}\right)° = \left(\frac{20}{\pi}\right)° \times \frac{\pi \text{ radians}}{180°} = \frac{1}{9}$ radians

17. $\frac{4}{3}\pi$ radians $= \frac{4\pi}{3} \times \frac{180°}{\pi \text{ radians}} = 240°$

19. $-\frac{2\pi}{3}$ radians $= -\frac{2\pi}{3}$ radians $\times \frac{180°}{\pi \text{ radians}} = -120°$

21. 3π radians $= 3\pi$ radians $\times \frac{180°}{\pi \text{ radians}} = 540°$

23. 4.52 radians $= 4.52$ radians $\times \frac{180°}{\pi \text{ radians}} \approx 259.0°$

25. $\frac{1}{\pi}$ radians $= \frac{1}{\pi}$ radians $\times \frac{180°}{\pi \text{ radians}} = \left(\frac{180}{\pi^2}\right)° \approx 18.2°$

Note: angle measurements have no dimensions; they do not appear when combined with other units.

27. a) $t = \frac{s}{r} = \frac{12 \text{ inches}}{6 \text{ inches}} = 2$ radians

b) $t = \frac{s}{r} = \frac{18.84 \text{ inches}}{6 \text{ inches}} = 3.14$ radians

29. a) $r = \frac{s}{t} = \frac{8.4 \text{ cm}}{2.8} = 3$ cm

b) $r = \frac{s}{t} = \frac{33 \text{ inches}}{6} = 5.5$ inches

31. a) $A = \frac{1}{2}r^2t = \frac{1}{2}(3 \text{ cm})^2(2.8) = 12.6$ cm^2

b) $A = \frac{1}{2}r^2t = \frac{1}{2}(5.5 \text{ in.})^2(6) = 90.75$ in^2

33. Since $\frac{\pi}{2} < 3 < \pi$, the point is in quadrant II.

35. Since $\pi < 4.7 < \frac{3\pi}{2}$, the point is in quadrant III.

37. $\frac{5\pi}{2} + 1 = 2\pi + \frac{\pi}{2} + 1$

The point traveled one revolution around the unit

circle, and then $\frac{\pi}{2} + 1$ additional units. Since
$$\frac{\pi}{2} < \frac{\pi}{2} + 1 < \pi,$$
the point is in quadrant II.

39. $\frac{100}{2\pi} \approx 15.9$. Thus
$$100 = 15(2\pi) + (100 - 15 \times 2\pi) \approx 15(2\pi) + 5.8$$
Since $\frac{3\pi}{2} < 5.8 < 2\pi$, the point is in quadrant IV.

41. $r = 8$ inches $\times \frac{1 \text{ ft}}{12 \text{ in}} = \frac{2}{3}$ ft

$\omega = \frac{4 \text{ rev}}{\sec} \times \frac{2\pi \text{ rad}}{\text{rev}} = \frac{8\pi \text{ rad}}{\sec}$

$v = r\omega = \frac{2}{3} \text{ ft} \times \frac{8\pi \text{ rad}}{\sec}$

$\quad = \frac{16}{3}\pi \frac{\text{ft}}{\sec} \approx 16.76 \frac{\text{ft}}{\sec}$

43. For the larger wheel,

$\omega = \frac{20 \text{ rev}}{\min} \times \frac{2\pi \text{ rad}}{\text{rev}} = \frac{40\pi \text{ rad}}{\min}$

$v = r\omega = 8 \text{ inches} \times \frac{40\pi \text{ rad}}{\min}$

$\quad = 320\pi \frac{\text{inches}}{\min} \approx 1005.3 \frac{\text{inches}}{\min}$

45. (a) $-1440° = -1440° \times \frac{\pi \text{ rad}}{180°} = -8\pi$ rad

(b) $2\frac{1}{2}$ rev $= \frac{5}{2}$ rev $\times \frac{2\pi \text{ rad}}{\text{rev}} = 5\pi$ rad

(c) $\left(\frac{60}{\pi}\right)° = \left(\frac{60}{\pi}\right)° \times \frac{\pi \text{ rad}}{180°} = \frac{1}{3}$ rad

47. (a) $s = rt = 4.25 \text{ cm} \times 6 \text{ rad} = 25.5$ cm

(b) First convert degrees to radians:

$\left(\frac{18}{13\pi}\right)° = \left(\frac{18}{13\pi}\right)° \times \frac{\pi \text{ rad}}{180°} = \frac{1}{130}$ rad

$s = rt = 4.25 \text{ cm} \times \frac{1}{130} \text{ rad} = .0327$ cm

(c) $s = rt = 4.25 \text{ cm} \times \frac{17\pi}{6} \text{ rad} = 37.83$ cm

49. Since the rear wheel sprocket has radius $\frac{1}{4}$ the size of the pedal sprocket, the rear wheel completes 4 revolutions for every one revolution of the pedal. Since the pedal completes 30 revolutions, the rear wheel completes 120 revolutions. The rear wheel, and therefore the bicycle, travels

$s = rt = 40 \text{ cm} \times \left(120 \text{ rev} \times \frac{2\pi \text{ rad}}{\text{rev}}\right)$

$\quad = 9600\pi \text{ cm} \approx 30{,}159$ cm

51. The earth completes one revolution around its axis in 24 hours.

$v = r\omega = 3960 \text{ mi} \times \left(\frac{1 \text{ rev}}{24 \text{ hr}} \times \frac{2\pi \text{ rad}}{\text{rev}}\right)$

$\quad = 330\pi \frac{\text{mi}}{\text{hr}} \approx 1037 \frac{\text{mi}}{\text{hr}}$

53. In the same amount of time, each circle will move through the same arc length. If the smaller circle moves through an angle θ_1 and sweeps out an area A while the larger circle moves through an an angle θ_2 and sweeps out an area A_2, then

$s = 6\theta_1 = 8\theta_2$

$\quad \frac{\theta_2}{\theta_1} = \frac{6}{8}$

and

$\frac{A_2}{A} = \frac{\frac{1}{2}\pi(8)^2\theta_2}{\frac{1}{2}\pi(6)^2\theta_1} = \frac{8^2}{6^2} \cdot \frac{\theta_2}{\theta_1} = \frac{8^2}{6^2} \cdot \frac{6}{8} = \frac{8}{6}$

$A_2 = \frac{8}{6}A = \frac{4}{3}A$

55.

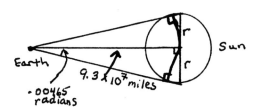

In the figure above we have assumed that the 93 million miles is the distance between the centers, and that the radius of the earth is negligible. Then

$\sin .00465 = \frac{r}{9.3 \times 10^7}$

$\quad r = (9.3 \times 10^7) \sin .00465 = 4.324 \times 10^5$

The diameter of the sun is twice the radius, r:

$\quad 2r = 8.6 \times 10^5$ miles

57. To find the distance s from Mr. Varberg's house to the North Pole, use

$t = 90° - 45° = 45° = \frac{\pi}{4}$ rad

$s = rt = 3960 \text{ miles} \times \frac{\pi}{4} \text{ rad} = 990\pi$ miles

Flying at 600 mi/hr it will take

$\frac{990\pi \text{ miles}}{600 \text{ mi/hr}} = \frac{33\pi}{20} \text{ hr} \approx 5.84$ hr

to fly to the North Pole.

59.

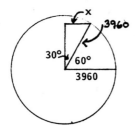

x is the radius of the 60th parallel (which is a circle of course). The hypotenuse is the earth's radius, 3960 mi.

$$\sin 30° = \frac{x}{3960}$$
$$x = 3960 \sin 30° = 1980 \text{ mi}$$

The angular distance between the two cities is

$$30° - 6° = 24° = 24° \times \frac{\pi \text{ rad}}{180°} = \frac{2\pi}{15} \text{ rad}$$

Hence the actual distance between the cities is

$$s = rt = 1980 \text{ mi} \times \frac{2\pi}{15}$$
$$= 264\pi \text{ mi} \approx 829 \text{ mi}$$

61.

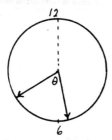

The minute hand moves at the rate of 360°/hr and the hour hand at $\frac{1}{12} \cdot 360° = 30°$/hr. Thus the minute hand moves 330°/hr faster. At 6 o'clock the two hands are 180° apart. At 5:40, $\frac{1}{3}$ hr earlier, they are $\frac{1}{3} \cdot 330° = 110°$ closer, or separated by

$$t = 70° \times \frac{\pi \text{ rad}}{180°} = \frac{7\pi}{18} \text{ rad}$$

The area of the pie-shaped region is

$$A = \frac{1}{2}r^2t = \frac{1}{2}(6 \text{ in})^2\frac{7\pi}{18} \text{ rad}$$
$$= 7\pi \text{ in}^2 \approx 21.99 \text{ in}^2$$

63. Let r be the radius of the smaller circle. Then

$$\frac{10 + r}{r} = \frac{18}{8}$$
$$18r = 80 + 8r$$
$$10r = 80$$

$$r = 8$$

This means that the angle at the center is 1 radian. The area of the polar rectangle is

$$A = \frac{1}{2}(18^2 - 8^2) \cdot 1 = 130 \text{ square units}$$

Problem Set 7.3

1.

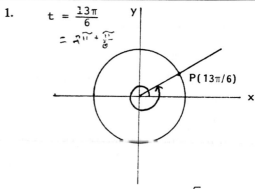

$$P\left(\frac{13\pi}{6}\right) = P\left(2\pi + \frac{\pi}{6}\right) = P\left(\frac{\pi}{6}\right) = \left(\frac{\sqrt{3}}{2}, \frac{1}{2}\right)$$

3.

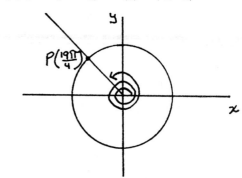

$$P\left(\frac{19\pi}{4}\right) = P\left(4\pi + \frac{3\pi}{4}\right) = P\left(\frac{3\pi}{4}\right) = \left(-\frac{\sqrt{2}}{2}, \frac{\sqrt{2}}{2}\right)$$

5.

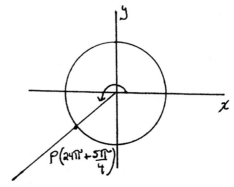

$$P\left(24\pi + \frac{5\pi}{4}\right) = P\left(12 \cdot 2\pi + \frac{5\pi}{4}\right) = P\left(\frac{5\pi}{4}\right) = \left(-\frac{\sqrt{2}}{2}, -\frac{\sqrt{2}}{2}\right)$$

7.

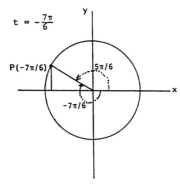

$$P\left(-\frac{7\pi}{6}\right) = P\left(-\frac{7\pi}{6} + 2\pi\right) = P\left(\frac{5\pi}{6}\right) = \left(-\frac{\sqrt{3}}{2}, \frac{1}{2}\right)$$

9.

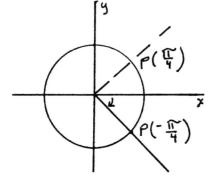

$P\left(\frac{\pi}{4}\right) = \left(\frac{\sqrt{2}}{2}, \frac{\sqrt{2}}{2}\right)$. Then $P\left(-\frac{\pi}{4}\right) = \left(\frac{\sqrt{2}}{2}, -\frac{\sqrt{2}}{2}\right)$ because of its position relative to P. Thus, $\sin\left(-\frac{\pi}{4}\right) = y$-coordinate of $P\left(-\frac{\pi}{4}\right) = -\frac{\sqrt{2}}{2}$.

11.

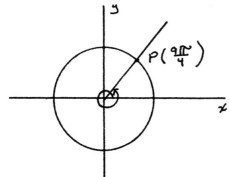

$$P\left(\frac{9\pi}{4}\right) = P\left(2\pi + \frac{\pi}{4}\right) = P\left(\frac{\pi}{4}\right) = \left(\frac{\sqrt{2}}{2}, \frac{\sqrt{2}}{2}\right)$$

Thus, $\sin\left(\frac{9\pi}{4}\right) = \frac{\sqrt{2}}{2}$

13.

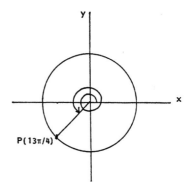

$$P\left(\frac{13\pi}{4}\right) = P\left(2\pi + \frac{5\pi}{4}\right) = P\left(\frac{5\pi}{4}\right) = \left(-\frac{\sqrt{2}}{2}, -\frac{\sqrt{2}}{2}\right)$$

$$\cos\left(\frac{13\pi}{4}\right) = x\text{-coordinate of } P\left(\frac{13\pi}{4}\right) = -\frac{\sqrt{2}}{2}$$

15.

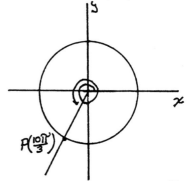

$$P\left(\frac{10\pi}{3}\right) = P\left(2\pi + \frac{4\pi}{3}\right) = P\left(\frac{4\pi}{3}\right) = \left(-\frac{1}{2}, -\frac{\sqrt{3}}{2}\right)$$

Thus, $\cos\left(\frac{10\pi}{3}\right) = -\frac{1}{2}$

17.

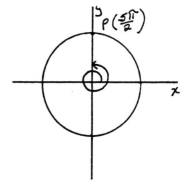

$$P\left(\frac{5\pi}{2}\right) = P\left(2\pi + \frac{\pi}{2}\right) = P\left(\frac{\pi}{2}\right) = (0, 1)$$

Thus, $\sin\left(\frac{5\pi}{2}\right) = 1$

19.

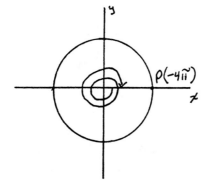

$P(-4\pi) = P(0) = (1, 0)$

Thus, $\sin(-4\pi) = 0$

21.

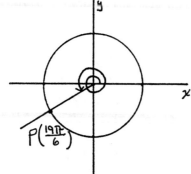

$P\left(\frac{19\pi}{6}\right) = P\left(2\pi + \frac{7\pi}{6}\right) = P\left(\frac{7\pi}{6}\right) = \left(-\frac{\sqrt{3}}{2}, -\frac{1}{2}\right)$

Thus, $\cos\left(\frac{19\pi}{6}\right) = -\frac{\sqrt{3}}{2}$

23.

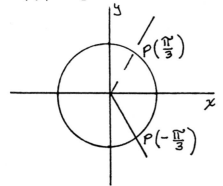

$P\left(\frac{\pi}{3}\right) = \left(\frac{1}{2}, \frac{\sqrt{3}}{2}\right)$. Then $P\left(-\frac{\pi}{3}\right) = \left(\frac{1}{2}, -\frac{\sqrt{3}}{2}\right)$

because of its position relative to P. Thus,

$\cos\left(-\frac{\pi}{3}\right) = \frac{1}{2}$

25.

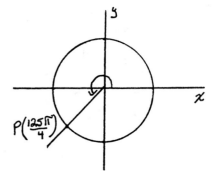

$P\left(\frac{125\pi}{4}\right) = P\left(30\pi + \frac{5\pi}{4}\right) = P\left(\frac{5\pi}{4}\right) = \left(-\frac{\sqrt{2}}{2}, -\frac{\sqrt{2}}{2}\right)$

Thus, $\cos\left(\frac{125\pi}{4}\right) = -\frac{\sqrt{2}}{2}$

27.

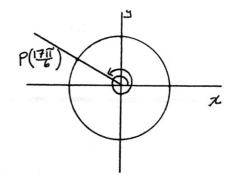

$510° = 510° \times \frac{\pi \text{ rad}}{180°} = \frac{17\pi}{6}$ rad

$P\left(\frac{17\pi}{6}\right) = P\left(2\pi + \frac{5\pi}{6}\right) = P\left(\frac{5\pi}{6}\right) = \left(-\frac{\sqrt{3}}{2}, \frac{1}{2}\right)$

Thus, $\sin(510°) = \frac{1}{2}$.

29.

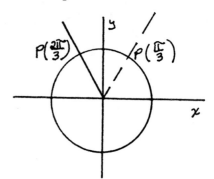

$840° = 840° \times \frac{\pi \text{ rad}}{180°} = \frac{14\pi}{3}$ rad

$P\left(\frac{14\pi}{3}\right) = P\left(4\pi + \frac{2\pi}{3}\right) = P\left(\frac{2\pi}{3}\right) = \left(-\frac{1}{2}, \frac{\sqrt{3}}{2}\right)$

Thus, $\cos 840° = -\frac{1}{2}$.

31.

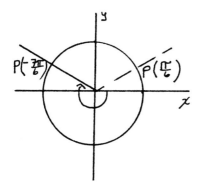

$$-210° = -210° \times \frac{\pi \text{ rad}}{180°} = -\frac{7\pi}{6} \text{ rad}$$

$$P\left(-\frac{7\pi}{6}\right) = P\left(2\pi - \frac{7\pi}{6}\right) = P\left(\frac{5\pi}{6}\right) = \left(-\frac{\sqrt{3}}{2}, \frac{1}{2}\right)$$

Thus, $\cos(-210°) = -\frac{\sqrt{3}}{2}$.

33. $\sin(-1.87) = -\sin(1.87) = -.95557$

$\cos(-1.87) = \cos(1.87) = -.29476$

35. a) Since the points $P(t)$ and $P(-t)$ are symmetric with respect to the x-axis, if

$$P(t) = \left(\frac{1}{\sqrt{5}}, -\frac{2}{\sqrt{5}}\right) \text{ then } P(-t) = \left(\frac{1}{\sqrt{5}}, \frac{2}{\sqrt{5}}\right).$$

b) $\sin(-t) = \frac{2}{\sqrt{5}}, \cos(-t) = \frac{1}{\sqrt{5}}$

37. Since the points $P(t)$ and $P(\pi + t)$ are symmetric with respect to the origin, if

$$P(t) = (x, y) \text{ then } P(\pi + t) = (-x, -y).$$

a) $\sin(t) = y$

$\sin(\pi + t) = -y = -\sin(t)$

b) $\cos(t) = x$

$\cos(\pi + t) = -x = -\cos(t)$

39. First note that since t is in quadrant IV, $\sin t$ will be a negative number.

$$\sin t = -\sqrt{1 - \cos^2 t}$$

$$= -\sqrt{1 - \left(\frac{3}{5}\right)^2}$$

$$= -\sqrt{1 - \frac{9}{25}}$$

$$= -\sqrt{\frac{16}{25}}$$

$$= -\frac{4}{5}$$

41. Since t is in quadrant II, $\cos t$ will be negative.

$$\cos t = -\sqrt{1 - \sin^2 t}$$

$$= -\sqrt{1 - \left(\frac{12}{13}\right)^2}$$

$$= -\sqrt{1 - \frac{144}{169}}$$

$$= -\sqrt{\frac{25}{169}}$$

$$= -\frac{5}{13}$$

From Problem 37, $\cos(\pi + t) = -\cos t = \frac{5}{13}$.

43. a) Since $\frac{\pi}{2} < 2 < \pi$, $P(2)$ is in quadrant II.

Thus $\cos 2$ is negative.

b) Since $-\pi < -3 < -\frac{\pi}{2}$, $P(-3)$ is in quadrant III.

Thus $\sin(-3)$ is negative.

c) Since $\frac{428°}{360°} = 1.19$ and $0 < .19 < .25$,

$P(428°)$ is in quadrant I.

Thus $\cos 428°$ is positive.

d) Since $\frac{21.4}{2\pi} = 3.41$ and $.25 < .41 < .5$,

$P(21.4)$ is in quadrant II.

Thus $\sin 21.4$ is positive.

e) Since $\frac{\pi}{2} < \frac{23\pi}{32} < \pi$, $P\left(\frac{23\pi}{32}\right)$ is in quadrant II.

Thus, $\sin\left(\frac{23\pi}{32}\right)$ is positive.

f) Since $\frac{-820°}{360°} = -2.28$ and $-.5 < -.28 < -.25$,

$P(-820°)$ is in quadrant III.

Thus, $\sin(-820°)$ is negative.

45. a) Since $P(t)$ lies on the unit circle,

$$x^2 + \left(-\frac{1}{2}\right)^2 = 1$$

$$x^2 + \frac{1}{4} = 1$$

$$x^2 = \frac{3}{4}$$

$$x = \pm\frac{\sqrt{3}}{2}$$

b) When $P(t) = \left(\frac{\sqrt{3}}{2}, -\frac{1}{2}\right)$, $t = \frac{11\pi}{6}$.

When $P(t) = \left(-\frac{\sqrt{3}}{2}, -\frac{1}{2}\right)$, $t = \frac{7\pi}{6}$.

47. $P\left(\frac{\pi}{6}\right) = \left(\frac{\sqrt{3}}{2}, \frac{1}{2}\right)$

$P\left(-\frac{3\pi}{4}\right) = P\left(\frac{5\pi}{4}\right) = \left(-\frac{\sqrt{2}}{2}, -\frac{\sqrt{2}}{2}\right)$

Use the distance formula

$$D = \sqrt{(x_2 - x_1)^2 + (y_2 - y_1)^2}$$
$$= \sqrt{\left(-\frac{\sqrt{2}}{2} - \frac{\sqrt{3}}{2}\right)^2 + \left(-\frac{\sqrt{2}}{2} - \frac{1}{2}\right)^2}$$
$$= \sqrt{\frac{2 + 2\sqrt{2}\sqrt{3} + 3 + 2 + 2\sqrt{2} + 1}{4}}$$
$$= \frac{\sqrt{8 + 2\sqrt{6} + 2\sqrt{2}}}{2}$$
$$\approx 1.98289$$

49. a) Let $P(t) = (x, y)$. Since $\sin t = \cos t$, $x = y$.

We also know that

$$x^2 + y^2 = 1$$

Substituting $y = x$,

$$x^2 + x^2 = 1$$
$$2x^2 = 1$$
$$x^2 = \frac{1}{2} = \frac{2}{4}$$
$$x^2 = \pm\frac{\sqrt{2}}{2}$$

Thus, $P(t) = \left(\frac{\sqrt{2}}{2}, \frac{\sqrt{2}}{2}\right)$ or $P(t) = \left(-\frac{\sqrt{2}}{2}, -\frac{\sqrt{2}}{2}\right)$.

$t = \frac{\pi}{4}$ or $\frac{5\pi}{4}$

b) $\frac{1}{2} < \sin t < \frac{\sqrt{3}}{2}$

$\sin t > 0$, so t must be in quadrants I or II. The given intervals include $\frac{\pi}{6} < t < \frac{\pi}{3}$ and $\frac{2\pi}{3} < t < \frac{5\pi}{6}$.

c) $\cos^2 t \geq .25$

$\cos t \geq .5$ or $\cos t \leq -.5$

These inequalities will be true when

$0 \leq t \leq \frac{\pi}{3}, \frac{2\pi}{3} \leq t \leq \frac{4\pi}{3}$, or $\frac{5\pi}{3} \leq t \leq 2\pi$

d) $\cos^2 t > \sin^2 t$

Substituting $\sin^2 t = 1 - \cos^2 t$,

$$\cos^2 t > 1 - \cos^2 t$$
$$2\cos^2 t > 1$$
$$\cos^2 t > \frac{1}{2} = \frac{2}{4}$$
$$\cos t > \frac{\sqrt{2}}{2} \text{ or } \cos t < -\frac{\sqrt{2}}{2}$$

These inequalities will be true when

$0 \leq t < \frac{\pi}{4}, \frac{3\pi}{4} < t < \frac{5\pi}{4}$, or $\frac{7\pi}{4} < t \leq 2\pi$.

51. a) Since θ is in the fourth quadrant,

$\cos\theta$ is positive.

$$\cos\theta = \sqrt{1 - \sin^2\theta}$$
$$= \sqrt{1 - \left(-\frac{4}{5}\right)^2}$$
$$= \sqrt{1 - \frac{16}{25}}$$
$$= \sqrt{\frac{9}{25}} = \frac{3}{5}$$

b) Since θ is in the fourth quadrant,

$\sin\theta$ is negative.

$$\sin\theta = -\sqrt{1 - \cos^2\theta}$$
$$= -\sqrt{1 - \left(\frac{24}{25}\right)^2}$$
$$= -\sqrt{1 - \frac{576}{625}}$$
$$= -\sqrt{\frac{49}{625}} = -\frac{7}{25}$$

53. $P(t) = \left(\frac{4}{5}, -\frac{3}{5}\right)$

a) $\sin(-t) = -\sin t = -\left(-\frac{3}{5}\right) = \frac{3}{5}$

b) $\sin\left(\frac{\pi}{2} - t\right) = \cos t = \frac{4}{5}$

c) $\cos(2\pi + t) = \cos t = \frac{4}{5}$

d) $\cos(2\pi - t) = \cos(-t) = \cos t = \frac{4}{5}$

e) From Problem 37,

$\sin(\pi + t) = -\sin t = -\left(-\frac{3}{5}\right) = \frac{3}{5}$

f) From Problem 38,

$\cos(\pi - t) = -\cos t = -\frac{4}{5}$

55. a) If x is nearest to n, then $x + 1$ is nearest to $n+1$.

$<x> = |n - x| = |(n + 1) - (x + 1)| = <x + 1>$

Thus $<x>$ is periodic with period 1.

b) $f(x) = <3x>$.

$f\left(x + \frac{1}{3}\right) = \left\langle 3\left(x + \frac{1}{3}\right)\right\rangle = <3x + 1> = <3x> = f(x)$

Thus, $f(x)$ is periodic with period $\frac{1}{3}$.

c) $[x]$ is not periodic.

d) $f(x) = x - [x]$

$$f(x+1) = x + 1 - [x+1]$$
$$= x + 1 - ([x] + 1)$$
$$= x - [x]$$
$$= f(x)$$

Thus, $f(x)$ is periodic with period 1.

57. $\sin 1° + \sin 2° + \sin 3° + \cdots$

$$+ \sin 357° + \sin 358° + \sin 359°$$
$$= [\sin 1° + \sin 359°] + [\sin 2° + \sin 358°]$$
$$+ [\sin 3° + \sin 357°] + \cdots + \sin 180°$$
$$= [\sin 1° + \sin(360° - 1°)] + [\sin 2° + \sin(360° - 2°)]$$
$$+ \sin 3° + \sin(360° - 3°)] + \cdots + \sin 180°$$
$$= [\sin 1° - \sin 1°] + [\sin 2° - \sin 2°] + [\sin 3° - \sin 3°]$$
$$+ \cdots + 0$$
$$= 0$$

Problem Set 7.4

1. $\sin t = \frac{4}{5}$, $\cos t = -\frac{3}{5}$

(a) $\tan t = \dfrac{\sin t}{\cos t} = \dfrac{\frac{4}{5}}{-\frac{3}{5}} = -\dfrac{4}{3}$

(b) $\cot t = \dfrac{\cos t}{\sin t} = \dfrac{-\frac{3}{5}}{\frac{4}{5}} = -\dfrac{3}{4}$

or

$\cot t = \dfrac{1}{\tan t} = \dfrac{1}{-\frac{4}{3}} = -\dfrac{3}{4}$

(c) $\sec t = \dfrac{1}{\cos t} = \dfrac{1}{-\frac{3}{5}} = -\dfrac{5}{3}$

(d) $\csc t = \dfrac{1}{\sin t} = \dfrac{1}{\frac{4}{5}} = \dfrac{5}{4}$

3. $\tan \theta = \dfrac{\sin \theta}{\cos \theta} = \dfrac{\frac{\sqrt{5}}{3}}{-\frac{2}{3}} = -\dfrac{\sqrt{5}}{2}$

$\csc \theta = \dfrac{1}{\sin \theta} = \dfrac{1}{\frac{\sqrt{5}}{3}} = \dfrac{3}{\sqrt{5}} = \dfrac{3\sqrt{5}}{5}$

5. $\tan\left(\dfrac{\pi}{6}\right) = \dfrac{\sin\left(\frac{\pi}{6}\right)}{\cos\left(\frac{\pi}{6}\right)} = \dfrac{\frac{1}{2}}{\frac{\sqrt{3}}{2}} = \dfrac{1}{\sqrt{3}} = \dfrac{\sqrt{3}}{3}$

7. $\sec\left(\dfrac{\pi}{6}\right) = \dfrac{1}{\cos\left(\frac{\pi}{6}\right)} = \dfrac{1}{\frac{\sqrt{3}}{2}} = \dfrac{2}{\sqrt{3}} = \dfrac{2\sqrt{3}}{3}$

9. $\cot\left(\dfrac{\pi}{4}\right) = \dfrac{\cos\left(\frac{\pi}{4}\right)}{\sin\left(\frac{\pi}{4}\right)} = \dfrac{\frac{\sqrt{2}}{2}}{\frac{\sqrt{2}}{2}} = 1$

11. $\csc\left(\dfrac{\pi}{3}\right) = \dfrac{1}{\sin\left(\frac{\pi}{3}\right)} = \dfrac{1}{\frac{\sqrt{3}}{2}} = \dfrac{2}{\sqrt{3}} = \dfrac{2\sqrt{3}}{3}$

▷ Note: If an angle is changed by $\pi = 180°$: the sine, cosine, secant, and cosecant change sign; the tangent and cotangent do not change.

13. $\sin\left(\dfrac{4\pi}{3}\right) = -\sin\left(\dfrac{\pi}{3}\right) = -\dfrac{\sqrt{3}}{2}$

15. $\tan\left(\dfrac{4\pi}{3}\right) = \dfrac{\sin\left(\frac{4\pi}{3}\right)}{\cos\left(\frac{4\pi}{3}\right)} = \dfrac{-\frac{\sqrt{3}}{2}}{-\frac{1}{2}} = \sqrt{3}$

17. $\tan \pi = \dfrac{\sin \pi}{\cos \pi} = \dfrac{0}{-1} = 0$

19. $\tan 330° = \dfrac{\sin 330°}{\cos 330°}$
$$= \dfrac{\sin(-30°)}{\cos(-30°)}$$
$$= \dfrac{\frac{1}{2}}{\frac{\sqrt{3}}{2}}$$
$$= -\dfrac{1}{\sqrt{3}}$$
$$= -\dfrac{\sqrt{3}}{3}$$

21. $\sec 600° = \dfrac{1}{\cos 600°}$
$$= \dfrac{1}{\cos 240°}$$
$$= \dfrac{1}{-\cos 60}$$
$$= \dfrac{1}{-\frac{1}{2}}$$
$$= -2$$

23. a) Since $\sec t = \dfrac{1}{\cos t}$, $\sec t$ is undefined when $\cos t = 0$; that is, when $t = \dfrac{\pi}{2}, \dfrac{3\pi}{2}, \dfrac{5\pi}{2}, \dfrac{7\pi}{2}$.

b) Since $\tan t = \dfrac{\sin t}{\cos t}$, $\tan t$ is undefined when $\cos t = 0$; that is, when $t = \dfrac{\pi}{2}, \dfrac{3\pi}{2}, \dfrac{5\pi}{2}, \dfrac{7\pi}{2}$.

c) Since $\csc t = \dfrac{1}{\sin t}$, $\csc t$ is undefined when $\sin t = 0$; that is, when $t = 0, \pi, 2\pi, 3\pi, 4\pi$.

d) Since $\cot t = \dfrac{\cos t}{\sin t}$, $\cot t$ is undefined when $\sin t = 0$; that is, when $t = 0, \pi, 2\pi, 3\pi, 4\pi$

25. a) Since t is in quadrant II, $\csc t$ is positive.
$$\begin{aligned}\csc t &= \sqrt{1 + \cot^2 t}\\ &= \sqrt{1 + (-\sqrt{6})^2}\\ &= \sqrt{1 + 6}\\ &= \sqrt{7}\end{aligned}$$

b) $\cos t = \dfrac{\cos t}{\sin t} \div \dfrac{1}{\sin t} = \dfrac{\cot t}{\csc t} = \dfrac{-\sqrt{6}}{\sqrt{7}} = \dfrac{\sqrt{42}}{7}$

c) $\cot\left(\dfrac{\pi}{2} - t\right) = \tan t = \dfrac{1}{\cot t} = \dfrac{1}{-\sqrt{6}} = -\dfrac{\sqrt{6}}{6}$

27. a) Since θ is in quadrant III, $\tan \theta$ is positive.
$$\tan \theta = \sqrt{\sec^2\theta - 1} = \sqrt{(-3)^2 - 1} = \sqrt{8} = 2\sqrt{2}$$

b) $\cot(-\theta) = -\cot\theta = -\dfrac{1}{\tan\theta} = -\dfrac{1}{2\sqrt{2}} = -\dfrac{\sqrt{2}}{4}$

c) $\sin(90° - \theta) = \cos\theta = \dfrac{1}{\sec\theta} = \dfrac{1}{-3} = -\dfrac{1}{3}$

29. If $(5, -12)$ is on the terminal side of the angle,
$$a = 5, \ b = -12$$
$$r = \sqrt{5^2 + (-12)^2} = \sqrt{169} = 13$$
$$\sin\theta = \dfrac{b}{r} = -\dfrac{12}{13}$$
$$\tan\theta = \dfrac{b}{a} = -\dfrac{12}{5}$$
$$\sec\theta = \dfrac{r}{a} = \dfrac{13}{5}$$

31. If $(-1, -2)$ is on the terminal side of the angle,
$$a = -1, \ b = -2$$
$$r = \sqrt{(-1)^2 + (-2)^2} = \sqrt{5}$$
$$\sin\theta = \dfrac{b}{r} = \dfrac{-2}{\sqrt{5}} = -\dfrac{2\sqrt{5}}{5}$$
$$\tan\theta = \dfrac{b}{a} = \dfrac{-2}{-1} = 2$$

$$\sec\theta = \dfrac{r}{a} = \dfrac{\sqrt{5}}{-1} = -\sqrt{5}$$

33. Since $(4, 3)$ is on the terminal side of the angle,
$$a = 4, \ b = 3$$
$$r = \sqrt{4^2 + 3^2} = \sqrt{25} = 5$$
$$\sin\theta = \dfrac{b}{r} = \dfrac{3}{5}$$
$$\sec\theta = \dfrac{r}{a} = \dfrac{5}{4}$$

35. $a = -12, \ b = 5, \ r = 13$
$$\cos\theta = \dfrac{a}{r} = -\dfrac{12}{13}$$
$$\cot\theta = \dfrac{a}{b} = -\dfrac{12}{5}$$

37. $r = \sqrt{5^2 + (-12)^2} = \sqrt{25 + 144} = \sqrt{169} = 13$
The required point is $\left(\dfrac{5}{r}, \dfrac{-12}{r}\right) = \left(\dfrac{5}{13}, -\dfrac{12}{13}\right)$.

39. $m = \tan\theta = \tan 150° = -\tan 30° = -\dfrac{\sqrt{3}}{3}$
Using $y - y_1 = m(x - x_1)$ with
$m = -\dfrac{\sqrt{3}}{3}$ and $(x_1, y_1) = (-3, 4)$
the equation of the line is
$$y - 4 = -\dfrac{\sqrt{3}}{3}(x + 3)$$
$$3y - 12 = -\sqrt{3}x - 3\sqrt{3}$$
$$\sqrt{3}x + 3y = 12 - 3\sqrt{3}$$

41.
$$5x + 2y = 6$$
$$2y = -5x + 6$$
$$y = -\dfrac{5}{2}x + 3$$
$$m = \tan\theta = -\dfrac{5}{2}$$

Use a calculator to find θ:
$5 \div 2 \boxed{+/-} = \boxed{2\text{nd}}\boxed{\text{TAN}} \rightarrow -68.2°$

Since the angle of inclination is positive, we add $180°$ to get $111.8°$.

43. a) $\sec\dfrac{7\pi}{6} = \dfrac{1}{\cos\dfrac{7\pi}{6}} = \dfrac{1}{-\dfrac{\sqrt{3}}{2}} = -\dfrac{2}{\sqrt{3}} = -\dfrac{2\sqrt{3}}{3}$

b) $\tan\left(-\dfrac{2\pi}{3}\right) = \tan\left(\dfrac{\pi}{3}\right) = \sqrt{3}$

c) $\csc\left(\dfrac{3\pi}{4}\right) = \dfrac{1}{\sin\left(\dfrac{3\pi}{4}\right)} = \dfrac{1}{\dfrac{\sqrt{2}}{2}} = \dfrac{2}{\sqrt{2}} = \sqrt{2}$

d) $\cot\left(\dfrac{11\pi}{4}\right) = \cot\left(2\pi + \dfrac{3\pi}{4}\right)$

$\qquad = \cot\left(\dfrac{3\pi}{4}\right)$

$\qquad = \dfrac{\cos\left(\dfrac{3\pi}{4}\right)}{\sin\left(\dfrac{3\pi}{4}\right)}$

$\qquad = \dfrac{-\dfrac{\sqrt{2}}{2}}{\dfrac{\sqrt{2}}{2}}$

$\qquad = -1$

e) $\csc 570° = \csc 210°$

$\qquad = -\csc 30°$

$\qquad = -\dfrac{1}{\sin 30°}$

$\qquad = -\dfrac{1}{\dfrac{1}{2}}$

$\qquad = -2$

f) $\tan(180{,}045°) = \tan[500(360°) + 45°] = \tan 45° = 1$

45. Using a, b, r, $\csc t = \dfrac{r}{b} = \dfrac{25}{24}$ so $r = 25$, $b = 24$.

Since $\cos t < 0$, a is negative.

$a = -\sqrt{25^2 - 24^2} = -\sqrt{625 - 576} = -\sqrt{49} = -7$

a) $\sin t = \dfrac{b}{r} = \dfrac{24}{25}$

b) $\cos t = \dfrac{a}{r} = -\dfrac{7}{25}$

c) $\tan t = \dfrac{b}{a} = -\dfrac{24}{7}$

d) $\sec\left(\dfrac{\pi}{2} - t\right) = \csc t = \dfrac{25}{24}$

e) $\cot\left(\dfrac{\pi}{2} - t\right) = \tan t = -\dfrac{24}{7}$

f) $\csc\left(\dfrac{\pi}{2} - t\right) = \sec t = \dfrac{r}{a} = -\dfrac{25}{7}$

47. a) Let $P(t) = (x, y)$.

$\tan t = -1$

$\dfrac{y}{x} = -1$

$y = -x$

$P(t) = \left(\dfrac{\sqrt{2}}{2}, -\dfrac{\sqrt{2}}{2}\right)$ or $P(t) = \left(-\dfrac{\sqrt{2}}{2}, \dfrac{\sqrt{2}}{2}\right)$

$t = \dfrac{3\pi}{4}$ or $t = \dfrac{7\pi}{4}$

b) $\sec t = \sqrt{2}$

$\cos t = = \dfrac{1}{\sqrt{2}} = \dfrac{\sqrt{2}}{2}$

$t = \dfrac{\pi}{4}$ or $t = \dfrac{7\pi}{4}$

c) $|\csc t| = 1$

$\left|\dfrac{1}{\sin t}\right| = 1$

$\sin t = \pm 1$

$t = \dfrac{\pi}{2}$ or $t = \dfrac{3\pi}{2}$

49. a) $\dfrac{\sec\theta\,\csc\theta}{\tan\theta + \cot\theta} = \dfrac{\dfrac{1}{\cos\theta}\cdot\dfrac{1}{\sin\theta}}{\dfrac{\sin\theta}{\cos\theta} + \dfrac{\cos\theta}{\sin\theta}}$

$\qquad = \dfrac{\dfrac{1}{\cos\theta}\cdot\dfrac{1}{\sin\theta}}{\dfrac{\sin\theta}{\cos\theta} + \dfrac{\cos\theta}{\sin\theta}}\cdot\dfrac{\cos\theta\,\sin\theta}{\cos\theta\,\sin\theta}$

$\qquad = \dfrac{1}{\sin^2\theta + \cos^2\theta}$

$\qquad = \dfrac{1}{1}$

$\qquad = 1$

b) $\tan\theta(\cos\theta - \csc\theta) = \dfrac{\sin\theta}{\cos\theta}\left(\cos\theta - \dfrac{1}{\sin\theta}\right)$

$\qquad = \sin\theta - \dfrac{1}{\cos\theta}$

c) $\dfrac{(1 + \tan\theta)^2}{\sec^2\theta} = \left(\dfrac{1 + \tan\theta}{\sec\theta}\right)^2$

$\qquad = \left(\dfrac{1 + \dfrac{\sin\theta}{\cos\theta}}{\dfrac{1}{\cos\theta}}\right)^2$

$\qquad = \left[\left(1 + \dfrac{\sin\theta}{\cos\theta}\right)\cos\theta\right]^2$

$\qquad = (\cos\theta + \sin\theta)^2$

$\qquad = \cos^2\theta + 2\sin\theta\cos\theta + \sin^2\theta$

$\qquad = 1 + 2\sin\theta\cos\theta$

d) $\dfrac{\sec\theta\,\cot\theta}{\sec^2\theta - \tan^2\theta} = \dfrac{\dfrac{1}{\cos\theta}\cdot\dfrac{\cos\theta}{\sin\theta}}{1} = \dfrac{1}{\sin\theta}$

e) $\dfrac{\cot\theta - \tan\theta}{\csc\theta - \sec\theta} = \dfrac{\dfrac{\cos\theta}{\sin\theta} - \dfrac{\sin\theta}{\cos\theta}}{\dfrac{1}{\sin\theta} - \dfrac{1}{\cos\theta}}$

$\qquad = \dfrac{\dfrac{\cos\theta}{\sin\theta} - \dfrac{\sin\theta}{\cos\theta}}{\dfrac{1}{\sin\theta} - \dfrac{1}{\cos\theta}}\cdot\dfrac{\cos\theta\,\sin\theta}{\cos\theta\,\sin\theta}$

$\qquad = \dfrac{\cos^2\theta - \sin^2\theta}{\cos\theta - \sin\theta}$

$\qquad = \dfrac{(\cos\theta + \sin\theta)(\cos\theta - \sin\theta)}{(\cos\theta - \sin\theta)}$

$\qquad = \cos\theta + \sin\theta$

f) $\tan^4\theta - \sec^4\theta = (\tan^2\theta - \sec^2\theta)(\tan^2\theta + \sec^2\theta)$
$$= -1\left(\frac{\sin^2\theta}{\cos^2\theta} + \frac{1}{\cos^2\theta}\right)$$
$$= -\frac{\sin^2\theta + 1}{\cos^2\theta}$$

51. a) $\tan(t + \pi) = \frac{\sin(t+\pi)}{\cos(t+\pi)} = \frac{-\sin t}{-\cos t} = \frac{\sin t}{\cos t} = \tan t$

b) $\cot(t + \pi) = \frac{\cos(t+\pi)}{\sin(t+\pi)} = \frac{-\cos t}{-\sin t} = \frac{\cos t}{\sin t} = \cot t$

c) $\sec(t + \pi) = \frac{1}{\cos(t+\pi)} = \frac{1}{-\cos t} = -\frac{1}{\cos t} = -\sec t$

d) $\csc(t + \pi) = \frac{1}{\sin(t+\pi)} = \frac{1}{-\sin t} = -\frac{1}{\sin t} = -\csc t$

53. $\tan\theta = \frac{5}{12} = \frac{b}{a}$. $b = 5$, $a = 12$
$$r = \sqrt{5^2 + 12^2} = \sqrt{169} = 13$$
$$\cos^2\theta - \sin^2\theta = \left(\frac{a}{r}\right)^2 - \left(\frac{b}{r}\right)^2$$
$$= \left(\frac{12}{13}\right)^2 - \left(\frac{5}{13}\right)^2$$
$$= \frac{144}{169} - \frac{25}{169}$$
$$= \frac{119}{169}$$

55. Let T be the speck and P be the point where the tangent line at T meets the x-axis. OPT is a right triangle and the angle at O is $\pi - \frac{2}{3}\pi = \frac{1}{3}\pi$.
$$\cos\frac{\pi}{3} = \frac{5}{OP}$$
$$\frac{1}{2} = \frac{5}{OP}$$
$$OP = 10$$
Because P is to the left of O, $x = -10$.

57. The angle at $(5, 0)$ between the chord and a tangent, which is vertical, is $150° - 90° = 60°$ so the arc measures $2 \cdot 60° - 120°$. Then
$$s = rt = 5 \times 120° \times \frac{\pi \text{ rad}}{180°} = \frac{10\pi}{3} \approx 10.47$$

59. a) $\frac{\theta}{90°} = \frac{9}{15}$
$$\theta = 54°$$
The minute hand has slope
$$-\tan 54° \approx -1.3764$$

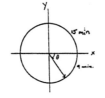

b) At 12:50,
$$\angle AOQ = \frac{5}{6}(30°) = 25°$$
$$\Rightarrow \angle ROQ = 65°$$
$$\angle AOP = 65°$$
$$\Rightarrow \angle ROP = 150°$$

The coordinates of P and Q are

$P(5 \cos 150°, 5 \sin 150°)$ and $(5 \cos 65°, 5 \sin 65°)$.

Therefore, the slope of PQ is
$$\frac{\sin 150° - \sin 65°}{\cos 150° - \cos 65°} \approx .3153$$

01. $\cos\theta = \frac{30}{100}$
$$\theta = 1.2661$$
$$\phi = \pi - \theta$$
$$x = \sqrt{100^2 - 30^2}$$
$$\approx 95.3939$$

The length of the belt is
$$2(\pi - \theta)50 + 2\theta(20) + 2x$$
$$= 100\pi - 60\theta + 2x$$
$$\approx 428.98 \text{ cm}$$

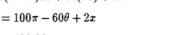

Problem Set 7.5

1. 1.38 is in quadrant I.
$$\sin 1.38 = .98185$$

3. 42.8° is in quadrant I.
$$\cos 42.8° = .7337$$

5. .82 is in quadrant I.
$$\cot .82 = .93309$$

7. 68.3° is in quadrant I.
$$\sin 68.3° = .9291$$

9. Since 1.84 is in quadrant II,
$$t_0 = 3.14 - 1.84 = 1.30$$

11. Since 3.54 is in quadrant III,
$$t_0 = 3.54 - 3.14 = .40$$

13. 5.18 is in quadrant IV.
$$t_0 = 6.28 - 5.18 = 1.10$$

150

15. Remove multiples of 2π from 10.48:

$10.48 - 1(6.28) = 4.20$

4.20 is in quadrant III.

$t_0 = 4.20 - 3.14 = 1.06$

17. -1.12 is in quadrant IV.

The acute angle between -1.12 and the x-axis is

$t_0 = 1.12$

19. -2.64 is in quadrant III.

$t_0 = 3.14 - 2.64 = .50$

21. Since $\frac{3\pi}{2} < \frac{13\pi}{8} < 2\pi$, $\frac{13\pi}{8}$ is in quadrant IV.

$t_0 = 2\pi - \frac{13\pi}{8} = \frac{16\pi}{8} - \frac{13\pi}{8} = \frac{3\pi}{8}$

23. Remove multiples of 2π:

$\left[\frac{40\pi}{3} \div 2\pi\right] = \left[\frac{20}{3}\right] = 6$

$\frac{40\pi}{3} - 6(2\pi) = \frac{40\pi}{3} - \frac{36\pi}{3} = \frac{4\pi}{3}$

$\frac{4\pi}{3}$ is in quadrant III.

$t_0 = \frac{4\pi}{3} - \pi = \frac{\pi}{3}$

25. Remove multiples of 2π:

$3\pi + .24 - 2\pi = \pi + .24$

$\pi + .24$ is in quadrant III

$t_0 = (\pi + .24) - \pi = .24$

27. Remove multiples of 2π:

$3\pi - .24 - 2\pi = \pi - .24$

$\pi - .24$ is in quadrant II.

$t_0 = \pi - (\pi - .24) = .24$

29. Remove multiples of 2π:

$\left[\frac{11\pi}{2} \div 2\pi\right] = \left[\frac{11}{4}\right] = 2$

$\frac{11\pi}{2} - 2(2\pi) = \frac{11\pi}{2} - \frac{8\pi}{2} = \frac{3\pi}{2}$

$\frac{3\pi}{2}$ is on the negative y-axis.

$t_0 = 2\pi - \frac{3\pi}{2} = \frac{\pi}{2}$

31. Since 1.42 is in quadrant I,

$\cos 1.42 = .15023$

33. Since 1.39 is in quadrant I,

$\tan 1.39 = 5.4707$

35. 2.14 is in quadrant II.

$t_0 = 3.14 - 2.14 = 1$

$\sin 2.14 = \sin 1 = .84147$

37. 5.62 is in quadrant IV.

$t_0 = 6.28 - 5.62 = .66$

$\cot 5.62 = -\cot .66 = -1.2885$

39. -2.54 is in quadrant III.

$t_0 = 3.14 - 2.54 = .60$

$\cos(-2.54) = -\cos .60 = -.82534$

41. Since $\sin t$ is positive, t must be in quadrants I or II. Find t_0:

$\sin t_0 = .94898$

$t_0 = 1.25$

So, $t = 1.25$ or $t = 3.14 - 1.25 = 1.89$

43. Since $\cos t$ is negative, t must be in quadrants II or III. Find t_0:

$\cos t_0 = .08071$

$t_0 = 1.49$

So, $t = 3.14 - 1.49 = 1.65$ or $t = 3.14 + 1.49 = 4.63$.

45. Since $\tan t$ is positive, t must be in quadrants I or III. Find t_0:

$\tan t_0 = 4.9131$

$t_0 = 1.37$

So, $t = 1.37$ or $t = 3.14 + 1.37 = 4.51$.

47. Since $\tan t$ is negative, t must be in quadrants II or IV. Find t_0:

$\tan t_0 = 3.6021$

$t_0 = 1.30$

So, $t = 3.14 - 1.30 = 1.84$ or $t = 6.28 - 1.30 = 4.98$.

49. $139.6°$ is in quadrant II.

$t_0 = 180° - 139.6° = 40.4°$

51. $348.7°$ is in quadrant IV.

$t_0 = 360° - 348.7° = 11.3°$

53. -99.8 is in quadrant III.

$t_0 = 180° - 99.8° = 80.2$

55. $156.1°$ is in quadrant II.

$t_0 = 180° - 156.1° = 23.9°$

$\sin 156.1° = \sin 23.9° = .4051$

57. $348.9°$ is in quadrant IV.

$t_0 = 360° - 348.9° = 11.1°$

$\tan 348.9° = -\tan 11.1° = -.1962$

59. $-66.1°$ is in quadrant IV.

$t_0 = 66.1°$

$\cos(-66.1°) = \cos 66.1° = .4051$

61. Remove multiples of $360°$:

$441.3° - 360° = 81.3°$

$81.3°$ is in quadrant I.

$t_0 = 81.3°$

$\cos 441.3° = \cos 81.3° = .15126$

63. $-134°$ is in quadrant III.

$t_0 = 180° - 134° = 46°$

$\cot(-134°) = \cot 46° = .9657$

65. Since $\sin \theta$ is positive, θ must be in quadrants I or II. Find θ_0:

$\sin \theta_0 = .3633$

$\theta_0 = 21.3°$

So, $\theta = 21.3°$ or $\theta = 180° - 21.3° = 158.7°$.

67. Since $\tan \theta$ is positive, θ must be in quadrants I or III. Find θ_0:

$\tan \theta_0 = .4942$

$\theta_0 = 26.3°$

So, $\theta = 26.3°$ or $\theta = 180° + 26.3° = 206.3°$.

69. Since $\cos \theta$ is negative, θ must be in quadrants II or III. Find θ_0:

$\cos \theta_0 = .9085$

$\theta_0 = 24.7°$

So, $\theta = 180° - 24.7° = 155.3°$

or $\theta = 180° + 24.7° = 204.7°$

71. a) 5.63 is in quadrant IV.

$t_0 = 6.28 - 5.63 = .65$

$\cos 5.63 = \cos .65 = .79608$

b) Remove multiples of 2π:

$10.34 - 6.28 = 4.06$

4.06 is in quadrant III.

$t_0 = 4.06 - 3.14 = .92$

$\sin 10.34 = -\sin .92 = -.79560$

c) Remove multiples of 2π:

$8.42 - 6.28 = 2.14$

2.14 is in quadrant II.

$t_0 = 3.14 - 2.14 = 1$

$\tan 8.42 = -\tan 1 = -1.5574$

d) $311.3°$ is in quadrant IV.

$\theta_0 = 360° - 311.3° = 48.7°$

$\sin 311.3° = -\sin 48.7° = -.7513$

e) $-411° + 360° = -51°$

$-51°$ is in quadrant IV.

$\theta_0 = 51°$

$\tan(-411°) = -\tan 51° = -1.2349$

f) $[1989° \div 360°] = 5$

$1989° - 5(360°) = 189°$

$189°$ is in quadrant III.

$\theta_0 = 189° - 180° = 9°$

$\cos 1989° = -\cos 9° = -.9877$

73. a) $\boxed{\text{DEG}}$ 2.42 $\boxed{\text{SIN}}$ $\boxed{\text{RAD}}$ $\boxed{\text{COS}}$ $\rightarrow .99108682$

b) $\boxed{\text{RAD}}$ 2.42 $\boxed{\text{SIN}}$ $\boxed{x^2}$ $\boxed{\text{COS}}$ $\boxed{y^x}$ $3 = \rightarrow .744399$

c) $\boxed{\text{RAD}}$ 4.21 $\boxed{\text{TAN}}$ $+ 7.12$ $\boxed{\text{SIN}}$ $\boxed{\text{LN}}$ $= \boxed{\sqrt{x}}$

$\rightarrow 1.233865$

75. a) Since $\sin t$ is positive, t is in quadrants I or II.

$\sin t_0 = .62879$

$t_0 = .679996$

So, $t = .679996$ or $t = \pi - .679996 = 2.461597$

b) Since $\cos t$ is positive, t is quadrants I or IV.

$\cos t_0 = .34176$

$t_0 = 1.222007$

So, $t = 1.222007$ or $t = 2\pi - 1.222007 = 5.061178$

c) Since tan t is negative, t is in quadrants II or IV.

$$\tan t_0 = 3.14159$$

$$t_0 = 1.262627$$

So, $t = \pi - 1.262627 = 1.878966$

or $t = 2\pi - 1.262627 = 5.020558$

77. a) $\theta = 180° + \phi$

θ is in quadrant III. The reference angle is

$\theta - 180° = (180° + \phi) - 180° = \phi$

b) $\theta = 270° - \phi$

θ is in quadrant III. The reference angle is

$\theta - 180° = (270° - \phi) - 180° = 90 - \phi$

c) $\theta = \phi - 90°$

θ is a negative angle in quadrant IV. The reference

angle is

$-\theta = -(\phi - 90°) = 90° - \phi$

79.

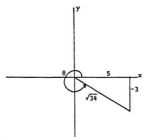

$3x + 5y = 0$

$y = -\frac{3}{5}x$, so slope is $-\frac{3}{5}$

$r = \sqrt{5^2 + (-3)^2} = \sqrt{25 + 9} = \sqrt{34}$

$\sin \theta = -\frac{3}{\sqrt{34}} = -.51450$

81. $\cos \theta = \frac{1}{\sec \theta} = \frac{1}{3}$

Using the calculator, we find the reference angle.

$$\theta_0 = 70.53°$$

Since θ is in the fourth quadrant,

$$\theta = 360° - 70.53° = 289.47°$$

83.

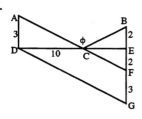

Extend BE 2 units to F and 3 units further to G.
$AC + CB = AC + CF$ has minimum length when
ACF is a straight line. Then
$\angle ACD = \angle FCE = \angle BCE = \angle EDG$
Call that angle θ.

$$\tan \theta = \frac{5}{10}$$

$$\theta = 26.565°$$

$$\phi = 180° - 2 \times 26.565° = 126.87°$$

Problem Set 7.6

1.

t	0	$\frac{\pi}{6}$	$\frac{\pi}{4}$	$\frac{\pi}{3}$	$\frac{\pi}{2}$	$\frac{3\pi}{4}$	π	$\frac{5\pi}{4}$	$\frac{3\pi}{4}$	$\frac{7\pi}{4}$	2π
$\cos t$	1	$\frac{\sqrt{3}}{2}$	$\frac{\sqrt{2}}{2}$	$\frac{1}{2}$	0	$-\frac{\sqrt{2}}{2}$	-1	$-\frac{\sqrt{2}}{2}$	0	$\frac{\sqrt{2}}{2}$	1

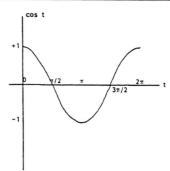

3. $y = \cot t = \dfrac{1}{\tan t}$

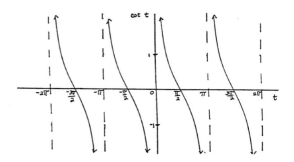

5. $\sec(t + 2\pi) = \dfrac{1}{\cos(t + 2\pi)} = \dfrac{1}{\cos t} = \sec t$

153

7. $\sec t = \dfrac{1}{\cos t}$

The domain of $y = \sec t$ is all values of t except those for which $\cos t = 0$, that is

$$\{t:\, t \neq \tfrac{\pi}{2} + k\pi,\ k \text{ any integer}\}$$

Since $|\cos t| \leq 1$, then $|\sec t| = \left|\dfrac{1}{\cos t}\right| \geq 1$.

That is, the range of $y = \sec t$ is

$$\{y:\, |y| \geq 1\}$$

9. The period of the cotangent is π:

$$\cot(t + \pi) = \dfrac{1}{\tan(t + \pi)} = \dfrac{1}{\tan t} = \cot t$$

The period of the secant is 2π, as shown in Problem 5 above.

11. $\cot(-t) = \dfrac{1}{\tan(-t)} = \dfrac{1}{-\tan t} = -\cot t$

13. $y = 3\cos t$

Amplitude: 3

Period: 2π

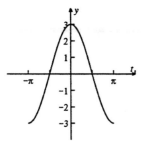

15. $y = -\sin t$

Amplitude: 1

Period: 2π

Reflect the graph of $y = \sin t$ in the x-axis.

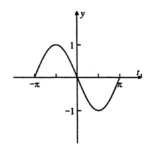

17. $y = \cos 4t$

Amplitude: 1

Period: $\dfrac{2\pi}{4} = \dfrac{\pi}{2}$

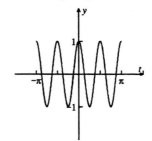

19. $y = 2\sin \tfrac{1}{2}t$

Amplitude: 2

Period: $\dfrac{2\pi}{\frac{1}{2}} = 4\pi$

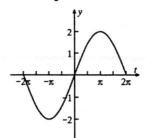

21. $y = 2\cos 3t$

Amplitude: 2

Period: $\dfrac{2\pi}{3}$

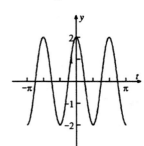

23. $y = 2\sin t$ has amplitude 2 and period 2π.

$y = \cos t$ has amplitude 1 and period 2π.

$y = 2\sin t + \cos t$ has amplitude

$$\sqrt{2^2 + 1^2} = \sqrt{5} \approx 2.24 \text{ and period } 2\pi.$$

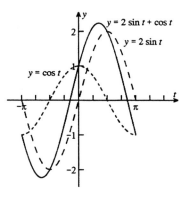

25. $y = \sin 2t$ has amplitude 1 and period $\frac{2\pi}{2} = \pi$.

$y = \cos t$ has amplitude 1 and period 2π.

$y = \sin 2t + \cos t$ has amplitude ≈ 1.760 and period 2π.

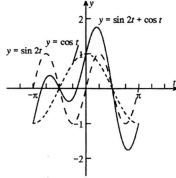

27. $y = \sin \frac{1}{2}t$ has amplitude 1 and period $\frac{2\pi}{1/2} = 4\pi$

$y = \frac{1}{2}\sin t$ has amplitude $\frac{1}{2}$ and period 2π

$y = \sin \frac{1}{2}t - \frac{1}{2}\sin t$ has amplitude ≈ 1.299 and period 4π.

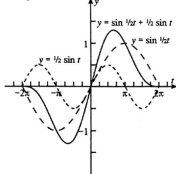

29. $y = -\cos t$, $-\pi \le t \le \pi$

The graph will be a reflection of the graph of $y = \cos t$ in the x-axis.

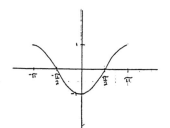

31. $y = \sin 4t$, $0 \le t \le \pi$

Amplitude: 1

Period: $\frac{2\pi}{4} = \frac{\pi}{2}$

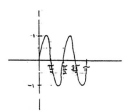

33. $y = -\cos t$

Amplitude: 1

Period: 2π

$y = \sin 4t$

Amplitude: 1

Period: $\frac{2\pi}{4} = \frac{\pi}{2}$

35. a) $y = \tan 2t$

Period: $\frac{\pi}{2}$

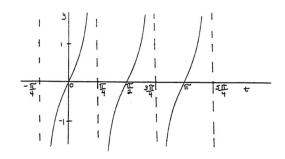

b) $y = 3\tan\left(\frac{t}{2}\right)$

Period: $\frac{\pi}{1/2} = 2\pi$

155

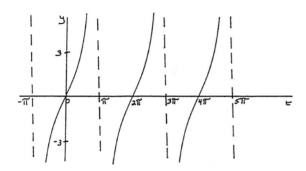

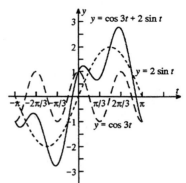

37. The graph of $y = 3 + \sin t$ will be the graph of $y = \sin t$ translated 3 units up.

The graph of $y = \sin\left(t - \frac{\pi}{4}\right)$ will be the graph of $y = \sin t$ translated $\frac{\pi}{4}$ units to the right.

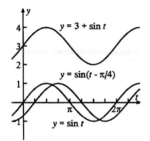

39. $f(x) = x^2$, $-1 < x \le 2$

f has amplitude $(4 - 0)/2 = 2$.

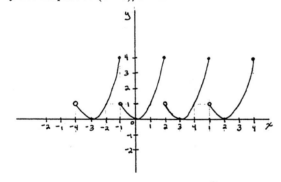

41. $y = \cos 3t$ has amplitude 1 and period $\frac{2\pi}{3}$.

$y = 2 \sin t$ has amplitude 2 and period 2π.
$y = \cos 3t + 2 \sin t$ has amplitude ≈ 2.779 and period 2π.

43. $y = t - \cos t$

t	0	.5	1	1.5	2	2.5	3
$t - \cos t$	-1	$-.38$	.45	1.43	2.42	3.30	3.99

t	3.5	4	4.5	5	5.5	6
$t - \cos t$	4.44	4.65	4.71	4.72	4.79	5.03

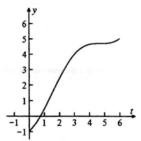

45. $I = 30 \sin(120\pi t)$

(a) Period: $\frac{2\pi}{120\pi} = \frac{1}{60}$ sec

(b) Cycles: 60 cycles/sec = 60 Hz

(c) Maximum strength: Occurs when I reaches its maximum amplitude. Hence it is 30 amperes.

47. (a) $y = 0$ whenever $\frac{1}{t} = k\pi$

$t = \frac{1}{k\pi}$, $k = 1, 2, 3, \ldots$

(b) $\sin \frac{1}{2/\pi} = 1$, $\sin \frac{1}{2/(3\pi)} = -1$, $\sin \frac{1}{2/(5\pi)} = 1, \ldots$

(c)

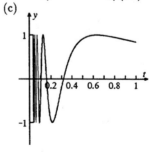

156

49. Press $\boxed{Y=}$

Press $\boxed{ZOOM}$ $\boxed{7}$ (7 is the Trig option)

Press $\boxed{GRAPH}$

Tick-marks on the x-axis are made at multiples of $\pi/2$. Looking at the graph, the period length is 2π. To find the amplitude, ZOOM in on a local maximum to find that $y = 2.736$.

51. Set standard range values.

Press $\boxed{Y=}$

Enter $x \wedge 2 - 2x$ as Y_1.

Enter $\sin x$ as Y_2.

Press $\boxed{GRAPH}$

ZOOM in on the leftmost intersection to find that $x = 0$; ZOOM in on the rightmost intersection to find that $x = 2.317$.

Chapter 7 Review Problem Set

1. T: $\sin\left(\frac{\pi}{3}\right) = \cos\left(\frac{\pi}{2} - \frac{\pi}{3}\right) = \cos\left(\frac{\pi}{6}\right)$

2. T: $135° = 135° \cdot \frac{\pi \text{ rad}}{180°} = \frac{3\pi}{4}$ rad

3. F: When $\theta = 180°$, $\sin\theta = 0$ but $\cos\theta = -1$.

4. F: $\cos(-t) = \cos t$ for all t

5. F: $\sin\beta = \sin(\alpha + \pi) = -\sin\alpha$

6. T:

7. F: Since $276°$ is in quadrant IV, the reference angle is $360° - 276° = 84°$

8. T:

9. F: The amplitude is 2.

10. T:

11.

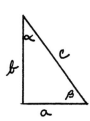

$a = 9,\ c = 15$

$b = \sqrt{15^2 - 9^2} = \sqrt{225 - 81} = \sqrt{144} = 12$

$\sin\alpha = \frac{9}{15} = .6$

$\alpha = 36.87°$

$\beta = 90° - 36.87° = 53.13°$

12. $\beta = 90° - 72.4° = 17.6°$

$\tan 72.4° = \frac{a}{29.6}$ so $a = 29.6 \times \tan 72.4° = 93.3$

$\cos 72.4° = \frac{29.6}{c}$ so $c = \frac{29.6}{\cos 72.4°} = 97.9$

13. $\sin\frac{3\pi}{2} = -1$

14. $\tan\frac{\pi}{6} = \frac{\sqrt{3}}{3}$

15. Since $\frac{5\pi}{3}$ is in quadrant IV, the reference angle is $2\pi - \frac{5\pi}{3} = \frac{\pi}{3}$. Secant is positive in quadrant IV.

$\sec\frac{5\pi}{3} = \frac{1}{\cos(5\pi/3)} = \frac{1}{\cos(\pi/3)} = \frac{1}{\frac{1}{2}} = 2$

16. $\cot 315° = -\cot 45° = -1$

17. Since $-225°$ is in quadrant II, the reference angle is $225° - 180° = 45°$. Sine is positive in quadrant II.

$\sin(-225°) = \sin 45° = \frac{\sqrt{2}}{2}$

18. $\cos 7\pi = \cos\pi = -1$

19. Since $\frac{3\pi}{4}$ is in quadrant II, the reference angle is $\pi - \frac{3\pi}{4} = \frac{\pi}{4}$. Tangent is negative in quadrant II.

$\tan\frac{3\pi}{4} = -\tan\frac{\pi}{4} = -1$

20. $\csc\frac{7\pi}{2} = -\csc\frac{\pi}{2} = -1$

21. Find the reference angle:

$\cos t_0 = \frac{1}{2}$

$t_0 = \frac{\pi}{3}$.

Since cosine is negative in quadrants II and III,

$t = \pi - \frac{\pi}{3} = \frac{2\pi}{3}$ or $t = \pi + \frac{\pi}{3} = \frac{4\pi}{3}$

22. Find the reference angle:

$\tan t_0 = 1$

$t_0 = \frac{\pi}{4}$

Since tangent is negative in quadrants II and IV,

$t = \pi - \frac{\pi}{4} = \frac{3\pi}{4}$ or $2\pi - \frac{\pi}{4} = \frac{7\pi}{4}$

23. Find the reference angle:

$\sin t_0 = \frac{\sqrt{2}}{2}$

$t_0 = \frac{\pi}{4}$.

Since sine is positive in quadrants I and II,

$t = \frac{\pi}{4}$ or $t = \pi - \frac{\pi}{4} = \frac{3\pi}{4}$

24. Find the reference angle:

$$\sec t_0 = 2$$
$$\cos t_0 = \frac{1}{2}$$
$$t_0 = \frac{\pi}{3}$$

Since cosine is positive in quadrants I and IV,

$$t = \frac{\pi}{3} \text{ or } t = 2\pi - \frac{\pi}{3} = \frac{5\pi}{3}$$

25. $\sin t = \frac{3}{7} = \frac{b}{r}$

$$a = \sqrt{r^2 - b^2} = \sqrt{7^2 - 3^2} = \sqrt{40} = 2\sqrt{10}$$

Since t is in quadrant II, $\tan t$ is negative.

$$\tan t = -\frac{b}{a} = \frac{-3}{2\sqrt{10}} = -\frac{3\sqrt{10}}{20}$$

26. $\cos \theta = \frac{1}{2}$

$$\theta_0 = 60°$$

Since cosine is positive in quadrants I and IV,
$\theta = 60°$ or $\theta = -60°$.

27. $\cos \theta = \frac{3}{5}$

a) $\cos(-\theta) = \cos \theta = \frac{3}{5}$

b) $\sin(\theta - 90°) = \sin[-(90° - \theta)]$
$$= -\sin(90° - \theta)$$
$$= -\cos \theta = -\frac{3}{5}$$

c) $\cos(\theta + 180°) = -\cos \theta = -\frac{3}{5}$

d) $\sin^2\theta = 1 - \cos^2\theta = 1 - \left(\frac{3}{5}\right)^2 = 1 - \frac{9}{25} = \frac{16}{25}$

Therefore, $|\sin \theta| = \sqrt{\frac{16}{25}} = \frac{4}{5}$

28. a) Since cosine is negative in quadrant II,
$$\cos \theta = -\sqrt{1 - \sin^2\theta}$$

b) $\csc \theta = \frac{1}{\sin \theta}$

c) $\tan \theta = \frac{\sin \theta}{\cos \theta} = \frac{\sin \theta}{-\sqrt{1 - \sin^2\theta}} = \frac{-\sin \theta}{\sqrt{1 - \sin^2\theta}}$

29. $a = -5$, $b = 12$

a) $\tan \theta = \frac{b}{a} = \frac{12}{-5} = -\frac{12}{5}$

b) $r = \sqrt{a^2 + b^2} = \sqrt{(-5)^2 + 12^2} = \sqrt{169} = 13$

$$\csc \theta = \frac{r}{b} = \frac{13}{12}$$

30. (a) $\sin t > 0$ when $0 < t < \pi$

(b) $\sin 2t > 0$ when
$$0 < 2t < \pi \text{ and } 2\pi < 2t < 3\pi$$
$$0 < t < \frac{\pi}{2} \text{ and } \pi < t < \frac{3\pi}{2}$$

31. $f(t) = 3 \cos t - 2$.

The smallest value is $3(-1) - 2 = -5$ and the largest value is $3(1) - 2 = 1$.

Thus the range of $f(t)$ is $\{y: -5 \le y \le 1\}$.

32. $\tan(-t) = \frac{\sin(-t)}{\cos(-t)} = \frac{-\sin t}{\cos t} = -\frac{\sin t}{\cos t} = -\tan t$

33. $y = a \sin x + b \cos x$

Replace (x, y) with $(\pi/2, 2)$ and then with $(\pi/3, 4)$, producing two equations in a and b.
$$\begin{cases} 2 = a \sin(\pi/2) + b \cos(\pi/2) \\ 4 = a \sin(\pi/3) + b \cos(\pi/3) \end{cases}$$
$$\begin{cases} 2 = a \\ 4 = \frac{a\sqrt{3}}{2} + b\left(\frac{1}{2}\right) \end{cases}$$

Let $a = 2$ in the second equation above:
$$4 = \frac{2\sqrt{3}}{2} + \frac{b}{2}$$
$$8 = 2\sqrt{3} + b$$
$$b = 8 - 2\sqrt{3}$$
Solution: $a = 2$, $b = 8 - 2\sqrt{3}$.

34. $f(t) = 2 + 4 \sin t$ for $0 \le t \le 2\pi$

(a) The maximum of $f(t) = 2 + 4(1) = 6$.
The minimum of $f(t) = 2 + 4(-1) = -2$

(b) $2 + 4 \sin t = 0$
$$\sin t = -\frac{2}{4} = -\frac{1}{2}$$
$$t_0 = \frac{\pi}{6}$$

Sine is negative in quadrants III and IV.

$$t = \pi + \frac{\pi}{6} = \frac{7\pi}{6} \text{ or } t = 2\pi - \frac{\pi}{6} = \frac{11\pi}{6}$$

35. a) $\cos(t + 4\pi) = \cos(t + 2 \cdot 2\pi) = \cos t$

b) $\cos(t + \pi) = -\cos t$

c) $\sin^2 t = 1 - \cos^2 t$

d) $\sin\left(t - \frac{\pi}{2}\right) = \sin\left[-\left(\frac{\pi}{2} - t\right)\right] = -\sin\left(\frac{\pi}{2} - t\right) = -\cos t$

36. $\tan \theta = \frac{3}{4} = \frac{b}{a}$. Since $\sin \theta < 0$, $a = -4$, $b = -3$.

$$c = \sqrt{(-4)^2 + (-3)^2} = \sqrt{25} = 5$$
$$\sec^2\theta - \sin^3\theta = \left(\frac{5}{-4}\right)^2 - \left(\frac{-3}{5}\right)^3 = \frac{25}{16} + \frac{27}{125} = 1.7785$$

37.

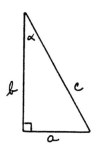

The area of the triangle is $\frac{1}{2}ab$. But

$$\frac{b}{a} = \cot \alpha$$
$$b = a \cot \alpha$$

So $\frac{1}{2}ab = \frac{1}{2}a(a \cot \alpha) = \frac{1}{2}a^2 \cot \alpha$

38. $y = 3 \cos 2t$ has period π and amplitude 3.

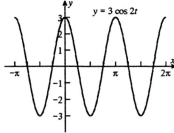

39. $y = \sin t$ has period 2π and amplitude 1.

$y = \sin 2t$ has period $\frac{2\pi}{2} = \pi$ and amplitude 1.

$y = \sin t + \sin 2t$ has period 2π and amplitude 1.760.

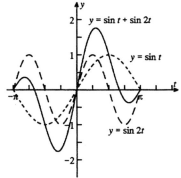

40. $\sin 2t = 1$
$$2t = \frac{\pi}{2}, \frac{5\pi}{2}$$
$$t = \frac{\pi}{4}, \frac{5\pi}{4}$$

41. a) $300 \frac{\deg}{\sec} = 300 \frac{\deg}{\sec} \times \frac{\pi \text{ rad}}{180°} = \frac{5\pi}{3} \frac{\text{rad}}{\sec}$

$10 \text{ inches} = 10 \text{ in.} \times \frac{1 \text{ ft}}{12 \text{ in.}} = \frac{5}{6} \text{ ft}$

$v = r\omega = \frac{5}{6} \text{ ft} \times \frac{5\pi}{3} \frac{\text{rad}}{\sec} = \frac{25\pi}{18} \frac{\text{ft}}{\sec} \approx 4.36 \text{ ft/sec}$

b) $2 \text{ min} = 2 \text{ min} \times \frac{60 \sec}{1 \text{ min}} = 120 \sec$

$d = \text{rate} \times \text{time}$
$$= \frac{25\pi}{18} \frac{\text{ft}}{\sec} \times 120 \sec$$
$$= \frac{500\pi}{3} \text{ ft}$$
$$\approx 523.6 \text{ ft}$$

42. If h is the height of the triangle, then
$$\tan 30° = \frac{h}{120}$$
$$h = 120 \tan 30° = 120 \cdot \frac{\sqrt{3}}{3} = 40\sqrt{3}$$

The area of the triangle is
$$\frac{1}{2} \cdot 120 \cdot 40\sqrt{3} = 2400\sqrt{3}$$

The area of the circular sector is
$$\frac{30°}{360°} \times \pi(120)^2 = 1200\pi$$

Thus, the area of the shaded region is
$$2400\sqrt{3} - 1200\pi \approx 387.0$$

43. Let y feet be the height of the wall.
$$\frac{y}{16} = \sin 55°$$
$$y = 16 \sin 55° \approx 16(.819152) \approx 13.11 \text{ ft}$$

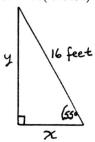

44. $x = 16 \cos 55° = 9.18$
If x is increased by 1, the new value of y is

$\sqrt{16^2 - 10.18^2} = 12.34$
The ladder has come down
$13.11 - 12.34 = .77$ ft

45. $\frac{y}{3} = \tan 38°$
$$y = 3 \tan 38° \approx 3(.781286) \approx 2.343857$$
$$\frac{x+y}{3} = \tan(11° + 38°)$$
$$x + y = 3 \tan 49°$$
$$x = -y + 3 \tan 49°$$
$$\approx -2.343857 + 3(1.150368)$$
$$\approx 1.107248$$
Solution: $x \approx 1.10725$, $y \approx 2.34386$

46. (a) $18° = 18° \times \frac{\pi \text{ rad}}{180°} = \frac{\pi}{10}$ rad

$s = 500 \cdot \frac{\pi}{10} = 50\pi \approx 157.08$ meters

(b) If d meters is the distance, then

$\sin 9° = \frac{d/2}{500}$

$d = 1000 \sin 9° \approx 156.43$ meters

47. For $0 \le t \le \pi$, $\sin(t/2)$ increases from 0 to 1.

For $\pi \le t \le 2\pi$, $\sin(t/2)$ decreases from 1 to 0.

48. Sketching the graphs of $y = \sin 2t$ and $y = -\cos t$ on the same coordinate axes, we find 4 intersections in the interval 0 to 2π. Using the formulas of Section 8.3, we find $t = \frac{\pi}{2}, \frac{7\pi}{6}, \frac{3\pi}{2}, \frac{11\pi}{6}$.

49.

Central angle AOB subtends an arc $\frac{1}{8}$ of the circumference of the circle. Therefore

$\angle \text{AOB} = \frac{1}{8}(360°) = 45°$

Since triangle AOB is isosceles,

$\angle \text{AOD} = \frac{1}{2}(45°) = 22.5°$

$\frac{\overline{\text{AD}}}{12} = \sin 22.5°$

$\overline{\text{AD}} = 12 \sin 22.5°$

The perimeter of the octagon is

$16 \, \overline{\text{AD}} = 16 \times 12 \sin 22.5°$

$= 192 \sin 22.5°$

≈ 73.4752 cm

50. Using the figure below we see that

$\tan 36° = \frac{921/2}{h}$

$h = \frac{921}{2 \tan 36°}$

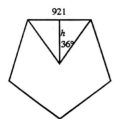

The pentagon, composed of 5 triangles, has area

$5 \cdot \frac{1}{2}(921)h = 5 \cdot \frac{1}{2}(921) \cdot \frac{921}{2 \tan 36°} \approx 1,459,379 \text{ ft}^2$

Chapter 8: Trigonometric Identities and Equations

1. a) $\cos^2 t = 1 - \sin^2 t$

   b) $\tan t \cos t = \dfrac{\sin t}{\cos t} \cdot \cos t = \sin t$

   c) $\dfrac{3}{\csc^2 t} + 2\cos^2 t - 2$

   $= 3\sin^2 t + 2(1 - \sin^2 t) - 2$

   $= 3\sin^2 t + 2 - 2\sin^2 t - 2$

   $= \sin^2 t$

   d) $\cot^2 t = \dfrac{\cos^2 t}{\sin^2 t} = \dfrac{1 - \sin^2 t}{\sin^2 t}$ or $\dfrac{1}{\sin^2 t} - 1$

3. a) $\cot^2 t = \dfrac{1}{\tan^2 t}$

   b) $\sec^2 t = 1 + \tan^2 t$

   c) $\sin t \sec t = \sin t \cdot \dfrac{1}{\cos t} = \dfrac{\sin t}{\cos t} = \tan t$

   d) $2\sec^2 t - 2\tan^2 t + 1$

   $= 2(1 + \tan^2 t) - 2\tan^2 t + 1$

   $= 2 + 2\tan^2 t - 2\tan^2 t + 1$

   $= 3$

5. $\cos t \sec t = \cos t \cdot \dfrac{1}{\cos t} = 1$

7. $\tan x \cot x = \tan x \cdot \dfrac{1}{\tan x} = 1$

9. $\cos y \csc y = \cos y \cdot \dfrac{1}{\sin y} = \dfrac{\cos y}{\sin y} = \cot y$

11. $\cot \theta \sin \theta = \dfrac{\cos \theta}{\sin \theta} \cdot \sin \theta = \cos \theta$

13. $\dfrac{\tan u}{\sin u} = \tan u \cdot \dfrac{1}{\sin u} = \dfrac{\sin u}{\cos u} \cdot \dfrac{1}{\sin u} = \dfrac{1}{\cos u}$

15. $(1 + \sin z)(1 - \sin z) = 1 - \sin^2 z = \cos^2 z = \dfrac{1}{\sec^2 z}$

17. $(1 - \sin^2 x)(1 + \tan^2 x) = \cos^2 x \sec^2 x$

   $= \cos^2 x \cdot \dfrac{1}{\cos^2 x}$

   $= 1$

19. $\sec t - \sin t \tan t = \dfrac{1}{\cos t} - \sin t \cdot \dfrac{\sin t}{\cos t}$

   $= \dfrac{1 - \sin^2 t}{\cos t}$

   $= \dfrac{\cos^2 t}{\cos t}$

   $= \cos t$

21. $\dfrac{\sec^2 t - 1}{\sec^2 t} = 1 - \dfrac{1}{\sec^2 t} = 1 - \cos^2 t = \sin^2 t$

23. $\cos t(\tan t + \cot t) = \cos t\left(\dfrac{\sin t}{\cos t} + \dfrac{\cos t}{\sin t}\right)$

   $= \cos t \cdot \dfrac{\sin^2 t + \cos^2 t}{\cos t \sin t}$

   $= \cos t \cdot \dfrac{1}{\cos t \sin t}$

   $= \dfrac{1}{\sin t}$

   $= \csc t$

25. $\dfrac{\sin^2 \theta}{\cos \theta} = \dfrac{1 - \cos^2 \theta}{\cos \theta} = \dfrac{1}{\cos \theta} - \dfrac{\cos^2 \theta}{\cos \theta} = \sec \theta - \cos \theta$

27. $\dfrac{\tan \theta - \cot \theta}{\sin \theta \cos \theta} = \dfrac{1}{\sin \theta \cos \theta}\left(\dfrac{\sin \theta}{\cos \theta} - \dfrac{\cos \theta}{\sin \theta}\right)$

   $= \dfrac{1}{\cos^2 \theta} - \dfrac{1}{\sin^2 \theta}$

   $= \sec^2 \theta - \csc^2 \theta$

29. $\dfrac{\sec t - 1}{\tan t} = \dfrac{\sec t - 1}{\tan t} \cdot \dfrac{\sec t + 1}{\sec t + 1}$

   $= \dfrac{\sec^2 t - 1}{\tan t(\sec t + 1)}$

   $= \dfrac{\tan^2 t}{\tan (\sec t + 1)}$

   $= \dfrac{\tan t}{\sec t + 1}$

31. $\dfrac{\tan^2 x}{\sec x + 1} = \dfrac{\sec^2 x - 1}{\sec x + 1}$

   $= \dfrac{(\sec x + 1)(\sec x - 1)}{\sec x + 1}$

   $= \sec x - 1$

   $= \dfrac{1}{\cos x} - 1$

   $= \dfrac{1 - \cos x}{\cos x}$

33. $\dfrac{\sin t + \cos t}{\tan^2 t - 1} = \dfrac{\sin t + \cos t}{\dfrac{\sin^2 t}{\cos^2 t} - 1}$

   $= \dfrac{\sin t + \cos t}{\dfrac{\sin^2 t}{\cos^2 t} - 1} \cdot \dfrac{\cos^2 t}{\cos^2 t}$

   $= \dfrac{(\sin t + \cos t)\cos^2 t}{\sin^2 t - \cos^2 t}$

   $= \dfrac{(\sin t + \cos t)\cos^2 t}{(\sin t + \cos t)(\sin t - \cos t)}$

   $= \dfrac{\cos^2 t}{\sin t - \cos t}$

35. $\cot \theta + \tan \theta = \dfrac{\cos \theta}{\sin \theta} + \dfrac{\sin \theta}{\cos \theta}$

   $= \dfrac{\cos^2 \theta + \sin^2 \theta}{\cos \theta \sin \theta}$

   $= \dfrac{1}{\cos \theta \sin \theta}$

   $= \sec \theta \csc \theta$

161

37. Since t is in quadrant II, $\sin t$ is positive.

$$\sin t = \sqrt{1 - \cos^2 t}$$

$$\tan t = \frac{\sin t}{\cos t} = \frac{\sqrt{1 - \cos^2 t}}{\cos t}$$

$$\cot t = \frac{\cos t}{\sin t} = \frac{\cos t}{\sqrt{1 - \cos^2 t}}$$

$$\sec t = \frac{1}{\cos t}$$

$$\csc t = \frac{1}{\sin t} = \frac{1}{\sqrt{1 - \cos^2 t}}$$

39. Since t is in quadrant II, $\cos t$ is negative.

$$\cos t = -\sqrt{1 - \sin^2 t} = -\sqrt{1 - \frac{16}{25}} = -\sqrt{\frac{9}{25}} = -\frac{3}{5}$$

$$\tan t = \frac{\sin t}{\cos t} = \frac{\frac{4}{5}}{-\frac{3}{5}} = -\frac{4}{3}$$

$$\cot t = \frac{1}{\tan t} = -\frac{3}{4}$$

$$\sec t = \frac{1}{\cos t} = -\frac{5}{3}$$

$$\csc t = \frac{1}{\sin t} = \frac{5}{4}$$

41. a) $[(\sin x + \cos x)^2 - 1]\sec x \csc^3 x$

$$= [\sin^2 + 2 \sin x \cos x + \cos^2 x - 1]\sec x \csc^3 x$$

$$= (1 + 2 \sin x \cos x - 1)\sec x \csc^3 x$$

$$= 2 \sin x \cos x (\sec x \csc^3 x)$$

$$= \frac{2 \sin x \cos x}{\cos x \sin^3 x}$$

$$= \frac{2}{\sin^2 x}$$

b) To express the above in terms of $\tan x$,

$$\frac{2}{\sin^2 x} = 2 \csc^2 x = 2(1 + \cot^2 x)$$

$$= 2\left(1 + \frac{1}{\tan^2 x}\right)$$

$$= 2 + \frac{2}{\tan^2 x}$$

43. $(1 + \tan^2 t)(\cos t + \sin t)$

$$= \sec^2 t(\cos t + \sin t)$$

$$= \sec t(\sec t \cos t + \sec t \sin t)$$

$$= \sec t\left(\frac{\cos t}{\cos t} + \frac{\sin t}{\cos t}\right)$$

$$= \sec t(1 + \tan t)$$

45. $2 \sec^2 y - 1 = \frac{2}{\cos^2 y} - 1$

$$= \frac{2 - \cos^2 y}{\cos^2 y}$$

$$= \frac{1 + 1 - \cos^2 y}{\cos^2 y}$$

$$= \frac{1 + \sin^2 y}{\cos^2 y}$$

47. $\dfrac{\sin z}{\sin z + \tan z} = \dfrac{\sin z}{\sin z + \dfrac{\sin z}{\cos z}}$

$$= \frac{\sin z}{\sin z\left(1 + \frac{1}{\cos z}\right)}$$

$$= \frac{1}{1 + \frac{1}{\cos z}}$$

$$= \frac{\cos z}{\cos z + 1}$$

49. $(\csc t + \cot t)^2 = \left(\dfrac{1}{\sin t} + \dfrac{\cos t}{\sin t}\right)^2$

$$= \frac{(1 + \cos t)^2}{\sin^2 t}$$

$$= \frac{(1 + \cos t)^2}{1 - \cos^2 t}$$

$$= \frac{(1 + \cos t)(1 + \cos t)}{(1 + \cos t)(1 - \cos t)}$$

$$= \frac{1 + \cos t}{1 - \cos t}$$

51. $\dfrac{\cos^4 u - \sin^4 u}{\cos u - \sin u} = \dfrac{(\cos^2 u - \sin^2 u)(\cos^2 u + \sin^2 u)}{\cos u - \sin u}$

$$= \frac{(\cos u + \sin u)(\cos u - \sin u)(1)}{\cos u - \sin u}$$

$$= \cos u + \sin u$$

53. $\dfrac{1 + \tan x}{1 - \tan x} = \dfrac{1 + \dfrac{\sin x}{\cos x}}{1 - \dfrac{\sin x}{\cos x}}$

$$= \frac{\cos x}{\cos x} \cdot \frac{1 + \frac{\sin x}{\cos x}}{1 - \frac{\sin x}{\cos x}}$$

$$= \frac{\cos x + \sin x}{\cos x - \sin x}$$

55. $(\sec t + \tan t)(\csc t - 1) = \left(\dfrac{1}{\cos t} + \dfrac{\sin t}{\cos t}\right)\left(\dfrac{1}{\sin t} - 1\right)$

$$= \frac{1 + \sin t}{\cos t} \cdot \frac{1 - \sin t}{\sin t}$$

$$= \frac{1 - \sin^2 t}{\cos t \sin t}$$

$$= \frac{\cos^2 t}{\cos t \sin t}$$

$$= \frac{\cos t}{\sin t}$$

$$= \cot t$$

57. $\dfrac{\cos^3 t + \sin^3 t}{\cos t + \sin t}$

$$= \frac{(\cos t + \sin t)(\cos^2 t - \sin t \cos t + \sin^2 t)}{\cos t + \sin t}$$

$$= \cos^2 t - \sin t \cos t + \sin^2 t$$

$$= 1 - \sin t \cos t$$

59. $\dfrac{1 - \cos \theta}{1 + \cos \theta} = \dfrac{1 - \cos \theta}{1 + \cos \theta} \cdot \dfrac{1 - \cos \theta}{1 - \cos \theta}$

$$= \frac{(1 - \cos \theta)^2}{1 - \cos^2 \theta}$$

$$= \left(\frac{1 - \cos \theta}{\sin \theta}\right)^2$$

61. $(\csc t - \cot t)^4 (\csc t + \cot t)^4$

$$= [(\csc t - \cot t)(\csc t + \cot t)]^4$$

$$= [\csc^2 t - \cot^2 t]^4$$

$$= [1]^4$$

$$= 1$$

63. $\sin^6 u + \cos^6 u$

$$= (\sin^2 u + \cos^2 u)(\sin^4 u - \sin^2 u \cos^2 u + \cos^4 u)$$

$$= \sin^4 u - \sin^2 u \cos^2 u + \cos^4 u$$

$$= \sin^4 u + 2 \sin^2 u \cos^2 u + \cos^4 u - 3 \sin^2 u \cos^2 u$$

$$= (\sin^2 u + \cos^2 u)^2 - 3 \sin^2 u \cos^2 u$$

$$= 1 - 3 \sin^2 u \cos^2 u$$

65. $\cot 3x = \tan\left(\dfrac{\pi}{2} - 3x\right)$

$$= \tan\left(\frac{3\pi}{2} - 3x\right)$$

$$= \tan 3\left(\frac{\pi}{2} - x\right)$$

$$= \frac{3 \tan\left(\frac{\pi}{2} - x\right) - \tan^3\left(\frac{\pi}{2} - x\right)}{1 - 3 \tan^2\left(\frac{\pi}{2} - x\right)}$$

$$= \frac{3 \cot x - \cot^3 x}{1 - 3 \cot^2 x}$$

67. Press $\boxed{Y=}$

Enter

$(\boxed{COS}\, x) \wedge 3 \times (1 - (\boxed{TAN}\, x) \wedge 4$

$+ (\boxed{COS}\, x) \wedge \boxed{(-)}4)$

Press $\boxed{ZOOM}\ \boxed{7}$ to get intervals of $\pi/2$ on the x-axis. The graph looks like the graph of $2 \cos x$.

$\cos^3(1 - \tan^4 x + \sec^4 x)$

$$= \cos^3 x[1 - \tan^4 + (1 + \tan^2 x)^2]$$

$$= \cos^3 x[1 - \tan^4 x + 1 + 2 \tan^2 x + \tan^4 x]$$

$$= \cos^3 x[2 + 2 \tan^2 x]$$

$$= 2 \cos^3 x[1 + \tan^2 x]$$

$$= 2 \cos^3 x[\sec^2 x]$$

$$= 2 \cos^3 x \cdot \frac{1}{\cos^2 x}$$

$$= 2 \cos x$$

Problem Set 8.2

1. a) $\sin \dfrac{\pi}{4} + \sin \dfrac{\pi}{6} = \dfrac{\sqrt{2}}{2} + \dfrac{1}{2} = \dfrac{\sqrt{2}+1}{2} \approx 1.21$

  b) $\sin\left(\dfrac{\pi}{4} + \dfrac{\pi}{6}\right) = \sin \dfrac{\pi}{4} \cos \dfrac{\pi}{6} + \cos \dfrac{\pi}{4} \sin \dfrac{\pi}{6}$

$$= \frac{\sqrt{2}}{2} \cdot \frac{\sqrt{3}}{2} + \frac{\sqrt{2}}{2} \cdot \frac{1}{2}$$

$$= \frac{\sqrt{6} + \sqrt{2}}{4}$$

$$\approx .97$$

3. a) $\cos \dfrac{\pi}{4} - \cos \dfrac{\pi}{6} = \dfrac{\sqrt{2}}{2} - \dfrac{\sqrt{3}}{2} = \dfrac{\sqrt{2} - \sqrt{3}}{2} \approx -.16$

  b) $\cos\left(\dfrac{\pi}{4} - \dfrac{\pi}{6}\right) = \cos \dfrac{\pi}{4} \cos \dfrac{\pi}{6} + \sin \dfrac{\pi}{4} \sin \dfrac{\pi}{6}$

$$= \frac{\sqrt{2}}{2} \cdot \frac{\sqrt{3}}{2} + \frac{\sqrt{2}}{2} \cdot \frac{1}{2}$$

$$= \frac{\sqrt{6} + \sqrt{2}}{4}$$

$$\approx .97$$

5. $\cos \dfrac{13\pi}{12} = \cos\left(\dfrac{\pi}{3} + \dfrac{3\pi}{4}\right)$

$$= \cos \frac{\pi}{3} \cos \frac{3\pi}{4} - \sin \frac{\pi}{3} \sin \frac{3\pi}{4}$$

$$= \frac{1}{2}\left(-\frac{\sqrt{2}}{2}\right) - \frac{\sqrt{3}}{2} \cdot \frac{\sqrt{2}}{2}$$

$$= -\frac{\sqrt{2} + \sqrt{6}}{2}$$

7. $\sin 165° = \sin(120° + 45°)$

$$= \sin 120° \cos 45° + \cos 120° \sin 45°$$

$$= \frac{\sqrt{3}}{2} \cdot \frac{\sqrt{2}}{2} + \left(-\frac{1}{2}\right)\frac{\sqrt{2}}{2}$$

$$= \frac{\sqrt{6} - \sqrt{2}}{4}$$

9. $\tan 75° = \tan(30° + 45°)$

$$= \frac{\tan 30° + \tan 45°}{1 - \tan 30° \tan 45°}$$

$$= \frac{\frac{\sqrt{3}}{3} + 1}{1 - \frac{\sqrt{3}}{3} \cdot 1}$$

$$= \frac{3 + \sqrt{3}}{3 - \sqrt{3}}$$

$$= \frac{3 + \sqrt{3}}{3 - \sqrt{3}} \cdot \frac{3 + \sqrt{3}}{3 + \sqrt{3}}$$

$$= \frac{12 + 6\sqrt{3}}{6}$$

$$= 2 + \sqrt{3}$$

11. $\sin(t + \pi) = \sin t \cos \pi + \cos t \sin \pi$

$$= (\sin t)(-1) + (\cos t)(0)$$

$$= -\sin t$$

13. $\sin\left(t + \frac{3\pi}{2}\right) = \sin t \cos \frac{3\pi}{2} + \cos t \sin \frac{3\pi}{2}$

$$= (\sin t)(0) + (\cos t)(-1)$$

$$= -\cos t$$

15. $\sin\left(t - \frac{\pi}{2}\right) = \sin t \cos \frac{\pi}{2} - \cos t \sin \frac{\pi}{2}$

$$= (\sin t)(0) - (\cos t)(1)$$

$$= -\cos t$$

17. $\cos\left(t + \frac{\pi}{3}\right) = \cos t \cos \frac{\pi}{3} - \sin t \sin \frac{\pi}{3}$

$$= (\cos t)\left(\frac{1}{2}\right) - (\sin t)\left(\frac{\sqrt{3}}{2}\right)$$

$$= \frac{1}{2}\cos t - \frac{\sqrt{3}}{2}\sin t$$

19. $\cos \frac{1}{2} \cos \frac{3}{2} - \sin \frac{1}{2} \sin \frac{3}{2} = \cos\left(\frac{1}{2} + \frac{3}{2}\right) = \cos 2$

21. $\sin \frac{7\pi}{8} \cos \frac{\pi}{8} + \cos \frac{7\pi}{8} \sin \frac{\pi}{8} = \sin\left(\frac{7\pi}{8} + \frac{\pi}{8}\right) = \sin \pi$

23. $\cos 33° \cos 27° - \sin 33° \sin 27° = \cos(33° + 27°)$

$$= \cos 60°$$

25. $\sin(\alpha + \beta) \cos \beta - \cos(\alpha + \beta) \sin \beta$

$$= \sin(\alpha + \beta - \beta)$$

$$= \sin \alpha$$

27. $\sin \alpha = -\frac{4}{5}$

$\cos \alpha = -\sqrt{1 - \left(-\frac{4}{5}\right)^2} = -\sqrt{1 - \frac{16}{25}} = -\sqrt{\frac{9}{25}} = -\frac{3}{5}$

$\cos \beta = -\frac{5}{13}$

$\sin \beta = -\sqrt{1 - \left(-\frac{5}{13}\right)^2} = -\sqrt{1 - \frac{25}{169}} = -\sqrt{\frac{144}{169}} = -\frac{12}{13}$

$\sin(\alpha + \beta) = \sin \alpha \cos \beta + \cos \alpha \sin \beta$

$$= \left(-\frac{4}{5}\right)\left(-\frac{5}{13}\right) + \left(-\frac{3}{5}\right)\left(-\frac{12}{13}\right)$$

$$= \frac{20}{65} + \frac{36}{65}$$

$$= \frac{56}{65}$$

$\cos(\alpha + \beta) = \cos \alpha \cos \beta - \sin \alpha \sin \beta$

$$= \left(-\frac{3}{5}\right)\left(-\frac{5}{13}\right) - \left(-\frac{4}{5}\right)\left(-\frac{12}{13}\right)$$

$$= \frac{15}{65} - \frac{48}{65}$$

$$= -\frac{33}{65}$$

Since $\sin(\alpha + \beta)$ is positive and $\cos(\alpha + \beta)$ is negative, $\alpha + \beta$ is in quadrant II.

29. $\sin \alpha = \frac{1}{\sqrt{10}}$

$\cos \alpha = \sqrt{1 - \left(\frac{1}{\sqrt{10}}\right)^2} = \sqrt{1 - \frac{1}{10}} = \sqrt{\frac{9}{10}} = \frac{3}{\sqrt{10}}$

$\cos \beta = -\frac{1}{2}$

$\sin \beta = \sqrt{1 - \left(-\frac{1}{2}\right)^2} = \sqrt{1 - \frac{1}{4}} = \sqrt{\frac{3}{4}} = \frac{\sqrt{3}}{2}$

$\sin(\alpha - \beta) = \sin \alpha \cos \beta - \cos \alpha \sin \beta$

$$= \left(\frac{1}{\sqrt{10}}\right)\left(-\frac{1}{2}\right) - \left(\frac{3}{\sqrt{10}}\right)\left(\frac{\sqrt{3}}{2}\right)$$

$$= -\frac{1}{2\sqrt{10}} - \frac{3\sqrt{3}}{2\sqrt{10}}$$

$$= -\frac{1 + 3\sqrt{3}}{2\sqrt{10}}$$

$\cos(\alpha - \beta) = \cos \alpha \cos \beta + \sin \alpha \sin \beta$

$$= \left(\frac{3}{\sqrt{10}}\right)\left(-\frac{1}{2}\right) + \left(\frac{1}{\sqrt{10}}\right)\left(\frac{\sqrt{3}}{2}\right)$$

$$= -\frac{3}{2\sqrt{10}} + \frac{\sqrt{3}}{2\sqrt{10}}$$

$$= \frac{\sqrt{3} - 3}{2\sqrt{10}}$$

Since $\sin(\alpha - \beta)$ is negative and $\cos(\alpha - \beta)$ is negative, $\alpha - \beta$ is in quadrant III.

31. $\tan(s-t) = \tan(s+(-t))$

$$= \frac{\tan s + \tan(-t)}{1 - \tan s \tan(-t)}$$

$$= \frac{\tan s - \tan t}{1 + \tan s \tan t}$$

33. $\tan\left(t + \frac{\pi}{4}\right) = \frac{\tan t + \tan \frac{\pi}{4}}{1 - \tan t \tan \frac{\pi}{4}}$

$$= \frac{\tan t + 1}{1 - (\tan t)(1)}$$

$$= \frac{1 + \tan t}{1 - \tan t}$$

35. a) $\sin\left(t - \frac{5}{6}\pi\right) = \sin t \cos \frac{5}{6}\pi - \cos t \sin \frac{5}{6}\pi$

$$= (\sin t)\left(-\frac{\sqrt{3}}{2}\right) - (\cos t)\left(\frac{1}{2}\right)$$

$$= -\frac{\sqrt{3} \sin t + \cos t}{2}$$

b) $\cos\left(\frac{\pi}{6} - t\right) = \cos \frac{\pi}{6} \cos t + \sin \frac{\pi}{6} \sin t$

$$= \frac{\sqrt{3}}{2} \cos t + \frac{1}{2} \sin t$$

$$= \frac{\sqrt{3} \cos t + \sin t}{2}$$

37. $\sin \alpha = \frac{2}{3}$, α is in quadrant I

$\cos \beta = -\frac{1}{3}$, β is in quadrant III

a) $\cos \alpha = \sqrt{1 - \left(\frac{2}{3}\right)^2} = \sqrt{1 - \frac{4}{9}} = \sqrt{\frac{5}{9}} = \frac{\sqrt{5}}{3}$

b) $\sin \beta = -\sqrt{1 - \left(-\frac{1}{3}\right)^2} = -\sqrt{1 - \frac{1}{9}} = -\sqrt{\frac{8}{9}} = -\frac{2\sqrt{2}}{2}$

c) $\cos(\alpha + \beta) = \cos \alpha \cos \beta - \sin \alpha \sin \beta$

$$= \left(\frac{\sqrt{5}}{3}\right)\left(-\frac{1}{3}\right) - \left(\frac{2}{3}\right)\left(-\frac{2\sqrt{2}}{3}\right)$$

$$= -\frac{\sqrt{5}}{9} + \frac{4\sqrt{2}}{9}$$

$$= \frac{4\sqrt{2} - \sqrt{5}}{9}$$

d) $\sin(\alpha - \beta) = \sin \alpha \cos \beta - \cos \alpha \cos \beta$

$$= \left(\frac{2}{3}\right)\left(-\frac{1}{3}\right) - \left(\frac{\sqrt{5}}{3}\right)\left(-\frac{2\sqrt{2}}{3}\right)$$

$$= -\frac{2}{9} + \frac{2\sqrt{10}}{9}$$

$$= \frac{-2 + 2\sqrt{10}}{9}$$

e) As in d) above, $\sin(\alpha + \beta) = \frac{-2 - 2\sqrt{10}}{9}$

$$\tan(\alpha + \beta) = \frac{\sin(\alpha + \beta)}{\cos(\alpha + \beta)}$$

$$= \frac{\dfrac{-2 - 2\sqrt{10}}{9}}{\dfrac{4\sqrt{2} - \sqrt{5}}{9}}$$

$$= \frac{-2 - 2\sqrt{10}}{4\sqrt{2} - \sqrt{5}}$$

$$= \frac{-2 - 2\sqrt{10}}{4\sqrt{2} - \sqrt{5}} \cdot \frac{4\sqrt{2} + \sqrt{5}}{4\sqrt{2} + \sqrt{5}}$$

$$= \frac{-8\sqrt{2} - 2\sqrt{5} - 8\sqrt{20} - 2\sqrt{50}}{32 - 5}$$

$$= \frac{-8\sqrt{2} - 2\sqrt{5} - 8 \cdot 2\sqrt{5} - 2 \cdot 5\sqrt{2}}{27}$$

$$= \frac{-18\sqrt{2} - 18\sqrt{5}}{27}$$

$$= -\frac{2}{3}(\sqrt{2} + \sqrt{5})$$

f) $\sin(2\beta) = \sin(\beta + \beta)$

$$= \sin \beta \cos \beta + \cos \beta \sin \beta$$

$$= 2 \sin \beta \cos \beta$$

$$= 2\left(-\frac{2\sqrt{2}}{3}\right)\left(-\frac{1}{3}\right)$$

$$= \frac{4\sqrt{2}}{9}$$

39. a) $\sin\left(t + \frac{\pi}{3}\right) \cos t - \cos\left(t + \frac{\pi}{3}\right) \sin t$

$$= \sin\left(t + \frac{\pi}{3} - t\right)$$

$$= \sin \frac{\pi}{3}$$

$$= \frac{\sqrt{3}}{2}$$

b) $\cos 175° \cos 25° + \sin 175° \sin 25°$

$$= \cos(175° - 25°)$$

$$= \cos 150°$$

$$= -\cos 30°$$

$$= -\frac{\sqrt{3}}{2}$$

c) $\sin t \cos(1 - t) + \cos t \sin(1 - t)$

$$= \sin(t + 1 - t)$$

$$= \sin 1 \approx .84147$$

41. a) $\sin(x + y) \sin(x - y)$

$$= (\sin x \cos y + \cos x \sin y)(\sin x \cos y - \cos x \sin y)$$

$$= \sin^2 x \cos^2 y - \cos^2 x \sin^2 y$$

$$= \sin^2 x (1 - \sin^2 y) - (1 - \sin^2 x)(\sin^2 y)$$

$$= \sin^2 x - \sin^2 x \sin^2 y - \sin^2 y + \sin^2 x \sin^2 y$$

$$= \sin^2 x - \sin^2 y$$

b) $\dfrac{\tan x + \tan y}{1 + \tan x \tan y} = \dfrac{\frac{\sin x}{\cos x} + \frac{\sin y}{\cos y}}{1 + \frac{\sin x}{\cos x} \cdot \frac{\sin y}{\cos y}}$

$= \dfrac{\frac{\sin x}{\cos x} + \frac{\sin y}{\cos y}}{1 + \frac{\sin x}{\cos x} \cdot \frac{\sin y}{\cos y}} \cdot \dfrac{\cos x \cos y}{\cos x \cos y}$

$= \dfrac{\sin x \cos y + \cos x \sin y}{\cos x \cos y + \sin x \sin y}$

$= \dfrac{\sin(x + y)}{\cos(x - y)}$

c) $\dfrac{\cos 5t}{\sin t} - \dfrac{\sin 5t}{\cos t} = \dfrac{\cos t \cos 5t - \sin t \sin 5t}{\sin t \cos t}$

$= \dfrac{\cos(t + 5t)}{\sin t \cos t}$

$= \dfrac{\cos 6t}{\sin t \cos t}$

43.

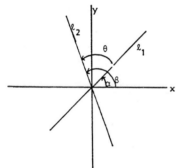

The slope of a line = tangent of smallest angle (called "angle of inclination") from positive x axis to line. Let α and β be the angles of inclination of ℓ_1 and ℓ_2, respectively. Then $\theta = \beta - \alpha$, $m_1 = \tan \alpha$, and $m_2 = \tan \beta$.

$\tan \theta = \tan(\beta - \alpha) = \dfrac{\tan \beta - \tan \alpha}{1 + \tan \beta \tan \alpha} = \dfrac{m_2 - m_1}{1 + m_1 m_2}$

45. $\tan \beta = \dfrac{\overline{AB}}{\overline{BC}} = \dfrac{\overline{AB}}{2\overline{AB}} = \dfrac{1}{2}$

$\tan \alpha = \dfrac{\overline{AB}}{\overline{BC} + \overline{CD}} = \dfrac{\overline{AB}}{2\overline{AB} + \overline{AB}} = \dfrac{1}{2+1} = \dfrac{1}{3}$

$\tan(\alpha + \beta) = \dfrac{\tan \alpha + \tan \beta}{1 - \tan \alpha \tan \beta} = \dfrac{\frac{2}{6} + \frac{3}{6}}{1 - \frac{1}{6}} = \dfrac{\frac{5}{6}}{\frac{5}{6}} = 1$

If $\tan(\alpha + \beta) = 1$, then $\alpha + \beta = 45°$.

47. a) $\frac{1}{2}[\cos(s + t) + \cos(s - t)]$

$= \frac{1}{2}[\cos s \cos t - \sin s \sin t$
$+ \cos s \cos t + \sin s \sin t]$

$= \frac{1}{2} \cdot 2 \cos s \cos t$

$= \cos s \cos t$

b) $-\frac{1}{2}[\cos(s + t) - \cos(s - t)]$

$= -\frac{1}{2}[\cos s \cos t - \sin s \sin t$
$- (\cos s \cos t + \sin s \sin t)]$

$= -\frac{1}{2}[-2 \sin s \sin t]$

$= \sin s \sin t$

c) $\frac{1}{2}[\sin(s + t) + \sin(s - t)]$

$= \frac{1}{2}[\sin s \cos t + \cos s \sin t$
$+ \sin s \cos t - \cos s \sin t]$

$= \frac{1}{2} \cdot 2 \sin s \cos t$

$= \sin s \cos t$

d) $\frac{1}{2}[\sin(s + t) - \sin(s - t)]$

$= \frac{1}{2}[\sin s \cos t + \cos s \sin t$
$- (\sin s \cos t - \cos s \sin t)]$

$= \frac{1}{2} \cdot 2 \cos s \sin t$

$= \cos s \sin t$

49. a) Use a result of Exercise 47:

$\cos 105° \cos 45° = \frac{1}{2}[\cos(105° + 45°) + \cos(105° - 45°]$

$= \frac{1}{2}[\cos 150° + \cos 60°]$

$= \frac{1}{2}\left(-\dfrac{\sqrt{3}}{2} + \dfrac{1}{2}\right)$

$= \dfrac{1 - \sqrt{3}}{4}$

b) Use a result of Exercise 48:

$\sin 15° - \sin 75° = 2 \cos \dfrac{15° + 75°}{2} \sin \dfrac{15° - 75°}{2}$

$= 2 \cos 45° \sin(-30°)$

$= 2\left(\dfrac{\sqrt{2}}{2}\right)\left(-\dfrac{1}{2}\right)$

$= -\dfrac{\sqrt{2}}{2}$

c) We know the exact values of $\cos 30°$, $\cos 45°$, and $\cos 60°$. Use a result of Exercise 48 to find

$\cos 15° + \cos 75° = 2 \cos \dfrac{15° + 75°}{2} \cos \dfrac{15° - 75°}{2}$

$= 2 \cos 45° \cos(-30°)$

$= 2 \cdot \dfrac{\sqrt{2}}{2} \cdot \dfrac{\sqrt{3}}{2}$

$= \dfrac{\sqrt{6}}{2}$

Therefore,

$\cos 15° + \cos 75° + \cos 30° + \cos 45° + \cos 60°$

$= \dfrac{\sqrt{6}}{2} + \dfrac{\sqrt{3}}{2} + \dfrac{\sqrt{2}}{2} + \dfrac{1}{2}$

$= \dfrac{\sqrt{6} + \sqrt{3} + \sqrt{2} + 1}{2}$

51. $\alpha + \beta = 45°$, from Exercise 45

$= \gamma$, since it is in an isosceles right triangle

53. Conjecture: $\sin t + \cos t = \sqrt{2}\sin\left(t + \frac{\pi}{4}\right)$

Proof:

$$\sqrt{2}\sin\left(t + \frac{\pi}{4}\right) = \sqrt{2}\left(\sin t \cos \frac{\pi}{4} + \cos t \sin \frac{\pi}{4}\right)$$

$$= \sqrt{2}\left[(\sin t)\left(\frac{\sqrt{2}}{2}\right) + (\cos t)\left(\frac{\sqrt{2}}{2}\right)\right]$$

$$= \frac{2}{2}\sin t + \frac{2}{2}\cos t$$

$$= \sin t + \cos t$$

Problem Set 8.3

1. $\cos t = \frac{4}{5}$, t is in quadrant I

$$\sin t = \sqrt{1 - \left(\frac{4}{5}\right)^2} = \sqrt{1 - \frac{16}{25}} = \sqrt{\frac{9}{25}} = \frac{3}{5}$$

a) $\sin 2t = 2\sin t \cos t = 2 \cdot \frac{3}{5} \cdot \frac{4}{5} = \frac{24}{25}$

b) $\cos 2t = 2\cos^2 t - 1 = 2\left(\frac{4}{5}\right)^2 - 1 = 2 \cdot \frac{16}{25} - \frac{25}{25} = \frac{7}{25}$

c) $\cos \frac{t}{2} = +\sqrt{\frac{1 - \frac{4}{5}}{2}} = \sqrt{\frac{9}{10}} = \frac{3}{\sqrt{10}} = \frac{3\sqrt{10}}{10}$

d) $\sin \frac{t}{2} = +\sqrt{\frac{1 + \frac{4}{5}}{2}} = \sqrt{\frac{1}{10}} = \frac{1}{\sqrt{10}} = \frac{\sqrt{10}}{10}$

3. $2\sin 5t \cos 5t = \sin(2 \cdot 5t) = \sin 10t$

5. $\cos^2\left(\frac{3t}{2}\right) - \sin^2\left(\frac{3t}{2}\right) = \cos\left(2 \cdot \frac{3t}{2}\right) = \cos 3t$

7. $2\cos^2\left(\frac{y}{4}\right) - 1 = \cos\left(2 \cdot \frac{y}{4}\right) = \cos \frac{y}{2}$

9. $1 - 2\sin^2(.6t) = \cos(2 \cdot .6t) = \cos 1.2t$

11. $\sin^2\left(\frac{\pi}{8}\right) - \cos^2\left(\frac{\pi}{8}\right) = -\left[\cos^2\left(\frac{\pi}{8}\right) - \sin^2\left(\frac{\pi}{8}\right)\right]$

$$= -\cos\left(2 \cdot \frac{\pi}{8}\right)$$

$$= -\cos \frac{\pi}{4}$$

13. $\frac{1 + \cos x}{2} = \cos^2\left(\frac{x}{2}\right)$

15. $\frac{1 - \cos 4\theta}{2} = \sin^2\left(\frac{4\theta}{2}\right) = \sin^2 2\theta$

17. a) $\sin \frac{\pi}{8} = \sin\left(\frac{1}{2} \cdot \frac{\pi}{4}\right)$

$$= \sqrt{\frac{1 - \cos\left(\frac{\pi}{4}\right)}{2}}$$

$$= \sqrt{\frac{1 - \frac{\sqrt{2}}{2}}{2}}$$

$$= \sqrt{\frac{2 - \sqrt{2}}{4}}$$

$$= \frac{1}{2}\sqrt{2 - \sqrt{2}}$$

$$\approx .3827$$

b) $\cos(112.5°) = \cos\left(\frac{1}{2} \cdot 225°\right)$

$$= -\sqrt{\frac{1 + \cos 225°}{2}}$$

$$= -\sqrt{\frac{1 + \left(-\frac{\sqrt{2}}{2}\right)}{2}}$$

$$= -\sqrt{\frac{2 - \sqrt{2}}{4}}$$

$$= -\frac{1}{2}\sqrt{2 - \sqrt{2}} \approx -.3827$$

21. Since $\frac{3\pi}{2} < u < 2\pi$, then $\frac{3\pi}{4} < \frac{u}{2} < \pi$.

$\frac{u}{2}$ is in quadrant II; $\cos\left(\frac{u}{2}\right)$ is negative.

Before we can find $\cos\left(\frac{u}{2}\right)$, we must find $\cos u$.

$$\cos u = \sqrt{1 - \left(-\frac{12}{13}\right)^2} = \sqrt{1 - \frac{144}{169}} = \sqrt{\frac{25}{169}} = \frac{5}{13}$$

$$\cos\left(\frac{u}{2}\right) = -\sqrt{\frac{1 + \cos u}{2}}$$

$$= -\sqrt{\frac{1 + \frac{5}{13}}{2}}$$

$$= -\sqrt{\frac{18}{26}}$$

$$= -\sqrt{\frac{9}{13}} \approx -.8321$$

23.

$$\cos 3t = \cos(2t + t)$$

$$= \cos 2t \cos t - \sin 2t \sin t$$

$$= (2\cos^2 t - 1)\cos t - (2\sin t \cos t)\sin t$$

$$= 2\cos^3 t - \cos t - 2\sin^2 t \cos t$$

$$= 2\cos^3 t - \cos t - 2(1 - \cos^2 t)\cos t$$

$$= 2\cos^3 t - \cos t - 2\cos t + 2\cos^3 t$$

$$= 4\cos^3 t - 3\cos t$$

25.

$$\csc 2t + \cot 2t = \frac{1}{\sin 2t} + \frac{\cos 2t}{\sin 2t}$$

$$= \frac{1 + \cos 2t}{\sin 2t}$$

$$= \frac{1 + 2\cos^2 t - 1}{2\sin t \cos t}$$

$$= \frac{2\cos^2 t}{2\sin t \cos t}$$

$$= \frac{\cos t}{\sin t} = \cot t$$

27.
$$\frac{\sin\theta}{1-\cos\theta} = \frac{\sin\left(2\cdot\frac{\theta}{2}\right)}{1-\cos\left(2\cdot\frac{\theta}{2}\right)}$$

$$= \frac{2\sin\frac{\theta}{2}\cos\frac{\theta}{2}}{1-\left[1-2\sin^2\left(\frac{\theta}{2}\right)\right]}$$

$$= \frac{2\sin\frac{\theta}{2}\cos\frac{\theta}{2}}{2\sin^2\left(\frac{\theta}{2}\right)}$$

$$= \frac{\cos\frac{\theta}{2}}{\sin\frac{\theta}{2}}$$

$$= \cot\frac{\theta}{2}$$

29.
$$\frac{2\tan\alpha}{1+\tan^2\alpha} = \frac{2\frac{\sin\alpha}{\cos\alpha}}{1+\frac{\sin^2\alpha}{\cos^2\alpha}}$$

$$= \frac{2\sin\alpha\cos\alpha}{\cos^2\alpha+\sin^2\alpha}$$

$$= 2\sin\alpha\cos\alpha$$

$$= \sin 2\alpha$$

31.
$$\sin 4\theta = \sin(2\cdot 2\theta)$$

$$= 2\sin 2\theta\cos 2\theta$$

$$= 2(2\sin\theta\cos\theta)(2\cos^2\theta-1)$$

$$= 4\sin\theta(2\cos^3\theta-\cos\theta)$$

33. $\tan 2t = \tan(t+t) = \dfrac{\tan t+\tan t}{1-\tan t\tan t} = \dfrac{2\tan t}{1-\tan^2 t}$

35. $\tan^2\left(\dfrac{t}{2}\right) = \dfrac{\sin^2\left(\frac{t}{2}\right)}{\cos^2\left(\frac{t}{2}\right)} = \dfrac{\frac{1-\cos t}{2}}{\frac{1+\cos t}{2}} = \dfrac{1-\cos t}{1+\cos t}$

Therefore, $\tan\dfrac{t}{2} = \pm\sqrt{\dfrac{1-\cos t}{1+\cos t}}$

37. a) $2\sin\dfrac{x}{2}\cos\dfrac{x}{2} = \sin\left(2\cdot\dfrac{x}{2}\right) = \sin x$

b) $\cos^2 3t-\sin^2 3t = \cos(2\cdot 3t) = \cos 6t$

c) $2\sin^2\left(\dfrac{y}{4}\right)-1 = -\left[1-2\sin^2\left(\dfrac{y}{4}\right)\right]$

$$= -\cos\left(2\cdot\dfrac{y}{4}\right)$$

$$= -\cos\dfrac{y}{2}$$

d) $\dfrac{\cos 4t-1}{2} = \dfrac{1-\cos 4t}{2}$

$$= -\sin^2\left(\dfrac{1}{2}\cdot 4t\right)$$

$$= -\sin^2 2t$$

e) $\dfrac{1-\cos 4t}{1+\cos 4t} = \dfrac{\sin^2\left(\frac{1}{2}\cdot 4t\right)}{\cos^2\left(\frac{1}{2}\cdot 4t\right)} = \tan^2 2t$

f)
$$\frac{\sin 6y}{1+\cos 6y} = \frac{2\sin 3y\cos 3y}{1+2\cos^2 3y-1}$$

$$= \frac{2\sin 3y\cos 3y}{2\cos^2 3y}$$

$$= \frac{\sin 3y}{\cos 3y}$$

$$= \tan 3y$$

39. $\cos t = -\dfrac{5}{13},\ \pi<t<\dfrac{3\pi}{2}$

$$\sin t = -\sqrt{1-\left(-\frac{5}{13}\right)^2} = -\sqrt{1-\frac{25}{169}} = -\sqrt{\frac{144}{169}} = -\frac{12}{13}$$

a) $\sin 2t = 2\sin t\cos t = 2\left(-\dfrac{12}{13}\right)\left(-\dfrac{5}{13}\right) = \dfrac{120}{169}$

b) Since $\pi<t<\dfrac{3\pi}{2}$, then $\dfrac{\pi}{2}<\dfrac{t}{2}<\dfrac{3\pi}{4}$.

Since $\dfrac{t}{2}$ is in quadrant II, $\cos\dfrac{t}{2}$ is negative.

$$\cos\frac{t}{2} = -\sqrt{\frac{1+\cos t}{2}}$$

$$= -\sqrt{\frac{1+\left(-\frac{5}{13}\right)}{2}}$$

$$= -\sqrt{\frac{13-5}{26}}$$

$$= -\sqrt{\frac{8}{26}}$$

$$= -\sqrt{\frac{4}{13}\cdot\frac{13}{13}}$$

$$= -\frac{2\sqrt{13}}{13}$$

c) $\tan\dfrac{t}{2} = \dfrac{\sin t}{1+\cos t} = \dfrac{-\frac{12}{13}}{1+\left(-\frac{5}{13}\right)} = \dfrac{-12}{13-5} = \dfrac{-12}{8} = -\dfrac{3}{2}$

41.
$$\cos^4 z-\sin^4 z = (\cos^2 z-\sin^2 z)(\cos^2 z+\sin^2 z)$$

$$= (\cos 2z)(1)$$

$$= \cos 2z$$

43. $1+\dfrac{1-\cos 8t}{1+\cos 8t} = 1+\tan^2\left(\dfrac{1}{2}\cdot 8t\right) = 1+\tan^2 4t = \sec^2 4t$

45.
$$\tan\frac{\theta}{2}-\sin\theta = \frac{\sin\theta}{1+\cos\theta}-\sin\theta$$

$$= \frac{\sin\theta-\sin\theta(1+\cos\theta)}{1+\cos\theta}$$

$$= \frac{\sin\theta-\sin\theta-\sin\theta\cos\theta}{1+\cos\theta}$$

$$= \frac{-\sin\theta\cos\theta}{1+\cos\theta}$$

$$= \frac{-\sin\theta\frac{\cos\theta}{\cos\theta}}{\frac{1}{\cos\theta}+\frac{\cos\theta}{\cos\theta}}$$

$$= \frac{-\sin\theta}{\sec\theta+1}$$

47. $3\cos 2t + 4\sin 2t$

$= 3(\cos^2 t - \sin^2 t) + 4(2\sin t \cos t)$

$= 3\cos^2 t + 8\sin t \cos t - 3\sin^2 t$

$= (3\cos t - \sin t)(\cos t + 3\sin t)$

49. $2(\cos 3x \cos x + \sin 3x \sin x)^2 = 2[\cos(3x - x)]^2$

$= 2\cos^2 2x$

$= 2 \cdot \dfrac{1 + \cos(2 \cdot 2x)}{2}$

$= 1 + \cos 4x$

51. $\tan 3t = \tan(2t + t)$

$= \dfrac{\tan 2t + \tan t}{1 - \tan 2t \tan t}$

$= \dfrac{\dfrac{2\tan t}{1 - \tan^2 t} + \tan t}{1 - \dfrac{2\tan^2 t}{1 - \tan^2 t}}$

$= \dfrac{2\tan t + \tan t(1 - \tan^2 t)}{1 - \tan^2 t - 2\tan^2 t}$

$= \dfrac{3\tan t - \tan^3 t}{1 - 3\tan^2 t}$

53. $\sin^4 u + \cos^4 u$

$= \sin^4 u + 2\sin^2 u \cos^2 u + \cos^4 u - 2\sin^2 u \cos^2 u$

$= (\sin^2 u + \cos^2 u)^2 - \dfrac{1}{2}(2\sin u \cos u)^2$

$= 1 - \dfrac{1}{2}(\sin 2u)^2$

$= 1 - \dfrac{1}{2} \cdot \dfrac{1 - \cos 4u}{2}$

$= 1 - \dfrac{1}{4} + \dfrac{1}{4}\cos 4u$

$= \dfrac{3}{4} + \dfrac{1}{4}\cos 4u$

55. $\cos^2 x + \cos^2 2x + \cos^2 3x$

$= \dfrac{1 + \cos 2x}{2} + \cos^2 2x + \dfrac{1 + \cos 6x}{2}$

$= 1 + \dfrac{1}{2}(\cos 2x + \cos 6x) + \cos^2 2x$

$= 1 + \cos 4x \cos 2x + \cos^2 2x$

$= 1 + \cos 2x(\cos 4x + \cos 2x)$

$= 1 + \cos 2x(2\cos 3x \cos x)$

$= 1 + 2\cos x \cos 2x \cos 3x$

57. $\alpha + \beta + \gamma = 180°$

$2\alpha + 2\beta + 2\gamma = 360°$

$2\gamma = 360° - 2\alpha - 2\beta$

$\sin 2\gamma = \sin(360° - 2\alpha - 2\beta)$

$= -\sin(2\alpha + 2\beta)$

$(\sin 2\alpha + \sin 2\beta) + \sin 2\gamma$

$= (\sin 2\alpha + \sin 2\beta) - \sin(2\alpha + 2\beta)$

$= 2\sin(\alpha + \beta)\cos(\alpha - \beta) - 2\sin(\alpha + \beta)\cos(\alpha + \beta)$

$= 2\sin(\alpha + \beta)[\cos(\alpha - \beta) - \cos(\alpha + \beta)]$

$= 2\sin(180° - \gamma)(-1)[\cos(\alpha + \beta) - \cos(\alpha - \beta)]$

$= 2\sin \gamma\,(-1)(-2)\sin \alpha \sin \beta$

$= 4\sin \alpha \sin \beta \sin \gamma$

59.

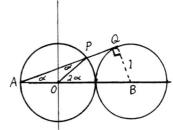

Let $\alpha = \angle PAB$. Then $\angle POB = 2\alpha$. Also $\overline{AB} = 3$ so

$\sin \alpha = \dfrac{\overline{BQ}}{\overline{AB}} = \dfrac{1}{3}$. Therefore

$\cos 2\alpha = 1 - 2\sin^2 \alpha = 1 - 2\left(\dfrac{1}{3}\right)^2 = 1 - \dfrac{2}{9} = \dfrac{7}{9}$

$\sin 2\alpha = \sqrt{1 - \left(\dfrac{7}{9}\right)^2} = \sqrt{1 - \dfrac{49}{81}} = \sqrt{\dfrac{32}{81}} = \dfrac{4\sqrt{2}}{9}$

The coordinates of P are $\left(\dfrac{7}{9}, \dfrac{4\sqrt{2}}{9}\right)$.

61. Press $\boxed{Y=}$

Enter 2 $\boxed{\text{COS}}$ $\boxed{\text{X|T}}$ -4 $\boxed{\text{SIN}}$ $\boxed{\text{X|T}}$ $\boxed{\text{SIN}}$ 2 $\boxed{\text{X|T}}$

Press $\boxed{\text{ZOOM}}$ $\boxed{7}$ to get intervals of $\dfrac{\pi}{2}$ on the x-axis.

It seems that there are 3 periods in each interval of 2π, so the period length is $\dfrac{2\pi}{3}$.

By moving the cursor to a maximum, the amplitude seems to be 2. In fact,

$2\cos x - 4\sin x \sin 2x$

$= 2\cos x - 4[-\dfrac{1}{2}\cos 3x - \dfrac{1}{2}\cos(-x)]$

$= 2\cos x + 2\cos 3x - 2\cos x$

$= 2\cos 3x$

Problem Set 8.4

1. $\sin^{-1}\left(\dfrac{\sqrt{3}}{2}\right) = \dfrac{\pi}{3}$

3. $\arcsin\left(\dfrac{\sqrt{2}}{2}\right) = \dfrac{\pi}{4}$

5. $\tan^{-1}(0) = 0$

7. $\tan^{-1}(\sqrt{3}) = \frac{\pi}{3}$

9. $\arccos\left(-\frac{1}{2}\right)$ is in quadrant II with reference angle $\frac{\pi}{3}$.

$\arccos\left(-\frac{1}{2}\right) = \frac{2\pi}{3}$

11. $\sec^{-1}(\sqrt{2}) = \cos^{-1}\left(\frac{1}{\sqrt{2}}\right) = \frac{\pi}{4}$

13. $\sin^{-1}(.21823) = .2200$
Set radian mode. .21823 $\boxed{\text{INV}}$ $\boxed{\text{SIN}}$ $\rightarrow$.2200

15. $\sin^{-1}(-.21823) = -.2200$
Set radian mode. $\boxed{+/-}$ .21823 $\boxed{\text{INV}}$ $\boxed{\text{SIN}}$ $\rightarrow$ $-.2200$

17. $\tan^{-1}(.20000) = .2037$
Set radian mode. .20660 $\boxed{\text{INV}}$ $\boxed{\text{TAN}}$ $\rightarrow$.2037

19. Set radian mode.
a) $\sec^{-1}(1.4263) = \cos^{-1}(1/1.4263) = .7938$
1.4263 $\boxed{1/x}$ $\boxed{\text{INV}}$ $\boxed{\text{COS}}$ $\rightarrow$.7938

b) $\sec^{-1}(-2.6715) = \cos^{-1}(1/-2.6715) = 1.9545$
2.6715 $\boxed{+/-}$ $\boxed{1/x}$ $\boxed{\text{INV}}$ $\boxed{\text{COS}}$ $\rightarrow$1.9545

21. $\sin t = .3416$
$t_0 = \sin^{-1}(.3416) = .3486$
So $t = .3486$ or $t = \pi - .3486 \approx 2.7930$

23. $\tan t = 3.345$
$t_0 = \tan^{-1}(3.345) = 1.2803$
So $t = 1.2803$ or $t = \pi + 1.2803 \approx 4.4219$

25. $\sin\left[\sin^{-1}\left(\frac{2}{3}\right)\right] = \frac{2}{3}$

27. $\tan[\tan^{-1}(10)] = 10$

29. $\sin^{-1}\left[\sin\left(\frac{\pi}{3}\right)\right] = \frac{\pi}{3}$

31. $\sin^{-1}\left[\cos\left(\frac{\pi}{4}\right)\right] = \sin^{-1}\left(\frac{\sqrt{2}}{2}\right) = \frac{\pi}{4}$

33. Let $\theta = \sin^{-1}\left(\frac{4}{5}\right)$. Then $\sin\theta = \frac{4}{5} = \frac{b}{c}$.
$a = \sqrt{c^2 - b^2} = \sqrt{5^2 - 4^2} = \sqrt{25 - 16} = \sqrt{9} = 3$
$\cos\left[\sin^{-1}\left(\frac{4}{5}\right)\right] = \cos\theta = \frac{a}{c} = \frac{3}{5}$

35. Let $\theta = \tan^{-1}\left(\frac{1}{2}\right)$. Then $\tan\theta = \frac{1}{2} = \frac{b}{a}$.
$c = \sqrt{a^2 + b^2} = \sqrt{2^2 + 1^2} = \sqrt{5}$
$\cos\left[\tan^{-1}\left(\frac{1}{2}\right)\right] = \cos\theta = \frac{a}{c} = \frac{2}{\sqrt{5}}$

37. Let $\theta = \sec^{-1}3$. Then $\sec\theta = 3$.
$\cos[\sec^{-1}(3)] = \cos\theta = \frac{1}{\sec\theta} = \frac{1}{3}$

39. $\sec^{-1}\left[\sec\left(\frac{2\pi}{3}\right)\right] = \frac{2\pi}{3}$

41. $\cos[\sin^{-1}(-.2564)] = .9666$:
Set radian mode. .2564 $\boxed{+/-}$ $\boxed{\text{INV}}$ $\boxed{\text{SIN}}$ $\boxed{\text{COS}}$ $\rightarrow$.9666

43. $\sin^{-1}(\cos 1.12) = .4508$:
Set radian mode. 1.12 $\boxed{\text{COS}}$ $\boxed{\text{INV}}$ $\boxed{\text{SIN}}$ $\rightarrow$.4508

45. $\tan[\sec^{-1}2.5] = \tan[\cos^{-1}(1/2.5)] = 2.2913$:
Set radian mode. 2.5 $\boxed{1/x}$ $\boxed{\text{INV}}$ $\boxed{\text{COS}}$ $\boxed{\text{TAN}}$ $\rightarrow$2.2913

47. Let $\theta = \cos^{-1}\left(\frac{3}{5}\right)$. Then $\cos\theta = \frac{3}{5}$.
$\sin\theta = \sqrt{1 - \left(\frac{3}{5}\right)^2} = \sqrt{1 - \frac{9}{25}} = \sqrt{\frac{16}{25}} = \frac{4}{5}$
$\sin\left[2\cos^{-1}\left(\frac{3}{5}\right)\right] = \sin 2\theta = 2\sin\theta\cos\theta = 2\cdot\frac{4}{5}\cdot\frac{3}{5} = \frac{24}{25}$

49. Let $\theta = \sin^{-1}\left(-\frac{3}{5}\right)$. Then $\sin\theta = -\frac{3}{5}$.
$\cos\left[2\sin^{-1}\left(-\frac{3}{5}\right)\right] = \cos 2\theta$
$= 1 - 2\sin^2\theta$
$= 1 - 2\left(-\frac{3}{5}\right)^2$
$= 1 - \frac{18}{25}$
$= \frac{7}{25}$

51. Let $\alpha = \cos^{-1}\frac{3}{5}$ and $\beta = \cos^{-1}\frac{5}{13}$.
Then $\cos\alpha = \frac{3}{5}$ and $\cos\beta = \frac{5}{13}$.
$\sin\alpha = \sqrt{1 - \left(\frac{3}{5}\right)^2} = \sqrt{1 - \frac{9}{25}} = \sqrt{\frac{16}{25}} = \frac{4}{5}$
$\sin\beta = \sqrt{1 - \left(\frac{5}{13}\right)^2} = \sqrt{1 - \frac{25}{169}} = \sqrt{\frac{144}{169}} = \frac{12}{13}$
$\sin\left[\cos^{-1}\left(\frac{3}{5}\right) + \cos^{-1}\left(\frac{5}{13}\right)\right]$
$= \sin(\alpha + \beta)$
$= \sin\alpha\cos\beta + \cos\alpha\sin\beta$
$= \frac{4}{5}\cdot\frac{5}{13} + \frac{3}{5}\cdot\frac{12}{13}$
$= \frac{20 + 36}{65}$
$= \frac{56}{65}$

53. Let $\alpha = \sec^{-1}\left(\frac{3}{2}\right)$. Then $\sec\alpha = \frac{3}{2}$ and $\cos\alpha = \frac{2}{3}$.
$\sin\alpha = \sqrt{1 - \left(\frac{2}{3}\right)^2} = \sqrt{1 - \frac{4}{9}} = \sqrt{\frac{5}{9}} = \frac{\sqrt{5}}{3}$
Let $\beta = \sec^{-1}\left(\frac{4}{3}\right)$. Then $\sec\beta = \frac{4}{3}$ and $\cos\beta = \frac{3}{4}$.
$\sin\beta = \sqrt{1 - \left(\frac{3}{4}\right)^2} = \sqrt{1 - \frac{9}{16}} = \sqrt{\frac{7}{16}} = \frac{\sqrt{7}}{4}$

$\cos\left[\sec^{-1}\left(\frac{3}{2}\right) - \sec^{-1}\left(\frac{4}{3}\right)\right]$

$= \cos(\alpha - \beta)$

$= \cos\alpha\cos\beta + \sin\alpha\sin\beta$

$= \frac{2}{3} \cdot \frac{3}{4} + \frac{\sqrt{5}}{3} \cdot \frac{\sqrt{7}}{4}$

$= \frac{6 + \sqrt{35}}{12}$

55. Let $\theta = \sin^{-1}x$. Then $\sin\theta = x$ and θ is in quadrant I or IV, so $\cos\theta = +\sqrt{1-x^2}$.

$\tan(\sin^{-1}x) = \tan\theta = \frac{\sin\theta}{\cos\theta} = \frac{x}{\sqrt{1-x^2}}$

57. Let $\theta = \tan^{-1}x$. Then $\tan\theta = x$.

$\tan(2\tan^{-1}x) = \tan 2\theta = \frac{2\tan\theta}{1-\tan^2\theta} = \frac{2x}{1-x^2}$

59. Let $\theta = \sec^{-1}x$. Then $\sec\theta = x$ and $\cos\theta = \frac{1}{x}$.

$\cos(2\sec^{-1}x) = \cos 2\theta$

$= 2\cos^2\theta - 1$

$= 2\left(\frac{1}{x}\right)^2 - 1$

$= \frac{2}{x^2} - 1$

61. a) $\sin x$ is negative is in quadrant IV.

$\arcsin\left(-\frac{\sqrt{3}}{2}\right) = -\arcsin\frac{\sqrt{3}}{2} = -\frac{\pi}{3}$

b) $\tan x$ is negative is in quadrant IV.

$\tan^{-1}(-\sqrt{3}) = -\tan^{-1}\sqrt{3} = -\frac{\pi}{3}$

c) $\cos x$ is negative is in quadrant II.

$\sec^{-1}(-2) = \cos^{-1}\left(-\frac{1}{2}\right) = \pi - \cos^{-1}\frac{1}{2}$

$= \pi - \frac{\pi}{3} = \frac{2\pi}{3}$

63. a) $\tan(\tan^{-1} 43) = 43$

b) Let $\theta = \sin^{-1}\left(\frac{5}{13}\right)$.

Then $\sin\theta = \frac{5}{13}$ and θ is in quadrant I.

$\cos\theta = +\sqrt{1-\left(\frac{5}{13}\right)^2} = \sqrt{1-\frac{25}{169}} = \sqrt{\frac{144}{169}} = \frac{12}{13}$

$\cos\left[\sin^{-1}\left(\frac{5}{13}\right)\right] = \cos\theta = \frac{12}{13}$

c) Let $\theta = \sin^{-1}(.8)$.

Then $\sin\theta = .8 = \frac{4}{5}$ and θ is in quadrant I.

$\cos\theta = +\sqrt{1-\left(\frac{4}{5}\right)^2} = \sqrt{1-\frac{16}{25}} = \sqrt{\frac{9}{25}} = \frac{3}{5}$

$\sin\left[\frac{\pi}{4} + \sin^{-1}(.8)\right] = \sin\left(\frac{\pi}{4} + \theta\right)$

$= \sin\frac{\pi}{4}\cos\theta + \cos\frac{\pi}{4}\sin\theta$

$= \frac{\sqrt{2}}{2} \cdot \frac{3}{5} + \frac{\sqrt{2}}{2} \cdot \frac{4}{5}$

$= \frac{7\sqrt{2}}{10}$

d) Let $\alpha = \sin^{-1}(.6)$.

Then $\sin\alpha = .6 = \frac{3}{5}$ and α is in quadrant I.

$\cos\alpha = +\sqrt{1-\left(\frac{3}{5}\right)^2} = \sqrt{1-\frac{9}{25}} = \sqrt{\frac{16}{25}} = \frac{4}{5}$

Let $\beta = \sec^{-1}(3)$. Then $\sec\beta = 3$, $\cos\beta = \frac{1}{3}$ and β is in quadrant I.

$\sin\beta = \sqrt{1-\left(\frac{1}{3}\right)^2} = \sqrt{1-\frac{1}{9}} = \sqrt{\frac{8}{9}} = \frac{2\sqrt{2}}{3}$

$\cos[\sin^{-1}(.6) + \sec^{-1}(3)]$

$= \cos(\alpha + \beta)$

$= \cos\alpha\cos\beta - \sin\alpha\sin\beta$

$= \frac{4}{5} \cdot \frac{1}{3} - \frac{3}{5} \cdot \frac{2\sqrt{2}}{3}$

$= \frac{4 - 6\sqrt{2}}{15}$

65. a) $\cos(\sin^{-1}x) = \frac{3}{4}$. Let $\sin^{-1}x = \theta$.

Then $\cos\theta = \frac{3}{4}$ and θ is in quadrant I or IV.

$x = \sin\theta = \pm\sqrt{1-\left(\frac{3}{4}\right)^2} = \pm\sqrt{\frac{7}{16}} = \pm\frac{\sqrt{7}}{4}$

b) $\sin(\cos^{-1}x) = \sqrt{.19}$. Let $\cos^{-1}x = \theta$.

Then $\sin\theta = \sqrt{.19}$ and θ is in quadrant I or II.

$x = \cos\theta$

$= \pm\sqrt{1-(\sqrt{.19})^2}$

$= \pm\sqrt{1-.19}$

$= \pm\sqrt{.81}$

$= \pm.9$

c) $\sin^{-1}(3x-5) = \frac{\pi}{6}$

$3x - 5 = \sin\frac{\pi}{6}$

$3x - 5 = \frac{1}{2}$

$3x = \frac{11}{2}$

$x = \frac{11}{6}$

d) $\tan^{-1}(x^2 - 3x + 3) = \frac{\pi}{4}$

$$x^2 - 3x + 3 = \tan\frac{\pi}{4}$$

$$x^2 - 3x + 3 = 1$$

$$x^2 - 3x + 2 = 0$$

$$(x-2)(x-1) = 0$$

$$x = 2 \text{ or } x = 1$$

67. a) $\sin^{-1}\left(\sin\frac{\pi}{2}\right) = \sin^{-1}(1) = \frac{\pi}{2}$

b) $\sin^{-1}\left(\sin\frac{3\pi}{4}\right) = \sin^{-1}\left(\frac{\sqrt{2}}{2}\right) = \frac{\pi}{4}$

c) $\sin^{-1}\left(\sin\frac{5\pi}{4}\right) = \sin^{-1}\left(-\frac{\sqrt{2}}{2}\right) = -\frac{\pi}{4}$

d) $\sin^{-1}\left(\sin\frac{3\pi}{2}\right) = \sin^{-1}(-1) = -\frac{\pi}{2}$

e) $\cos^{-1}(\cos 3\pi) = \cos^{-1}(-1) = \pi$

f) $\tan^{-1}\left(\tan\frac{13\pi}{4}\right) = \tan^{-1}(1) = \frac{\pi}{4}$

69. a) $\theta = \sin^{-1}\left(\frac{x}{5}\right)$

b) $\theta = \tan^{-1}\left(\frac{x}{3}\right)$

c) $\theta = \sin^{-1}\left(\frac{3}{x}\right)$ or $\theta = \csc^{-1}\left(\frac{x}{3}\right)$

d) $\theta = \tan^{-1}\left(\frac{3}{x}\right) - \tan^{-1}\left(\frac{1}{x}\right)$

or $\theta = \cot^{-1}\left(\frac{x}{3}\right) - \cot^{-1}x$

71. a) $\sin^{-1}(.6) = \tan^{-1}\left(\dfrac{.6}{\sqrt{1 - (.6)^2}}\right) = .6435$ radians

b) $\sin^{-1}(-.3) = \tan^{-1}\left(\dfrac{-.3}{\sqrt{1 - (-.3)^2}}\right) = -.3047$ rad

c) $\cos^{-1}(.8) = \frac{\pi}{2} - \tan^{-1}\left(\dfrac{.8}{\sqrt{1 - (.8)^2}}\right) = .6435$ rad

d) $\cos^{-1}(-.9) = \frac{\pi}{2} - \tan^{-1}\left(\dfrac{-.9}{\sqrt{1 - (-.9)^2}}\right) = 2.6907$

73. $\tan\left(2\tan^{-1}\frac{1}{5}\right) = \dfrac{2 \cdot \frac{1}{5}}{1 - \left(\frac{1}{5}\right)^2} = \dfrac{\frac{2}{5}}{\frac{24}{25}} = \dfrac{5}{12}$

$\tan\left(4\tan^{-1}\frac{1}{5}\right) = \dfrac{2 \cdot \frac{5}{12}}{1 - \left(\frac{5}{12}\right)^2} = \dfrac{\frac{5}{6}}{\frac{119}{144}} = \dfrac{120}{119}$

$\tan\left(4\tan^{-1}\frac{1}{5} - \frac{\pi}{4}\right) = \dfrac{\frac{120}{119} - 1}{1 + \frac{120}{119} \cdot 1} = \dfrac{\frac{1}{119}}{\frac{239}{119}} = \dfrac{1}{239}$

Therefore $4\tan^{-1}\frac{1}{5} - \frac{\pi}{4} = \tan^{-1}\frac{1}{239}$.

75.

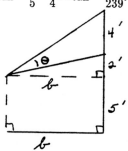

a) $\theta = \tan^{-1}\left(\frac{6}{b}\right) - \tan^{-1}\left(\frac{2}{b}\right)$

b) When $b = 8$,

$\theta = \tan^{-1}\left(\frac{6}{8}\right) - \tan^{-1}\left(\frac{2}{8}\right) \approx 36.87° - 14.04° = 22.83°$

c) $\qquad 30° = \tan^{-1}\left(\frac{6}{b}\right) - \tan^{-1}\left(\frac{2}{b}\right)$

$\tan 30° = \tan\left[\tan^{-1}\left(\frac{6}{b}\right) - \tan^{-1}\left(\frac{2}{b}\right)\right]$

$\dfrac{1}{\sqrt{3}} = \dfrac{\frac{6}{b} - \frac{2}{b}}{1 + \frac{6}{b} \cdot \frac{2}{b}} = \dfrac{\frac{4}{b}}{1 + \frac{12}{b^2}} = \dfrac{4b}{b^2 + 12}$

$b^2 + 12 = 4\sqrt{3}b$

$b^2 - 4\sqrt{3}b + 12 = 0$

$(b - 2\sqrt{3})^2 = 0$

$b = 2\sqrt{3}$

77. a) Press $\boxed{Y=}$

Enter $\quad \boxed{\text{2nd}}\ [\text{COS}^{-1}]\ \boxed{\text{COS}}\ \boxed{\text{X|T}}$

Press $\quad \boxed{\text{ZOOM}}\ \boxed{7}$
to get intervals of $\pi/2$ on the x-axis.

The period of $\cos^{-1}(\cos x)$ is 2π.
Zoom in on a local maximum to find that the maximum height is 3.14 or π. Since the amplitude is half the difference of the maximum and minimum, the amplitude is $\pi/2$.

b) Press $\boxed{Y=}$

Enter $\quad \boxed{\text{2nd}}\ [\text{SIN}^{-1}]\ \boxed{\text{SIN}}\ \boxed{\text{X|T}}$

Press $\quad \boxed{\text{ZOOM}}\ \boxed{7}$
to get intervals of $\pi/2$ on the x-axis.

The period of $\sin^{-1}(\sin x)$ is 2π.
Zoom in on a local maximum to find that the amplitude is 1.57 or $\pi/2$.

172

1. $\sin t = 0$

$t = 0, \pi$

3. $\sin t = -1$

$t = \dfrac{3\pi}{2}$

5. $\sin t = 2$

No solution: for any t, $|\sin t| \le 1$

7. $2 \cos x + \sqrt{3} = 0$

$\cos x = -\dfrac{\sqrt{3}}{2}$

The reference angle is $\dfrac{\pi}{6}$, since $\cos \dfrac{\pi}{6} = \dfrac{\sqrt{3}}{2}$.

$x = \pi - \dfrac{\pi}{6} = \dfrac{5\pi}{6}$ or $x = \pi + \dfrac{\pi}{6} = \dfrac{7\pi}{6}$.

9. $\tan^2 x = 1$

$\tan x = 1$ $\qquad\qquad \tan x = -1$

$x = \dfrac{\pi}{4}$ $\qquad\qquad x = \pi - \dfrac{\pi}{4} = \dfrac{3\pi}{4}$

or $\quad x = \pi + \dfrac{\pi}{4} = \dfrac{5\pi}{4}$ $\quad$ or $x = 2\pi - \dfrac{\pi}{4} = \dfrac{7\pi}{4}$

Solution set: $\left\{ \dfrac{\pi}{4}, \dfrac{3\pi}{4}, \dfrac{5\pi}{4}, \dfrac{7\pi}{4} \right\}$

11. $(2 \cos \theta + 1)(2 \sin \theta - \sqrt{2}) = 0$

$2 \cos \theta + 1 = 0$ $\qquad\qquad 2 \sin \theta - \sqrt{2} = 0$

$\cos \theta = -\dfrac{1}{2}$ $\qquad\qquad \sin \theta = \dfrac{\sqrt{2}}{2}$

$\theta_0 = \dfrac{\pi}{3}$ $\qquad\qquad\qquad \theta_0 = \dfrac{\pi}{4}$

$\theta = \pi - \dfrac{\pi}{3} = \dfrac{2\pi}{3}$ $\qquad\quad \theta = \dfrac{\pi}{4}$

or $\quad \theta = \pi + \dfrac{\pi}{3} = \dfrac{4\pi}{3}$ $\qquad \theta = \pi - \dfrac{\pi}{4} = \dfrac{3\pi}{4}$

Solution set: $\left\{ \dfrac{\pi}{4}, \dfrac{2\pi}{3}, \dfrac{3\pi}{4}, \dfrac{4\pi}{3} \right\}$

13. $\sin^2 x + \sin x = 0$

$\sin x (\sin x + 1) = 0$

$\sin x = 0$ $\qquad\qquad \sin x = -1$

$x = 0, \pi$ $\qquad\qquad x = \dfrac{3\pi}{2}$

Solution set: $\left\{ 0, \pi, \dfrac{3\pi}{2} \right\}$

15. $\tan^2 \theta = \sqrt{3} \tan \theta$

$\tan^2 \theta - \sqrt{3} \tan \theta = 0$

$\tan \theta (\tan \theta - \sqrt{3}) = 0$

$\tan \theta = 0$ $\qquad\qquad \tan \theta = \sqrt{3}$

$\theta = 0, \pi$ $\qquad\qquad \theta = \dfrac{\pi}{3}, \dfrac{4\pi}{3}$

Solution set: $\left\{ 0, \dfrac{\pi}{3}, \pi, \dfrac{4\pi}{3} \right\}$

17. $2 \sin^2 x = 1 + \cos x$

$2(1 - \cos^2 x) = 1 + \cos x$

$2 - 2 \cos^2 x = 1 + \cos x$

$2 \cos^2 x + \cos x - 1 = 0$

$(2 \cos x - 1)(\cos x + 1) = 0$

$\cos x = \dfrac{1}{2}$ $\qquad\qquad \cos x = -1$

$x = \dfrac{\pi}{3}, \dfrac{5\pi}{3}$ $\qquad\qquad x = \pi$

Solution set: $\left\{ \dfrac{\pi}{3}, \pi, \dfrac{5\pi}{3} \right\}$

19. $\tan^2 x - 3 \tan x + 1 = 0$

Using the quadratic formula:

$\tan x = \dfrac{3 \pm \sqrt{9 - 4}}{2} = \dfrac{3 \pm \sqrt{5}}{2}$

$\tan x = 2.61803$ $\qquad\qquad \tan x = .381966$

$x = 1.20593$ $\qquad\qquad\quad x = .36486$

or $\quad x = \pi + 1.20593$ $\qquad$ or $x = \pi + .36486$

$\quad\quad = 4.34753$ $\qquad\qquad\qquad = 3.50646$

Solution set: $\{.36486, 1.20593, 3.50646, 4.34753\}$

21. $\sin t + \cos t = 1$

$(\sin t + \cos t)^2 = 1$

$\sin^2 + 2 \sin t \cos t + \cos^2 t = 1$

$1 + 2 \sin t \cos t = 1$

$2 \sin t \cos t = 0$

$\sin t = 0$ $\qquad\qquad \cos t = 0$

$t = 0, \pi$ $\qquad\qquad t = \dfrac{\pi}{2}, \dfrac{3\pi}{2}$

But π and $\dfrac{3\pi}{2}$ make the left side of the original equation -1.

Solution set: $\left\{ 0, \dfrac{\pi}{2} \right\}$

23. $\sqrt{3}(1 - \sin t) = \cos t$

$3(1 - \sin t)^2 = \cos^2 t$

$3(1 - 2 \sin t + \sin^2 t) = \cos^2 t$

$3 - 6 \sin t + 3 \sin^2 t = 1 - \sin^2 t$

$4 \sin^2 t - 6 \sin t + 2 = 0$

$2 \sin^2 t - 3 \sin t + 1 = 0$

$(2 \sin t - 1)(\sin t - 1) = 0$

$\sin t = \dfrac{1}{2}$ $\qquad\qquad \sin t = 1$

$t = \dfrac{\pi}{6}, \dfrac{5\pi}{6}$ $\qquad\qquad t = \dfrac{\pi}{2}$

$\dfrac{5\pi}{6}$ does not satisfy the original equation.

Solution set: $\left\{ \dfrac{\pi}{6}, \dfrac{\pi}{2} \right\}$

25.
$$\sec t + \tan t = 1$$
$$\sec t = 1 - \tan t$$
$$\sec^2 t = (1 - \tan t)^2$$
$$1 + \tan^2 t = 1 - 2 \tan t + \tan^2 t$$
$$2 \tan t = 0$$
$$\tan t = 0$$
$$t = 0, \pi$$

$t = \pi$ does not satisfy the original equation.
Solution set: $\{0\}$

27. $\sin t = \frac{1}{2}$

In the period $0 \leq t < 2\pi$ there are two solutions:
$t = \frac{\pi}{6}$ and $t = \frac{5\pi}{6}$. The entire solution set is thus
$\left\{\frac{\pi}{6} + 2\pi k, \frac{5\pi}{6} + 2\pi k; k \text{ is an integer}\right\}$

29. $\tan t = 0$
In the period $0 \leq t < \pi$ there is one solution:
$t = 0$. The entire solution set is $\{k\pi; k \text{ is an integer}\}$

31. $\sin^2 t = \frac{1}{4}$
$$\sin t = \pm \frac{1}{2}$$

In the period $0 \leq t < 2\pi$ there are 4 solutions:
$t = \frac{\pi}{6}, \frac{5\pi}{6}, \frac{7\pi}{6}, \frac{11\pi}{6}$.
Notice that $\frac{7\pi}{6} = \frac{\pi}{6} + \pi$ and $\frac{11\pi}{6} = \frac{5\pi}{6} + \pi$.

We can thus describe the entire solution set as
$\left\{\frac{\pi}{6} + k\pi, \frac{5\pi}{6} + k\pi; k \text{ is an integer}\right\}$

33. $\sin 2t = 0$
$$2t = 0, \pi, 2\pi, 3\pi$$
$$t = 0, \frac{\pi}{2}, \pi, \frac{3\pi}{2}$$

Notice that we let $2t$ take on values between 0 and
4π so that t would fall between 0 and 2π.

35. $\sin 4t = 1$
$$4t = \frac{\pi}{2}, \frac{5\pi}{2}, \frac{9\pi}{2}, \frac{13\pi}{2}$$
$$t = \frac{\pi}{8}, \frac{5\pi}{8}, \frac{9\pi}{8}, \frac{13\pi}{8}$$

Notice that we let $4t$ take on values between 0 and
8π so that t would fall between 0 and 2π.

37. $\tan 2t = -1$
$$2t = \frac{3\pi}{4}, \frac{7\pi}{4}, \frac{11\pi}{4}, \frac{15\pi}{4}$$
$$t = \frac{3\pi}{8}, \frac{7\pi}{8}, \frac{11\pi}{8}, \frac{15\pi}{8}$$

Notice that we let $2t$ take on values between 0 and

4π so that t would fall between 0 and 2π.

39.
$$2 \sin^2 x = \sin x$$
$$2 \sin^2 x - \sin x = 0$$
$$\sin x (2 \sin x - 1) = 0$$

$$\sin x = 0 \qquad\qquad \sin x = \frac{1}{2}$$
$$x = 0, \pi \qquad\qquad x = \frac{\pi}{6}, \frac{5\pi}{6}$$
Solution set: $\left\{0, \frac{\pi}{6}, \frac{5\pi}{6}, \pi\right\}$

41. $\cos^2 x = \frac{1}{3}$
$$\cos x = \pm \sqrt{\frac{1}{3}}$$
$$x_0 = \cos^{-1}\left(\sqrt{\frac{1}{3}}\right) \approx .9553$$
$$x = .9553$$

or $x = \pi - .9553 = 2.1863$

or $x = \pi + .9553 = 4.0969$

or $x = 2\pi - .9553 = 5.3279$

43.
$$2 \tan x - \sec^2 x = 0$$
$$2 \tan x - (\tan^2 x + 1) = 0$$
$$\tan^2 x - 2 \tan x + 1 = 0$$
$$(\tan x - 1)^2 = 0$$
$$\tan x - 1 = 0$$
$$\tan x = 1$$
$$x = \frac{\pi}{4}, \frac{5\pi}{4}$$

45.
$$\tan 2x = 3 \tan x$$
$$\frac{2 \tan x}{1 - \tan^2 x} = 3 \tan x$$
$$2 \tan x = 3 \tan x (1 - \tan^2 x)$$
$$2 \tan x = 3 \tan x - 3 \tan^3 x$$
$$3 \tan^3 x - \tan x = 0$$
$$\tan x (3 \tan^2 x - 1) = 0$$

$$\tan x = 0 \qquad\qquad \tan^2 x = \frac{1}{3}$$
$$\qquad\qquad\qquad\qquad \tan x = \pm \sqrt{\frac{1}{3}}$$
$$x = 0, \pi \qquad\qquad x = \frac{\pi}{6}, \frac{5\pi}{6}, \frac{7\pi}{6}, \frac{11\pi}{6}$$
Solution set: $\left\{0, \frac{\pi}{6}, \frac{5\pi}{6}, \pi, \frac{7\pi}{6}, \frac{11\pi}{6}\right\}$

47. $\sin^2 x + 3 \sin x - 1 = 0$

Using the quadratic formula,

$$\sin x = \frac{-3 \pm \sqrt{9+4}}{2} = \frac{-3 \pm \sqrt{13}}{2}$$

$$\sin x = -3.3028 \qquad \sin x = .3028$$

No solution: $|\sin x| \le 1 \qquad x = .3076$

$$\text{or } x = \pi - .3076 = 2.8340$$

Solution set: $\{.3076, 2.8340\}$

49.
$$\sin 2x + \sin x + 4 \cos x = -2$$

$$2 \sin x \cos x + \sin x + 4 \cos x + 2 = 0$$

$$\sin x(2 \cos x + 1) + 2(2 \cos x + 1) = 0$$

$$(\sin x + 2)(2 \cos x + 1) = 0$$

$$\sin x = -2 \qquad \cos x = -\frac{1}{2}$$

No solution: $|\sin x| \le 1 \qquad x = \frac{2\pi}{3}, \frac{4\pi}{3}$

51.
$$\sin x \cos x = -\frac{\sqrt{3}}{4}$$

$$2 \sin x \cos x = -\frac{\sqrt{3}}{2}$$

$$\sin 2x = -\frac{\sqrt{3}}{2}$$

$$2x = \frac{4\pi}{3}, \frac{5\pi}{3}, \frac{10\pi}{3}, \frac{11\pi}{3}$$

$$x = \frac{2\pi}{3}, \frac{5\pi}{6}, \frac{5\pi}{3}, \frac{11\pi}{6}$$

Notice that we let $2x$ take on values between 0 and 4π so that x would fall between 0 and 2π.

53. We show how the addition law can be used to avoid extraneous solutions. $1^2 + 2^2 = 5$. Let

$$\cos \theta = \frac{1}{\sqrt{5}}, \sin \theta = \frac{2}{\sqrt{5}} = \cos\left(\frac{\pi}{2} - \theta\right)$$

$$\cos x - 2 \sin x = 2$$

$$\frac{1}{\sqrt{5}} \cos x - \frac{2}{\sqrt{5}} \sin x = \frac{2}{\sqrt{5}}$$

$$\cos x \cos \theta - \sin x \sin \theta = \frac{2}{\sqrt{5}}$$

$$\cos(x + \theta) = \cos\left(\frac{\pi}{2} - \theta\right)$$

$$x + \theta = 2\pi + \left(\frac{\pi}{2} - \theta\right) \qquad x + \theta = 2\pi - \left(\frac{\pi}{2} - \theta\right)$$

$$= \frac{5\pi}{2} - \theta \qquad = \frac{3\pi}{2} + \theta$$

$$x = \frac{5\pi}{2} - 2\theta \qquad x = \frac{3\pi}{2}$$

$$= \frac{5\pi}{2} - 2 \cos^{-1}\left(\frac{1}{\sqrt{5}}\right)$$

$$= 5.6397$$

55.
$$\cos^8 x - \sin^8 x = 0$$

$$\sin^8 x = \cos^8 x$$

$$\left(\frac{\sin x}{\cos x}\right)^8 = 1$$

$$\tan^8 x = 1$$

$$\tan x = \pm 1$$

$$x = \frac{\pi}{4}, \frac{3\pi}{4}, \frac{5\pi}{4}, \frac{7\pi}{4}$$

57. a) Since the two triangles are similar,

$$\frac{10}{x} = \frac{30}{60 - x}$$

Therefore,

$$30x = 600 - 10x$$

$$40x = 600$$

$$x = 15 \text{ inches}$$

b) $$\tan \theta = \frac{10}{15} = \frac{2}{3}$$

$$\theta = \tan^{-1} \frac{2}{3} = 33.69°$$

59.
$$\sin 2t \ge \cos t$$

$$2 \sin t \cos t \ge \cos t$$

$$2 \sin t \cos t - \cos t \ge 0$$

$$2 \cos t\left(\sin t - \frac{1}{2}\right) \ge 0$$

Find the split points:

$$\cos t = 0 \qquad\qquad \sin t = \frac{1}{2}$$

$$t = \frac{\pi}{2}, \frac{3\pi}{2} \qquad\qquad t = \frac{\pi}{6}, \frac{5\pi}{6}$$

Interval	$\cos t$	$\sin t - \frac{1}{2}$	product
$0 < t < \frac{\pi}{6}$	$+$	$-$	$-$
$\frac{\pi}{6} < t < \frac{\pi}{2}$	$+$	$+$	$+$
$\frac{\pi}{2} < t < \frac{5\pi}{6}$	$-$	$+$	$-$
$\frac{5\pi}{6} < t < \frac{3\pi}{2}$	$-$	$-$	$+$
$\frac{3\pi}{2} < t < 2\pi$	$+$	$-$	$-$

$2 \cos t(\sin t - \frac{1}{2}) \ge 0$, and hence $\sin 2t \ge \cos t$ for

$$\left\{t: \frac{\pi}{6} \le t \le \frac{\pi}{2} \text{ or } \frac{5\pi}{6} \le t \le \frac{3\pi}{2}\right\}$$

61. a) $$\tan \alpha = \frac{1}{2}$$

$$t = 26.6°$$

b) $$\tan(\alpha + \theta) = \frac{15}{20}$$

$$\alpha + \theta = 36.9°$$

$$\theta = 36.9° - \alpha = 36.9° - 26.6° = 10.3°$$

63.
$$\sin 4t + \sin 3t + \sin 2t = 0$$
$$[\sin 4t + \sin 2t] + \sin 3t = 0$$
$$2\sin\left(\frac{4t+2t}{2}\right)\cos\left(\frac{4t-2t}{2}\right) + \sin 3t = 0$$
$$2\sin 3t \cos t + \sin 3t = 0$$
$$\sin 3t(2\cos t + 1) = 0$$

$$\sin 3t = 0$$
$$3t = 0, \pm\pi, \pm2\pi, \pm3\pi, \pm4\pi, \dots$$
$$t = 0, \pm\frac{\pi}{3}, \pm\frac{2\pi}{3}, \pm\frac{3\pi}{3}, \pm\frac{4\pi}{3}, \dots$$
or
$$\cos t = -\frac{1}{2}$$
$$t = \pm\frac{2\pi}{3}, \pm\frac{4\pi}{3}, \pm\frac{8\pi}{3}, \pm\frac{10\pi}{3}, \dots$$
Solution set: $\left\{\frac{k\pi}{3}, k \text{ is an integer}\right\}$

65. Let $\cos 2u = x$. Then $\cos^2 u = \frac{1+x}{2}$, $\sin^2 u = \frac{1-x}{2}$.
$$\cos^8 u + \sin^8 u = \frac{41}{128}$$
$$\left(\frac{1+x}{2}\right)^4 + \left(\frac{1-x}{2}\right)^4 = \frac{41}{128}$$
$$\frac{1+4x+6x^2+4x^3+x^4}{16} + \frac{1-4x+6x^2-4x^3+x^4}{16}$$
$$= \frac{41}{128}$$
$$\frac{2(1+6x^2+x^4)}{16} = \frac{41}{128}$$
$$x^4 + 6x^2 + 1 = \frac{41}{16}$$
$$x^4 + 6x^2 + 9 = \frac{41}{16} + 8$$
$$(x^2+3)^2 = \frac{169}{16}$$
$$x^2 + 3 = \frac{13}{4}$$
$$x^2 = \frac{1}{4}$$

$$\begin{array}{ll} x = \frac{1}{2} & x = -\frac{1}{2} \\ \cos 2u = \frac{1}{2} & \cos 2u = -\frac{1}{2} \\ 2u = \frac{\pi}{3}, \frac{5\pi}{3} & 2u = \frac{2\pi}{3}, \frac{4\pi}{3} \\ u = \frac{\pi}{6}, \frac{5\pi}{6} & u = \frac{\pi}{3}, \frac{2\pi}{3} \end{array}$$

67. Press $\boxed{Y=}$
Enter   X as Y_1
Enter   2 $\boxed{\text{2nd}}$ [TAN^{-1}] as Y_2
Press $\boxed{\text{GRAPH}}$

There appear to be 3 points at which the two curves intersect. One is clearly $x = 0$.
Zoom in on the point of intersection in quadrant III, to find that $x = -2.331$. By the symmetry of the graph, we would expect that the point of intersection in quadrant I occurs at $x = 2.331$. This can be checked by zooming in on that point.

Chapter 8 Review Problem Set

1. T
$$\sin\theta \cos\theta \tan\theta = \sin\theta \cos\theta \frac{\sin\theta}{\cos\theta}$$
$$= \sin^2\theta$$
$$= 1 - \cos^2\theta$$

2. False.
Consider $\sin 2\theta = \frac{1}{2}$ which has 4 solutions in $[0, 2\pi]$.

3. False.
Consider $\sin\theta = 0$. $\theta = k\pi$ where k is any integer.

4. T

5. T

6. False.
Consider $\alpha = 3k\pi$ where k is any integer

7. T
$$\cos 75° \sin 15° - \sin 75° \cos 15°$$
$$= \sin 15° \cos 75° - \cos 15° \sin 75°$$
$$= \sin(15° - 75°)$$
$$= \sin(-60°)$$
$$= -\frac{\sqrt{3}}{2}$$

8. T

9. False. Consider $\sin^{-1}\left(\sin\frac{4\pi}{3}\right) = -\frac{\pi}{3}$

10. False. Consider $f\left(\frac{\pi}{3}\right) = f\left(-\frac{\pi}{3}\right)$.

11. a) $2 - 3\sin^2 t = 2 - 3(1 - \cos^2 t) = -1 + 3\cos^2 t$

b) $\dfrac{\sin^2 t \sec t}{1 + \sec t} = \dfrac{(1 - \cos^2 t) \cdot \frac{1}{\cos t}}{1 + \frac{1}{\cos t}}$
$$= \frac{1 - \cos^2 t}{\cos t + 1}$$
$$= \frac{(1 + \cos t)(1 - \cos t)}{1 + \cos t}$$
$$= 1 - \cos t$$

c) $\sin^2 2t \cos^2\left(\frac{t}{2}\right) = 4\sin^2 t \cos^2 t\left(\frac{1 - \cos t}{2}\right)$
$$= 2(1 - \cos^2 t)(\cos^2 t)(1 + \cos t)$$

12. $\frac{\pi}{2} < t < \pi$. $\tan t = \frac{3}{-4} = \frac{b}{a}$. $r = \sqrt{3^2 + 4^2} = \sqrt{25} = 5$
a) $\sin t = \frac{b}{r} = \frac{3}{5}$
b) $\cos t = \frac{a}{r} = \frac{-4}{5}$
c) $\sec t = \frac{r}{a} = \frac{5}{-4} = -\frac{5}{4}$

13.
$$\cot\theta\cos\theta = \frac{\cos\theta}{\sin\theta}\cdot\cos\theta$$
$$= \frac{\cos^2\theta}{\sin\theta}$$
$$= \frac{1-\sin^2\theta}{\sin\theta}$$
$$= \frac{1}{\sin\theta}-\sin\theta$$
$$= \csc\theta-\sin\theta$$

14.
$$\sec t-\cos t = \frac{1}{\cos t}-\cos t$$
$$= \frac{1-\cos^2 t}{\cos t}$$
$$= \frac{\sin^2 t}{\cos t}$$
$$= \sin t\cdot\frac{\sin t}{\cos t}$$
$$= \sin t\tan t$$

15. $\left(\cos\frac{t}{2}+\sin\frac{t}{2}\right)^2 = \cos^2\left(\frac{t}{2}\right)+\sin^2\left(\frac{t}{2}\right)+2\cos\frac{t}{2}\sin\frac{t}{2}$
$$= 1+\sin 2\left(\frac{t}{2}\right)$$
$$= 1+\sin t$$

16.
$$\sec^4\theta-\sec^2\theta = \sec^2\theta(\sec^2\theta-1)$$
$$= (\tan^2\theta+1)\tan^2\theta$$
$$= \tan^4\theta+\tan^2\theta$$

17.
$$\tan u+\cot u = \frac{\sin u}{\cos u}+\frac{\cos u}{\sin u}$$
$$= \frac{\sin^2 u+\cos^2 u}{\cos u\sin u}$$
$$= \frac{1}{\cos u\sin u}$$
$$= \sec u\csc u$$

18.
$$\frac{1-\cos x}{\sin x} = \frac{1-\cos x}{\sin x}\cdot\frac{1+\cos x}{1+\cos x}$$
$$= \frac{1-\cos^2 x}{\sin x(1+\cos x)}$$
$$= \frac{\sin^2 x}{\sin x(1+\cos x)}$$
$$= \frac{\sin x}{1+\cos x}$$

19. $\cos 153°\cos 33°+\sin 153°\sin 33° = \cos(153°-33°)$
$$= \cos 120°$$
$$= -\frac{1}{2}$$

20. $\sin\frac{\pi}{8}\cos\frac{3\pi}{8}+\cos\frac{\pi}{8}\sin\frac{3\pi}{8} = \sin\left(\frac{\pi}{8}+\frac{3\pi}{8}\right) = \sin\frac{\pi}{2}=1$

21. $2\sin^2 112.5°-1 = -(1-2\sin^2 112.5°)$
$$= -\cos 2\cdot 112.5°$$
$$= -\cos 225°$$
$$= -\left(-\frac{\sqrt{2}}{2}\right)$$
$$= \frac{\sqrt{2}}{2}$$

22. $\frac{\tan 20°+\tan 25°}{1-\tan 20°\tan 25°} = \tan(20°+25°) = \tan 45° = 1$

23.
$$\tan\theta\tan 2\theta = \frac{\sin\theta}{\cos\theta}\cdot\frac{\sin 2\theta}{\cos 2\theta}$$
$$= \frac{\sin\theta(2\sin\theta\cos\theta)}{\cos\theta(1-2\sin^2\theta)}$$
$$= \frac{2\sin^2\theta}{1-2\sin^2\theta}$$

24.
$$\cos 4t = \cos 2\cdot 2t$$
$$= 2\cos^2 2t-1$$
$$= 2(1-2\sin^2 t)^2-1$$
$$= 2(1-4\sin^2 t+4\sin^4 t)-1$$
$$= 8\sin^4 t-8\sin^2 t+1$$

25. $\sin t = -\frac{12}{13},\ \pi<t<\frac{3\pi}{2}$

a) $\cos t = -\sqrt{1-\left(-\frac{12}{13}\right)^2} = -\sqrt{\frac{25}{169}} = -\frac{5}{13}$

b) $\sin 2t = 2\sin t\cos t = 2\left(-\frac{12}{13}\right)\left(-\frac{5}{13}\right) = \frac{120}{169}$

c) $\cos 2t = 1-2\sin^2 t$
$$= 1-2\left(-\frac{12}{13}\right)^2$$
$$= 1-2\cdot\frac{144}{169}$$
$$= \frac{169-288}{169}$$
$$= -\frac{119}{169}$$

d) $\tan\frac{t}{2} = \frac{\sin t}{1+\cos t} = \frac{-\frac{12}{13}}{1+\left(-\frac{5}{13}\right)} = \frac{-12}{8} = -\frac{3}{2}$

26. a) $\cos\alpha = -\sqrt{1-\left(\frac{2}{3}\right)^2} = -\sqrt{1-\frac{4}{9}} = -\sqrt{\frac{5}{9}} = -\frac{\sqrt{5}}{3}$

b) $\cos\beta = -\sqrt{1-\left(\frac{4}{5}\right)^2} = -\sqrt{1-\frac{16}{25}} = -\sqrt{\frac{9}{25}} = -\frac{3}{5}$

c) $\sin(\alpha+\beta) = \sin\alpha\cos\beta+\cos\alpha\sin\beta$
$$= \frac{2}{3}\left(-\frac{3}{5}\right)+\left(-\frac{\sqrt{5}}{3}\right)\left(\frac{4}{5}\right)$$
$$= \frac{-6-4\sqrt{5}}{15}$$

27. $\cos\left(u+\frac{\pi}{3}\right)\cos(\pi-u)-\sin\left(u+\frac{\pi}{3}\right)\sin(\pi-u)$
$$= \cos\left[\left(u+\frac{\pi}{3}\right)+(\pi-u)\right]$$
$$= \cos\frac{4\pi}{3}$$
$$= -\frac{1}{2}$$

28. $\dfrac{\cos 5t}{\sin t} - \dfrac{\sin 5t}{\cos t} = \dfrac{\cos 5t \cos t - \sin 5t \sin t}{\sin t \cos t}$

$$= \dfrac{\cos(5t + t)}{\frac{1}{2}\sin 2t}$$

$$= \dfrac{2\cos 6t}{\sin 2t}$$

29. $\csc 2t + \cot 2t = \dfrac{1}{\sin 2t} + \dfrac{\cos 2t}{\sin 2t}$

$$= \dfrac{1 + \cos 2t}{\sin 2t}$$

$$= \dfrac{1}{\dfrac{\sin 2t}{1 + \cos 2t}}$$

$$= \dfrac{1}{\tan t}$$

$$= \cot t$$

30. $\sin 3\theta = \sin(\theta + 2\theta)$

$$= \sin \theta \cos 2\theta + \cos \theta \sin 2\theta$$

$$= \sin \theta(1 - 2\sin^2\theta) + \cos \theta \cdot 2 \sin \theta \cos \theta$$

$$= \sin \theta - 2\sin^3\theta + 2\sin \theta(1 - \sin^2\theta)$$

$$= 3\sin \theta - 4\sin^3\theta$$

31. $\dfrac{\sin(\alpha - \beta)}{\cos \alpha \cos\beta} = \dfrac{\sin \alpha \cos \beta - \cos \alpha \sin \beta}{\cos \alpha \cos \beta}$

$$= \dfrac{\sin \alpha \cos \beta}{\cos \alpha \cos \beta} - \dfrac{\cos \alpha \sin \beta}{\cos \alpha \cos \beta}$$

$$= \dfrac{\sin \alpha}{\cos \alpha} - \dfrac{\sin \beta}{\cos \beta}$$

$$= \tan \alpha - \tan \beta$$

32. $\dfrac{1 - \tan^2\left(\frac{u}{2}\right)}{1 + \tan^2\left(\frac{u}{2}\right)} = \dfrac{1 - \dfrac{\sin^2(u/2)}{\cos^2(u/2)}}{1 + \dfrac{\sin^2(u/2)}{\cos^2(u/2)}}$

$$= \dfrac{\cos^2\left(\frac{u}{2}\right) - \sin^2\left(\frac{u}{2}\right)}{\cos^2\left(\frac{u}{2}\right) + \sin^2\left(\frac{u}{2}\right)}$$

$$= \dfrac{\cos 2 \cdot \frac{u}{2}}{1}$$

$$= \cos u$$

33. a) $\sin^{-1}\left(\dfrac{\sqrt{2}}{2}\right) = \dfrac{\pi}{4}$

b) $\cos^{-1}\left(-\dfrac{1}{2}\right) = \dfrac{2\pi}{3}$

c) $\tan^{-1}(-1) = -\dfrac{\pi}{4}$

d) $\sec^{-1}(\sqrt{2}) = \cos^{-1}\left(\dfrac{1}{\sqrt{2}}\right) = \dfrac{\pi}{4}$

34. a) $\sec(\sec^{-1} 2.5) = 2.5$

b) $\sin^{-1}\left(\sin \dfrac{3\pi}{4}\right) = \sin^{-1}\left(\dfrac{\sqrt{2}}{2}\right) = \dfrac{\pi}{4}$

c) $\csc\left[\sin^{-1}(.6)\right] = \dfrac{1}{\sin[\sin^{-1}(.6)]} = \dfrac{1}{.6} = \dfrac{5}{3}$

d) $\sec[\cos^{-1}(-.2)] = \dfrac{1}{\cos[\cos^{-1}(-.2)]} = \dfrac{1}{-.2} = -5$

35. a) Let $\theta = \cos^{-1}(.8)$.

Then $\cos \theta = .8$ and θ is in quadrant I.

$\sin \theta = \sqrt{1 - (.8)^2} = \sqrt{1 - .64} = \sqrt{.36} = .6$

$\sin[2 \cos^{-1}(.8)] = \sin 2\theta$

$$= 2 \sin \theta \cos \theta$$

$$= 2(.6)(.8)$$

$$= .96$$

b) Let $\theta = \cos^{-1}(.7)$. Then $\cos \theta = .7$

$\cos[2 \cos^{-1}(.7)] = \cos 2\theta$

$$= 2\cos^2\theta - 1$$

$$= 2(.7)^2 - 1$$

$$= .98 - 1$$

$$= -.02$$

36. a) $\sin[\cos^{-1}(.6) + \cos^{-1}(.5)]$

$= \sin[\cos^{-1}(.6) \cos[\cos^{-1}(.5)]$

$\qquad\qquad + \cos[\cos^{-1}(.6)] \sin[\cos^{-1}(.5)]$

$= \sqrt{1 - (.6)^2}(.5) + (.6)\sqrt{1 - (.5)^2}$

$= (.8)(.5) + (.6)\dfrac{\sqrt{3}}{2}$

$= .4 + .3\sqrt{3}$

b) $\tan(\tan^{-1} 1 + \tan^{-1} 2) = \dfrac{1 + 2}{1 - 1 \cdot 2} = -3$

37. $\tan t = -\dfrac{\sqrt{3}}{3}$

The reference angle is $t_0 = \tan^{-1}\left(\dfrac{\sqrt{3}}{3}\right) = \dfrac{\pi}{6}$

$t = \pi - t_0 = \dfrac{5\pi}{6}$ or $t = 2\pi - t_0 = \dfrac{11\pi}{6}$

Solution set: $\left\{\dfrac{5\pi}{6}, \dfrac{11\pi}{6}\right\}$

38. $\qquad 2\cos^2 t - \cos t = 0$

$\qquad \cos t(2 \cos t - 1) = 0$

$\cos t = 0 \qquad\qquad \cos t = \dfrac{1}{2}$

$t = \dfrac{\pi}{2}, \dfrac{3\pi}{2} \qquad\qquad t = \dfrac{\pi}{3}, \dfrac{5\pi}{3}$

Solution set: $\left\{\dfrac{\pi}{3}, \dfrac{\pi}{2}, \dfrac{3\pi}{2}, \dfrac{5\pi}{3}\right\}$

39. $\qquad \sin^2 t - 2 \sin t - 3 = 0$

$\qquad (\sin t - 3)(\sin t + 1) = 0$

$\sin t = 3 \qquad\qquad \sin t = -1$

No solution $\qquad\qquad t = \dfrac{3\pi}{2}$

Solution set: $\left\{\dfrac{3\pi}{2}\right\}$

40. $\sec^4 t - 3\sec^2 t + 2 = 0$

$(\sec^2 t - 1)(\sec^2 t - 2) = 0$

$\sec^2 t = 1 \qquad\qquad \sec^2 t = 2$

$\sec t = \pm 1 \qquad\qquad \sec t = \pm\sqrt{2}$

$t = 0,\ \pi \qquad\qquad t = \dfrac{\pi}{4}, \dfrac{3\pi}{4}, \dfrac{5\pi}{4}, \dfrac{7\pi}{4}$

41. $\sin 2t = \dfrac{1}{2}$

$2t = \dfrac{\pi}{6}, \dfrac{5\pi}{6}, \dfrac{13\pi}{6}, \dfrac{17\pi}{6}$

$t = \dfrac{\pi}{12}, \dfrac{5\pi}{12}, \dfrac{13\pi}{12}, \dfrac{17\pi}{12}$

Notice that we let $2t$ take on values between 0 and 4π so that t would fall between 0 and 2π.

42. $\cos 4t = -\dfrac{1}{2}$

$4t = \dfrac{2\pi}{3}, \dfrac{4\pi}{3}, \dfrac{8\pi}{3}, \dfrac{10\pi}{3}, \dfrac{14\pi}{3}, \dfrac{16\pi}{3}, \dfrac{20\pi}{3}, \dfrac{22\pi}{3}$

$t = \dfrac{\pi}{6}, \dfrac{\pi}{3}, \dfrac{2\pi}{3}, \dfrac{5\pi}{6}, \dfrac{7\pi}{6}, \dfrac{4\pi}{3}, \dfrac{5\pi}{3}, \dfrac{11\pi}{6}$

Notice that we let $4t$ take on values between 0 and 8π so that t would fall between 0 and 2π.

43. $3 + \cos 2t = 5\cos t$

$3 + 2\cos^2 t - 1 = 5\cos t$

$2\cos^2 t - 5\cos t + 2 = 0$

$(2\cos t - 1)(\cos t - 2) = 0$

$\cos t = \dfrac{1}{2} \qquad\qquad \cos t = 2$

$t = \dfrac{\pi}{3}, \dfrac{5\pi}{3} \qquad\qquad$ No solution

Solution set: $\left\{\dfrac{\pi}{3}, \dfrac{5\pi}{3}\right\}$

44. $\tan^3 t + \tan^2 t - 3\tan t = 3$

$\tan^3 t + \tan^2 t - 3\tan t - 3 = 0$

$\tan^2 t(\tan t + 1) - 3(\tan t + 1) = 0$

$(\tan t + 1)(\tan^2 t - 3) = 0$

$\tan t = -1 \qquad\qquad \tan t = \pm\sqrt{3}$

$t = \dfrac{3\pi}{4}, \dfrac{7\pi}{4} \qquad\qquad t = \dfrac{\pi}{3}, \dfrac{2\pi}{3}, \dfrac{4\pi}{3}, \dfrac{5\pi}{3}$

45. $y = f(x) = \cos^{-1} x$

Domain $= \{x : -1 \le x \le 1\}$

Range: $= \{y : 0 \le y \le \pi\}$

46. $y = \sin 2x$ is one-to-one if $-\dfrac{\pi}{2} \le 2x \le \dfrac{\pi}{2},\ -\dfrac{\pi}{4} \le x \le \dfrac{\pi}{4}$

$2x = \sin^{-1} y$

$x = \dfrac{1}{2}\sin^{-1} y$

$f^{-1}(x) = \dfrac{1}{2}\sin^{-1} x$

47. Let $\theta = \sin^{-1} x$.

Then $\sin\theta = x$ and θ is in quadrant I or IV.

$\cos\theta = \sqrt{1 - x^2}$

$\cot\theta = \dfrac{\cos\theta}{\sin\theta} = \dfrac{\sqrt{1 - x^2}}{x}$

48. $0 < \tan^{-1} 2 + \tan^{-1} 3 < \dfrac{\pi}{2} + \dfrac{\pi}{2} = \pi$

$\tan(\tan^{-1} 2 + \tan^{-1} 3) = \dfrac{2 + 3}{1 - 2\cdot 3} = -1$

Therefore $\tan^{-1} 2 + \tan^{-1} 3 = \dfrac{3\pi}{4}$

49. $\tan(\tan^{-1} 2 + \tan^{-1} 3) = \dfrac{2 + 3}{1 - 2\cdot 3} = \dfrac{5}{-5} = -1$

$\tan[(\tan^{-1} 2 + \tan^{-1} 3) + \tan^{-1} 5] = \dfrac{-1 + 5}{1 - (-1)5}$

$= \dfrac{4}{6}$

$= \dfrac{2}{3}$

50. $\tan t = .5$

$\tan 2t = \dfrac{2\tan t}{1 - \tan^2 t} = \dfrac{2(.5)}{1 - (.5)^2} = \dfrac{1}{.75} = \dfrac{4}{3}$

$\tan 3t = \tan(t + 2t) = \dfrac{\frac{1}{2} + \frac{4}{3}}{1 - \frac{1}{2}\cdot\frac{4}{3}} = \dfrac{\frac{11}{6}}{\frac{1}{3}} = \dfrac{11}{2} = 5.5$

Chapter 9: Applications of Trigonometry

1. $\alpha = 42.6°$, $\beta = 81.9°$, $a = 14.3$
    Find γ: $\gamma = 180° - (42.6° + 81.9°) = 55.5°$

    Find b:
    $$\frac{a}{\sin \alpha} = \frac{b}{\sin \beta}$$
    $$\frac{14.3}{\sin 42.6°} = \frac{b}{\sin 81.9°}$$
    $$b = \frac{14.3 \times \sin 81.9°}{\sin 42.6°} \approx 20.9$$

    Find c:
    $$\frac{a}{\sin \alpha} = \frac{c}{\sin \gamma}$$
    $$\frac{14.3}{\sin 42.6°} = \frac{c}{\sin 55.5°}$$
    $$c = \frac{14.3 \times \sin 55.5°}{\sin 42.6°} \approx 17.4$$

3. $\alpha = \gamma = 62°$, $b = 50$

    Find β: $\beta = 180° - 2 \times 62° = 56°$

    Find a:
    $$\frac{a}{\sin \alpha} = \frac{b}{\sin \beta}$$
    $$\frac{a}{\sin 62°} = \frac{50}{\sin 56°}$$
    $$a = \frac{50 \times \sin 62°}{\sin 56°} \approx 53.3$$

    Find c: Since $\alpha = \gamma$, the triangle is isosceles, with $a = c$. $c = 53.3$

Note. To ensure that the final answer is as accurate as the given, intermediate results require an extra digit.

5. $\alpha = 115°$, $a = 46$, $b = 34$
    β must be an acute angle.

    Find β:
    $$\frac{a}{\sin \alpha} = \frac{b}{\sin \beta}$$
    $$\frac{46}{\sin 115°} = \frac{34}{\sin \beta}$$
    $$\sin \beta = \frac{34 \times \sin 115°}{46} \approx .66988$$
    $$\beta = 42.1° \approx 42°$$

    Find γ: $\gamma = 180° - (115° + 42.1°) = 22.9° \approx 23°$

    Find c:
    $$\frac{a}{\sin \alpha} = \frac{c}{\sin \gamma}$$
    $$\frac{46}{\sin 115°} = \frac{c}{\sin 22.9°}$$
    $$c = \frac{46 \times \sin 22.9°}{\sin 115°} = 19.75 \approx 20$$

7. $\alpha = 30°$, $a = 8$, $b = 5$
    Since $a \geq b$, there is a unique triangle, and $\beta < \alpha$.

    Find β:
    $$\frac{a}{\sin \alpha} = \frac{b}{\sin \beta}$$
    $$\frac{8}{\sin 30°} = \frac{5}{\sin \beta}$$
    $$\sin \beta = \frac{5 \times \sin 30°}{8} = .3125$$

$\beta = 18.2° \approx 18°$

Find γ: $\gamma = 180° - (30° + 18.2°) = 131.8° \approx 132°$

Find c:
$$\frac{a}{\sin \alpha} = \frac{c}{\sin \gamma}$$
$$\frac{8}{\sin 30°} = \frac{c}{\sin 131.8°}$$
$$c = \frac{8 \times \sin 131.8°}{\sin 30°} = 11.9 \approx 12$$

9. $\alpha = 30°$, $a = 5$, $b = 8$
    Since $a < b$, there may be 1, 2, or no triangles.

    Find β:
    $$\frac{a}{\sin \alpha} = \frac{b}{\sin \beta}$$
    $$\frac{5}{\sin 30°} = \frac{8}{\sin \beta}$$
    $$\sin \beta = \frac{8 \times \sin 30°}{5} = .8$$

    There are two triangles.
    $$\beta_1 = 53.1° \approx 53°$$
    $$\beta_2 = 180° - 53.1° = 126.9° \approx 127°$$

    Find γ_1 and c_1:
    $\gamma_1 = 180° - (30° + 53.1°) = 96.9° \approx 97°$
    $$\frac{a}{\sin \alpha} = \frac{c_1}{\sin \gamma_1}$$
    $$\frac{5}{\sin 30°} = \frac{c_1}{\sin 96.9°}$$
    $$c_1 = \frac{5 \times 96.9°}{\sin 30°} = 9.93 \approx 9.9$$

    Find γ_2 and c_2:
    $\gamma_2 = 180° - (30° + 126.9°) = 23.1° \approx 23°$
    $$\frac{5}{\sin 30°} = \frac{c_2}{\sin 23.1°}$$
    $$c_2 = \frac{5 \times \sin 23.1°}{\sin 30°} = 3.92 \approx 3.9$$

11.

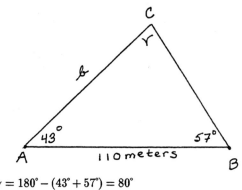

$\gamma = 180° - (43° + 57°) = 80°$
$$\frac{110}{\sin 80°} = \frac{b}{\sin 57°}$$
$$b = \frac{110 \times \sin 57°}{\sin 80°} = 93.7 \text{ meters}$$

Problem Set 9.1

13.

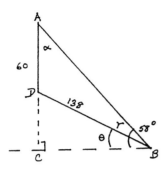

Because ABC is a right triangle, we know that
$\alpha = 90° - 58° = 32°$
We find γ by applying the law of sines to Δ ADB.

$$\frac{60}{\sin \gamma} = \frac{138}{\sin 32°}$$

$$\sin \gamma = \frac{60 \times \sin 32°}{138} = .2304$$

$$\gamma = 13.3°$$

$$\theta = 58° - 13.3° = 44.7°$$

15. $A = \frac{1}{2}bc \sin \alpha = \frac{1}{2} \times 20 \times 30 \times \sin 40° = 192.8$

17. We must find the length of another side of the triangle, say a. First, find γ:
$\gamma = 180° - (25.3° + 112.2°) = 42.5°$

$$\frac{a}{\sin \alpha} = \frac{c}{\sin \gamma}$$

$$\frac{a}{\sin 25.3°} = \frac{30.1}{\sin 42.5°}$$

$$a = \frac{30.1 \times \sin 25.3°}{\sin 42.5°} = 19.04$$

$A = \frac{1}{2}ac \sin \beta = \frac{1}{2} \times 19.04 \times 30.1 \times \sin 112.2° = 265.3$

19.

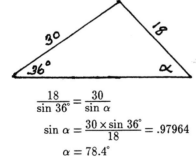

$$\frac{18}{\sin 36°} = \frac{30}{\sin \alpha}$$

$$\sin \alpha = \frac{30 \times \sin 36°}{18} = .97964$$

$$\alpha = 78.4°$$

21.

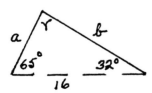

Find γ: $\gamma = 180° - (65° + 32°) = 83°$

Find a: $\quad \frac{a}{\sin 32°} = \frac{16}{\sin 83°}$

$$a = \frac{16 \times \sin 32°}{\sin 83°} = 8.54 \text{ feet}$$

Find b: $\quad \frac{b}{\sin 65°} = \frac{16}{\sin 83°}$

$$b = \frac{16 \times \sin 65°}{\sin 83°} = 14.61 \text{ feet}$$

The area of the ceiling is the sum of the areas of two rectangles, one 30 feet by 8.54 feet and the other 30 feet by 14.61 feet.

Area $= 30 \times 8.54 + 30 \times 14.61 = 694.5$ square feet

23.

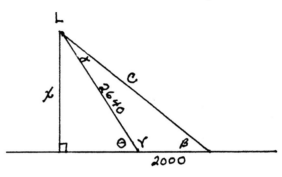

If the beam of light completes 1 revolution (360°) every minute (60 seconds), then it rotates $\frac{360°}{60} = 6°$ degrees per second. In 3 seconds it rotates 18°. Therefore $\alpha = 18°$.

Find β: $\quad \frac{2000}{\sin 18°} = \frac{2640}{\sin \beta}$

$$\sin \beta = \frac{2640 \times \sin 18°}{2000} = .4079$$

Since β is an angle of a right triangle,

$$\beta = 24.07°$$

From geometry, we know that
$\theta = \alpha + \beta = 18° + 24.07° = 42.07°$
Let x feet be the shortest distance from the lighthouse to the shore.

$$\frac{x}{2640} = \sin 42.07$$

$$x = 2640 \times \sin 42.07 = 1769 \text{ feet}$$

181

25.

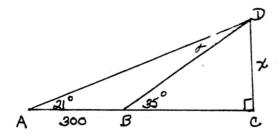

From geometry, we know that

$\alpha = 35° - 21° = 14°$

Find $\overline{BD}$:
$$\frac{\overline{BD}}{\sin 21°} = \frac{300}{\sin 14°}$$

$$\overline{BD} = \frac{300 \times \sin 21°}{\sin 14°} = 444.4 \text{ feet}$$

Let x feet be the height of the tower.

$$\frac{x}{\overline{BD}} = \sin 35°$$

$$x = 444.4 \times \sin 35° = 255 \text{ feet}$$

27.

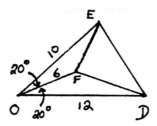

Area $\triangle DEF =$
$\qquad$ Area $\triangle OED -$ Area $\triangle OFE -$ Area $\triangle OFD$

Area $\triangle OED = \frac{1}{2} \times 12 \times 10 \times \sin 40° = 38.57$

Area $\triangle OFE = \frac{1}{2} \times 6 \times 10 \times \sin 20° = 10.26$

Area $\triangle OFD = \frac{1}{2} \times 6 \times 12 \times \sin 20° = 12.31$

Area $\triangle DEF = 38.57 - 10.26 - 12.31 \approx 16$ square units

29.

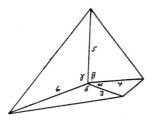

$A = \frac{1}{2}(3 \cdot 4 \sin \alpha + 4 \cdot 5 \sin \beta + 5 \cdot 6 \sin \gamma + 6 \cdot 3 \sin \delta)$

This is a maximum when $\alpha = \beta = \gamma = \delta = 90°$.

Then $A = \frac{1}{2}(12 + 20 + 30 + 18) = 40$

31.

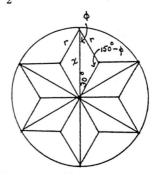

The required area is six times the sum of the areas of an isosceles triangle of side r and an equilateral triangle of side $2x$.

$$\frac{x}{r} = \sin \phi$$

$$x = r \sin \phi$$

$$\text{Area} = 6[\tfrac{1}{2}r^2 \sin 2\phi + \tfrac{1}{2}(2r \sin \phi)^2 \sin 60°]$$

$$= 3r^2 \sin 2\phi + 6\sqrt{3}r^2 \sin^2 \phi$$

$$= 6r^2 \sin \phi \cos \phi + 6\sqrt{3}r^2 \sin^2 \phi$$

33. Set new range values:
$\qquad$ $X_{min} = 0$
$\qquad$ $X_{max} = 60$
$\qquad$ $Y_{min} = 0$
$\qquad$ $Y_{max} = 20$
$\qquad$ Press $\boxed{Y=}$
$\qquad$ Enter: $\boxed{\text{2nd}}$ [ABS] $\boxed{(}$10 $\boxed{\text{SIN}}$ $\boxed{(}$$\pi$X$\boxed{\div}30\boxed{)}$

$\qquad\qquad$ $- 7.5$ $\boxed{\text{SIN}}$ $\boxed{(}$ πX $\boxed{\div}$ 360 $\boxed{)}$

$\qquad\qquad$ $- 12$ $\boxed{\text{SIN}}$ $\boxed{(}$11 π X $\boxed{\div}$ 360 $\boxed{)}$ $\boxed{)}$

$\qquad$ Press: $\boxed{\text{GRAPH}}$

Zoom in on the highest point to find that $x \approx 42.4$ there. This represents 42.4 minutes past 12 o'clock or, 12:42.4.

Problem Set 9.2

1. $\alpha = 60°$, $b = 14$, $c = 10$

$\qquad$ Find a: $\quad a^2 = b^2 + c^2 - 2bc \cos \alpha$
$\qquad\qquad\qquad\qquad = 14^2 + 10^2 - 2 \times 14 \times 10 \times \cos 60°$
$\qquad\qquad\qquad\qquad = 156$

$\qquad\qquad\qquad a = \sqrt{156} = 12.5$

Find γ: (Since $c < a$, then $\gamma < \alpha$)

$$\frac{10}{\sin \gamma} = \frac{12.5}{\sin 60°}$$

$$\sin \gamma = \frac{10 \times \sin 60°}{12.5} = .6928$$

$$\gamma = 44°$$

Then $\quad \beta = 180° - (44° + 60°) = 76°$

3. $\gamma = 120°$, $a = 8$, $b = 10$

Find c: $\quad c^2 = a^2 + b^2 - 2ab \cos \gamma$
$$= 8^2 + 10^2 - 2 \times 8 \times 10 \times \cos 120°$$
$$= 244$$

$$c = \sqrt{244} = 15.6$$

Find α: (Since γ is obtuse, α will be acute.)

$$\frac{8}{\sin \alpha} = \frac{15.6}{\sin 120°}$$

$$\sin \alpha = \frac{8 \times \sin 120°}{15.6} = .4441$$

$$\alpha = 26°$$

Then $\beta = 180° - (26° + 120°) = 34°$

5. $a = 5$, $b = 6$, $c = 7$

Find γ (the largest angle): $c^2 = a^2 + b^2 - 2ab \cos \gamma$

$$\cos \gamma = \frac{a^2 + b^2 - c^2}{2ab} = \frac{5^2 + 6^2 - 7^2}{2 \times 5 \times 6} = .2$$

$$\gamma = 78.5°$$

Find α: (Since $a < c$, then $\alpha < \gamma$.)

$$\frac{5}{\sin \alpha} = \frac{7}{\sin 78.5°}$$

$$\sin \alpha = \frac{5 \times \sin 78.5°}{7} = .6999$$

$$\alpha = 44.4°$$

Then $\beta = 180° - (44.4° + 78.5°) = 57.1°$

7. $a = 12.2$, $b = 19.1$, $c = 23.8$

Find γ (the largest angle):

$$\cos \gamma = \frac{a^2 + b^2 - c^2}{2ab}$$

$$= \frac{12.2^2 + 19.1^2 - 23.8^2}{2 \times 12.2 \times 19.1}$$

$$= -.1133$$

$$\gamma = 96.5°$$

Find α: (Since $a < c$, then $\alpha < \gamma$.)

$$\frac{12.2}{\sin \alpha} = \frac{23.8}{\sin 96.5°}$$

$$\sin \alpha = \frac{12.2 \times \sin 96.5°}{23.8} = .5093$$

$$\alpha = 30.6°$$

Then $\beta = 180° - \left(30.6° + 96.5°\right) = 52.9°$

9. Let the third side of the field be c meters long.

$$c^2 = 100^2 + 120^2 - 2 \times 100 \times 120 \times \cos 52.4° = 9756$$

$$c = 98.8 \text{ meters}$$

11. At 3:00 PM, the first runner is $6 \times 3 = 18$ miles north of the starting point, and the second runner is $8 \times 3 = 24$ miles away from the starting point on a line 68° east of north. Let the distance between the two runners be c miles.

$$c^2 = 18^2 + 24^2 - 2 \times 18 \times 24 \times \cos 68° = 576.34$$

$$c = 24 \text{ miles}$$

13. The largest angle of the triangle will be opposite the longest side of the triangle.
Let $c = 60$, $a = 40$, $b = 35$. We want to find γ.

$$\cos \gamma = \frac{a^2 + b^2 - c^2}{2ab} = \frac{40^2 + 35^2 - 60^2}{2 \times 40 \times 35} = -.27679$$

$$\gamma = 106°$$

15. Let $a = 3$, $b = 4$, $c = 5$. $s = \dfrac{3 + 4 + 5}{2} = 6$

$$A = \sqrt{s(s-a)(s-b)(s-c)}$$

$$= \sqrt{6(6-3)(6-4)(6-5)}$$

$$= \sqrt{6(3)(2)(1)}$$

$$= \sqrt{36}$$

$$= 6$$

17. Let $a = 5.9$, $b = 6.7$, $c = 10.3$.

$$s = \frac{5.9 + 6.7 + 10.3}{2} = 11.45$$

$$A = \sqrt{s(s-a)(s-b)(s-c)}$$

$$= \sqrt{11.45(11.45 - 5.9)(11.45 - 6.7)(11.45 - 10.3)}$$

$$= \sqrt{11.45 \times 5.55 \times 4.75 \times 1.15}$$

$$= \sqrt{347.13}$$

$$= 18.63$$

19. The smallest angle is opposite the shortest side.
Let $a = 50$, $b = 63$, $c = 42$. Find γ.

$$\cos \gamma = \frac{a^2 + b^2 - c^2}{2ab} = \frac{50^2 + 63^2 - 42^2}{2 \times 50 \times 63} = .7468$$

$$\gamma = 41.68°$$

21. After 36 minutes, the two cars have traveled

$$\frac{36}{60} \times 55 = 33 \text{ miles and } \frac{36}{60} \times 65 = 39 \text{ miles,}$$

respectively. Let the cars be c miles apart after 36 minutes.

$$c^2 = 33^2 + 39^2 - 2 \times 33 \times 39 \times \cos 72° = 1814.6$$

$$c = 42.60 \text{ miles}$$

23. Let the sides of the triangle be $3a$, $2a$, and $2a$ meters. Let s be the semiperimeter of the garden.

$$s = \frac{3a + 2a + 2a}{2} = \frac{7a}{2}$$

Using Heron's formula to compute the area of the garden, we get the equation

$$200 = \sqrt{\frac{7a}{2}\left(\frac{7a}{2} - 3a\right)\left(\frac{7a}{2} - 2a\right)\left(\frac{7a}{2} - 2a\right)}$$

$$= \sqrt{\frac{7a}{2} \cdot \frac{a}{2} \cdot \frac{3a}{2} \cdot \frac{3a}{2}}$$

$$= \frac{3a^2}{4}\sqrt{7}$$

$$a^2 = \frac{800}{3\sqrt{7}} = 100.79$$

$$a = 10.04$$

The longer side is $3 \times 10.04 = 30.12$ meters long, and each of the two shorter sides is $2 \times 10.04 = 20.08$ meters.

25.

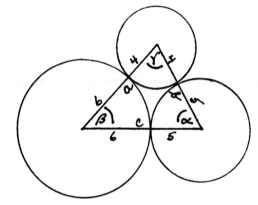

a) $c = 6 + 5 = 11$, $b = 4 + 5 = 9$, $a = 6 + 4 = 10$

$$\cos \gamma = \frac{a^2 + b^2 - c^2}{2ab} = \frac{10^2 + 9^2 - 11^2}{2 \times 10 \times 9} = \frac{1}{3}$$

$$\gamma = 70.53°$$

$$\cos \alpha = \frac{b^2 + c^2 - a^2}{2bc} = \frac{9^2 + 11^2 - 10^2}{2 \times 9 \times 11} = \frac{51}{99}$$

$$\alpha = 58.99°$$

$$\beta = 180° - (70.53° + 58.99°) = 50.48°$$

b) The area of the white region between the circles is the area A_T of the triangle, minus the areas A_α, A_β, A_γ, of the sectors of the circles with central angles α, β, and γ.

Using Heron's formula with $s = \frac{9 + 10 + 11}{2} = 15$,

$$A_T = \sqrt{15(15 - 9)(15 - 10)(15 - 11)}$$

$$= \sqrt{15 \times 6 \times 5 \times 4}$$

$$= \sqrt{1800} = 42.43$$

$$A_\gamma = \frac{\pi(4)^2(70.53°)}{360°} = 9.85$$

$$A_\alpha = \frac{\pi(5)^2(58.99°)}{360°} = 12.87$$

$$A_\beta = \frac{\pi(6)^2(50.48°)}{360°} = 15.86$$

The area of the white region is
$42.43 - 9.85 - 12.87 - 15.86 = 3.85$ square units

27.

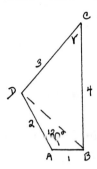

$$\overline{DB} = \sqrt{\overline{AB}^2 + \overline{AD}^2 - 2(\overline{AB})(\overline{AD})\cos 120°}$$

$$= \sqrt{1^2 + 2^2 - 2 \times 1 \times 2 \times \cos 120°}$$

$$= \sqrt{7}$$

$$\cos \gamma = \frac{\overline{DC}^2 + \overline{BC}^2 - \overline{DB}^2}{2(\overline{DC})(\overline{BC}^2)} = \frac{3^2 + 4^2 - (\sqrt{7})^2}{2 \times 3 \times 4} = \frac{3}{4}$$

$$\gamma = 41.41°$$

$$\sin \gamma = \sqrt{1 - \left(\frac{3}{4}\right)^2} = \sqrt{1 - \frac{9}{16}} = \sqrt{\frac{7}{16}} = \frac{\sqrt{7}}{4}$$

Area of Q = area of $\triangle ABD$ + area of $\triangle BCD$

$$= \frac{1}{2} \times 2 \times 1 \times \sin 120° + \frac{1}{2} \times 3 \times 4 \times \sin \gamma$$

$$= 1 \times \frac{\sqrt{3}}{2} + 6 \times \frac{\sqrt{7}}{4}$$

$$= \frac{1}{2}(\sqrt{3} + 3\sqrt{7})$$

29.

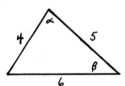

$$\cos \alpha = \frac{4^2 + 5^2 - 6^2}{2 \times 4 \times 5} = \frac{1}{8}$$

$$\cos \beta = \frac{5^2 + 6^2 - 4^2}{2 \times 5 \times 6} = \frac{3}{4}$$

$$\cos 2\beta = 2\cos^2\beta - 1 = 2 \cdot \frac{9}{16} - 1 = \frac{1}{8} = \cos \alpha$$

Therefore, $\alpha = 2\beta$.

31.

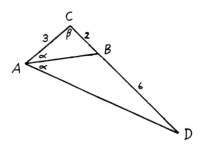

Because ∠CBA and ∠DBA are supplementary, $\sin\angle CBA = \sin\angle DBA$. Applying the law of sines to $\triangle CBA$ and $\triangle DBA$, we get

$$\frac{3}{2} = \frac{\sin\angle CBA}{\sin\alpha} = \frac{\sin\angle DBA}{\sin\alpha} = \frac{\overline{AD}}{6}$$

$$\overline{AD} = \frac{6 \times 3}{2} = 9$$

Applying the law of cosines to $\triangle ACD$, we get

$$\cos\beta = \frac{\overline{CD}^2 + \overline{AC}^2 - \overline{AD}^2}{2(CD)(AC)} = \frac{8^2 + 3^2 - 9^2}{2(8)(3)} = -\frac{1}{6}$$

Applying the law of cosines to $\triangle ACB$, we get

$$\overline{AB}^2 = \overline{AC}^2 + \overline{BC}^2 - 2\,\overline{AC}\,\overline{BC}\cos\beta$$
$$= 3^2 + 2^2 - 2(3)(2)\left(-\frac{1}{6}\right)$$
$$= 15$$
$$\overline{AB} = \sqrt{15}$$

33. Set new range:
Xmin $= 0$
Xmax $= 70$
Ymin $= 0$
Ymax $= 20$

Press $\boxed{Y=}$

Enter $\boxed{\sqrt{}}\boxed{(}(\,41 - 40\,\boxed{COS}\boxed{(}(\,\pi\,X\,\boxed{\div}\,30\,\boxed{)}\boxed{)}$
$\quad + \boxed{\sqrt{}}\boxed{(}(\,34 - 30\,\boxed{COS}\boxed{(}(\,\pi\,X\,\boxed{\div}\,360\,\boxed{)}\boxed{)}$
$\quad + \boxed{\sqrt{}}\boxed{(}(\,25 - 24\,\boxed{COS}\boxed{(}(\,11\,\pi\,X\,\boxed{\div}\,360\,\boxed{)}\boxed{)}$

Press $\boxed{GRAPH}$

Zoom in on the highest point of the graph. At that point, $x = 31.5$, which means that the perimeter is maximum 31.5 minutes after 12:00, at 12:31.5.

Problem Set 9.3

1.

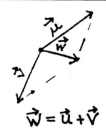

$$\vec{W} = \vec{u} + \vec{v}$$

$$\vec{w} = \vec{u} + \vec{v}$$

3.

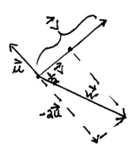

$$\vec{W} = -2\vec{u} + \tfrac{1}{2}\vec{v}$$

$$\mathbf{w} = -2\mathbf{u} + \tfrac{1}{2}\mathbf{v}$$

5. $\mathbf{w} = \dfrac{\mathbf{u} + \mathbf{v}}{2} = \tfrac{1}{2}\mathbf{u} + \tfrac{1}{2}\mathbf{v}$

7. $\|\mathbf{w}\|^2 = 1^2 + 1^2 - 2(1)(1)\cos 60° = 2 - 2(.5) = 1$

$\quad \|\mathbf{w}\| = 1$

9.

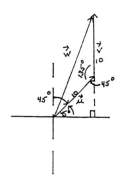

$$\mathbf{w} = \mathbf{u} + \mathbf{v}$$

$$\|\mathbf{w}\|^2 = 10^2 + 10^2 - 2(10)(10)\cos 135°$$
$$= 200 - 200\left(-\frac{\sqrt{2}}{2}\right)$$
$$= 200 + 100\sqrt{2}$$
$$= 100(2 + \sqrt{2})$$
$$\|\mathbf{w}\| = 10\sqrt{2 + \sqrt{2}} \approx 18.48$$

11.

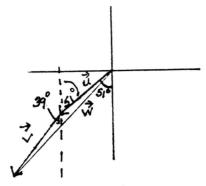

The angle between **u** and **v** is
$39° + (180° − 51°) = 168°$.

$$\|\mathbf{w}\|^2 = \|\mathbf{u}\|^2 + \|\mathbf{v}\|^2 - 2\|\mathbf{u}\|\|\mathbf{v}\|\cos 168°$$

$$= 100^2 + 145^2 - 2 \times 100 \times 145 \times \cos 168°$$

$$\approx 59391$$

$\|\mathbf{w}\| \approx 243.7$ kilometers

Let α be the angle between **u** and **w**.

$$\frac{\|\mathbf{v}\|}{\sin \alpha} = \frac{\|\mathbf{w}\|}{\sin 168°}$$

$$\sin \alpha = \frac{145 \times \sin 168°}{243.7} \approx .1237$$

$$\alpha \approx 7.1°$$

The bearing of the plane is
$S(51° − 7.1°)W = S43.9W$

13.

$$\frac{\|\mathbf{w}\|}{\sin 73°} = \frac{6}{\sin 17°}$$

$$\|\mathbf{w}\| = \frac{6 \times \sin 73°}{\sin 17°} \approx 19.6 \text{ mph}$$

$$t = \frac{d}{r} = \frac{\frac{1}{2}}{19.6} = .026 \text{ hour}$$

15.

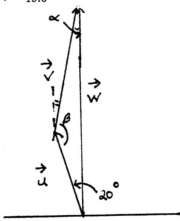

Let α be the angle between **v** and **w**.

$$\frac{\|\mathbf{u}\|}{\sin \alpha} = \frac{\|\mathbf{v}\|}{\sin 20°}$$

$$\sin \alpha = \frac{\|\mathbf{u}\|\sin 20°}{\|\mathbf{v}\|} = \frac{58 \sin 20°}{425} \approx .04668$$

$$\alpha \approx 2.68°$$

Let β be the angle between **u** and **v**.
$\beta = 180° − (20° + 2.68°) = 157.32°$

$$\frac{\|\mathbf{w}\|}{\sin \beta} = \frac{\|\mathbf{v}\|}{\sin 20°}$$

$$\|\mathbf{w}\| = \frac{425 \sin 157.32°}{\sin 20°} \approx 479$$

The plane should be flying at 479 mph, at a bearing of N 2.68° E.

17. Let $\mathbf{p} = \mathbf{u} + \mathbf{v}$.

$$\|\mathbf{p}\|^2 = \|\mathbf{u}\|^2 + \|\mathbf{v}\|^2 - 2\|\mathbf{u}\|\|\mathbf{v}\|\cos 105°$$
$$= 10^2 + 10^2 - 2 \times 10 \times 10 \times \cos 105°$$
$$\approx 251.76$$

$\|\mathbf{p}\| \approx 15.9$

Since the triangle created by **u**, **v**, and **p** is isosceles, the angle between **p** and **v** is $\frac{1}{2}(180° − 105°) = 37.5°$. Since **v** is 45° east of north, **p** is $(45° − 37.5°) = 7.5°$ east of north. Therefore, $\mathbf{w} = −\mathbf{p}$ has magnitude 15.9 and direction 7.5° west of south or S 7.5° W.

19.

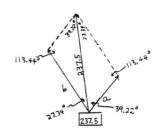

$$180° − (27.34° + 39.22°) = 113.44°$$

$$\frac{237.5}{\sin 113.44°} = \frac{a}{\sin 27.34°}$$

$$a = \frac{237.5 \times \sin 27.34°}{\sin 113.4} \approx 118.9 \text{ pounds}$$

$$\frac{237.5}{\sin 113.44°} = \frac{b}{\sin 39.22°}$$

$$b = \frac{237.5 \times \sin 39.22°}{\sin 113.44°} \approx 163.7 \text{ pounds}$$

21.

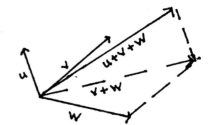

Problem Set 9.3

23. 0

25. a) $\overrightarrow{BD} = \overrightarrow{AD} - \overrightarrow{AB}$

b) $\overrightarrow{AF} = \frac{1}{2}\overrightarrow{AC} = \frac{1}{2}(\overrightarrow{AD} + \overrightarrow{AB})$

c) $\overrightarrow{DE} = \overrightarrow{DC} + \frac{1}{2}\overrightarrow{CB} = \overrightarrow{AB} - \frac{1}{2}\overrightarrow{AD}$

d) $\overrightarrow{AF} - \overrightarrow{DE} = \frac{1}{2}(\overrightarrow{AD} + \overrightarrow{AB}) - (\overrightarrow{AB} - \frac{1}{2}\overrightarrow{AD})$

$\qquad = \overrightarrow{AD} - \frac{1}{2}\overrightarrow{AB}$

27.

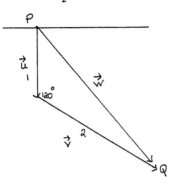

$\|\mathbf{w}\|^2 = 1^2 + 2^2 - 2(1)(2)\cos 120° = 7$

$\|\mathbf{w}\| = \sqrt{7}$

$r = \frac{d}{t} = \frac{\sqrt{7}}{2} \approx 1.32$ mph

29.

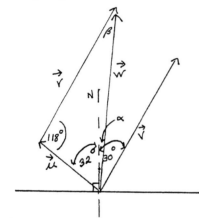

$\|\mathbf{w}\|^2 = \|\mathbf{u}\|^2 + \|\mathbf{v}\|^2 - 2\|\mathbf{u}\|\|\mathbf{v}\|\cos 118°$

$\qquad = 50^2 + 100^2 - 2 \times 50 \times 100 \times \cos 118°$

$\qquad \approx 17195$

$\|\mathbf{w}\| \approx 131.1$ pounds

$\frac{\|\mathbf{u}\|}{\sin \beta} = \frac{\|\mathbf{w}\|}{\sin 118°}$

$\sin \beta = \frac{50 \times \sin 118°}{131.1} \approx .3367$

$\beta = 19.7°$

$\alpha = 30° - \beta = 30° - 19.7° = 10.3°$

The object is moving N 10.3° E.

31. $\|\mathbf{v}\|^2$

$= \|\mathbf{u}\|^2 + \|\mathbf{w}\|^2 - 2\|\mathbf{u}\|\|\mathbf{w}\|\cos168°$

$= 56^2 + 600^2 - 2(56)(600)\cos 168°$

≈ 428867.5

$\|\mathbf{v}\| \approx 654.88$ mph

$\frac{\|\mathbf{w}\|}{\sin \alpha} = \frac{\|\mathbf{v}\|}{\sin 168°}$

$\sin \alpha = \frac{600 \times \sin 168°}{654.88} \approx .1905$

$\alpha = 10.981°$

$\beta = 180° - 168° - 10.981°$

$\qquad = 1.019°$

The speed of the plane must be 654.88 mph, heading N 1.019° W.

33.

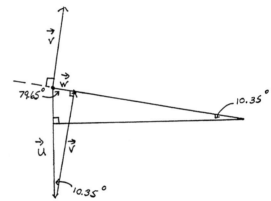

Two forces only act on the car; the downward pull of gravity $\mathbf{u}$ ($\|\mathbf{u}\| = 3625$) and the outward push of the hill $\mathbf{v}$. (We are assuming there is no frictional force.) $\mathbf{v}$ is at right angles to the hill, but its magnitude is unknown. The resultant $\mathbf{w} = \mathbf{u} + \mathbf{v}$ is the force propelling the car down the hill.

$\|\mathbf{w}\| = \|\mathbf{u}\|\sin 10.35° = 3625 \sin 10.35° = 651.3$

A force of 651.3 pounds exerted in the opposite direction of $\mathbf{w}$ would be needed to keep the car from rolling down the hill.

35.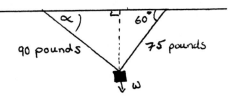

The magnitude of the leftward force must equal the magnitude of the rightward force:

$$90 \cos \alpha = 75 \cos 60°$$
$$\cos \alpha = \frac{75(\frac{1}{2})}{90} = \frac{5}{12}$$
$$\sin \alpha = \sqrt{1 - \left(\frac{5}{12}\right)^2} = \sqrt{\frac{119}{144}} = \frac{\sqrt{119}}{12}$$
$$\alpha = 65.00°$$

The downward force must balance the upward force:

$$w = 90 \sin \alpha + 75 \sin 60°$$
$$= 90\left(\frac{\sqrt{119}}{12}\right) + 75\left(\frac{\sqrt{3}}{2}\right)$$
$$\approx 146.77 \text{ pounds}$$

Problem Set 9.4

1. $3\mathbf{u} - \mathbf{v} = 3(3\mathbf{i} - 4\mathbf{j}) - (5\mathbf{i} + 12\mathbf{j})$
$$= 9\mathbf{i} - 12\mathbf{j} - 5\mathbf{i} - 12\mathbf{j}$$
$$= 4\mathbf{i} - 24\mathbf{j}$$

$$\mathbf{u} \cdot \mathbf{v} = 3(5) + (-4)12 = 15 - 48 = -33$$
$$\cos \theta = \frac{\mathbf{u} \cdot \mathbf{v}}{\|\mathbf{u}\|\|\mathbf{v}\|} = \frac{-33}{\sqrt{9+16}\sqrt{25+144}} = \frac{-33}{5 \cdot 13} = -\frac{33}{65}$$

3. $3\mathbf{u} - \mathbf{v} = 3(2\mathbf{i} - \mathbf{j}) - (3\mathbf{i} - 4\mathbf{j})$
$$= 6\mathbf{i} - 3\mathbf{j} - 3\mathbf{i} + 4\mathbf{j}$$
$$= 3\mathbf{i} + \mathbf{j}$$

$$\mathbf{u} \cdot \mathbf{v} = 2(3) + (-1)(-4) = 6 + 4 = 10$$
$$\cos \theta = \frac{\mathbf{u} \cdot \mathbf{v}}{\|\mathbf{u}\|\|\mathbf{v}\|} = \frac{10}{\sqrt{4+1}\sqrt{9+16}} = \frac{10}{\sqrt{5}(5)} = \frac{2}{\sqrt{5}}$$

5. $\mathbf{u} = 14.1\mathbf{i} + 32.7\mathbf{j}, \ \mathbf{v} = 19.2\mathbf{i} - 13.3\mathbf{j}$
$$\cos \theta = \frac{\mathbf{u} \cdot \mathbf{v}}{\|\mathbf{u}\|\|\mathbf{v}\|}$$
$$= \frac{(14.1)(19.2) + (32.7)(-13.3)}{\sqrt{14.1^2 + 32.7^2}\sqrt{19.2^2 + 13.3^2}}$$
$$\approx -.1974$$
$$\theta \approx 101.39°$$

7. $\mathbf{u} = (6-1)\mathbf{i} + (3-1)\mathbf{j} = 5\mathbf{i} + 2\mathbf{j}$
$$\mathbf{v} = (5-1)\mathbf{i} + (-2-1)\mathbf{j} = 4\mathbf{i} - 3\mathbf{j}$$
$$\mathbf{u} \cdot \mathbf{v} = (5)(4) + (2)(-3) = 20 - 6 = 14$$

9. $\mathbf{u} = (-3-1)\mathbf{i} + (-4-1)\mathbf{j} = -4\mathbf{i} - 5\mathbf{j}$
$$\mathbf{v} = (-5-1)\mathbf{i} + (6-1)\mathbf{j} = -6\mathbf{i} + 5\mathbf{j}$$
$$\mathbf{u} \cdot \mathbf{v} = (-4)(-6) + (-5)(5) = 24 - 25 = -1$$

11.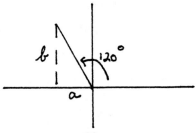

$$a = 10 \cos 120° = -10\left(\frac{1}{2}\right) = -5$$
$$b = 10 \sin 120° = 10\left(\frac{\sqrt{3}}{2}\right) = 5\sqrt{3}$$
$$\mathbf{u} = -5\mathbf{i} + 5\sqrt{3}\mathbf{j}$$

13. $(x\mathbf{i} + \mathbf{j}) \cdot (3\mathbf{i} - 4\mathbf{j}) = 0$
$$3x - 4 = 0$$
$$x = \frac{4}{3}$$

15. $\dfrac{\mathbf{u}}{\|\mathbf{u}\|} = \dfrac{3\mathbf{i} - 4\mathbf{j}}{\sqrt{9+16}} = \dfrac{3\mathbf{i} - 4\mathbf{j}}{5} = \dfrac{3}{5}\mathbf{i} - \dfrac{4}{5}\mathbf{j}$

17. $-2\mathbf{u} = -2(2\mathbf{i} - 5\mathbf{j}) = -4\mathbf{i} + 10\mathbf{j}$

19. Let $\mathbf{u} = a\mathbf{i} + b\mathbf{j}$
$$\mathbf{u} \cdot \mathbf{u} = a \cdot a + b \cdot b = a^2 + b^2 = \|\mathbf{u}\|^2$$

21.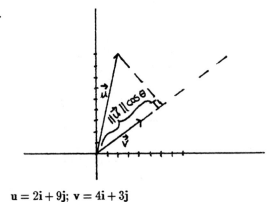

$$\mathbf{u} = 2\mathbf{i} + 9\mathbf{j}; \ \mathbf{v} = 4\mathbf{i} + 3\mathbf{j}$$

The scalar projection of $\mathbf{u}$ on $\mathbf{v}$ is
$$\frac{\mathbf{u} \cdot \mathbf{v}}{\|\mathbf{v}\|} = \frac{(2)(4) + (9)(3)}{\sqrt{16+9}} = \frac{35}{5} = 7$$

23.

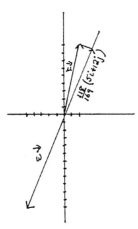

$\mathbf{u} = 2\mathbf{i} + 9\mathbf{j};\ \mathbf{w} = -5\mathbf{i} - 12\mathbf{j}$

The vector projection of $\mathbf{u}$ on $\mathbf{w}$ is

$$\frac{\mathbf{u} \cdot \mathbf{w}}{\mathbf{w} \cdot \mathbf{w}}\mathbf{w} = \frac{2(-5) + 9(-12)}{(-5)^2 + (-12)^2}(-5\mathbf{i} - 12\mathbf{j})$$
$$= \frac{-118}{169}(-5\mathbf{i} - 12\mathbf{j})$$
$$= \frac{118}{169}(5\mathbf{i} + 12\mathbf{j})$$

25.

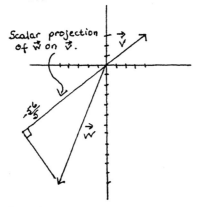

Scalar projection of $\vec{w}$ on $\vec{v}$.

$\mathbf{w} = -5\mathbf{i} - 12\mathbf{j};\ \mathbf{v} = 4\mathbf{i} + 3\mathbf{j}$

The scalar projection of $\mathbf{w}$ on $\mathbf{v}$ is

$$\frac{\mathbf{w} \cdot \mathbf{v}}{\|\mathbf{v}\|} = \frac{(-5)(4) + (-12)(3)}{\sqrt{16 + 9}} = -\frac{56}{5}$$

27. Work $= \mathbf{F} \cdot \mathbf{D} = (3\mathbf{i} + 10\mathbf{j})(10\mathbf{j}) = 100$

29. $\mathbf{F} = (50\cos 45°)\mathbf{i} + (50\sin 45°)\mathbf{j} = 25\sqrt{2}\mathbf{i} + 25\sqrt{2}\mathbf{j}$

$\mathbf{D} = (6 - 1)\mathbf{i} + (9 - 1)\mathbf{j} = 5\mathbf{i} + 8\mathbf{j}$

Work $= \mathbf{F} \cdot \mathbf{D}$
$$= (25\sqrt{2})(5) + (25\sqrt{2})(8)$$
$$= 325\sqrt{2} \text{ dyne-cm}$$

31. $\mathbf{u} = 2\mathbf{i} + 3\mathbf{j};\ \mathbf{v} = 3\mathbf{i} + 4\mathbf{j}$

$3\mathbf{u} - \mathbf{v} = 3(2\mathbf{i} + 3\mathbf{j}) - (3\mathbf{i} + 4\mathbf{j}) = 3\mathbf{i} + 5\mathbf{j}$

$\|3\mathbf{u} - \mathbf{v}\| = \sqrt{3^2 + 5^2} = \sqrt{34}$

33. $\pm(4\mathbf{i} + 3\mathbf{j})$ is perpendicular to $3\mathbf{i} - 4\mathbf{j}$ since

$$\pm(4\mathbf{i} + 3\mathbf{j}) \cdot (3\mathbf{i} - 4\mathbf{j}) = \pm[(4)(3) + (3)(-4)] = 0$$

The required unit vectors are
$$\frac{\pm(4\mathbf{i} + 3\mathbf{j})}{\|4\mathbf{i} + 3\mathbf{j}\|} = \frac{\pm(4\mathbf{i} + 3\mathbf{j})}{\sqrt{4^2 + 3^2}} = \frac{\pm(4\mathbf{i} + 3\mathbf{j})}{5} = \pm\left(\tfrac{4}{5}\mathbf{i} + \tfrac{3}{5}\mathbf{j}\right)$$

35. Let $\mathbf{u} = 5\mathbf{i} + 3\mathbf{j}$ and $\mathbf{v} = 3\mathbf{i} - 4\mathbf{j}$.

The vector projection of $\mathbf{u}$ on $\mathbf{v}$ is
$$\frac{\mathbf{u} \cdot \mathbf{v}}{\mathbf{v} \cdot \mathbf{v}}\mathbf{v} = \frac{5(3) + 3(-4)}{3(3) + (-4)(-4)}(3\mathbf{i} - 4\mathbf{j})$$
$$= \frac{3}{25}(3\mathbf{i} - 4\mathbf{j})$$
$$= \frac{9}{25}\mathbf{i} - \frac{12}{25}\mathbf{j}$$

Let θ be the angle between $\mathbf{u}$ and $\mathbf{v}$.
$$\cos\theta = \frac{\mathbf{u} \cdot \mathbf{v}}{\|\mathbf{u}\|\|\mathbf{v}\|} = \frac{5(3) + 3(-4)}{\sqrt{25 + 9}\sqrt{9 + 16}} = \frac{3}{5\sqrt{34}} \approx .1029$$
$$\theta \approx 84.1°$$

37. $\mathbf{u} = 20\mathbf{i}$

$\mathbf{v} = 15(\cos 30°)\mathbf{i} + 15(\sin 30°)\mathbf{j} = \dfrac{15\sqrt{3}}{2}\mathbf{i} + \dfrac{15}{2}\mathbf{j}$

$\mathbf{w} = 20(\cos 45°)\mathbf{i} + 20(\sin 45°)\mathbf{j} = 10\sqrt{2}\mathbf{i} + 10\sqrt{2}\mathbf{j}$

$\mathbf{p} = 20(\cos 120°)\mathbf{i} + 20(\sin 120°)\mathbf{j} = -10\mathbf{i} + 10\sqrt{3}\mathbf{j}$

$\mathbf{q} = 10\mathbf{j}$

$\mathbf{u} + \mathbf{v} + \mathbf{w} + \mathbf{p} + \mathbf{q}$

$$= \left(20 + \frac{15\sqrt{3}}{2} + 10\sqrt{2} - 10\right)\mathbf{i}$$
$$+ \left(\frac{15}{2} + 10\sqrt{2} + 10\sqrt{3} + 10\right)\mathbf{j}$$

$$\approx 37.1325\mathbf{i} + 48.9626\mathbf{j}$$

39. Let θ measure the angle counterclockwise from the x-axis to the minute hand. Then

$$\theta = -\left(t \min \cdot \frac{2\pi \text{ rad}}{60 \min}\right) = -\frac{\pi t}{30}$$

The minute hand is the vector

$$r(\cos\theta)\mathbf{i} + r(\sin\theta)\mathbf{j} = 3\cos\left(-\frac{\pi t}{30}\right)\mathbf{i} + 3\sin\left(-\frac{\pi t}{30}\right)\mathbf{j}$$
$$= \left(3\cos\frac{\pi t}{30}\right)\mathbf{i} - \left(3\sin\frac{\pi t}{30}\right)\mathbf{j}$$

41. Let θ be the angle between $\mathbf{u}$ and $\mathbf{v}$.

$|\mathbf{u} \cdot \mathbf{v}| = \|\mathbf{u}\|\|\mathbf{v}\||\cos\theta| \le \|\mathbf{u}\|\|\mathbf{v}\|$

since $0 \le |\cos\theta| \le 1$. Equality holds when $\cos\theta = \pm 1$, that is, when $\theta = 0°$ or $180°$.

43.
$$(\mathbf{u} + \mathbf{v}) \cdot (\mathbf{u} - \mathbf{v}) = 0$$
$$\mathbf{u} \cdot \mathbf{u} - \mathbf{v} \cdot \mathbf{v} = 0$$
$$\|\mathbf{u}\|^2 - \|\mathbf{v}\|^2 = 0$$
$$\|\mathbf{u}\|^2 = \|\mathbf{v}\|^2$$
$$\|\mathbf{u}\| = \|\mathbf{v}\|$$

45. $\cos 54° = \sin 36° = \sin(2 \cdot 18) = 2 \sin 18°\cos 18°$

$\cos 54° = \cos(36° + 18°)$
$\quad = \cos 36° \cos 18° - \sin 36° \sin 18°$
$\quad = \cos(2 \cdot 18°)\cos 18° - \sin(2 \cdot 18°)\sin 18°$
$\quad = (1 - 2 \sin^2 18°)\cos 18° - 2 \sin^2 18°\cos 18°$
$\quad = \cos 18°(1 - 4 \sin^2 18°)$

Equating these two expressions,

$2 \sin 18°\cos 18° = \cos 18°(1 - 4 \sin^2 18°)$

Since $\cos 18° \neq 0$, we can divide by it:

$$2 \sin 18° = 1 - 4 \sin^2 18°$$

$$4 \sin^2 18° + 2 \sin 18° - 1 = 0$$

$$\sin 18° = \frac{-2 \pm \sqrt{4 - 4(4)(-1)}}{2(4)}$$

$$= \frac{-2 \pm \sqrt{20}}{8}$$

$$= \frac{-1 \pm \sqrt{5}}{4}$$

Since $\sin 18° > 0$, $\sin 18° = \frac{-1 + \sqrt{5}}{4}$.

47. Set new range values:
Xmin $= 0$
Xmax $= 120$
Ymin $= 0$
Ymax $= 10$

Press ⬛Y=⬛

Enter ☑ ⬛(⬛ $25 - 24$ ⬛COS⬛ ⬛(⬛ π X ⬛÷⬛ 30 ⬛)⬛ ⬛)⬛

Press ⬛GRAPH⬛

We can see two periods in the 120 units shown on the screen. Therefore the period is 60. Zoom in on the highest point of the graph, to see that at that point, $y = 7$. Since the minimum value of y is 1, the amplitude of the graph is $\frac{1}{2}(7 - 1) = 3$.

1. a) $y = \cos t$

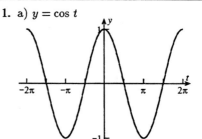

b) $y = \cos 2t$

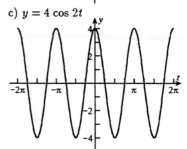

c) $y = 4 \cos 2t$

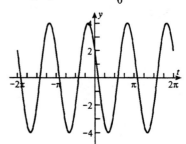

d) $y = 4 \cos\left(2t + \frac{\pi}{3}\right) = 4 \cos 2\left(t + \frac{\pi}{6}\right)$

The graph is shifted $\frac{\pi}{6}$ units to the left.

3. a) $y = 4 \sin 2t$

Amplitude: 4

Period: $\dfrac{2\pi}{2} = \pi$

Phase shift: 0

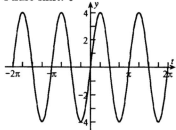

b) $y = 3 \cos\left(t + \dfrac{\pi}{8}\right)$

Amplitude: 3

Period: 2π

Phase shift: $-\dfrac{\pi}{8}$

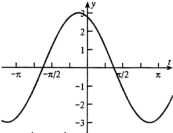

c) $\sin\left(4t + \dfrac{\pi}{8}\right)$

Amplitude: 1

Period: $\dfrac{2\pi}{4} = \dfrac{\pi}{2}$

Phase shift: $\dfrac{-\pi/8}{4} = -\dfrac{\pi}{32}$

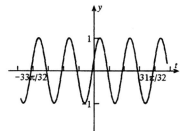

d) $y = 3 \cos\left(3t - \dfrac{\pi}{2}\right)$

Amplitude: 3

Period: $\dfrac{2\pi}{3}$

Phase shift: $\dfrac{\pi/2}{3} = \dfrac{\pi}{6}$

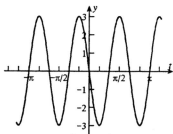

5. $y = 3 + 2 \cos\left(\dfrac{1}{2}t - \dfrac{\pi}{16}\right)$

Amplitude: 2

Period: $\dfrac{2\pi}{1/2} = 4\pi$

Phase shift: $\dfrac{\pi/16}{1/2} = \dfrac{\pi}{8}$

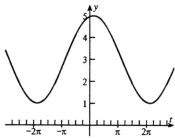

7. $y = -2 \sin 3t$

Amplitude: $|-2| = 2$ Period: $\dfrac{2\pi}{3}$

Phase shift: 0

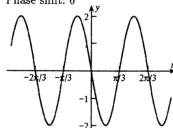

9. $y = -\sin\left(2t + \dfrac{\pi}{3}\right)$

Amplitude: $|-1| = 1$ Period: $\dfrac{2\pi}{2} = \pi$

Phase shift: $-\dfrac{\pi/3}{2} = -\dfrac{\pi}{6}$

11. $y = -2\cos\left(t - \frac{1}{6}\right)$

Amplitude: $|-2| = 2$

Period: 2π

Phase shift: $\frac{1}{6}$

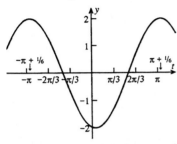

13.

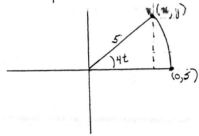

After t seconds, the hole has rotated through an angle of $4t$ radians.

$$(x, y) = (5\cos 4t, 5\sin 4t)$$

15. From Problem 13, we know that at time t the hole is at position $(5\cos 4t, 5\sin 4t)$. The point P will always have the same x-coordinate as the hole, and will always be 8 units below it. The coordinates of P are

$$(5\cos 4t, -8 + 5\sin 4t)$$

17. a) $y = \sin 5t$

Amplitude: 1

Period: $\frac{2\pi}{5}$

Phase shift: 0

b) $y = \frac{3}{2}\cos\frac{1}{2}t$

Amplitude: $\frac{3}{2}$

Period: $\frac{2\pi}{1/2} = 4\pi$

Phase shift: 0

c) $y = 2\cos(4t - \pi)$

Amplitude: 2

Period: $\frac{2\pi}{4} = \frac{\pi}{2}$

Phase shift: $\frac{\pi}{4}$

d) $y = -4\sin\left(3t + \frac{3\pi}{4}\right)$

Amplitude: $|-4| = 4$

Period: $\frac{2\pi}{3}$

Phase shift: $\dfrac{-3\pi/4}{3} = -\frac{\pi}{4}$

19. $y = 8 + 4\cos\left(\frac{\pi}{2}t + \frac{\pi}{4}\right)$

The closest the weight gets to the ceiling is 4 feet. This happens when

$$\cos\left(\frac{\pi}{2}t + \frac{\pi}{4}\right) = -1$$
$$\frac{\pi}{2}t + \frac{\pi}{4} = \pi$$
$$\frac{\pi}{2}t = \frac{3\pi}{4}$$
$$t = \frac{3\pi}{4} \cdot \frac{2}{\pi} = \frac{3}{2} = 1.5 \text{ sec}$$

21. After t seconds, P has rotated through an angle of t radians. The coordinates of P are $(\cos t, \sin t)$. Let $R = (0, \sin t)$ be the point on the y-axis at the same height as P. Then PQR is a right triangle.

$$\overline{RQ} = \sqrt{25 - \cos^2 t}$$

Therefore the y-coordinate of Q is

$$\sin t + \sqrt{25 - \cos^2 t}$$

23. After t seconds, $E = 156\sin 110\pi t$. The maximum voltage drop is the amplitude of the sine curve: 156. The period of the curve is $\frac{2\pi}{110\pi} = \frac{1}{55}$ sec, so 55 cycles occur in one second.

25. $y(t) = 55[1 + .02\sin 2400\pi t]\sin(2 \times 10^5\pi t)$

a) $\sin(2 \times 10^5 \cdot 3\pi) = \sin(6 \times 10^5\pi) = 0$, so $y(3) = 0$.

b) $\sin(2 \times 10^5 \cdot .03216\pi) = \sin(2 \times 3216\pi) = 0$, so $y(.03216) = 0$

c) $y(.0000321)$
$= 55[1 + .02\sin(2400 \times .000321\pi)]\sin(2 \times 3.21\pi)$
$\approx 55 \times 1.004794 \times .9686$
≈ 53.53

27. $R = 1000 + 150\sin 2t$

Amplitude: 150

Period: $\frac{2\pi}{2} = \pi$

Phase shift: 0

$C = 200 + 50\sin(2t - .7)$

Amplitude: 50

Period: $\frac{2\pi}{2} = \pi$

Phase shift: $\frac{.7}{2} = .35$

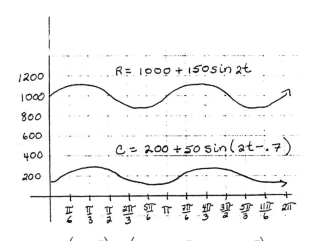

29. a) $4\sin\left(2t - \frac{\pi}{4}\right) = 4\left(\sin 2t \cos\frac{\pi}{4} - \cos 2t \sin\frac{\pi}{4}\right)$

$\qquad = 4 \cdot \frac{\sqrt{2}}{2}\sin 2t - 4 \cdot \frac{\sqrt{2}}{2}\cos 2t$

$\qquad = 2\sqrt{2}\sin 2t - 2\sqrt{2}\cos 2t$

b) $3\cos\left(3t + \frac{\pi}{3}\right) = 3\left(\cos 3t \cos\frac{\pi}{3} - \sin 3t \sin\frac{\pi}{3}\right)$

$\qquad = 3 \cdot \frac{1}{2}\cos 3t - 3 \cdot \frac{\sqrt{3}}{2}\sin 3t$

$\qquad = -\frac{3\sqrt{3}}{2}\sin 3t + \frac{3}{2}\cos 3t$

31. Let $A = \sqrt{A_1{}^2 + A_2{}^2}$ and let $P(C) = \left(\frac{A_1}{A}, \frac{A_2}{A}\right)$.

Then $\cos C = \frac{A_1}{A}$ and $\sin C = \frac{A_2}{A}$. Because A_1 and

A_2 are positive, $C = \tan^{-1}\frac{A_2}{A_1}$. Also

$A_1 \sin Bt + A_2 \cos Bt$

$= A\left(\frac{A_1}{A}\sin Bt + \frac{A_2}{A}\cos Bt\right)$

$= A(\sin Bt \cos C + \cos Bt \sin C)$

$= A \sin(Bt + C)$

33. We practice the method.

a) $A = \sqrt{3^2 + 4^2} = 5$, $\cos C = \frac{3}{5}$, $\sin C = \frac{4}{5}$,

$C = \tan^{-1}\left(\frac{4}{3}\right)$

$4\cos 2t + 3\sin 2t = 5\left(\frac{3}{5}\sin 2t + \frac{4}{5}\cos 2t\right)$

$\qquad = 5(\sin 2t \cos C + \cos 2t \sin C)$

$\qquad = 5\sin(2t + C)$

$\qquad = 5\sin\left[2t + \tan^{-1}\left(\frac{4}{3}\right)\right]$

b) $A = \sqrt{9 + 3} = \sqrt{12} = 2\sqrt{3}$

$\cos C = \frac{3}{2\sqrt{3}} = \frac{\sqrt{3}}{2}$, $\sin C = \frac{-\sqrt{3}}{2\sqrt{3}} = -\frac{1}{2}$, $C = \frac{11\pi}{6}$

$3\sin 4t - \sqrt{3}\cos 4t$

$= 2\sqrt{3}\left(\frac{3}{2\sqrt{3}}\sin 4t - \frac{\sqrt{3}}{2\sqrt{3}}\cos 4t\right)$

$= 2\sqrt{3}\left(\sin 4t \cos\frac{11\pi}{6} + \cos 4t \sin\frac{11\pi}{6}\right)$

$= 2\sqrt{3}\sin\left(4t + \frac{11\pi}{6}\right)$

35. $\cos t \pm \sin t = \sqrt{2}\left(\frac{1}{\sqrt{2}}\cos t \pm \frac{1}{\sqrt{2}}\sin t\right)$

$\qquad = \sqrt{2}\cos\left(t \mp \frac{\pi}{4}\right)$

The maximum is $\sqrt{2}$ and the minimum is $-\sqrt{2}$.

37. Set new range values:
Xmin = 0
Xmax = 32
Enter 3 [SIN] [(] .5X + π [÷] 8 [)] as Y_1

$\quad$ −5 [COS] [(] .25X − .1 [)] as Y_2

Press [GRAPH]
Zooming in on the leftmost point of intersection, we find $x \approx 6.039$. Zooming in on the rightmost point of intersection, we find $x \approx 21.724$. The period of $y = -5\cos\left(\frac{1}{4}t - \frac{1}{10}\right)$ is $\frac{2\pi}{1/4} = 8\pi$. The period of $y = 3\sin\left(\frac{1}{2}t + \frac{\pi}{8}\right)$ is $\frac{2\pi}{1/2} = 4\pi$. The graph of the two functions will repeat every 8π units. Thus the complete solution set of the equation is

$\{6.039 + 8\pi k, \ 21.724 + 8\pi k, \text{ where } k \text{ is any integer}\}$

Problem Set 9.6

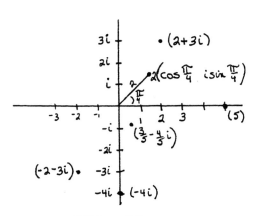

13. 1) $|2 + 3i| = \sqrt{4 + 9} = \sqrt{13}$

$\quad$ 3) $|-2 - 3i| = \sqrt{4 + 9} = \sqrt{13}$

$\quad$ 5) $|5| = 5$

$\quad$ 7) $|-4i| = 4$

9) $\left|\frac{3}{5} - \frac{4}{5}i\right| = \sqrt{\frac{9}{25} + \frac{16}{25}} = 1$

11) $\left|2\left(\cos \frac{\pi}{4} + i \sin \frac{\pi}{4}\right)\right| = 2$

Note: $r = 2$

15. $4\left(\cos \frac{3\pi}{2} + i \sin \frac{3\pi}{2}\right) = 4(0 - i) = -4i$

17. $2(\cos 225° + i \sin 225°) = 2\left(-\frac{\sqrt{2}}{2} - \frac{\sqrt{2}}{2}i\right)$
$= -\sqrt{2} - \sqrt{2}i$

19. $r = 4,\ \theta = \pi$
$-4 = 4(\cos \pi + i \sin \pi)$

21. $r = 5,\ \theta = 270°$
$-5i = 5(\cos 270° + i \sin 270°)$

23. $r = \sqrt{(2)^2 + (-2)^2} = \sqrt{8} = 2\sqrt{2}$
$2 - 2i$ is in quadrant IV;
$\cos \theta = \frac{2}{2\sqrt{2}} = \frac{1}{\sqrt{2}} = \frac{\sqrt{2}}{2}$
$\theta = 315°$
$2 - 2i = 2\sqrt{2}(\cos 315° + i \sin 315°)$

25. $r = \sqrt{(2\sqrt{3})^2 + (2)^2} = \sqrt{12 + 4} = \sqrt{16} = 4$
$2\sqrt{3}i$ is in quadrant I;
$\cos \theta = \frac{2\sqrt{3}}{4} = \frac{\sqrt{3}}{2}$
$\theta = 30° = \frac{\pi}{6}$
$2\sqrt{3} + 2i = 4\left(\cos \frac{\pi}{6} + i \sin \frac{\pi}{6}\right)$

27. $r = \sqrt{5^2 + 4^2} = \sqrt{41} \approx 6.403$
$5 + 4i$ is in quadrant I;
$\cos \theta = \frac{5}{\sqrt{41}} \approx .781$
$\theta \approx 38.66° \approx .675$ rad
$5 + 4i = 6.403(\cos .675 + i \sin .675)$

29. $uv = 2 \cdot 3[\cos(140° + 70°) + i \sin(140° + 70°]$
$= 6[\cos 210° + i \sin 210°]$

31. $vw = 3 \cdot \frac{1}{2}[\cos(70° + 55°) + i \sin(70° + 55°)]$
$= \frac{3}{2}[\cos 125° + i \sin 125°]$

33. $\frac{u}{v} = \frac{2}{3}[\cos(140° - 70°) + i \sin(140° - 70°)]$
$= \frac{2}{3}[\cos 70° + i \sin 70°]$

35. $\frac{1}{w} = \frac{1[\cos 360° + i \sin 360°}{\frac{1}{2}[\cos 55° + i \sin 55°]}$
$= 2[\cos(360° - 55°) + i \sin(360° - 55°)]$
$= 2[\cos 305° + i \sin 305°]$

37. $(4 - 4i)(2 + 2i) = (8 + 8) + (8 - 8)i = 16$

$4 - 4i$ is in quadrant IV. Change to polar form:
$r = 4\sqrt{1^2 + (-1)^2} = 4\sqrt{2}$
$\cos \theta = \frac{4}{4\sqrt{2}} = \frac{1}{\sqrt{2}}$
$\theta = 315°$

$2 + 2i$ is in quadrant I. Change to polar form:
$r = 2\sqrt{1^2 + 1^2} = 2\sqrt{2}$
$\cos \theta = \frac{2}{2\sqrt{2}} = \frac{1}{\sqrt{2}}$
$\theta = 45°$

$2 + 2i = 2\sqrt{2}(\cos 45° + i \sin 45°)$

$(4 - 4i)(2 + 2i)$
$= (4\sqrt{2})(2\sqrt{2})[\cos(315° + 45°) + i \sin(315° + 45°)]$
$= 16[\cos 360° + i \sin 360°]$
$= 16[1 + 0i]$
$= 16$

39. $(1 + \sqrt{3}i)(1 + \sqrt{3}i) = (1 - 3) + (\sqrt{3} + \sqrt{3})i$
$= -2 + 2\sqrt{3}i$

$1 + \sqrt{3}i$ is in quadrant I. Change to polar form:
$r = \sqrt{1 + 3} = 2$
$\cos \theta = \frac{1}{2}$
$\theta = 60°$

$1 + \sqrt{3}i = 2(\cos 60° + i \sin 60°)$
$(1 + \sqrt{3}i)(1 + \sqrt{3}i)$
$= 2 \cdot 2[\cos(60° + 60°) + i \sin(60° + 60°)]$
$= 4[\cos 120° + i \sin 120°]$
$= 4\left[-\frac{1}{2} + i\frac{\sqrt{3}}{2}\right]$
$= -2 + 2\sqrt{3}i$

41. Change $4i$ to polar form:
$r = 4 \qquad \theta = 90°$
$4i = 4[\cos 90° + i \sin 90°]$

$2\sqrt{3} - 2i$ is in quadrant IV. Change to polar form:
$r = \sqrt{(2\sqrt{3})^2 + (-2)^2} = \sqrt{12 + 4} = \sqrt{16} = 4$
$\cos \theta = \frac{2\sqrt{3}}{4} = \frac{\sqrt{3}}{2}$
$\theta = 330°$
$2\sqrt{3} - 2i = 4[\cos 330° + i \sin 330°]$

$2\sqrt{3} - 2i = 4[\cos 330° + i \sin 330°]$

$4i(2\sqrt{3} - 2i)$

$\quad = 4 \cdot 4[\cos(90° + 330°) + i \sin(90° + 330°)]$

$\quad = 16[\cos 420° + i \sin 420°]$

$\quad = 16[\cos 60° + i \sin 60°]$

43. From Problem 41, we know

$4i = 4[\cos 90° + i \sin 90°]$

$2\sqrt{3} - 2i = 4[\cos 330° + i \sin 330°]$

$\dfrac{4i}{2\sqrt{\ } - 2i} = \dfrac{4}{4}[\cos(90° - 330°) + i \sin (90° - 330°)]$

$\quad = 1[\cos(-240°) + i \sin(-240°)]$

$\quad = 1[\cos 120° + i \sin 120°]$

45. $2\sqrt{2} - 2\sqrt{2}i$ is in quadrant IV. Change to polar form:

$r = \sqrt{(2\sqrt{2})^2 + (-2\sqrt{2})^2} = \sqrt{8 + 8} = 4$

$\cos \theta = \dfrac{2\sqrt{2}}{4} = \dfrac{\sqrt{2}}{2}$

$\theta = 315°$

$2\sqrt{2} - 2\sqrt{2}\mathbf{i} = 4[\cos 315° + \mathbf{i} \sin 315°]$

$(2\sqrt{2} - 2\sqrt{2}i)(2\sqrt{2} - 2\sqrt{2}i)$

$\quad = 4 \cdot 4[\cos(315° + 315°) + i \sin(315° + 315°)]$

$\quad = 16[\cos 630° + i \sin 630°]$

$\quad = 16[\cos 270° + i \sin 270°]$

47.

a) $|-5 + 12i| = \sqrt{(-5)^2 + 12^2} = \sqrt{169} = 13$

b) $|-4i| = 4$

c) $|5(\cos 60° + i \sin 60°)| = 5$

Note: $r = 5$

49. a) $\quad r = 12 \qquad \theta = 0$

$\quad 12 = 12(\cos 0 + i \sin 0)$

b) $-\sqrt{2} + \sqrt{2}i$ is in quadrant II.

$r = \sqrt{(-\sqrt{2})^2 + (\sqrt{2})^2} = \sqrt{2 + 2} = 2$

$\cos \theta = \dfrac{-\sqrt{2}}{2}$

$\theta = 135°$

$-\sqrt{2} + \sqrt{2}i = 2(\cos 135° + i \sin 135°)$

c) $\quad r = 3 \qquad \theta = 270°$

$\quad -3i = 3(\cos 270° + i \sin 270°)$

d) $2 - 2\sqrt{3}i$ is in quadrant IV.

$r = \sqrt{(2)^2 + (-2\sqrt{3})^2} = \sqrt{4 + 12} = 4$

$\cos \theta = \dfrac{2}{4} = \dfrac{1}{2}$

$\theta = 300°$

$2 - 2\sqrt{3}i = 4(\cos 300° + i \sin 300°)$

e) $4\sqrt{3} + 4i$ is in quadrant I.

$r = \sqrt{(4\sqrt{3})^2 + 4^2} = \sqrt{48 + 16} = 8$

$\cos \theta = \dfrac{4\sqrt{3}}{8} = \dfrac{\sqrt{3}}{2}$

$\theta = 30°$

$4\sqrt{3} + 4i = 8(\cos 30° + i \sin 30°)$

f) $\quad 2(\cos 45° - i \sin 45°) = 2[\cos(-45°) + i \sin(-45°)]$

$\quad\quad\quad\quad\quad\quad\quad\quad\quad = 2[\cos 315° + i \sin 315°]$

51. a) The result is

$1.5 \cdot 4 \cdot 2[\cos(110° + 30° + 20°) + i \sin(110° + 30° + 20°)]$

$= 12[\cos 160° + i \sin 160°]$

b) The result is

$\dfrac{12}{4}[\cos(115° - 55° - 20°) + i \sin(115° - 55° - 20°)]$

$= 3[\cos 40° + i \sin 40°]$

c) From the results of Problem 49, we can express the quotient in polar form:

$\dfrac{2(\cos 135° + i \sin 135°)4(\cos 300° + i \sin 300°)}{8(\cos 30° + i \sin 30°)}$

$= \dfrac{8}{8}[\cos(135° + 300° - 30°) + i \sin(135° + 300° - 30°)]$

$= \cos 405° + i \sin 405°$

$= \cos 45° + i \sin 45°$

Problem Set 9.6

53. a) Let $z = a + 5i$.
$$\sqrt{a^2 + 25} = 13$$
$$a^2 + 25 = 169$$
$$a^2 = 144$$
$$a = \pm 12$$
$z = 12 + 5i$ or $z = -12 + 5i$

b) If z is in quadrant I,
$$z = 8(\cos 45° + i \sin 45°)$$
$$= 8\left(\frac{\sqrt{2}}{2} + \frac{\sqrt{2}}{2}i\right)$$
$$= 4\sqrt{2} + 4\sqrt{2}$$

If z is in quadrant III, then
$$z = -4\sqrt{2} - 4\sqrt{2}i$$

55. $u = r(\cos \theta + i \sin \theta)$

a) u^3
$$= r(\cos \theta + i \sin \theta) \cdot r(\cos \theta + i \sin \theta)$$
$$\cdot r(\cos \theta + i \sin \theta)$$
$$= r^3(\cos 3\theta + i \sin 3\theta)$$

b) $\bar{u} = r(\cos \theta - i \sin \theta) = r[\cos(-\theta) + i \sin(-\theta)]$

c) $u\bar{u} = |u|^2 = r^2 = r^2[\cos 0 + i \sin 0]$

d) $\frac{1}{u} = \frac{1(\cos 0 + i \sin 0)}{r(\cos \theta + i \sin \theta)} = \frac{1}{r}[\cos(-\theta) + i \sin(-\theta)]$

e) u^{-2}
$$= \left(\frac{1}{u}\right)\left(\frac{1}{u}\right)$$
$$= \frac{1}{r}[\cos(-\theta) + i \sin(-\theta)] \cdot \frac{1}{r}[\cos(-\theta) + i \sin(-\theta)]$$
$$= \frac{1}{r^2}[\cos(-2\theta) + i \sin(-2\theta)]$$

f) $-u = r(-\cos \theta - i \sin \theta)$
$$= r[\cos(\theta + \pi) + i \sin(\theta + \pi)]$$

57. As in Problem 55a,
$$\frac{U^2}{V^3} = \frac{r^2[\cos 2\alpha + i \cos 2\alpha]}{s^3[\cos 3\beta + i \sin 3\beta]}$$
$$= \frac{r^2}{s^3}[\cos(2\alpha - 3\beta) + i \sin(2\alpha - 3\beta)]$$

59. a) Let $U = a + bi$ and $V = c + di$. Then
$$|U - V| = |(a - c) + (b - d)i| = \sqrt{(a - c)^2 + (b - d)^2}$$
which is the distance between U and V in the complex plane.

b) The angle of $U - V$ is the angle from the positive x-axis to the line joining U and V.

61. a) $z_1 z_2 z_3 \cdots z_8$
$$= 2^8\left[\cos\left(\frac{\pi}{4} + \frac{2\pi}{4} + \cdots + \frac{8\pi}{4}\right)\right.$$
$$\left. + i \sin\left(\frac{\pi}{4} + \frac{2\pi}{4} + \cdots + \frac{8\pi}{4}\right)\right]$$
$$= 2^8\left[\cos \frac{\pi}{4}(1 + 2 + 3 + \cdots + 8)\right.$$
$$\left. i \sin\frac{\pi}{4}(1 + 2 + 3 + \cdots + 8)\right]$$
$$= 2^8\left[\cos \frac{36\pi}{4} + i \sin \frac{36\pi}{4}\right]$$
$$= 2^8[\cos 9\pi + i \sin 9\pi]$$
$$= 2^8[-1 + 0]$$
$$= -256$$

b) From the graph, we see that we can interpret z_1, z_2, ..., z_8 as uniformly distributed vectors of equal magnitude. We conclude that their sum is 0.

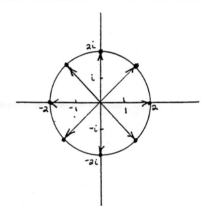

Problem Set 9.7

1. $\left[2\left(\cos \frac{\pi}{4} + i \sin \frac{\pi}{4}\right)\right]^3 = 2^3\left(\cos \frac{3\pi}{4} + i \sin \frac{3\pi}{4}\right)$
$$= 8\left(\cos \frac{3\pi}{4} + i \sin \frac{3\pi}{4}\right)$$

3. $\left[\sqrt{5}\left(\cos 11° + i \sin 11°\right)\right]^6$
$$= (\sqrt{5})^6(\cos 66° + i \sin 66°)$$
$$= 125(\cos 66° + i \sin 66°)$$

5. Find the polar form of $1 + i$.
$$r = \sqrt{1 + 1} = \sqrt{2}$$
$$\cos \theta = \frac{1}{\sqrt{2}}$$
$$\theta = 45°$$
$$(1 + i)^8 = [\sqrt{2}(\cos 45° + i \sin 45°)]^8$$
$$= (\sqrt{2})^8(\cos 360° + i \sin 360°)$$
$$= 16(\cos 0 + i \sin 0)$$

196

7. $(\cos 36° + i \sin 36°)^{10} = (\cos 360° + i \sin 360°)$
$$= 1 + 0i$$

9. Find the polar form of $\sqrt{3} + i$:
$$r = \sqrt{(\sqrt{3})^2 + 1^2} = \sqrt{3+1} = 2$$
$$\cos \theta = \frac{\sqrt{3}}{2}$$
$$\theta = 30°$$
$$(\sqrt{3} + i)^5 = [2(\cos 30° + i \sin 30°)]^5$$
$$= 2^5 (\cos 150° + i \sin 150°)$$
$$= 32\left(-\frac{\sqrt{3}}{2} + \frac{1}{2}i\right)$$
$$= -16\sqrt{3} + 16i$$

11. $u = 125(\cos 45° + i \sin 45°)$

The 3 third roots of u are:
$$u_0 = \sqrt[3]{125}\left[\cos\left(\frac{45°}{3}\right) + i \sin\left(\frac{45°}{3}\right)\right]$$
$$= 5(\cos 15° + i \sin 15°)$$
$$u_1 = 5\left[\cos\left(\frac{45° + 360°}{3}\right) + i \sin\left(\frac{45° + 360°}{3}\right)\right]$$
$$= 5(\cos 135° + i \sin 135°)$$
$$u_2 = 5\left[\cos\left(\frac{45° + 720°}{3}\right) + i \sin\left(\frac{45° + 720°}{3}\right)\right]$$
$$= 5(\cos 255° + i \sin 255°)$$

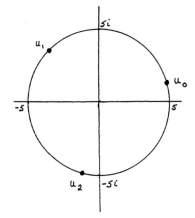

13. $u = 64\left(\cos \frac{\pi}{2} + i \sin \frac{\pi}{2}\right)$

The first sixth root of u is:
$$u_0 = \sqrt[6]{64}\left(\cos \frac{1}{6} \cdot \frac{\pi}{2} + i \sin \frac{1}{6} \cdot \frac{\pi}{2}\right)$$
$$= 2\left(\cos \frac{\pi}{12} + i \sin \frac{\pi}{12}\right)$$

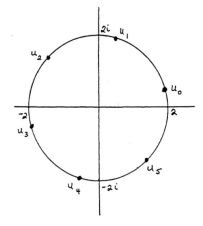

We increase θ successively by $\frac{2\pi}{6} = \frac{4\pi}{12}$ to find the other 5 roots.

$$\theta_1 = \frac{\pi}{12} + \frac{4\pi}{12} = \frac{5\pi}{12}$$
$$u_1 = 2\left(\cos \frac{5\pi}{12} + i \sin \frac{5\pi}{12}\right)$$
$$\theta_2 = \frac{5\pi}{12} + \frac{4\pi}{12} = \frac{9\pi}{12} = \frac{3\pi}{4}$$
$$u_2 = 2\left(\cos \frac{3\pi}{4} + i \sin \frac{3\pi}{4}\right)$$
$$\theta_3 = \frac{9\pi}{12} + \frac{4\pi}{12} = \frac{13\pi}{12}$$
$$u_3 = 2\left(\cos \frac{13\pi}{12} + i \sin \frac{13\pi}{12}\right)$$
$$\theta_4 = \frac{13\pi}{12} + \frac{4\pi}{12} = \frac{17\pi}{12}$$
$$u_4 = 2\left(\cos \frac{17\pi}{12} + i \sin \frac{17\pi}{12}\right)$$
$$\theta_5 = \frac{17\pi}{12} + \frac{4\pi}{12} = \frac{21\pi}{12} = \frac{7\pi}{4}$$
$$u_5 = 2\left(\cos \frac{7\pi}{4} + i \sin \frac{7\pi}{4}\right)$$

15. $u = 4(\cos 112° + i \sin 112°)$

The first fourth root of u is
$$u_0 = \sqrt[4]{4}\left(\cos \frac{112°}{4} + i \sin \frac{112°}{4}\right)$$
$$= \sqrt{2}(\cos 28° + i \sin 28°)$$

We increase θ successively by $\frac{360°}{4} = 90°$ to find the other 3 roots.

$$u_1 = \sqrt{2}(\cos 118° + i \sin 118°)$$
$$u_2 = \sqrt{2}(\cos 208° + i \sin 208°)$$
$$u_3 = \sqrt{2}(\cos 298° + i \sin 298°)$$

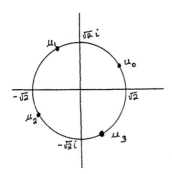

17. $u = 16 = 16(\cos 0° + i \sin 0°)$

The first fourth root of u is

$u_0 = \sqrt[4]{16}\left(\cos \frac{0°}{4} + i \sin \frac{0°}{4}\right) = 2(\cos 0° + i \sin 0°) = 2$

We increase θ successively by $\frac{360°}{4} = 90°$ to find the other 3 roots.

$u_1 = 2(\cos 90° + i \sin 90°) = 2(0 + 1i) = 2i$

$u_2 = 2(\cos 180° + i \sin 180°) = 2(-1 + 0i) = -2$

$u_3 = 2(\cos 270° + i \sin 270°) = 2(0 - 1i) = -2i$

19. $u = 4i = 4(\cos 90° + i \sin 90°)$

The 2 square roots of u are:

$u_0 = \sqrt{4}\left(\cos \frac{90°}{2} + i \sin \frac{90°}{2}\right)$

$= 2(\cos 45° + i \sin 45°)$

$= 2\left(\frac{\sqrt{2}}{2} + i \frac{\sqrt{2}}{2}\right)$

$= \sqrt{2} + \sqrt{2}i$

$u_1 = -u_0 = -\sqrt{2} - \sqrt{2}i$

21. $u = -4 + 4\sqrt{3}i$

Find the polar form of u:

$r = \sqrt{(-4)^2 + (4\sqrt{3})^2} = \sqrt{16 + 48} = \sqrt{64} = 8$

θ is in quadrant II

$\cos \theta = \frac{-4}{8} = -\frac{1}{2}$

$\theta = 120°$

$u = 8(\cos 120° + i \sin 120°)$

The 2 square roots of u are:

$u_0 = \sqrt{8}\left(\cos \frac{120°}{2} + i \sin \frac{120°}{2}\right)$

$= 2\sqrt{2}(\cos 60° + i \sin 60°)$

$= 2\sqrt{2}\left(\frac{1}{2} + \frac{\sqrt{3}}{2}i\right)$

$= \sqrt{2} + \sqrt{6}i$

$u_1 = -u_0 = -\sqrt{2} - \sqrt{6}i$

23. $u = 1 = \cos 0° + i \sin 0°$

The first fourth root of u is

$u_0 = \cos 0° + i \sin 0° = 1$

We increase θ successively by $\frac{360°}{4} = 90°$ to find the other 3 roots.

$u_1 = \cos 90° + i \sin 90° = i$

$u_2 = \cos 180° + i \sin 180° = -1$

$u_3 = \cos 270° + i \sin 270° = -i$

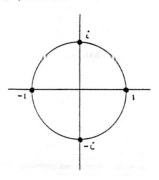

25. $u = 1 = \cos 0° + i \sin 0°$

The first tenth root of u is

$u_0 = \cos 0° + i \sin 0° = 1$

We increase θ successively by $\frac{360°}{10} = 36°$ to find the other 9 roots.

$u_1 = \cos 36° + i \sin 36°$
$u_2 = \cos 72° + i \sin 72°$
$u_3 = \cos 108° + i \sin 108°$
$u_4 = \cos 144° + i \sin 144°$
$u_5 = \cos 180° + i \sin 180°$
$u_6 = \cos 216° + i \sin 216°$
$u_7 = \cos 252° + i \sin 252°$
$u_8 = \cos 288° + i \sin 288°$
$u_9 = \cos 324° + i \sin 324°$

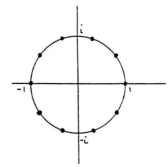

Problem Set 9.7

27. a) $[3(\cos 20° + i \sin 20°)]^4$

$= 3^4[\cos(4 \cdot 20°) + i \sin(4 \cdot 20°)]$

$= 81(\cos 80° + i \sin 80°)$

b) $[2.46(\cos 1.54 + i \sin 1.54)]^5$

$= 2.46^5[\cos 5(1.54) + i \sin 5(1.54)]$

$\approx 90.09[\cos 7.70 + i \sin 7.70]$

c) $[2(\cos 50° + i \sin 50°)(\cos 30° + i \sin 30°)]^3$

$= [2(\cos 80° + i \sin 80°0)]^3$

$= 2^3[\cos(3 \cdot 80°) + i \sin(3 \cdot 80°)]$

$= 8[\cos 240° + i \sin 240°\}$

d) $\left[\dfrac{8\left(\cos \frac{2\pi}{3} + i \sin \frac{2\pi}{3}\right)}{4\left(\cos \frac{\pi}{4} + i \sin \frac{\pi}{4}\right)}\right]^4$

$= \left\{2\left[\cos\left(\frac{2\pi}{3} - \frac{\pi}{4}\right) + i \sin\left(\frac{2\pi}{3} - \frac{\pi}{4}\right)\right]\right\}^4$

$= 2^4\left[\cos 4\left(\frac{5\pi}{12}\right) + i \sin 4\left(\frac{5\pi}{12}\right)\right]$

$= 16\left(\cos \frac{5\pi}{3} + i \sin \frac{5\pi}{3}\right)$

29. $u = 32(\cos 255° + i \sin 255°)$

The first fifth root of u is

$u_0 = \sqrt[5]{32}\left(\cos \frac{255°}{5} + i \sin \frac{255°}{5}\right)$

$= 2(\cos 51° + i \sin 51°)$

We increase θ successively by $\frac{360°}{5} = 72°$ to find the other 4 roots.

$u_1 = 2(\cos 123° + i \sin 123°)$

$u_2 = 2(\cos 195° + i \sin 195°)$

$u_3 = 2(\cos 267° + i \sin 267°)$

$u_4 = 2(\cos 339° + i \sin 339°)$

31. $u = 1 = \cos 0° + i \sin 0°$

The first eighth root of u is

$u_0 = \cos 0° + i \sin 0° = 1$

We increase θ successively by $\frac{360°}{8} = 45°$ to find the other 7 roots.

$u_1 = \cos 45° + i \sin 45° = \frac{\sqrt{2}}{2} + \frac{\sqrt{2}}{2}i$

$u_2 = \cos 90° + i \sin 90° = 0 + i$

$u_3 = \cos 135° + i \sin 135° = -\frac{\sqrt{2}}{2} + \frac{\sqrt{2}}{2}i$

$u_4 = \cos 180° + i \sin 180° = -1 + 0i$

$u_5 = \cos 225° + i \sin 225° = -\frac{\sqrt{2}}{2} - \frac{\sqrt{2}}{2}i$

$u_6 = \cos 270° + i \sin 270° = 0 - i$

$u_7 = \cos 315° + i \sin 315° = \frac{\sqrt{2}}{2} - \frac{\sqrt{2}}{2}i$

Sum $= 0$

Product $= \cos 1260° + i \sin 1260°$

$= \cos 180° + i \sin 180°$

$= -1$

Looking ahead to to Section 10-2, Problem 69, these are the roots of $x^8 - 1 = 0$, $n = 8$, $a_8 = 1$, $a_7 = 0$, $a_0 = -1$. The sum is $-a_7/a_0 = 0$ and the product is $(-1)^8 a_0/a_8 = -1$.

33. $x^5 + \sqrt{2} - \sqrt{2}i = 0$

$x^5 = -\sqrt{2} + \sqrt{2}i$

$x = (-\sqrt{2} + \sqrt{2}i)^{1/5}$

To find the fifth roots of $-\sqrt{2} + \sqrt{2}i$, which is in quadrant II, change to polar form:

$r = \sqrt{(-\sqrt{2})^2 + (\sqrt{2})^2} = \sqrt{4} = 2$

$\cos \theta = \frac{-\sqrt{2}}{2}$

$\theta = 135°$

$\theta_0 = \frac{135°}{5} = 27°$, and increases by $\frac{360°}{5} = 72°$.

$\theta_1 = 99°$, $\theta_2 = 171°$, $\theta_3 = 243°$, and $\theta_4 = 315°$.

The root with the largest real part is the one with the largest value of $\cos \theta$, that is, with θ in quadrants I or IV and having the smallest reference angle. Thus, the root with the largest real part is

$\sqrt[5]{2}(\cos 27° + i \sin 27°) \approx 1.0235 + .5215i$

The exact value of $\cos 27°$ can be obtained from the result of Problem 45 of Section 9-4.

35. Method I.

$$x^6 - 1 = 0$$

$$(x^3)^2 - 1 = 0$$

$$(x^3 + 1)(x^3 - 1) = 0$$

$$(x + 1)(x^2 - x + 1)(x - 1)(x^2 + x + 1) = 0$$

Set each factor to 0.

$x + 1 = 0 \Rightarrow x = -1$

$x - 1 = 0 \Rightarrow x = 1$

$x^2 - x + 1 = 0 \Rightarrow x = \frac{1 \pm \sqrt{1 - 4}}{2} = \frac{1 \pm \sqrt{3}i}{2}$

$x^2 + x + 1 = 0 \Rightarrow x = \frac{-1 \pm \sqrt{1 - 4}}{2} = \frac{-1 \pm \sqrt{3}i}{2}$

Solution set: $\left\{1, -1, \frac{1}{2} \pm \frac{\sqrt{3}}{2}i, -\frac{1}{2} \pm \frac{\sqrt{3}}{2}i\right\}$

199

Method II.

To solve $x^6 = 1$, find the 6 roots of unity.

$\theta_0 = 0$ and increases by $\frac{360°}{6} = 60°$.

$x_0 = \cos 0° + i \sin 0° = 1$

$x_1 = \cos 60° + i \sin 60° = \frac{1}{2} + \frac{\sqrt{3}}{2}i$

$x_2 = \cos 120° + i \sin 120° = -\frac{1}{2} + \frac{\sqrt{3}}{2}i$

$x_3 = \cos 180° + i \sin 180° = -1$

$x_4 = \cos 240° + i \sin 240° = -\frac{1}{2} - \frac{\sqrt{3}}{2}i$

$x_5 = \cos 300° + i \sin 300° = \frac{1}{2} - \frac{\sqrt{3}}{2}i$

37. $(x^2 + 1)(x^4 + 1) = 0$

$x^2 + 1 = 0 \qquad\qquad x^4 + 1 = 0$

$x^2 = -1 \qquad\qquad x^4 = -1$

$x = \pm\sqrt{-1} = \pm i$

Find the 4 fourth roots of -1.

$x^4 = -1 = \cos 180° + i \sin 180°$

$\theta_0 = \frac{180°}{4} = 45°$ and increases by $\frac{360°}{4} = 90°$.

$x = \cos 45° + i \sin 45° = \frac{\sqrt{2}}{2} + i\frac{\sqrt{2}}{2}$

$x = \cos 135° + i \sin 135° = -\frac{\sqrt{2}}{2} + i\frac{\sqrt{2}}{2}$

$x = \cos 225° + i \sin 225° = -\frac{\sqrt{2}}{2} - i\frac{\sqrt{2}}{2}$

$x = \cos 315° + i \sin 315° = \frac{\sqrt{2}}{2} - i\frac{\sqrt{2}}{2}$

39. The real 8th roots of $15 = 15(\cos 0° + i \sin 0°)$ are found from the values of k for which

$\sin \frac{0° + k(360°)}{8} = \sin 45°k, \ k = 0, 1, 2, \ldots, 7$

will be 0. This occurs only when k is 0 or 4. Thus there are two real 8th roots of 15.

The real 15th roots of $-8 = 8(\cos 180° + i \sin 180°)$ are found from the values of k for which

$\sin \frac{180° + k(360°)}{15} = 12° + 24°k, \ k = 0, 1, 2, \ldots, 14$

will be 0. This occurs only when $k = 7$. Thus there is only one real 15th root of -8.

41. a) Find the polar form of $1 + \sqrt{3}i$:

$r = \sqrt{1 + 3} = 2$

$\cos \theta = \frac{1}{2}$

$\theta = 60°$

$1 + \sqrt{3}i = 2(\cos 60° + i \sin 60°)$

The square root of $1 + \sqrt{3}i$ with positive real part:

$\sqrt{2}\left(\cos \frac{60°}{2} + i \sin \frac{60°}{2}\right) = \sqrt{2}(\cos 30° + i \sin 30°)$

$\qquad = \sqrt{2}\left(\frac{\sqrt{3}}{2} + \frac{1}{2}i\right)$

$\qquad = \frac{\sqrt{6}}{2} + \frac{\sqrt{2}}{2}i$

b) From the result in the statement of the problem,

$\sqrt{-1 + \sqrt{3}i} = \sqrt{\frac{-4 + 4\sqrt{3}i}{4}}$

$\qquad = \frac{\sqrt{2} + \sqrt{6}i}{2}$

$\qquad = \frac{\sqrt{2}}{2} + \frac{\sqrt{6}}{2}i$

43. Note that

$[\cos \theta - i \sin \theta]^n = [\cos(-\theta) + i \sin(-\theta)]^n$

$\qquad = \cos(-n\theta) + i \sin(-n\theta)$

$\qquad = \cos n\theta - i \sin n\theta$

$-1 + \sqrt{3}i = 2[\cos 120° + i \sin 120°]$

$-1 - \sqrt{3}i = 2[\cos 120° - i \sin 120°]$

$(-1 + \sqrt{3}i)^n + (-1 - \sqrt{3}i)^n$

$\qquad = 2^n[\cos 120°n + i \sin 120°n]$
$\qquad\quad + 2^n[\cos 120°n - i \sin 120°n]$

$\qquad = 2 \cdot 2^n \cos 120°n$

Since n is not divisible by 3, the possibilities for $\cos 120°n$ are $\cos(120°) = -\frac{1}{2}$ and $\cos(240°) = -\frac{1}{2}$.

Therefore, the answer is $2 \cdot 2^n(-\frac{1}{2}) = -2^n$.

Chapter 9 Review Problem Set

1. True

2. True

3. False. It is both associative and commutative.

4. True

5. True

6. True

7. False

The phase shift is $\frac{-\pi/2}{3/2} = -\frac{\pi}{3}; \frac{\pi}{3}$ units to the left.

8. False. $u = \cos \theta + i \sin \theta$ for any θ.

9. True

10. False. None of the 8th roots of i is real.

11. $\gamma = 180° - (30° + 45°) = 105°$

Use the law of sines:

$$\frac{b}{\sin \beta} = \frac{c}{\sin \gamma}$$

$$\frac{b}{\sin 45°} = \frac{10}{\sin 105°}$$

$$b = \frac{10 \sin 45°}{\sin 105°} \approx 7.32$$

$$\frac{a}{\sin \alpha} = \frac{c}{\sin \gamma}$$

$$\frac{a}{\sin 30°} = \frac{10}{\sin 105°}$$

$$a = \frac{10 \sin 30°}{\sin 105°} \approx 5.18$$

12. Use the law of cosines:

$$\cos \alpha = \frac{b^2 + c^2 - a^2}{2bc} = \frac{3^2 + 4^2 - 2^2}{2(3)(4)} = \frac{7}{8}$$

$$\alpha = 28.96°$$

$$\cos \beta = \frac{a^2 + c^2 - b^2}{2ac} = \frac{2^2 + 4^2 - 3^2}{2(2)(4)} = \frac{11}{16}$$

$$\beta = 46.56°$$

$$\gamma = 180° - (28.96° + 46.56°) = 104.48°$$

13. Use the law of sines:

$$\frac{a}{\sin \alpha} = \frac{b}{\sin \beta}$$

$$\frac{67}{\sin \alpha} = \frac{94}{\sin 142°}$$

$$\sin \alpha = \frac{67 \sin 142°}{94} \approx .4388$$

$$\alpha = 26.03°$$

$$\gamma = 180° - (26.03° + 142°) = 11.97°$$

$$\frac{c}{\sin \gamma} = \frac{b}{\sin \beta}$$

$$\frac{c}{\sin 11.97°} = \frac{94}{\sin 142°}$$

$$c = \frac{94 \sin 11.97°}{\sin 142°} \approx 31.67$$

14. Use the law of cosines:

$$c^2 = a^2 + b^2 - 2ab \cos \gamma$$

$$= 11.6^2 + 20.3^2 - 2 \times 11.6 \times 20.3 \times \cos 37.6°$$

$$= 173.51$$

$$c = 13.17 \approx 13.2$$

Use the law of sines to find the smallest angle:

$$\frac{a}{\sin \alpha} = \frac{c}{\sin \gamma}$$

$$\frac{11.6}{\sin \alpha} = \frac{13.2}{\sin 37.6°}$$

$$\sin \alpha = \frac{11.6 \sin 37.6°}{13.17} = .5374$$

$$\alpha = 32.5°$$

$$\beta = 180° - (37.6° + 32.6°) = 109.9°$$

15. Area $= \frac{1}{2}ab \sin \gamma = \frac{1}{2}(11.6)(20.3)\sin 37.6° = 71.84$

16. Use Heron's formula with $s = \frac{7 + 8 + 9}{2} = 12$

$$A = \sqrt{12(12-7)(12-8)(12-9)}$$

$$= \sqrt{12 \cdot 5 \cdot 4 \cdot 3}$$

$$= \sqrt{720}$$

$$= 26.8$$

17.

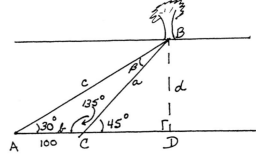

a) $\beta = 180° - (30° + 135°) = 15°$

Let $\overline{AB} = c$ and use the law of sines:

$$\frac{c}{\sin 135°} = \frac{100}{\sin 15°}$$

$$c = \frac{100 \sin 135°}{\sin 15°} = 273.2 \text{ yards}$$

b) Let d be the width of the river. In right $\triangle$ ABD,

$$\frac{d}{c} = \sin 30°$$

$$d = 273.2 \sin 30° = 136.6 \text{ yards}$$

18. $\mathbf{u} + \frac{4}{5}\mathbf{v} = 6(-\sin 60°\mathbf{i} + \cos 60°\mathbf{j}) + \frac{4}{5}(-10\mathbf{i} + 15\mathbf{j})$

$$= 6\left(-\frac{\sqrt{3}}{2}\mathbf{i} + \frac{1}{2}\mathbf{j}\right) - 8\mathbf{i} + 12\mathbf{j}$$

$$= -(3\sqrt{3} + 8)\mathbf{i} + 15\mathbf{j}$$

19. $\mathbf{u} = 5\mathbf{i} - 12\mathbf{j}$, $\mathbf{v} = 24\mathbf{i} + 7\mathbf{j}$

a) $\|\mathbf{u}\| = \sqrt{5^2 + (-12)^2} = \sqrt{25 + 144} = \sqrt{169} = 13$

b) $\|\mathbf{v}\| = \sqrt{24^2 + 7^2} = \sqrt{625} = 25$

c) $\mathbf{u} \cdot \mathbf{v} = 5 \cdot 24 + (-12)(7) = 120 - 84 = 36$

d) $\cos \theta = \frac{\mathbf{u} \cdot \mathbf{v}}{\|\mathbf{u}\|\|\mathbf{v}\|} = \frac{36}{13 \cdot 25} = .11077$

$$\theta = 83.64°$$

e) A unit length vector in the same directions as $\mathbf{v}$:

$$\frac{\mathbf{v}}{\|\mathbf{v}\|} = \frac{24\mathbf{i} + 7\mathbf{j}}{25} = \frac{24}{25}\mathbf{i} + \frac{7}{25}\mathbf{j}$$

20. a) The scalar projection of **u** on **v** is $\frac{\mathbf{u} \cdot \mathbf{v}}{\|\mathbf{v}\|} = \frac{36}{25}$

b) The vector projection of **u** on **v** is

$\frac{\mathbf{u} \cdot \mathbf{v}}{\mathbf{v} \cdot \mathbf{v}}\mathbf{v} = \frac{36}{625}(24\mathbf{i} + 7\mathbf{j}) = \frac{864}{625}\mathbf{i} + \frac{252}{625}\mathbf{j}$

21. $\mathbf{v} = 90\mathbf{j}$

Let θ be the angle between **u** and the **j** axis. Then

$$\theta = \cos^{-1}\left(\frac{4}{5}\right)$$

$\cos\theta = \frac{4}{5}$ and θ is in quadrant I

$$\sin\theta = \sqrt{1 - \left(\frac{4}{5}\right)^2} = \sqrt{\frac{9}{25}} = \frac{3}{5}$$

$$\mathbf{u} = 120(\sin\theta)\mathbf{i} + 120(\cos\theta)\mathbf{j}$$

$$= 120\left(\frac{3}{5}\right)\mathbf{i} + 120\left(\frac{4}{5}\right)\mathbf{j}$$

$$= 72\mathbf{i} + 96\mathbf{j}$$

$\mathbf{w} = \mathbf{u} + \mathbf{v} = (72\mathbf{i} + 96\mathbf{j}) + 90\mathbf{j} = 72\mathbf{i} + 186\mathbf{j}$

22. a)   work $= \mathbf{F} \cdot \mathbf{D}$

$$= \mathbf{w} \cdot 4\frac{\mathbf{w}}{\|\mathbf{w}\|}$$

$$= 4\|\mathbf{w}\|$$

$$= 4\sqrt{72^2 + 186^2}$$

$$= 797.8 \text{ foot-pounds}$$

b)   work $= \mathbf{w} \cdot 4\frac{\mathbf{u}}{\|\mathbf{u}\|}$

$$= 4\frac{(72\mathbf{i} + 186\mathbf{j}) \cdot (72\mathbf{i} + 96\mathbf{j})}{120}$$

$$= \frac{23040}{30}$$

$$= 768 \text{ foot-pounds}$$

23. $y = \cos t$

Amplitude: 1

Period: $\frac{2\pi}{2} = \pi$

Phase shift: 0

24. $y = 3\cos 4t$

Amplitude: 3

Period: $\frac{2\pi}{4} = \frac{\pi}{2}$

Phase shift: 0

25. $y = 2\sin\left(3t - \frac{\pi}{2}\right)$

Amplitude: 2

Period: $\frac{2\pi}{3}$

Phase shift: $\frac{\pi/2}{3} = \frac{\pi}{6}$

26. $y = -2\sin(\frac{1}{2}t + \pi)$

Amplitude: 2

Period: $\frac{2\pi}{1/2} = 4\pi$

Phase shift: $\frac{-\pi}{1/2} = -2\pi$

27. $y = \cos 2t$

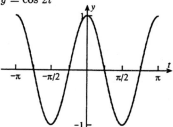

28. $y = 3\cos 4t$

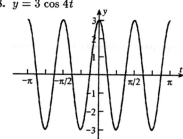

29. $y = 2\sin\left(3t - \frac{\pi}{2}\right)$

Amplitude: 2

Period: $\frac{2\pi}{3}$

Phase shift: $\frac{\pi/2}{3} = \frac{\pi}{6}$

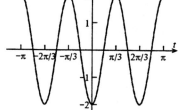

30. $y = -2\sin(\frac{1}{2}t - \pi)$

Amplitude: 2

Period: $\frac{2\pi}{1/2} = 4\pi$

Phase shift: $\frac{\pi}{1/2} = 2\pi$

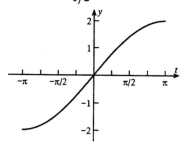

31. a) After t seconds, the speck has rotated through an angle of $\frac{5\pi}{6}t$ radians. The coordinates of the speck are
$$\left(3\cos\frac{5\pi}{6}t,\ 3\sin\frac{5\pi}{6}t\right)$$
b) The speck will be at $(-3,0)$ when
$$\frac{5\pi}{6}t = \pi$$
$$t = \frac{6}{5}\text{ seconds}$$

32. After t seconds, $x = 3\sin\left(5t+\frac{\pi}{6}\right)$
a) period $= \frac{2\pi}{5}$ seconds
b) $x(0) = 3\sin\frac{\pi}{6} = 3\cdot\frac{1}{2} = \frac{3}{2}$
c) $x = 0$ for the first time when
$$5t + \frac{\pi}{6} = \pi$$
$$5t = \frac{5\pi}{6}$$
$$t = \frac{\pi}{6}\text{ seconds}$$

33. $\dfrac{\left(\frac{13\pi}{6}-\frac{\pi}{6}\right)\text{ seconds}}{\frac{2\pi}{5}\text{ seconds/period}} = \dfrac{\frac{2\pi}{\ }}{\frac{2\pi}{5}}$ period $= 5$ periods

Since the particle moves 12 meters in each period (from -3 to 3 and back again) the particle moves a total of $12(5) = 60$ meters in the time interval.

34.

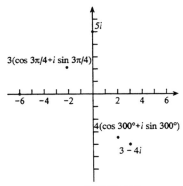

35. a) $|3-4i| = \sqrt{3^2+(-4)^2} = \sqrt{25} = 5$
b) $|-6| = 6$
c) $|5i| = 5$
d) $\left|3\left(\cos\frac{3\pi}{4}+i\sin\frac{3\pi}{4}\right)\right| = 3$
e) $\left|4(\cos 300° + i\sin 300°)\right| = 4$

36. $4\sqrt{2}\left(\cos\frac{7\pi}{4}+i\sin\frac{7\pi}{4}\right) = 4\sqrt{2}\left(\frac{1}{\sqrt{2}}-\frac{1}{\sqrt{2}}i\right) = 4-4i$

37. $-4 = 4(\cos 180° + i\sin 180°)$

38. $9i = 9(\cos 90° + i\sin 90°)$

39. $2+2i$ is in quadrant I
$$r = \sqrt{2^2+2^2} = \sqrt{8} = 2\sqrt{2}$$
$$\cos\theta = \frac{2}{2\sqrt{2}} = \frac{1}{\sqrt{2}}$$
$$\theta = 45°$$
$$2+2i = 2\sqrt{2}(\cos 45° + i\sin 45°)$$

40. $2-2\sqrt{3}i$ is in quadrant IV
$$r = \sqrt{2^2+(2\sqrt{3})^2} = \sqrt{4+12} = \sqrt{16} = 4$$
$$\cos\theta = \frac{2}{4} = \frac{1}{2}$$
$$\theta = 300°$$
$$2-2\sqrt{3}i = 4(\cos 300° + i\sin 300°)$$

41. $u = r(\cos t + i\sin t)$
a) $u^3 = r^3(\cos 3t + i\sin 3t)$
b) $\frac{1}{u} = \frac{1(\cos 0 + i\sin 0)}{r(\cos t + i\sin t)} = \frac{1}{r}[\cos(-t) + i\sin(-t)]$
c) $u\left(\cos\frac{\pi}{3}+i\sin\frac{\pi}{3}\right)$
$$= r(\cos t + i\sin t)\left(\cos\frac{\pi}{3}+i\sin\frac{\pi}{3}\right)$$
$$= r\left[\cos\left(t+\frac{\pi}{3}\right)+i\sin\left(t+\frac{\pi}{3}\right)\right]$$

42. $u = 6(\cos 145° + i\sin 145°)$, $v = 3(\cos 65° + i\sin 65°)$
a) $v^2 = 3^2[\cos(2\cdot 65°)+i\sin(2\cdot 65°)]$
$$= 9(\cos 130° + i\sin 130°)$$
$$uv^2 = 6\cdot 9[\cos(145°+130°)+i\sin(145°+130°)]$$
$$= 54(\cos 275° + i\sin 275°)$$
b) $\frac{2u}{v} = \frac{12}{3}[\cos(145°-65°)+i\sin(145°-65°)]$
$$= 4(\cos 80° + i\sin 80°)$$
c) $\frac{u^2}{v^3} = \frac{36(\cos 290° + i\sin 290°)}{27(\cos 195° + i\sin 195°)}$
$$= \frac{4}{3}(\cos 95° + i\sin 95°)$$

43. Let $u = 2^8(\cos 96° + i\sin 96°)$
$$r = \sqrt[4]{2^8} = 2^2 = 4$$
$$\theta_0 = \frac{96°}{4} = 24°\text{ and increases by }\frac{360°}{4} = 90°.$$
$$u_0 = 4(\cos 24° + i\sin 24°)$$
$$u_1 = 4(\cos 114° + i\sin 114°)$$
$$u_2 = 4(\cos 204° + i\sin 204°)$$
$$u_3 = 4(\cos 294° + i\sin 294°)$$

44. Let $u = -i = \cos 270° + i \sin 270°$

$\theta_0 = \dfrac{270°}{6} = 45°$ and increases by $\dfrac{360°}{6} = 60°$.

$u_0 = \cos 45° + i \sin 45°$

$u_1 = \cos 105° + i \sin 105°$

$u_2 = \cos 165° + i \sin 165°$

$u_3 = \cos 225° + i \sin 225°$

$u_4 = \cos 285° + i \sin 285°$

$u_5 = \cos 345° + i \sin 345°$

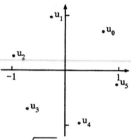

45. $u = \sqrt{1 - i}$

$u^6 = (\sqrt{1-i})^6 = (1-i)^3$

$v = \sqrt[4]{1+i}$

$v^{12} = (\sqrt[4]{1+i})^{12} = (1+i)^3$

$u^6 v^{12} = (1-i)^3(1+i)^3 = [(1-i)(1+i)]^3 = 2^3 = 8$

46. $u = a + bi$ and $v = c + di$

$uv = (ac - bd) + (ad + bc)i$

$|uv| = \sqrt{(ac-bd)^2 + (ad+bc)^2}$

$\quad = \sqrt{a^2c^2 - 2abcd + b^2d^2 + a^2d^2 + 2abcd + b^2c^2}$

$\quad = \sqrt{a^2c^2 + b^2d^2 + a^2d^2 + b^2c^2}$

$|u||v| = \sqrt{a^2 + b^2}\sqrt{c^2 + d^2}$

$\quad = \sqrt{(a^2+b^2)(c^2+d^2)}$

$\quad = \sqrt{a^2c^2 + b^2d^2 + a^2d^2 + b^2c^2}$

$\quad = |uv|$

47. $x^3 + 8i = 0$

$\qquad x^3 = -8i = 8(\cos 270° + i \sin 270°)$

$r = \sqrt[3]{8} = 2$

$\theta_0 = \dfrac{270°}{3} = 90°$ and increases by $\dfrac{360°}{3} = 120°$

$x_0 = 2(\cos 90° + i \sin 90°) = 2(0 + i) = 2i$

$x_1 = 2(\cos 210° + i \sin 210°)$

$\quad = 2\left[-\dfrac{\sqrt{3}}{2} + i\left(-\dfrac{1}{2}\right)\right]$

$\quad = -\sqrt{3} - \mathbf{i}$

$x_2 = 2(\cos 330° + i \sin 330°)$

$\quad = 2\left[\dfrac{\sqrt{3}}{2} + i\left(-\dfrac{1}{2}\right)\right]$

$\quad = \sqrt{3} - i$

48. a) $(u^n)^6 = 1$ when $6n$ is a multiple of 12, that is when $n = 2, 4, 6, 8, 10, 12$.

b) $(u^n)^8 = 1$ when $8n$ is a multiple of 12, that is when $n = 3, 6, 9, 12$.

49. Use the law of sines to find β.

$\dfrac{2}{\sin \beta} = \dfrac{1.5}{\sin 40°}$

$\sin \beta = \dfrac{2 \sin 40°}{1.5} = .85705$

$\beta_1 = 58.99°$

$\beta_2 = 180° - 58.99° = 121.01°$

In the acute triangle,

$\gamma_1 = 180° - (40° + 58.99°) = 81.01°$

$\dfrac{c_1}{\sin 81.01°} = \dfrac{1.5}{\sin 40°}$

$c_1 = \dfrac{1.5 \sin 81.01°}{\sin 40°} = 2.30$

In the obtuse triangle,

$\gamma_2 = 180° - (40° + 121.01°) = 18.99°$

$\dfrac{c_2}{\sin 18.99°} = \dfrac{1.5}{\sin 40°}$

$c_2 = \dfrac{1.5 \sin 18.99}{\sin 40°} = .76$

50. Area sector $-$ area triangle

$= \dfrac{36°}{360°}\pi(10^2) - \dfrac{1}{2}(10^2)\sin 36°$

$= 10\pi - 50 \sin 36°$

$= 2.027$ square units

Chapter 10: Theory of Polynomial Equations

Problem Set 10.1

1.

$$x^2 - 2x + 3 \;\big)\; \overline{x^3 - x^2 + x + 3} \quad \Big\uparrow x+1$$

$$
\begin{array}{r}
x + 1 \\
x^2 - 2x + 3 \;\overline{)\; x^3 - \;\; x^2 + \;\; x + 3} \\
x^3 - 2x^2 + 3x \\
\hline
x^2 - 2x + 3 \\
x^2 - 2x + 3 \\
\hline
0
\end{array}
$$

Quotient: $x + 1$
Remainder: 0

3.

$$
\begin{array}{r}
3x - 1 \\
2x^2 + 3x - 5 \;\overline{)\; 6x^3 + 7x^2 - 18x + 15} \\
6x^3 + 9x^2 - 15x \\
\hline
-2x^2 - \;\; 3x + 15 \\
-2x^2 - \;\; 3x + \;\; 5 \\
\hline
10
\end{array}
$$

Quotient: $3x - 1$
Remainder: 10

5.

$$
\begin{array}{r}
2x^2 + x + 3 \\
2x^2 - x - 3 \;\overline{)\; 4x^4 \qquad\quad - \;\; x^2 - 6x - 9} \\
4x^4 - 2x^3 - 6x^2 \\
\hline
2x^3 + 5x^2 - 6x \\
2x^3 - \;\; x^2 - 3x \\
\hline
6x^2 - 3x - 9 \\
6x^2 - 3x - 9 \\
\hline
0
\end{array}
$$

Quotient: $2x^2 + x + 3$
Remainder: 0

7.

$$
\begin{array}{r}
x^2 + 4 \\
2x^3 - 2x^2 + x - 4 \;\overline{)\; 2x^5 - 2x^4 + 9x^3 - 12x^2 + 4x - 16} \\
2x^5 - 2x^4 + \;\; x^3 - 4x^2 \\
\hline
8x^3 - 8x^2 + 4x - 16 \\
8x^3 - 8x^2 + 4x - 16 \\
\hline
0
\end{array}
$$

Quotient: $x^2 + 4$
Remainder: 0

9.

$$
\begin{array}{r|rrrr}
1 & 2 & -1 & 1 & -4 \\
  &   & 2 & 1 & 2 \\
\hline
  & 2 & 1 & 2 & \boxed{-2}
\end{array}
$$

Quotient: $2x^2 + x + 2$
Remainder: -2

11.

$$
\begin{array}{r|rrrr}
1 & 3 & 5 & 2 & -10 \\
  &   & 3 & 8 & 10 \\
\hline
  & 3 & 8 & 10 & \boxed{0}
\end{array}
$$

Quotient: $3x^2 + 8x + 10$
Remainder: 0

13.

$$
\begin{array}{r|rrrrr}
3 & 1 & 0 & -2 & 0 & -1 \\
  &   & 3 & 9 & 21 & 63 \\
\hline
  & 1 & 3 & 7 & 21 & \boxed{62}
\end{array}
$$

Quotient: $x^3 + 3x^2 + 7x + 21$
Remainder: 62

15.

$$
\begin{array}{r|rrrr}
-1 & 1 & 2 & -3 & 2 \\
   &   & -1 & -1 & 4 \\
\hline
   & 1 & 1 & -4 & \boxed{6}
\end{array}
$$

Quotient: $x^2 + x - 4$
Remainder: 6

17.

$$
\begin{array}{r|rrrrr}
-\frac{1}{2} & 2 & 1 & 4 & 7 & 4 \\
             &   & -1 & 0 & -2 & -\frac{5}{2} \\
\hline
             & 2 & 0 & 4 & 5 & \boxed{\frac{3}{2}}
\end{array}
$$

Quotient: $2x^3 + 4x + 5$
Remainder: $\frac{3}{2}$

19.

$$
\begin{array}{r|rrrr}
2 - 3i & 1 & -2 & 5 & 30 \\
       &   & (2-3i) & (-9-6i) & -26 \\
\hline
       & 1 & -3i & (-4-6i) & \boxed{4}
\end{array}
$$

Quotient: $x^2 - 3ix - 4 - 6i$
Remainder: 4

21.

$$
\begin{array}{r|rrrrr}
2i & 1 & 0 & 0 & 0 & -17 \\
   &   & 2i & -4 & -8i & 16 \\
\hline
   & 1 & 2i & -4 & -8i & \boxed{-1}
\end{array}
$$

Quotient: $x^3 + 2ix^2 - 4x - 8i$
Remainder: -1

23. $\dfrac{x^3 + 2x^2 + 5}{x^2} = \dfrac{x^3}{x^2} + \dfrac{2x^2}{x^2} + \dfrac{5}{x^2} = x + 2 + \dfrac{5}{x^2}$

25.

$$
\begin{array}{r}
x - 1 \\
x^2 + x - 2 \;\overline{)\; x^3 \qquad\quad - 4x + 5} \\
x^3 + x^2 - 2x \\
\hline
-x^2 - 2x + 5 \\
-x^2 - \;\; x + 2 \\
\hline
-x + 3
\end{array}
$$

$$\frac{x^3 - 4x + 5}{x^2 + x - 2} = x - 1 + \frac{(-x + 3)}{x^2 + x - 2}$$

27.

$$
\begin{array}{r}
2 \\
x^2 + 1 \;\overline{)\; 2x^2 - 4x + 5} \\
2x^2 \qquad\;\; + 2 \\
\hline
-4x + 3
\end{array}
$$

$$\frac{2x^2 - 4x + 5}{x^2 + 1} = 2 + \frac{3 - 4x}{x^2 + 1}$$

29.

$$
\begin{array}{r}
2x-1 \\
x^3+1\ \overline{\smash{\big)}\ 2x^4 - x^3 - x^2\quad\ \ - 2} \\
\underline{2x^4\qquad\qquad\ +2x} \\
-x^3 - x^2 - 2x - 2 \\
\underline{-x^3\qquad\qquad - 1} \\
-x^2 - 2x - 1
\end{array}
$$

$$\frac{2x^4 - x^3 - x^2 - 2}{x^3 + 1} = 2x - 1 + \frac{(-x^2 - 2x - 1)}{x^3 + 1}$$

$$= 2x - 1 + \frac{-(x+1)^2}{(x+1)(x^2 - x + 1)}$$

$$= 2x - 1 + \frac{(-x-1)}{x^2 - x + 1}$$

31. a) $\dfrac{2x^0 + 3x^2 - 11x + 9}{x^2}$

$= 2x + 3$ with remainder of $(-11x + 9)$

b) $\dfrac{2(x+3)^2 + 10(x+3) - 14}{x + 3}$

$= 2(x + 3) + 10$; remainder is -14.

c) $\dfrac{(x-4)^5 + x^2 + x + 1}{(x-4)^3} = (x-4)^2$

with remainder of $(x^2 + x + 1)$

d) $\dfrac{(x^2+3)^3 + 2x(x^2+3) + 4x - 1}{x^2 + 3}$

$= (x^2 + 3)^2 + 2x$ with remainder of $(4x - 1)$
(Quotient here may also be written as $x^4 + 6x^2 + 2x + 9$.)

33. a) Divide $x^5 + x^4 - 16x - 16$ by $x - 2$:

$$
\begin{array}{r|rrrrr}
2 & 1 & 1 & 0 & 0 & -16 & -16 \\
  &   & 2 & 6 & 12 & 24 & 16 \\
\hline
  & 1 & 3 & 6 & 12 & 8 & 0
\end{array}
$$

Since the remainder is 0, $x - 2$ is a factor. The other factor is $x^4 + 3x^3 + 6x^2 + 12x + 8$.

b) Divide $x^5 + 32$ by $x + 2$:

$$
\begin{array}{r|rrrrrr}
-2 & 1 & 0 & 0 & 0 & 0 & 32 \\
   &   & -2 & 4 & -8 & 16 & -32 \\
\hline
   & 1 & -2 & 4 & -8 & 16 & 0
\end{array}
$$

Since the remainder is 0, $x + 2$ is a factor. The other factor is $x^4 - 2x^3 + 4x^2 - 8x + 16$.

c) Divide $x^4 - \frac{3}{2}x^3 + 3x^2 + 6x + 2$ by $x + \frac{1}{2}$:

$$
\begin{array}{r|rrrrr}
-\frac{1}{2} & 1 & -\frac{3}{2} & 3 & 6 & 2 \\
             &   & -\frac{1}{2} & 1 & -2 & -2 \\
\hline
             & 1 & -2 & 4 & 4 & 0
\end{array}
$$

Since the remainder is 0, $x + \frac{1}{2}$ is a factor. The other factor is $x^3 - 2x^2 + 4x + 4$.

d) Divide $x^3 - 2ix^2 + x - 2i$ by $x - 2i$:

$$
\begin{array}{r|rrrr}
2i & 1 & -2i & 1 & -2i \\
   &   & 2i & 0 & 2i \\
\hline
   & 1 & 0 & 1 & 0
\end{array}
$$

Since the remainder is 0, $x - 2i$ is a factor. The other factor is $x^2 + 1$.

35. $x^6 + 2x^5 - x^4 - 10x^3 - 16x^2 + 8x + 16$
$$= (x^2 - 1)(x+2)h(x)$$
$$= (x^3 + 2x^2 - x - 2)h(x)$$

$$
\begin{array}{r}
x^3 - 8 \\
x^3 + 2x^2 - x - 2\ \overline{\smash{\big)}\ x^6 + 2x^5 - x^4 - 10x^3 - 16x^2 + 8x + 16} \\
\underline{x^6 + 2x^5 - x^4 -\ \ 2x^3} \\
-8x^3 - 16x^2 + 8x + 16 \\
\underline{-8x^3 - 16x^2 + 8x + 16} \\
0
\end{array}
$$

$h(x) = x^3 - 8$

37. a) Divide $x^3 + x^2 - 10x + k$ by $x - 4$:

$$
\begin{array}{r|rrrr}
4 & 1 & 1 & -10 & k \\
  &   & 4 & 20 & 40 \\
\hline
  & 1 & 5 & 10 & (40 + k)
\end{array}
$$

Since we want $x - 4$ to be a factor, we want the remainder to equal 0:

$$40 + k = 0$$
$$k = -40$$

b) Divide $x^4 + kx + 10$ by $x + 2$:

$$
\begin{array}{r|rrrrr}
-2 & 1 & 0 & 0 & k & 10 \\
   &   & -2 & 4 & -8 & -2(k-8) \\
\hline
   & 1 & -2 & 4 & (k-8) & 10 - 2(k-8)
\end{array}
$$

Since we want $x + 2$ to be a factor, we want the remainder to equal 0.

$$10 - 2(k - 8) = 0$$
$$10 - 2k + 16 = 0$$
$$2k = 26$$
$$k = 13$$

c) Divide $k^2x^3 - 4kx + 4$ by $x - 1$:

$$
\begin{array}{r|rrrr}
1 & k^2 & 0 & -4k & 4 \\
  &   & k^2 & k^2 & k^2 - 4k \\
\hline
  & k^2 & k^2 & (k^2 - 4k) & k^2 - 4k + 4
\end{array}
$$

Since we want $x - 1$ to be a factor, we want the remainder to equal 0.

$$k^2 - 4k + 4 = 0$$
$$(k - 2)^2 = 0$$
$$k - 2 = 0$$
$$k = 2$$

39. Divide $x^4 + ax^3 + bx^2 + cx - 4$ by $x - 1$:

```
1 | 1    a        b          c          -4
  |      1       a+1       a+b+1     a+b+c+1
    ----------------------------------------------
    1   a+1     a+b+1     a+b+c+1   |a+b+c-3
```

If $(x-1)$ is a factor, then
$$a + b + c - 3 = 0, \text{ or}$$
$$a + b + c = 3 \quad \textbf{(1)}$$

Divide the quotient of the previous division by $x - 1$:

```
1 | 1   a+1     a+b+1      a+b+c+1
  |     1        a+2       2a+b+3
    ----------------------------------------
    1   a+2     2a+b+3    |3a+2b+c+4
```

If $(x-1)$ is a factor twice, then
$$3a + 2b + c + 4 = 0, \text{ or}$$
$$3a + 2b + c = -4$$

Multiply by -1:
$$-3a - 2b - c = 4 \quad \textbf{(2)}$$

Divide the quotient of the second division by $x - 1$:

```
1 | 1   a+2      2a+b+3
  |     1         a+3
    --------------------------
    1   a+3      |3a+b+6
```

If $(x-1)$ is a factor three times, then
$$3a + b + 6 = 0$$
$$3a + b = -6$$
$$b = -6 - 3a \quad \textbf{(3)}$$

With three equations we can solve for 3 variables. First add (1) and (2) to obtain (4):

$$
\begin{aligned}
a + b + c &= 3 \quad \textbf{(1)}\\
-3a - 2b - c &= 4 \quad \textbf{(2)}\\
\hline
-2a - b \quad &= 7 \quad \textbf{(4)}
\end{aligned}
$$

Substitute (3) into (4):
$$-2a - (-6 - 3a) = 7$$
$$-2a + 6 + 3a = 7, \ a = 1$$

Substituting into (4) again,
$$-2(1) - b = 7$$
$$-2 - 7 = b, \ b = -9$$

Now use (1) again:
$$1 - 9 + c = 3$$
$$c = 11$$
$$a = 1, \ b = -9, \ c = 11$$

41. Set new range values:
Xmin: -8
Xmax: 8
Ymin: -10
Ymax: 80
Press $\boxed{Y=}$
Enter as Y_1:
$\boxed{(}2X \char`\^ 4 - 4X \char`\^ 3 - 8X + 5 \boxed{)} \boxed{\div} \boxed{(} X \char`\^ 2 - 2X + 3 \boxed{)}$
Enter $2X \char`\^ 2 - 6$ as Y_2.
Press $\boxed{\text{GRAPH}}$.
The conjecture is that as $|x|$ gets very large, $f(x) - g(x)$ approaches 0.

Problem Set 10.2

1.
```
2 | 2   -5    3    -4
  |      4   -2    2
    --------------------
    2   -1    1   |-2
```
$P(2) = -2$.
By substitution,
$$P(2) = 2(2^3) - 5(2^2) + 3(2) - 4$$
$$= 16 - 20 + 6 - 4 = -2$$

3.
```
1/2 | 8   0    -3     0      -2
    |     4     2    -1/2    -1/4
      ---------------------------------
      8   4    -1    -1/2   |-9/4
```
$P\left(\frac{1}{2}\right) = -\frac{9}{4}$.
By substitution,
$$P\left(\tfrac{1}{2}\right) = 8\left(\tfrac{1}{2}\right)^4 - 3\left(\tfrac{1}{2}\right)^2 - 2$$
$$= \tfrac{8}{16} - \tfrac{3}{4} - 2$$
$$= \tfrac{2}{4} - \tfrac{3}{4} - \tfrac{8}{4}$$
$$= -\tfrac{9}{4}$$

5. The remainder equals
$$P(1) = (1)^{10} - 15(1) + 8 = -6.$$

7. The remainder equals $P\left(-\frac{1}{2}\right) = 64\left(\frac{1}{2}\right)^6 + 13$
$$= \tfrac{64}{64} + 13 = 14$$

9. The zeros of $(x-1)(x+2)(x-3)$ are $x = 1$, $x = -2$, and $x = 3$, each of multiplicity 1.

11. The zeros of $(2x-1)(x-2)^2 x^3$ are $x = \frac{1}{2}$ (multiplicity 1); $x = 2$ (multiplicity 2); and $x = 0$ (multiplicity 3).

13. The zeros of $3(x - 1 - 2i)(x + \frac{2}{3})$ are $1 + 2i$ and $-\frac{2}{3}$, each of multiplicity 1.

15. $P(1) = 2(1)^3 - 7(1) + 9(1) - 4 = 2 - 7 + 9 - 4 = 0.$
Since $P(1) = 0$, 1 is a zero of $P(x)$, so $x - 1$ is a factor of $P(x)$.

17. $P(3) = (3)^3 - 7(3)^2 + 16(3) - 12$
$$= 27 - 63 + 48 - 12$$
$$= 0$$
Since $P(3) = 0$, 3 is a zero of $P(x)$, so $x - 3$ is a factor of $P(x)$.

19. $P(x) = 2x^3 - 7x^2 + 9x - 4$; $c = 1$. From Problem 15, we know that 1 is a zero.

$$\begin{array}{r|rrrr} 1 & 2 & -7 & 9 & -4 \\ & & 2 & -5 & 4 \\ \hline & 2 & -5 & 4 & \boxed{0} \end{array}$$

We know $(x-1)$ is a factor and $2x^2 - 5x + 4$ is another. To find the other zeros, set $2x^2 - 5x + 4$ equal to zero and use the quadratic formula.
$2x^2 - 5x + 4 = 0$; $a = 2$, $b = -5$, $c = 4$

$$x = \frac{-b \pm \sqrt{b^2 - 4ac}}{2a}$$
$$= \frac{-(-5) \pm \sqrt{(-5)^2 - 4(2)(4)}}{2(2)}$$
$$= \frac{5 \pm \sqrt{25 - 32}}{4} = \frac{5 \pm \sqrt{-7}}{4}$$
$$= \frac{5 \pm i\sqrt{7}}{4}$$

21. $P(x) = x^3 - 7x^2 + 16x - 12$.
From Problem 17, we know that 3 is a zero.

$$\begin{array}{r|rrrr} 3 & 1 & -7 & 16 & -12 \\ & & 3 & -12 & 12 \\ \hline & 1 & -4 & 4 & \boxed{0} \end{array}$$

The synthetic division shows us that
$P(x) = (x-3)(x^2 - 4x + 4) = (x-3)(x-2)^2$
Therefore, 2 is the other zero of $P(x)$. It is of multiplicity 2.

23. $x^2 - 5x + 6 = (x-3)(x-2)$

25. $x^4 - 5x^2 + 4 = (x^2 - 4)(x^2 - 1)$
$\qquad = (x+2)(x-2)(x+1)(x-1)$

27. $P(x) = x^3 - 3x^2 - 28x + 60$.
We know that 2 is a zero of $P(x)$.

$$\begin{array}{r|rrrr} 2 & 1 & -3 & -28 & 60 \\ & & 2 & -2 & -60 \\ \hline & 1 & -1 & -30 & \boxed{0} \end{array}$$

The synthetic division shows us that
$P(x) = (x-2)(x^2 - x - 30) = (x-2)(x-6)(x+5)$.

29. $P(x) = x^3 + 3x^2 - 10x - 12$.
We know that -1 is a zero of $P(x)$.

$$\begin{array}{r|rrrr} -1 & 1 & 3 & -10 & -12 \\ & & -1 & -2 & 12 \\ \hline & 1 & 2 & -12 & \boxed{0} \end{array}$$

The synthetic division shows us that
$P(x) = (x+1)(x^2 + 2x - 12)$. $x^2 + 2x - 12$ is not factorable over the integers. We will need to find its zeros using the quadratic formula.

$x^2 + 2x - 12 = 0$

$$x = \frac{-2 \pm \sqrt{(-2)^2 - 4(1)(-12)}}{2(1)}$$
$$= \frac{-2 \pm \sqrt{52}}{2}$$
$$= \frac{-2 \pm 2\sqrt{13}}{2}$$
$$= -1 \pm \sqrt{13}$$

Since $-1 + \sqrt{13}$ and $-1 - \sqrt{13}$ are zeros of $x^2 + 2x - 12$, $x - (-1 + \sqrt{13})$ and $x - (-1 - \sqrt{13})$ are factors. Thus
$P(x) = (x+1)(x+1-\sqrt{13})(x+1+\sqrt{13})$.

31. 2, 1, and -4 are simple zeros, so
$$P(x) = (x-2)(x-1)(x+4)$$
$$= (x-2)(x^2 + 3x - 4)$$
$$= x^3 + 3x^2 - 4x - 2x^2 - 6x + 8$$
$$= x^3 + x^2 - 10x + 8$$

33. $\frac{1}{2}$ and $-\frac{5}{6}$ are simple zeros, so $P(x) = a(x - \frac{1}{2})(x + \frac{5}{6})$. Choose $a = 12$ to eliminate fractions.
$$P(x) = 12(x - \tfrac{1}{2})(x + \tfrac{5}{6})$$
$$= (2x - 1)(6x + 5)$$
$$= 12x^2 + 4x - 5$$

35. 2, $\sqrt{5}$, $-\sqrt{5}$ are simple zeros, so
$$P(x) = a(x-2)(x-\sqrt{5})(x+\sqrt{5})$$
$$= a(x-2)(x^2 - 5)$$
$$= a(x^3 - 2x^2 - 5x + 10)$$
Choose $a = 1$, so that $P(x) = x^3 - 2x^2 - 5x + 10$.

37. $\frac{1}{2}$ is a zero of multiplicity 2, and -2 is a zero of multiplicity 3, so $P(x) = a(x - \frac{1}{2})^2 (x+2)^3$.
Choose $a = 4$ to clear fractions.
$$P(x) = 4(x - \tfrac{1}{2})^2 (x+2)^3$$
$$= (2x-1)^2 (x+2)^3$$
$$= (4x^2 - 4x + 1)(x^3 + 6x^2 + 12x + 8)$$
$$= 4x^5 + 24x^4 + 48x^3 + 32x^2$$
$$\quad - 4x^4 - 24x^3 - 48x^2 - 32x$$
$$\quad + x^3 + 6x^2 + 12x + 8$$
$$= 4x^5 + 20x^4 + 25x^3 - 10x^2 - 20x + 8$$

39. $P(x) = (x-2)(x+2)(x-i)(x+i)$
$$= (x^2 - 4)(x^2 + 1)$$
$$= x^4 - 3x^2 - 4$$

41. $P(x) = (x-2)(x+5)(x-2-3i)(x-2+3i)$
$$= (x^2 + 3x - 10)[(x-2) - 3i][(x-2) + 3i]$$
$$= (x^2 + 3x - 10)[(x-2)^2 + 9]$$
$$= (x^2 + 3x - 10)(x^2 - 4x + 13)$$
$$= x^4 - 4x^3 + 13x^2$$
$$\quad + 3x^3 - 12x^2 + 39x$$
$$\quad - 10x^2 + 40x - 130$$
$$= x^4 - x^3 - 9x^2 + 79x - 130$$

43. Use synthetic division three times:

$$
\begin{array}{r|rrrrrr}
1 & 1 & 2 & -6 & -4 & 13 & -6 \\
  &   & 1 & 3 & -3 & -7 & 6 \\
\hline
1 & 1 & 3 & -3 & -7 & 6 & \lfloor 0 \\
  &   & 1 & 4 & 1 & -6 & \\
\hline
1 & 1 & 4 & 1 & -6 & \lfloor 0 & \\
  &   & 1 & 5 & 6 & & \\
\hline
  & 1 & 5 & 6 & \lfloor 0 & & \\
\end{array}
$$

Last quotient is $x^2 + 5x + 6 = (x+2)(x+3)$

$x^5 + 2x^4 - 6x^3 - 4x^2 + 13x - 6$
$$= (x-1)^3(x+2)(x+3)$$

and set of zeros is $\{1, 1, 1, -2, -3\}$.

45. Use synthetic division 4 times:

$$
\begin{array}{r|rrrrrrr}
-1 & 1 & -8 & 7 & 32 & 31 & 40 & 25 \\
   &   & -1 & 9 & -16 & -16 & -15 & -25 \\
\hline
-1 & 1 & -9 & 16 & 16 & 15 & 25 & \lfloor 0 \\
   &   & -1 & 10 & -26 & 10 & -25 & \\
\hline
5 & 1 & -10 & 26 & -10 & 25 & \lfloor 0 & \\
  &   & 5 & -25 & 5 & -25 & & \\
\hline
5 & 1 & -5 & 1 & -5 & \lfloor 0 & & \\
  &   & 5 & 0 & 5 & & & \\
\hline
  & 1 & 0 & 1 & \lfloor 0 & & & \\
\end{array}
$$

The synthetic division shows us that
$P(x) = (x+1)^2(x-5)^2(x^2+1)$
$$= (x+1)^2(x-5)^2(x+i)(x-i)$$
$x = i$ and $x = -i$ are the remaining zeros of $P(x)$.

47. $P(x) = x^3 - 2ix^2 - 9x + 18i$

$$
\begin{array}{r|rrrr}
2i & 1 & -2i & -9 & 18i \\
   &   & 2i & 0 & -18i \\
\hline
   & 1 & 0 & -9 & \lfloor 0 \\
\end{array}
$$

The synthetic division shows us that
$P(x) = (x - 2i)(x^2 - 9) = (x - 2i)(x + 3)(x - 3)$.

49. $P(x) = x^3 + (1-i)x^2 - (1+2i)x - 1 - i$

$$
\begin{array}{r|rrrr}
1+i & 1 & 1-i & -1-2i & -1-i \\
    &   & 1+i & 2+2i & 1+i \\
\hline
    & 1 & 2 & 1 & \lfloor 0 \\
\end{array}
$$

The synthetic division shows us that
$P(x) = [x - (1+i)][x^2 + 2x + 1]$
$$= (x - 1 - i)(x + 1)^2$$

51. a) By the remainder theorem,
$R = P(-1) = 3(-1)^{44} + 5(-1)^{41} + 4$
$$= 3 - 5 + 4$$
$$= 2$$

b) $R = P(1) = 1988(1)^3 - 1989(1)^2 + 1990(1) - 1991$
$$= -2$$

c) $R = P(-2) = (-2)^9 + 512$
$$= -512 + 512$$
$$= 0$$

53. a) $P(x) = (x^2 - 4)^3$
$$= [(x+2)(x-2)]^3 = (x+2)^3(x-2)^3$$
The zeros of $P(x)$ are -2 and 2, each of multiplicity 3.

b) $P(x) = (x^2 - 3x + 2)^2$
$$= [(x-2)(x-1)]^2 = (x-2)^2(x-1)^2$$
The zeros of $P(x)$ are 2 and 1, each of multiplicity 2.

c) $P(x) = (x^2 + 2x - 4)^3(x+2)^4$
Find the zeros of $x^2 + 2x - 4$ using the quadratic formula:

$$
\begin{aligned}
x &= \frac{-2 \pm \sqrt{4 - 4(1)(-4)}}{2(1)} \\
&= \frac{-2 \pm \sqrt{20}}{2} \\
&= \frac{-2 \pm 2\sqrt{5}}{2} \\
&= -1 \pm \sqrt{5}
\end{aligned}
$$

The zeros of $P(x)$ are $-1 + \sqrt{5}$ and $-1 - \sqrt{5}$, each of multiplicity 3, and -2, which is of multiplicity 4.

55. Let c be a positive number. Then
$$
\begin{aligned}
P(-c) &= 3(-c)^{31} - 2(-c)^{18} + 4(-c)^3 - (-c)^2 - 4 \\
&= -3c^{31} - 2c^{18} - 4c^3 - c^2 - 4 \\
&< 0,
\end{aligned}
$$
since each term in the polynomial is less than 0. Since $P(-c) \neq 0$, $-c$ is not a zero.

57. a)
$$
\begin{array}{r|rrrr}
\frac{1}{2} & 12 & 4 & -3 & -1 \\
            &   & 6 & 5 & 1 \\
\hline
            & 12 & 10 & 2 & \lfloor 0 \\
\end{array}
$$

$12x^3 + 4x^2 - 3x - 1 = (x - \frac{1}{2})(12x^2 + 10x + 2)$
$$= 2(x - \frac{1}{2})(6x^2 + 5x + 1)$$
$$= (2x - 1)(3x + 1)(2x + 1)$$

b)
$$
\begin{array}{r|rrrr}
\frac{1}{2} & 2 & -1 & -4 & 2 \\
            &   & 1 & 0 & -2 \\
\hline
            & 2 & 0 & -4 & \lfloor 0 \\
\end{array}
$$

$2x^3 - x^2 - 4x + 2 = (x - \frac{1}{2})(2x^2 - 4)$
$$= 2(x - \frac{1}{2})(x^2 - 2)$$
$$= (2x - 1)(x + \sqrt{2})(x - \sqrt{2})$$

c)
$$
\begin{array}{r|rrrr}
\frac{1}{2} & 2 & -1 & 2 & -1 \\
            &   & 1 & 0 & 1 \\
\hline
            & 2 & 0 & 2 & \lfloor 0 \\
\end{array}
$$

$2x^3 - x^2 + 2x - 1 = (x - \frac{1}{2})(2x^2 + 2)$
$$= 2(x - \frac{1}{2})(x^2 + 1)$$

$$= (2x - 1)(x + i)(x - i)$$

59. $y = x^3 + 4x^2 - 2x$
$$= x(x^2 + 4x - 2)$$

To sketch the graph, find the x-intercepts.
Clearly, $x = 0$ is one. To find the others, solve $x^2 + 4x - 2 = 0$.

$$x = \frac{-4 \pm \sqrt{(4)^2 - 4(1)(-2)}}{2}$$
$$= \frac{-4 \pm \sqrt{24}}{2}$$
$$= \frac{-4 \pm 2\sqrt{6}}{2}$$
$$= -2 \pm \sqrt{6}$$

The remaining x-intercepts are
$$x \approx .45 \text{ and } x \approx -4.45$$

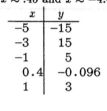

x	y
-5	-15
-3	15
-1	5
0.4	-0.096
1	3

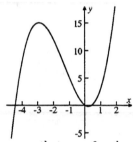

From the graph, it appears that $y = 8$ when $x = -4$. To check this, and to find the other values of x at which $y = 8$, consider the polynomial
$$P(x) = x^3 + 4x^2 - 2x - 8$$

The zeros of $P(x)$ will be the values of x at which $x^3 + 4x^2 - 2x = 8$.

$$\begin{array}{r|rrrr} -4 & 1 & 4 & -2 & -8 \\ & & -4 & 0 & 8 \\ \hline & 1 & 0 & -2 & \;\;\underline{|\;0} \end{array}$$

$$P(x) = (x + 4)(x^2 - 2)$$
$$= (x + 4)(x + \sqrt{2})(x - \sqrt{2})$$

The remaining zeros of $P(x)$, and hence the values of x at which $y = 8$, are $x = \sqrt{2}$ and $x = -\sqrt{2}$.

61. Using the formula $V = lwh$, with $l = 16 - 2x$, $w = 16 - 2x$, and $h = x$, we want
$$(16 - 2x)(16 - 2x)x = 300$$
$$(8 - x)(8 - x)x = 75$$
$$(64 - 16x + x^2)x = 75$$
$$x^3 - 16x^2 + 64x - 75 = 0$$

Substituting $x = 1$, $x = 2$, $x = 3$, we find that $x = 3$ is a root of the equation. We can use synthetic division to find the other roots.

$$\begin{array}{r|rrrr} 3 & 1 & -16 & 64 & -75 \\ & & 3 & -39 & 75 \\ \hline & 1 & -13 & 25 & \;\;\underline{|\;0} \end{array}$$

$$x^3 - 16x^2 + 64x - 75 = (x - 3)(x^2 - 13x + 25)$$

Solve $x^2 - 13x + 25 = 0$ using the quadratic formula.

$$x = \frac{13 \pm \sqrt{(-13)^2 - 4(1)(25)}}{2}$$
$$= \frac{13 \pm \sqrt{69}}{2}$$
$$x \approx 10.7 \text{ or } x = 2.35$$

The values of x that make sense in the physical problem are $x = 3$ inches or $x = 2.35$ inches.

63. By the Complete Factorization Theorem,
$$P(x) = a(x - \tfrac{1}{2})^4.$$
Substituting $x = 0$,
$$P(0) = a(0 - \tfrac{1}{2})^4 = 1$$
$$a(\tfrac{1}{2})^4 = 1$$
$$a = 2^4$$

Therefore,
$$P(x) = 2^4(x - \tfrac{1}{2})^4 = [2(x - \tfrac{1}{2})]^4 = (2x - 1)^4$$

Using the binomial theorem,
$$(2x - 1)^4$$
$$= (2x)^4 + 4(2x)^3(-1) + 6(2x)^2(-1)^2$$
$$+ 4(2x)(-1)^3 + (-1)^4$$
$$= 16x^4 - 32x^3 + 24x^2 - 8x + 1$$

65. Let c_1, c_2, ..., c_n, c_{n+1} be the $n + 1$ distinct zeros of $P(x)$. Using just the first n of these, by the Complete Factorization Theorem, we can say that
$$P(x) = a_n(x - c_1)(x - c_2) \cdots (x - c_n). \text{ Let } x = c_{n+1}.$$
$$P(c_{n+1}) = a_n(c_{n+1} - c_1)(c_{n+1} - c_2) \cdots (c_{n+1} - c_n)$$
$$= 0$$

Since c_{n+1} is distinct from all the other zeros,
$$c_{n+1} - c_1 \neq 0, \quad c_{n+1} - c_2 \neq 0, \quad ..., \quad c_{n+1} - c_n \neq 0.$$
For the product to equal 0, $a_n = 0$. Therefore,
$$P(x) = a_{n-1}x^{n-1} + a_{n-2}x^{n-2} + \cdots + a_1 x + a_0.$$
$P(x)$ still has $n + 1$ distinct roots. The same argument as used above will show that $a_{n-1} = 0$, and then $a_{n-2} = 0$; in fact all the coefficients must equal 0.

67. Let $P(x) = x^6 - 5x^5 + 3x^4 + 7x^3 + 3x^2 - 5x + 1$.
It is given that
$$P(c) = c^6 - 5c^5 + 3c^4 + 7c^3 + 3c^2 - 5c + 1 = 0.$$
Clearly $c \neq 0$ and so $\frac{1}{c}$ is a number.

$$P\left(\tfrac{1}{c}\right) = \frac{1}{c^6} - \frac{5}{c^5} + \frac{3}{c^4} + \frac{7}{c^3} + \frac{3}{c^2} - \frac{5}{c} + 1$$
$$= \frac{1}{c^6}(1 - 5c + 3c^2 + 7c^3 + 3c^4 - 5c^5 + c^6)$$
$$= \frac{1}{c^6}(P(c)) = 0$$

Therefore, $\frac{1}{c}$ is a solution of $P(x) = 0$.

69. $a_n x^n + a_{n-1} x^{n-1} + a_{n-2} x^{n-2} + \cdots + a_1 x + a_0$
$= a_n(x - c_1)(x - c_2) \cdots (x - c_n)$

a) When this is multiplied out, the coefficient of x^{n-1} is $a_n(-c_1 - c_2 - \cdots - c_n)$. This must equal a_{n-1}. Therefore, $c_1 + c_2 + \cdots + c_n = \dfrac{-a_{n-1}}{a_n}$.

b) The coefficient of x^{n-2} is
$a_n(c_1 c_2 + c_1 c_3 + \cdots + c_1 c_n + c_2 c_3 + c_2 c_4 + \cdots$
$\hspace{5cm} + c_{n-1} c_n)$
This must equal a_{n-2}. Therefore,
$c_1 c_2 + \cdots + c_1 c_n + c_2 c_3 + \cdots + c_2 c_n + \cdots + c_{n-1} c_n$
$= \dfrac{a_{n-2}}{a_n}$

c) The constant term a_0 must equal
$a_n(-c_1)(-c_2) \cdots (-c_n) = (-1)^n a_n c_1 c_2 \cdots c_n$.
Therefore, $c_1 c_2 \cdots c_n = \dfrac{a_0}{(-1)^n a_n} = (-1)^n \dfrac{a_0}{a_n}$.

71. Set new range values:
Ymax: 350
Press $\boxed{Y=}$.
Enter X $\boxed{(}$ 16 $-$ 2X $\boxed{)}$ ^ 2
Press $\boxed{\text{GRAPH}}$.
Zoom in on the local maximum near $x = 2.8$. The value of x that maximizes V is close to 2.67; $V = 303.4$.

Problem Set 10.3

1. $\overline{2 + 3i} = 2 - 3i$

3. $\overline{4i} = -4i$

5. $\overline{4 + \sqrt{6}} = 4 + \sqrt{6}$

7. $\overline{(2 - 3i)^8} = (\overline{2 - 3i})^8 = (2 + 3i)^8$

9. $\overline{2(1 + 2i)^3 - 3(1 + 2i)^2 + 5}$
$= \overline{2(1 + 2i)^3} - \overline{3(1 + 2i)^2} + \overline{5}$
$= \overline{2}\,\overline{(1 + 2i)^3} - \overline{3}\,\overline{(1 + 2i)^2} + \overline{5}$
$= 2(\overline{1 + 2i})^3 - 3(\overline{1 + 2i})^2 + 5$
$= 2(1 - 2i)^3 - 3(1 - 2i)^2 + 5$

11. Since zeros come in conjugate pairs, the other zero must be $5 + i$.

13. Since zeros come in conjugate pairs, the other two zeros must be $3 + 2i$ and $5 - 4i$.

15. Since i is one solution, we know that $-i$ is another.

$$
\begin{array}{r|rrrr}
i & 2 & -1 & 2 & -1 \\
& & 2i & -2 - i & 1 \\
\hline
-i & 2 & -1 + 2i & -i & \boxed{0} \\
& & -2i & i & \\
\hline
& 2 & -1 & \boxed{0} &
\end{array}
$$

The synthetic division shows us that
$2x^3 - x^2 + 2x - 1 = (x - i)(x + i)(2x - 1)$.
Therefore, the remaining solution of the equation is found by solving $2x - 1 = 0$
$$x = \tfrac{1}{2}$$
The three roots of the equation are $x = i$, $x = -i$, and $x = \tfrac{1}{2}$.

17. Since $1 + 3i$ is a solution, we know that $1 - 3i$ is another.

$$
\begin{array}{r|rrrrr}
1+3i & 1 & 1 & 6 & 26 & 20 \\
& & 1 + 3i & -7 + 9i & -28 + 6i & -20 \\
\hline
1-3i & 1 & 2 + 3i & -1 + 9i & -2 + 6i & \boxed{0} \\
& & 1 - 3i & 3 - 9i & 2 - 6i & \\
\hline
& 1 & 3 & 2 & \boxed{0} &
\end{array}
$$

The synthetic division shows us that
$x^4 + x^3 + 6x^2 + 26x + 20$
$\hspace{1cm} = [x - (1 + 3i)][x - (1 - 3i)][x^2 + 3x + 2]$
The remaining roots of the equation are found by solving $x^2 + 3x + 2 = 0$
$$(x + 2)(x + 1) = 0$$
$$x = -2 \text{ or } x = -1$$
The 4 roots of the equation are $x = 1 + 3i$, $x = 1 - 3i$, $x = -2$, and $x = -1$.

19. The second zero must be $\overline{2 + 5i} = 2 - 5i$.
The polynomial is
$P(x) = [x - (2 + 5i)][x - (2 - 5i)]$
$\hspace{1cm} = [(x - 2) - 5i][(x - 2) + 5i]$
$\hspace{1cm} = (x - 2)^2 - (5i)^2$
$\hspace{1cm} = x^2 - 4x + 4 - 25i^2$
$\hspace{1cm} = x^2 - 4x + 4 + 25$
$\hspace{1cm} = x^2 - 4x + 29$

21. Since $2i$ is a zero, we know that $-2i$ must also be a zero.
$P(x) = (x - 2i)(x + 2i)(x + 3)$
$\hspace{1cm} = (x^2 + 4)(x + 3)$
$\hspace{1cm} = x^3 + 3x^2 + 4x + 12$

23. Since $3i$ is a zero of multiplicity 2, we know that $-3i$ must also be a zero of multiplicity 2.
$P(x) = (x - 3i)^2(x + 3i)^2(x - 2)$
$\hspace{1cm} = (x - 3i)(x + 3i)(x - 3i)(x + 3i)(x - 2)$
$\hspace{1cm} = (x^2 + 9)(x^2 + 9)(x - 2)$
$\hspace{1cm} = (x^4 + 18x^2 + 81)(x - 2)$
$\hspace{1cm} = x^5 - 2x^4 + 18x^3 - 36x^2 + 81x - 162$

25. $x^3 - 3x^2 - x + 3 = 0$

The only possible rational roots are $\frac{c}{d}$ where $c = \pm 1, \pm 3$ and $d = \pm 1$.

Therefore, $\frac{c}{d} = \pm 1$ or ± 3. We try by synthetic division:

$$\begin{array}{r|rrrr} 1 & 1 & -3 & -1 & 0 \\ & & 1 & -2 & -3 \\ \hline & 1 & -2 & -3 & \boxed{0} \end{array}$$

So 1 is a solution. We could continue by trial and error, but the quotient is $x^2 - 2x - 3$ so we factor: $(x-3)(x+1)$, yielding $x = -1, +3$.

The solution set is $\{1, -1, +3\}$.

27. $2x^3 + 3x^2 - 4x + 1 = 0$

The only possible rational roots are $\frac{c}{d}$ where $c = \pm 1$ and $d = \pm 1, \pm 2$.

Therefore, $\frac{c}{d} = \pm 1, \pm \frac{1}{2}$. Trying these in order from left to right using synthetic division, the first root we come to is $x = \frac{1}{2}$:

$$\begin{array}{r|rrrr} \frac{1}{2} & 2 & 3 & -4 & 1 \\ & & 1 & 2 & -1 \\ \hline & 2 & 4 & -2 & \boxed{0} \end{array}$$

Therefore,
$$2x^3 + 3x^2 - 4x + 1 = (x - \tfrac{1}{2})(2x^2 + 4x - 2)$$
$$= 2(x - \tfrac{1}{2})(x^2 + 2x - 1)$$

To find the additional roots, solve $x^2 + 2x - 1 = 0$.
$$x = \frac{-2 \pm \sqrt{4+4}}{2}$$
$$= \frac{-2 \pm 2\sqrt{2}}{2}$$
$$= -1 \pm \sqrt{2}$$

The solution set is $\{\tfrac{1}{2}, -1 \pm \sqrt{2}\}$.

29. $\frac{1}{3}x^3 - \frac{1}{2}x^2 - \frac{1}{6}x + \frac{1}{6} = 0$

Multiply by 6 to clear fractions.
$$2x^3 - 3x^2 - x + 1 = 0$$

The only possible rational roots are $\frac{c}{d}$ where $c = \pm 1$ and $d = \pm 1, \pm 2$.

Therefore, $\frac{c}{d} = \pm 1, \pm \frac{1}{2}$. Trying these in order from left to right using synthetic division, the first root we come to is $x = \frac{1}{2}$:

$$\begin{array}{r|rrrr} \frac{1}{2} & 2 & -3 & -1 & 1 \\ & & 1 & -1 & -1 \\ \hline & 2 & -2 & -2 & \boxed{0} \end{array}$$

Therefore,
$$2x^3 - 3x^2 - x + 1 = (x - \tfrac{1}{2})(2x^2 - 2x - 2)$$
$$= 2(x - \tfrac{1}{2})(x^2 - x - 1)$$

To find the additional roots, solve $x^2 - x - 1 = 0$.
$$x = \frac{1 \pm \sqrt{1+4}}{2}$$
$$= \frac{1 \pm \sqrt{5}}{2}$$

The solution set is $\left\{\frac{1}{2}, \frac{1 \pm \sqrt{5}}{2}\right\}$.

31. If $2 + i$ is a solution to $x^4 - 3x^3 + 2x^2 + x + 5 = 0$, then $\overline{2+i} = 2 - i$ is also a solution.

$$\begin{array}{r|rrrrr} 2+i & 1 & -3 & 2 & 1 & 5 \\ & & 2+i & -3+i & -3+i & i^2-4 \\ \cline{2-6} 2-i & 1 & -1+i & -1+i & -2+i & \boxed{0} \\ & & 2-i & 2-i & 2-i & \\ \hline & 1 & 1 & 1 & \boxed{0} & \end{array}$$

The quotient is $x^2 + x + 1$. Using the quadratic formula to get the other two zeros,
$$x = \frac{-1 \pm \sqrt{1-4}}{2} = \frac{-1 \pm \sqrt{-3}}{2}$$
$$= \frac{-1 \pm i\sqrt{3}}{2}$$

So the solutions are $2 - i$, $2 + i$, and $\dfrac{-1 \pm i\sqrt{3}}{2}$.

33. $x^4 - 3x^3 - 20x^2 - 24x - 8 = 0$

Look for rational solutions first. Any rational solution must be of the form $\frac{c}{d}$, where $c = \pm 1, \pm 2, \pm 4, \pm 8$ and $d = \pm 1$.

Therefore, $\frac{c}{d} = \pm 1, \pm 2, \pm 4, \pm 8$. Trying these in order from left to right using synthetic division, the first root we come to is $x = -1$:

$$\begin{array}{r|rrrrr} -1 & 1 & -3 & -20 & -24 & -8 \\ & & -1 & 4 & 16 & 8 \\ \hline & 1 & -4 & -16 & -8 & \boxed{0} \end{array}$$

Continuing to test $\frac{c}{d}$ on the quotient obtained above, being sure to retest $x = -1$, the next root we come to is $x = -2$:

$$\begin{array}{r|rrrr} -2 & 1 & -4 & -16 & -8 \\ & & -2 & 12 & 8 \\ \hline & 1 & -6 & -4 & \boxed{0} \end{array}$$

Since the quotient is quadratic, the remaining roots can be found by solving $x^2 - 6x - 4 = 0$:
$$x = \frac{6 \pm \sqrt{36+16}}{2}$$
$$= \frac{6 \pm \sqrt{52}}{2}$$
$$= 3 \pm \sqrt{13}$$

The solution set is $\{-1, -2, 3 \pm \sqrt{13}\}$.

35. $2x^5 - 3x^4 + 13x^3 - 22x^2 - 24 + 16 = 0$ will have rational solutions of the form $\frac{c}{d}$, where $c = \pm 1$, ± 2, ± 4, ± 8, ± 16 and $d = \pm 1$, ± 2.

Therefore, $\frac{c}{d} = \pm 1$, ± 2, ± 4, ± 8, ± 16, $\pm \frac{1}{2}$. Trying these in order from left to right, the first root we come to is $x = -1$:

$$\begin{array}{r|rrrrrr} -1 & 2 & -3 & 13 & -22 & -24 & 16 \\ & & -2 & 5 & -18 & 40 & -16 \\ \hline & 2 & -5 & 18 & -40 & 16 & \underline{0} \end{array}$$

Continuing to test $\frac{c}{d}$ on the quotient above, being sure to retest $x = -1$, the next root we come to is $x = 2$:

$$\begin{array}{r|rrrrr} 2 & 2 & -5 & 18 & -40 & 16 \\ & & 4 & -2 & 32 & -16 \\ \hline & 2 & -1 & 16 & -8 & \underline{0} \end{array}$$

Continuing to test $\frac{c}{d}$ on the quotient obtained above, being sure to retest $x = 2$, but noting that it is not necessary to test ± 16, the next root we come to is $x = \frac{1}{2}$:

$$\begin{array}{r|rrrr} \frac{1}{2} & 2 & -1 & 16 & -8 \\ & & 1 & 0 & 8 \\ \hline & 2 & 0 & 16 & \underline{0} \end{array}$$

At this point, the quotient is quadratic, so the remaining roots can be found by solving

$$2x^2 + 16 = 0$$
$$x^2 = -8$$
$$x = \pm\sqrt{-8}$$
$$= \pm 2i\sqrt{2}$$

The solution set is $\{-1, 2, \frac{1}{2}, \pm 2i\sqrt{2}\}$.

37. $x^5 + 6x^4 - 34x^3 + 56x^2 - 39x + 10 = 0$

First, look for rational roots. Any rational root will be of the form $\frac{c}{d}$, where $c = \pm 1$, ± 2, ± 5, ± 10 and $d = \pm 1$.

Therefore, $\frac{c}{d} = \pm 1$, ± 2, ± 5, ± 10. Trying these in order from left to right, the first root we come to is $x = 1$:

$$\begin{array}{r|rrrrrr} 1 & 1 & 6 & -34 & 56 & -39 & 10 \\ & & 1 & 7 & -27 & 29 & -10 \\ \hline & 1 & 7 & -27 & 29 & -10 & \underline{0} \end{array}$$

Continuing to test the possibilities on the quotient obtained above, we find that $x = 1$ is again a root:

$$\begin{array}{r|rrrrr} 1 & 1 & 7 & -27 & 29 & -10 \\ & & 1 & 8 & -19 & 10 \\ \hline & 1 & 8 & -19 & 10 & \underline{0} \end{array}$$

Trying the possibilities again on the previous quotient, we find that $x = 1$ is again a root:

$$\begin{array}{r|rrrr} 1 & 1 & 8 & -19 & 10 \\ & & 1 & 9 & -10 \\ \hline & 1 & 9 & -10 & \underline{0} \end{array}$$

At this point, the quotient is quadratic, so we can find any remaining roots by solving

$$x^2 + 9x - 10 = 0$$
$$(x + 10)(x - 1) = 0$$
$$x = -10 \text{ or } x = 1$$

The roots are $x = 1$ (of multiplicity 4) and $x = -10$.

39. Any polynomial equation with real coefficients and with a complex root $a + bi$ also has the conjugate $a - bi$ as a root. Hence, such an equation has an even number of complex solutions. But an equation of degree n has n roots. If n is odd, there must therefore be (at least) one real solution.

41. Let $u = a + bi$, where a and b are real. Then $\bar{u} = a - bi$.

$u + \bar{u} = (a + bi) + (a - bi) = 2a$, a real number.

$$\begin{aligned} u\bar{u} &= (a + bi)(a - bi) \\ &= a^2 - abi + abi - b^2i^2 \\ &= a^2 + b^2, \text{ a real number.} \end{aligned}$$

43. Finding the roots of the equation $x^4 + 3x^3 + 3x^2 - 3x - 4 = 0$ will allow us to write the polynomial in factored form. Look for rational roots first. Any rational root will be of the form $\frac{c}{d}$, where $c = \pm 1$, ± 2, ± 4 and $d = \pm 1$.

Therefore, $\frac{c}{d} = \pm 1$, ± 2, ± 4. Trying the possibilities in order from left to right, the first root we come to is $x = 1$:

$$\begin{array}{r|rrrrr} 1 & 1 & 3 & 3 & -3 & -4 \\ & & 1 & 4 & 7 & 4 \\ \hline & 1 & 4 & 7 & 4 & \underline{0} \end{array}$$

Trying the possibilities again on the quotient obtained above, the next root we come to is $x = -1$:

$$\begin{array}{r|rrrr} -1 & 1 & 4 & 7 & 4 \\ & & -1 & -3 & -4 \\ \hline & 1 & 3 & 4 & \underline{0} \end{array}$$

Since the quotient is quadratic, at this point we can write the polynomial as the product of linear and quadratic factors.

$$\begin{aligned} x^4 + 3x^3 + 3x^2 - 3x - 4 \\ = (x - 1)(x + 1)(x^2 + 3x + 4) \end{aligned}$$

45. $x^8 - 1$
$$\begin{aligned} &= (x^4 - 1)(x^4 + 1) \\ &= (x^2 - 1)(x^2 + 1)(x^4 + 2x^2 + 1 - 2x^2) \\ &= (x - 1)(x + 1)(x^2 + 1)[(x^2 + 1)^2 - (\sqrt{2}x)^2] \\ &= (x - 1)(x + 1)(x^2 + 1)(x^2 + 1 - \sqrt{2}x)(x^2 + 1 + \sqrt{2}x) \\ &= (x - 1)(x + 1)(x^2 + 1)(x^2 - \sqrt{2}x + 1)(x^2 + \sqrt{2}x + 1) \end{aligned}$$

47. Let $r = \sqrt{5} - 2$ and $s = \sqrt{5} + 2$.
Then $x = r^{1/3} - s^{1/3}$, and
$$x^3 = r - 3r^{2/3}s^{1/3} + 3r^{1/3}s^{2/3} - s$$
$$= (r - s) - 3r^{1/3}s^{1/3}(r^{1/3} - s^{1/3})$$
$$= -4 - 3r^{1/3}s^{1/3}x$$
Now, since $rs = 5 - 4 = 1$, $r^{1/3}s^{1/3} = 1$.
Therefore, $x^3 = -4 - 3x$, and so
$$x^3 + 3x + 4 = 0$$
$$(x + 1)(x^2 - x + 4) = 0$$
Since x is real, it has to equal -1. (The zeros of $x^2 - x + 4$ are not real.)

49. The graph touches but does not cross the x-axis at a real zero of even multiplicity; it is tangent to and crosses the x-axis at a real 0 of odd multiplicity greater than 1.

Problem Set 10.4

1. $P(x) = 2x^2 - 5x + 6$
$P'(x) = 2(2)x - 5 = 4x - 5$
Slope of tangent line at $x = 1$ is
$P'(1) = 4 \cdot 1 - 5 = -1$

3. $P(x) = 2x^2 + x - 2$
$P'(x) = 2(2)x + 1 = 4x + 1$
Slope of tangent line at $x = 1$ is
$P'(1) = 4 \cdot 1 + 1 = 5$

5. $P(x) = 2x^5 + x^4 - 2x^3 + 8x - 4$
$P'(x) = 5(2)x^4 + 4x^3 + 3(-2)x^2 + 2(0)x + 8$
$= 10x^4 + 4x^3 - 6x^2 + 8$
Slope of tangent line at $x = 1$ is
$P'(1) = 10(1)^4 + 4(1)^3 - 6(1)^2 + 8 = 16$

7. $y = x^3 + 2x - 5$

x	y
-2	-17
-1	-8
0	-5
1	-2
2	7

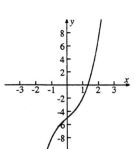

A first guess at a solution might be $x = 1.4$. Calculate values of y for x near 1.4 until a change of sign in y occurs.

x	y
1.3	-0.203
1.4	0.544

Pretend that the graph is a straight line in the interval $1.3 \le x \le 1.4$.

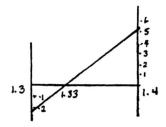

The straight line seems to cross the x-axis near 1.33. Use this as a second approximation to the solution. Calculate values of of y for x near 1.33 until a change of sign in y occurs.

x	y
1.32	-0.060
1.33	0.013

Pretend that the graph is a straight line in the interval $1.32 \le x \le 1.33$.

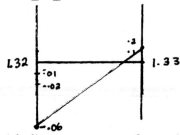

The straight line seems to cross the x-axis at 1.328, which is a fairly good approximation to the solution. Notice that $(1.328)^3 + 2(1.328) - 5 = .0019$.

9. $y = x^3 - 3x - 10$

x	y
-2	-12
-1	-8
0	-10
1	-12
2	-8
3	8

A first good guess at a solution might be $x = 2.5$. Calculate values of y for x near 2.5 until a change of sign in y occurs.

x	y
2.5	-1.875
2.6	-0.224
2.7	1.583

Pretend that the graph is a straight line in the interval $2.6 \leq x \leq 2.7$.

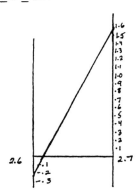

The straight line seems to cross the x-axis at 2.61. Use this as a second approximation to the solution. Calculate values of y for x near 2.61 until a change of sign in y occurs.

x	y
2.61	-0.050
2.62	0.125

Pretend that the graph is a straight line in the interval $2.61 \leq x \leq 2.62$.

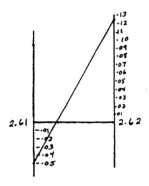

The straight line seems to cross the x-axis at 2.613, which is a fairly good approximation to the solution. Notice that
$$(2.613)^3 - 3(2.613) - 10 = .00196.$$

▷ It can be shown that Newton's method has an error-squaring property, so that if an approximation agrees with the previous one to n decimal places, as indicated by the underscores, it is accurate to about $2n$ decimal places. To avoid loss of accuracy, use your calculator's memory to store each approximation, then recall it from memory when calculating the next one.
We use Min and MR to store and retrieve; your calculator may use STO and RCL or x→M and RM.

11. $P(x) = x^3 + 2x - 5$
$P'(x) = 3x^2 + 2$
$x_1 = 1.4$ Min
$x_2 = x_1 - \dfrac{P(x_1)}{P'(x_1)}$. This is calculated as

MR $-$ (MR y^x 3 + 2 × MR $-$ 5)
  $\div$ (3 × MR x^2 + 2) $=$
The calculator will display 1.330964467. Press Min and repeat the calculation to get $x_3 = 1.\underline{32}827282$. Press Min and repeat the calculation to get $x_4 = 1.\underline{328}238856$. $x \approx 1.32826885$, to 8 decimals.

13. $P(x) = x^3 - 3x - 10$
$P'(x) = 3x^2 - 3$
$x_1 = 2.5$ Min
$x_2 = x_1 - \dfrac{P(x_1)}{P'(x_1)}$. This is calculated as

MR $-$ (MR y^x 3 $-$ 3 × MR $-$ 10)
  $\div$ (3 × MR x^2 $-$ 3) $=$
The calculator will display 2.619047619. Press Min and repeat the calculation to get $x_3 = 2.\underline{612}904811$. Press Min and repeat the calculation to get $x_4 = 2.\underline{612}887865$. $x \approx 2.612887$, to 6 decimals.

15. $y = P(x) = x^4 + x^3 - 3x^2 + 4x - 28$
$P'(x) = 4x^3 + 3x^2 - 6x + 4$

x	y
-4	100
-3	-13
-2	-40
-1	-35
0	-28
1	-25
2	-8
3	65

It is not really necessary to sketch the graph to see that there will be two solutions: one in the interval $-4 < x < -3$ and one in the interval $2 < x < 3$.
To find the first solution, a reasonable choice for an initial guess is -3.5.
$x_1 = 3.5$ +/− Min
$x_2 = x_1 - \dfrac{P(x_1)}{P'(x_1)}$. This is calculated as

MR $-$
(MR y^x 4 + MR y^x 3 $-$ 3 × MR x^2 + 4 × MR $-$ 2)
  $\div$ (4 × MR y^x 3 + 3 × MR x^2 $-$ 6 × MR + 4) $=$
The calculator will display -3.240888383. Press Min and repeat the calculation to get $x_3 = -3.19401118$. Press Min and repeat the calculation to get $x_4 = -3.\underline{1925}837$.
$x \approx 3.1925$, to 4 decimals $= 3.193$ to 3 decimals.

To find the second solution, a reasonable choice for a first guess is 2.5.

$x_1 = 2.5$ $\boxed{\text{Min}}$ Repeat the calculation as above to get $x_2 = 2.244661922$. Press $\boxed{\text{Min}}$ and repeat:
$x_3 = 2.194365201$. Press $\boxed{\text{Min}}$ and repeat:
$x_4 = 2.\underline{192}584573$.
$x \approx 2.1925$, to 4 decimals $= 2.193$ to 3 decimals.

17. $y = P(x) = x^3 - 3x + 1$
$P'(x) = 3x^2 - 3$

x	y
-3	-17
-2	-1 }
-1	3 }
0	1 }
1	-1 }
2	3 }

It is not really necessary to sketch the graph to see that three solutions must occur: one in the interval $-2 < x < -1$, one in the interval $0 < x < 1$, and one in the interval $1 < x < 2$.

To find the first solution, our initial guess is -1.5.
$x_1 = 1.5$ $\boxed{+/-}$ $\boxed{\text{Min}}$
$x_2 = x_1 - \dfrac{P(x_1)}{P'(x_1)}$. This is calculated as

$\boxed{\text{MR}} - (\boxed{\text{MR}}\boxed{y^x}3 - 3\times\boxed{\text{MR}}+1)$
$\div(3\times\boxed{\text{MR}}\boxed{x^2}-3)\boxed{=}$

The calculator will display -2.066666667. Since this is outside the interval $-2 < x < -1$, we'll start again, with $x_1 = -2$:
$x_1 = \boxed{+/-}\boxed{\text{Min}}$ Repeat the calculation as above, to get $x_2 = -1.888888889$. Press $\boxed{\text{Min}}$ and repeat:
$x_3 = -1.\underline{879}451567$. Press $\boxed{\text{Min}}$ and repeat:
$x_4 = -1.\underline{879}385245$.
$x \approx -1.879385$, to 6 places $= -1.879$ to 3 places.
To find the second solution, a good initial guess is $x_1 = .5$ $\boxed{\text{Min}}$ Repeat the calculation as above, to get $x_2 = 0.333333333$. Press $\boxed{\text{Min}}$ and repeat:
$x_3 = 0.\underline{347}666666$. Press $\boxed{\text{Min}}$ and repeat:
$x_4 = 0.\underline{3472}963532$.
$x \approx 0.347296$, to 6 decimals $= 0.347$ to 3 decimals.
To find the third solution, a good initial guess is $x_1 = 1.5$ $\boxed{\text{Min}}$ Repeat the calculation as above, to get $x_2 = 1.\underline{5}33333333$. Press $\boxed{\text{Min}}$ and repeat:
$x_3 = 1.\underline{532}090643$.
$x \approx 1.5320$, to 4 decimals $= 1.532$ to 3 decimals.

19. Solve $y = x + 3$ and $y = x^3 - 3x + 4$ simultaneously. Equate values of y:
$x^3 - 3x + 4 = x + 3$
$x^3 - 4x + 1 = 0$
$P(x) = x^3 - 4x + 1$
$P'(x) = 3x^2 - 4$

x	y
-4	-47
-3	-14 }
-2	1 }
-1	4
0	1 }
1	-2 }
2	1 }

The 3 roots are in the intervals $-3 < x < -2$, $0 < x < 1$, and $1 < x < 2$.
The first root is, ostensibly, very close to -2, so we'll choose -2 as our initial guess.
$x_1 = 2$ $\boxed{+/-}\boxed{\text{Min}}$
$x_2 = x_1 - \dfrac{P(x_1)}{P'(x_1)}$. This is calculated as

$\boxed{\text{MR}} - (\boxed{\text{MR}}\boxed{y^x}3 - 4\times\boxed{\text{MR}}+1)$
$\div(3\times\boxed{\text{MR}}\boxed{x^2}-4)\boxed{=}$

The calculator will display -2.125. Press $\boxed{\text{Min}}$ and repeat the calculation to get $x_3 = -2.\underline{11}4975450$.
$x = -2.11$ to 2 decimal places.
For the second root, a good initial guess is $x_1 = .5$ $\boxed{\text{Min}}$ Repeat the calculation as above, to get $x_2 = 0.2307692308$. Press $\boxed{\text{Min}}$ and repeat:
$x_3 = 0.\underline{25}40002371$. $x = 0.25$ to 2 decimal places.
For the third root, our initial guess is $x_1 = 1.5$ $\boxed{\text{Min}}$ Repeat the calculation to get $x_2 = 2.090909091$. Since this is outside the interval $1 < x < 2$, we'll choose 2 as x_1:
$x_1 = 2$ $\boxed{\text{Min}}$ Repeat the calculation:
$x_2 = 1.875$. Press $\boxed{\text{Min}}$ and repeat:
$x_3 = 1.\underline{86}097852$. $x = 1.86$ to 2 decimal places.

21. $P(x) = x^4 - 4x^3 - 2x^2 + 4x + 5$
$P'(x) = 4x^3 - 12x^2 - 4x + 4$
Let x_1 be 4 $\boxed{\text{Min}}$
$x_2 = x_1 - \dfrac{P(x_1)}{P'(x_1)}$. This is calculated as

$\boxed{\text{MR}} -$
$(\boxed{\text{MR}}\boxed{y^x}4 - 4\times\boxed{\text{MR}}\boxed{y^x}3 - 2\times\boxed{\text{MR}}\boxed{x^2}+4\times\boxed{\text{MR}}+5)$
$\div(4\times\boxed{\text{MR}}\boxed{y^x}3 - 12\times\boxed{\text{MR}}\boxed{x^2}-4\times\boxed{\text{MR}}+4)\boxed{=}$
The calculator will display 4.211538462. Press $\boxed{\text{Min}}$ and repeat:
$x_3 = 4.181802223$. Press $\boxed{\text{Min}}$ and repeat:
$x_4 = 4.\underline{181}125791$.
$x \approx 4.181125$. $50x = 50(4.181125) = 209.05625$.

▷ Do not confuse the i in Problem 23 with the imaginary unit.

23. With $x = 1 + i$, solve $5x^3 - 2x^2 - 2x - 2 = 0$.

$P(x) = 5x^3 - 2x^2 - 2x - 2$

$P'(x) = 15x^2 - 4x - 2$

10% is a good initial guess for interest rate problems, so let

$x_1 = 1.10$ [Min]

$x_2 = x_1 - \dfrac{P(x_1)}{P'(x_1)}$. This is calculated as

[MR] $- (5 \times$ [MR] [y^x] $3 - 2 \times$ [MR] [x^2] $- 2 \times$ [MR] $- 2)$

$\div (15 \times$ [MR] [x^2] $- 4 \times$ [MR] $- 2)$ [=]

The calculator will display 1.097021277. Press [Min] and repeat:

$x_3 = 1.\underline{097010258}$

$x \approx 1.0970102 = 1 + i$

$.0970102 = i$

$i \approx 9.701\%$

25. If $f(x) = 3x^2$, then $f'(x) = 6x$.

The slope of the tangent line at $(2, 12)$ is $m = f'(2) = 6(2) = 12$. The tangent line at $(2, 12)$ passes through $(2, 12)$ and has slope 12, so its equation is $y - 12 = 12(x - 2)$.

27. a) $P(x) = 2x^3 - 3x^2 - 36x + 10$

$P'(x) = 6x^2 - 6x - 36$

The tangent line will be horizontal when its slope is 0. Solve

$6x^2 - 6x - 36 = 0$

$x^2 - x - 6 = 0$

$(x - 3)(x + 2) = 0$

$x = 3$ or $x = -2$

The tangent line is horizontal at $x = -2$ and $x = 3$.

b) $P(x) = 3x^4 - 8x^3 - 6x^2 + 9$

$P'(x) = 12x^3 - 24x^2 - 12x$

The tangent line will be horizontal when its slope is 0. Solve

$12x^3 - 24x^2 - 12x = 0$

$12x(x^2 - 2x - 1) = 0$

$12x = 0$ or $x^2 - 2x - 1 = 0$

$x = \dfrac{2 \pm \sqrt{4 + 4}}{2}$

$x = 1 \pm \sqrt{2}$

The tangent line is horizontal at $x = 0$, $x = 1 + \sqrt{2}$, and $x = 1 - \sqrt{2}$.

29. a) The coordinates of P and Q are $(a, 2a^3)$ and $(a + h, 2(a + h)^3)$

$m_h = \dfrac{y_2 - y_1}{x_2 - x_1} = \dfrac{2(a + h)^3 - 2a^3}{h}$

b) $ = \dfrac{2[a^3 + 3a^2h + 3ah^2 + h^3 - a^3]}{h}$

$ = 6a^2 + 6ah + 2h^2$

c) As $h \to 0$, $m_h \to 6a^2 + 6a(0) + 2(0)^2 = 6a^2$

d) According to the slope theorem, if $y = P(x) = 2x^3$, then the slope of the tangent line at x is $P'(x) = 6x^2$. Thus, the value of the slope at $x = a$ is $P'(a) = 6a^2$.

This agrees with the answer to part (c).

31. Set standard range values.

Press [Y=].

Enter $\mathrm{X} \wedge 5 - 4\,\mathrm{X} \wedge 4 + 2\,\mathrm{X} \wedge 3 + 3\,\mathrm{X} \wedge 2 + \mathrm{X} + 6$

Press [GRAPH].

Zoom in on the leftmost x-intercept, to find that it is close to $x = -1.04$.

Reset standard range values, and zoom in on the first x-intercept to the right of the origin, to find that it is close to $x = 2.46$.

Without having to reset the range, move the cursor to the last x-intercept appearing on the screen. The graph is magnified enough for us to be able to see that it is close to $x = 2.59$.

The graph has one "hill" and one "valley"—one local maximum and one local minimum.

Chapter 10 Review Problem Set

1. True. The remainder can be found by evaluating $P(1) = 1 - 2 + 3 - 4 = -2$

2. False. If $x - a$ is a factor, then $P(a) = 0$.

3. False. $(x - 1)^3$ has 1 as its only zero.

4. False. The graph of $y = (x - 2)^3(x + 1)^2$ crosses the x-axis only at $x = 2$, although it touches, but doesn't cross, the x-axis again at $x = -1$.

5. True. If $x^n - a^n$ is divided by $x - a$, the remainder will be $P(a) = a^n - a^n = 0$.

6. True. There are 13 solutions, an even number of which are imaginary. Therefore, there is at least one real solution.

7. False. The polynomial must have real coefficients for the statement to be true.

8. False. $(x^2 + 2x - 3)^2$ can be further factored as $(x - 1)^2(x + 3)^2$, so the number 1 is actually a zero of multiplicity 6.

9. True. The only possible rational solutions are 1 and -1, and $P(1) \neq 0$ and $p(-1) \neq 0$.

10. True. (It may have any odd number of real zeros between 2 and 3.)

11.
$$x^2 + x + 1 \overline{\smash{\big)}\, 2x^3 + x^2 - 3x + 4}$$
$$\underline{2x^3 + 2x^2 + 2x}$$
$$-x^2 - 5x + 4$$
$$\underline{-x^2 - x - 1}$$
$$-4x + 5$$
Quotient: $2x - 1$
Remainder: $-4x + 5$

12.
$$2x^2 + 1 \overline{\smash{\big)}\, 2x^4 - 5x^2 + 2x - 3}$$
$$\underline{2x^4 + x^2}$$
$$-6x^2 + 2x - 3$$
$$\underline{-6x^2 - 3}$$
$$2x$$
Quotient: $x^2 - 3$
Remainder: $2x$

13.
$$x^3 + 4x^2 + 5x + 4 \overline{\smash{\big)}\, 2x^7 + 8x^6 + 10x^5 + 8x^4 + x^3 + 4x^2 + 5x + 6}$$
$$\underline{2x^7 + 8x^6 + 10x^5 + 8x^4}$$
$$x^3 + 4x^2 + 5x + 6$$
$$\underline{x^3 + 4x^2 + 5x + 4}$$
$$2$$
Quotient: $2x^4 + 1$
Remainder: 2

14.
$$x^2 + ix + 1 \overline{\smash{\big)}\, x^3 + 3ix^2 - x + 2i}$$
$$\underline{x^3 + ix^2 + x}$$
$$2ix^2 - 2x + 2i$$
$$\underline{2ix^2 - 2x + 2i}$$
$$0$$
Quotient: $x + 2i$
Remainder: 0

15.

2	1	-1	2	3	5	-1
		2	2	8	22	54
	1	1	4	11	27	53

Quotient: $x^4 + x^3 + 4x^2 + 11x + 27$
Remainder: 53

16.

$\frac{3}{2}$	2	-5	5	-4
		3	-3	3
	2	-2	2	-1

Quotient: $2x^2 - 2x + 2$
Remainder: -1

17.

$\sqrt{2}-1$	1	0	-2	5	-7
		$\sqrt{2}-1$	$3-2\sqrt{2}$	$3\sqrt{2}-5$	$6-3\sqrt{2}$
	1	$\sqrt{2}-1$	$1-2\sqrt{2}$	$3\sqrt{2}$	$-1-3\sqrt{2}$

Quotient: $x^3 + (\sqrt{2}-1)x^2 + (1-2\sqrt{2})x + 3\sqrt{2}$
Remainder: $-1 - 3\sqrt{2}$

18.

$2i$	2	-4	1	-7
		$4i$	$-8-8i$	$16-14i$
	2	$-4+4i$	$-7-8i$	$9-14i$

Quotient: $2x^2 + (-4+4i)x - 7 - 8i$
Remainder: $9 - 14i$

19.

$2-3i$	1	$-2+3i$	-4	$16-12i$	$-16+24i$
		$2-3i$	0	$-8+12i$	$16-24i$
	1	0	-4	8	0

Quotient: $x^3 - 4x + 8$
Remainder: 0

20. $R = P(-1) = (-1)^{18} - 2(-1)^5 + 3 = 6$

21. $R = P(\sqrt{2}) = 2(\sqrt{2})^6 + (\sqrt{2})^4 - 5(\sqrt{2})^2 - 7$
$$= 2 \cdot 8 + 4 - 5 \cdot 2 - 7$$
$$= 3$$

22. $R = P(i) = i^4 + 2i^3 + 3i^2 + 4i + 5$
$$= 1 - 2i - 3 + 4i + 5$$
$$= 3 + 2i$$

23. $P(x) = (x-2)^2(2x-3)(x+4)^3$. The zeros are $x = 2$ (multiplicity 2); $x = \frac{3}{2}$; and $x = -4$ (multiplicity 3).

24. $P(x) = (x^2-4)^3(x^2+1)$
$$= [(x+2)(x-2)]^3(x+i)(x-i)$$
$$= (x+2)^3(x-2)^3(x+i)(x-i)$$
The zeros are $x = -2$ (multiplicity 3); $x = 2$ (multiplicity 3); $x = -i$; and $x = i$

25. $P(x) = (x^2-5x+6)^2(x^2-9)$
$$= [(x-3)(x-2)]^2(x+3)(x-3)$$
$$= (x-3)^2(x-2)^2(x+3)(x-3)$$
$$= (x-3)^3(x-2)^2(x+3)$$
The zeros are $x = 3$ (multiplicity 3); $x = 2$ (multiplicity 2); and $x = -3$.

26. $P(x) = x^2(x^2-6x+10)^3(x+3\sqrt{2})$
First find the zeros of $(x^2-6x+10)^3$. They are the zeros of $x^2 - 6x + 10$, with a multiplicity of 3.
$$x^2 - 6x + 10 = 0$$
$$x = \frac{6 \pm \sqrt{36-40}}{2} = \frac{6 \pm i\sqrt{4}}{2} = 3 \pm i$$
Two zeros are $3+i$ and $3-i$ each of multiplicity 3. x^2 has the zero 0 (multiplicity 2); the zero of $x + 3\sqrt{2}$ is $-3\sqrt{2}$ (multiplicity 1).

27. $P(x) = x^3 - 2x^2 - 4x + 8$. Since 2 is a zero, $x - 2$ is a factor. Use synthetic division to find the others:

$$\begin{array}{r|rrrr} 2 & 1 & -2 & -4 & 8 \\ & & 2 & 0 & -8 \\ \hline & 1 & 0 & -4 & \boxed{0} \end{array}$$

$$\begin{aligned} P(x) &= (x-2)(x^2-4) \\ &= (x-2)(x+2)(x-2) \\ &= (x-2)^2(x+2) \end{aligned}$$

28. $P(x) = x^3 - 5x^2 + 2x + 8$. Since -1 is a zero, $x - (-1)$ is a factor. Use synthetic division:

$$\begin{array}{r|rrrr} -1 & 1 & -5 & 2 & 8 \\ & & -1 & 6 & -8 \\ \hline & 1 & -6 & 8 & \boxed{0} \end{array}$$

$$P(x) = (x+1)(x^2-6x+8) = (x+1)(x-2)(x-4)$$

29. $P(x) = x^3 - (2+3i)x^2 + (-8+6i)x + 24i$. Since $3i$ is a zero, $x - 3i$ is a factor. Use synthetic division:

$$\begin{array}{r|rrrr} 3i & 1 & -2-3i & -8+6i & 24i \\ & & 3i & -6i & -24i \\ \hline & 1 & -2 & -8 & \boxed{0} \end{array}$$

$$P(x) = (x-3i)(x^2-2x-8) = (x-3i)(x-4)(x+2)$$

30. $P(x) = 2x^3 + (1-2\pi)x^2 - (\pi+1)x + \pi$. Since π is a zero, $x - \pi$ is a factor. Use synthetic division:

$$\begin{array}{r|rrrr} \pi & 2 & 1-2\pi & -\pi-1 & \pi \\ & & 2\pi & \pi & -\pi \\ \hline & 2 & 1 & -1 & \boxed{0} \end{array}$$

$$\begin{aligned} P(x) &= (x-\pi)(2x^2+x-1) \\ &= (x-\pi)(2x-1)(x+1) \end{aligned}$$

31. $P(x) = a(x+4)(x-1)^2$. Letting $a = 1$,

$$\begin{aligned} P(x) &= (x+4)(x-1)^2 \\ &= (x+4)(x^2-2x+1) \\ &= x^3-2x^2+x+4x^2-8x+4 \\ &= x^3+2x^2-7x+4 \end{aligned}$$

32. $P(x) = a(x-\sqrt{2})(x+\sqrt{2})(x-\frac{3}{4})(x+\frac{2}{3})$. Let $a = 12$:

$$\begin{aligned} P(x) &= 12(x-\sqrt{2})(x+\sqrt{2})(x-\tfrac{3}{4})(x+\tfrac{2}{3}) \\ &= (x^2-2)(4x-3)(3x+2) \\ &= (x^2-2)(12x^2-x-6) \\ &= 12x^4-x^3-6x^2-24x^2+2x+12 \\ &= 12x^4-x^3-30x^2+2x+12 \end{aligned}$$

33. $P(x) = a(x-i)(x+i)[x-(2-i)][x-(2+i)]$. Letting $a = 1$,

$$\begin{aligned} P(x) &= (x-i)(x+i)[(x-2)+i][(x-2)-i] \\ &= (x^2+1)[(x-2)^2+1] \\ &= (x^2+1)(x^2-4x+5) \\ &= x^4-4x^3+5x^2+x^2-4x+5 \\ &= x^4-4x^3+6x^2-4x+5 \end{aligned}$$

34. Divide $x^4 - 4x^3 + 5x^2 - 4x + 4$ by $x - 2$ and then do it again for the quotient:

$$\begin{array}{r|rrrrr} 2 & 1 & -4 & 5 & -4 & 4 \\ & & 2 & -4 & 2 & -4 \\ \cline{2-6} 2 & 1 & -2 & 1 & -2 & \boxed{0} \\ & & 2 & 0 & 2 & \\ \cline{2-6} & 1 & 0 & 1 & \boxed{0} & \end{array}$$

Hence, $x^4 - 4x^3 + 5x^2 - 4x + 4 = (x-2)^2(x^2+1)$.

35. Let $P(x) = a(x-1)(x+4)(x+\frac{2}{3})$.
Let $x = -1$:

$$P(-1) = a(-2)(3)(-\tfrac{1}{3}) = 18$$
$$2a = 18, \ a = 9$$

Therefore,

$$\begin{aligned} P(x) &= 9(x-1)(x+4)(x+\tfrac{2}{3}) \\ &= (x-1)(x+4)(9x+6) \\ &= (x^2+3x-4)(9x+6) \\ &= 9x^3+6x^2+27x^2+18x-36x-24 \\ &= 9x^3+33x^2-18x-24 \end{aligned}$$

36. We need $(\sqrt{3})^3 - 2(\sqrt{3})^2 - 3\sqrt{3} + k = 0$

$$3\sqrt{3} - 6 - 3\sqrt{3} + k = 0$$
$$-6 + k = 0, \ k = 6$$

37. If $5 + 2i$ is a zero, $\overline{5+2i} = 5 - 2i$ is a zero.

$$P(x) = a(x+3)[x-(5+2i)][x-(5-2i)]$$

Letting $a = 1$,

$$\begin{aligned} P(x) &= (x+3)[(x-5)-2i][(x-5)+2i] \\ &= (x+3)[(x-5)^2+4] \\ &= (x+3)(x^2-10x+29) \\ &= x^3-10x^2+29x+3x^2-30x+87 \\ &= x^3-7x^2-x+87 \end{aligned}$$

38. Since $1 + 5i$ is a solution, so is $1 - 5i$.

$$\begin{aligned} &(x-(1+5i))(x-(1-5i)) \\ &= [(x-1)-5i][(x-1)+5i] \\ &= [(x-1)^2+25] \\ &= x^2-2x+26 \text{ is a factor of} \end{aligned}$$

$2x^4 - 5x^3 + 53x^2 - 24x - 26$. Apply long division to find the other factor:

$$\begin{array}{r} 2x^2 - x - 1 \\ x^2-2x+26 \overline{\smash{\big)}\ 2x^4 - 5x^3 + 53x^2 - 24x - 26} \\ \underline{2x^4 - 4x^3 + 52x^2 } \\ -x^3 + x^2 - 24x \\ \underline{-x^3 + 2x^2 - 26x } \\ -x^2 + 2x - 26 \\ \underline{-x^2 + 2x - 26} \\ 0 \end{array}$$

Therefore, the other solutions come from

$$2x^2 - x - 1 = 0$$
$$(2x+1)(x-1) = 0$$
$$x = -\tfrac{1}{2} \text{ or } x = 1$$

The solutions are $1 + 5i$, $1 - 5i$, $-\frac{1}{2}$, and 1.

39. $3x^3 + 4x^2 - 7x + 2 = 0$

Any rational solution must be of the form $\frac{c}{d}$, where $c = \pm 1, \pm 2$ and $d = \pm 1, \pm 3$. Therefore, $\frac{c}{d} = \pm 1, \pm 2, \pm\frac{1}{3}, \pm\frac{2}{3}$. Using synthetic division to test the possibilities from left to right, the first root we come to is $x = \frac{2}{3}$:

$$
\begin{array}{r|rrrr}
\frac{2}{3} & 3 & 4 & -7 & 2 \\
  &   & 2 & 4 & -2 \\
\hline
  & 3 & 6 & -3 & \underline{|\,0}
\end{array}
$$

The remaining solutions can be found by solving the quadratic equation

$$3x^2 + 6x - 3 = 0$$
$$x^2 + 2x - 1 = 0$$
$$x = \frac{-2 \pm \sqrt{4+4}}{2}$$
$$= \frac{-2 \pm 2\sqrt{2}}{2}$$
$$= -1 \pm \sqrt{2}$$

Solution set: $\left\{\frac{2}{3}, -1 + \sqrt{2}, -1 - \sqrt{2}\right\}$

40. Since imaginary solutions occur in conjugate pairs, this odd degree polynomial equation has at least one real solution. The possible rational solutions are $\pm 1, \pm 3$. Trying each of these, we see that none of them work; thus, any real solution must be irrational.

41. $P(x) = x^3 - 8x - 4$

x	y
-2	4
-1	3
0	-4
1	-11
2	-12
3	-1
4	28

It is not necessary to sketch the graph to see that a zero must occur in the positive interval $3 < x < 4$. Choose several values of x near $x = 3$, looking for a sign change in y.

x	y
3	-1
3.1	0.99

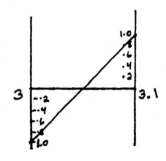

The straight line seems to cross the x-axis at $x = 3.05$. Choose values of x near 3.05 looking for a sign change in y.

x	y
3.04	-0.23
3.05	-0.03
3.06	0.17

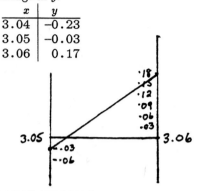

The straight line seems to cross the x-axis at $x = 3.051$.

▷ See note preceding Problem 11 in the solutions for Problem Set 7.4.

42. $P(x) = x^3 - 8x - 4$. $P'(x) = 3x^2 - 8$.
$x_1 = 3$ [Min]
$x_2 = x_1 - \dfrac{P(x_1)}{P'(x_1)}$. This is calculated as

[MR] $-$ ([MR][y^x]$3 - 8 \times$[MR]$- 4$)
$\div (3 \times$[MR]$[x^2] - 8)$[=]

The calculator will display 3.052631579. Press [Min] and repeat the calculation to get $x_3 = 3.\underline{051}374967$. Press [Min] and repeat the calculation to get $x_4 = 3.\underline{051374}242$.
(This is accurate to all 9 decimal places.)

43. $y = P(x) = x^4 - 2x^2 - 3x + 5$
$$P'(x) = 4(1)x^3 - 2(2)x - 3$$
$$= 4x^3 - 4x - 3$$
The slope of the tangent line at $(1,1)$
$$= P'(1) = 4 - 4 - 3 = -3.$$
The equation of the line passing through $(1,1)$ with slope -3 is
$$y - y_1 = m(x - x_1)$$
$$y - 1 = -3(x - 1)$$
$$y - 1 = -3x + 3$$
$$3x + y = 4$$

44. Let x centimeters be the length of a side of the original cube. The volume of the remaining chunk is given by $V = x^2(x-2) = 384$
$$x^3 - 2x^2 - 384 = 0$$
In the hopes that x is rational, try the possible rational solutions to this equation. Such candidates will be of the form $\pm p$ where p is a factor of $384 = 128 \cdot 3 = 2^7 \cdot 3$. Furthermore, we know that $x > 2$, so we try 4, 8, 16, ..., 128, 3, 6, 12, ..., 384. Trying these, we find that $x = 8$ is a solution. Since there is only one cube that will yield the resulting volume, the original cube of cheese was 8 centimeters on a side. (The other two solutions must be imaginary.)

45. Use the formula $V = lwh$, with $l = 40 - 2x$, $w = 24 - 2x$, and $h = x$.
$$(40 - 2x)(24 - 2x)x = 2000$$
$$(20 - x)(12 - x)x = 500$$
$$240x - 32x^2 + x^3 = 500$$
$$x^3 - 32x^2 + 240x - 500 = 0$$
$$y = P(x) = x^3 - 32x^2 + 240x - 500$$
$$P'(x) = 3x^2 - 64x + 240$$

x	y
0	-500
1	-291
2	-140
3	-41
4	396
5	25
6	4
7	-44

The table shows that there is a solution in the interval $3 < x < 4$, and a solution in the interval $6 < x < 7$.
Since we are interested in the larger solution, let
$x_1 = 6$ [Min]
$x_2 = x_1 - \dfrac{P(x_1)}{P'(x_1)}$. This is calculated as

[MR] $-$ ([MR]$\boxed{y^x}$3 $- 3 \times$ [MR]$\boxed{x^2}$ + 240 $\times$ [MR] $- 500$)
$\div$ ($3 \times$ [MR]$\boxed{x^2}$ $- 64 \times$ [MR] + 240) [=]
The calculator will display 6.111111111.
Press [Min] and repeat the calculation to get
$x_3 = 6.\underline{1}06722837$. Press [Min] and repeat:
$x_4 = 6.\underline{106}716078$. $x \approx 6.106716078$,
to 8 decimals places $= 6.107$ to 3 decimals places.

46. We need to find the smallest positive t such that $3t^3 - 4t^2 + 5t - 6 = 0$. We start by looking at possible positive rational solutions: 1, 2, 3, 6, $\frac{1}{3}$, $\frac{2}{3}$. None of them work, so we use Newton's Method.

$P(t) = 3t^3 - 4t^2 + 5t - 6$; $P'(t) = 9t^2 - 8t + 5$

t	$P(t)$
0	-6
1	-4
2	12

$t_1 = 1$ [Min]
$t_2 = t_1 - \dfrac{P(t_1)}{P'(t_1)}$. This is calculated as

[MR] $-$ ($3 \times$ [MR]$\boxed{y^x}$3 $- 4 \times$ [MR]$\boxed{x^2}$ + 5 $\times$ [MR] $- 6$)
$\div$ ($9 \times$ [MR]$\boxed{x^2}$ $- 8 \times$ [MR] + 5) [=]
The calculator will display 1.33333333. Press [Min] and repeat the calculation to get
$t_3 = 1.268817204$. Press [Min] and repeat:
$t_4 = 1.\underline{265}337747$.
The object hit the origin after approximately 1.2653 seconds, to 4 decimal places.

47. $P(x) = ax^2 + bx + c$
By the Complete Factorization Theorem, we can also write
$$\begin{aligned} P(x) &= a(x - r_1)(x - r_2) \\ &= a(x^2 - r_1x - r_2x + r_1r_2) \\ &= ax^2 - ax(r_1 + r_2) + ar_1r_2 \end{aligned}$$
Equating the coefficients in the two forms of $P(x)$, we see that $b = -a(r_1 + r_2)$
$$-\frac{b}{a} = r_1 + r_2$$
and
$$c = ar_1r_2$$
$$\frac{c}{a} = r_1r_2$$

48. $P(x) = ax^3 + bx^2 + cx + d$
Start by dividing $P(x)$ by a:
$$P(x) = x^3 + \frac{b}{a}x^2 + \frac{c}{a}x + \frac{d}{a} \text{ which has factors}$$
$$(x - r_1)(x - r_2)(x - r_3)$$
$$= x^3 - (r_1 + r_2 + r_3)x^2 + (r_1r_2 + r_1r_3 + r_2r_3)x$$
$$- r_1r_2r_3$$
Comparing coefficients, $r_1 + r_2 + r_3 = -\dfrac{b}{a}$ and
$$r_1r_2r_3 = -\frac{d}{a}.$$

Chapter 11: Systems of Equations and Inequalities

1. $\begin{cases} 2x - 3y = 7 \\ \quad\quad y = -1 \end{cases}$

$y = -1$, which we substitute back into the first equation:

$$2x - 3(-1) = 7$$
$$2x + 3 = 7$$
$$2x = 7 - 3 = 4$$
$$x = 2$$

Solution: $(x, y) = (2, -1)$

3. $\begin{cases} x \quad\quad = -2 \\ 2x + 7y = 24 \end{cases}$

$x = -2$, which we substitute into the second equation:

$$2(-2) + 7y = 24$$
$$-4 + 7y = 24$$
$$7y = 24 + 4 = 28$$
$$y = 4$$

Solution: $(x, y) = (-2, 4)$

5. $\begin{cases} x - 3y = 7 \\ 4x + y = 2 \end{cases}$

Add -4 times the first equation to the second:

$$\begin{cases} x - 3y = 7 \\ \quad 13y = -26 \end{cases}$$
$$y = -2$$

Substitute in the first equation:

$$x - 3(-2) = 7$$
$$x + 6 = 7$$
$$x = 1$$

Solution: $(x, y) = (1, -2)$

7. $\begin{cases} 2x - y + 3z = -6 \\ \quad 2y - z = 2 \\ \quad\quad\quad z = -2 \end{cases}$

Already in triangular form, we use back substitution.

$$2y - (-2) = 2$$
$$2y = 0$$
$$y = 0$$

Substitute in first equation:

$$2x - 0 + 3(-2) = -6$$
$$2x - 6 = -6$$
$$2x = 0$$
$$x = 0$$

Solution: $(x, y, z) = (0, 0, -2)$

9. $\begin{cases} 3x - 2y + 5z = -10 \\ \quad y - 4z = 8 \\ \quad 2y + z = 7 \end{cases}$

Add -2 times the second equation to the third:

$$\begin{cases} 3x - 2y + 5z = -10 \\ \quad y - 4z = 8 \\ \quad\quad 9z = -9 \\ \quad\quad z = -1 \end{cases}$$

Substitute in the second equation:

$$y - 4(-1) = 8$$
$$y + 4 = 8$$
$$y = 4$$

Substitute in the first equation:

$$3x - 2(4) + 5(-1) = -10$$
$$3x - 8 - 5 = -10$$
$$3x - 13 = -10$$
$$3x = 3$$
$$x = 1$$

Solution: $(x, y, z) = (1, 4, -1)$

11. $\begin{cases} x + 2y + z = 8 \\ 2x - y + 3z = 15 \\ -x + 3y - 3z = -11 \end{cases}$

Add -2 times the first equation to the second, and 1 times the first equation to the third:

$$\begin{cases} x + 2y + z = 8 \\ \quad -5y + z = -1 \\ \quad 5y - 2z = -3 \end{cases}$$

Add the second equation to the third:

$$\begin{cases} x + 2y + 2z = 8 \\ \quad -5y + z = -1 \\ \quad\quad -z = -4 \\ \quad\quad z = 4 \end{cases}$$

Use back substitution:

$$-5y + 4 = -1$$
$$-5y = -5$$
$$y = 1$$
$$x + 2(1) + 4 = 8$$
$$x = 2$$

Solution: $(x, y, z) = (2, 1, 4)$

13. $\begin{cases} x - 2y + 3z = 0 \\ 2x - 3y - 4z = 0 \\ x + y - 4z = 0 \end{cases}$

Add -2 times the first equation to the second, and -1 times the first equation to the third:

$$\begin{cases} x - 2y + 3z = 0 \\ \quad y - 10z = 0 \\ \quad 3y - 7z = 0 \end{cases}$$

Add -3 times the second equation to the third:

$$\begin{cases} x - 2y + 3z = 0 \\ \quad y - 10z = 0 \\ \quad\quad 23z = 0 \\ \quad\quad z = 0 \end{cases}$$

Use back substitution:

$$y - 10(0) = 0$$
$$y = 0$$
$$x = 0$$

Solution: $(x, y, z) = (0, 0, 0)$

15.
$$\begin{cases} x + y + z + w = 10 \\ y + 3z - w = 7 \\ x + y + 2z \quad = 11 \\ x - 3y \quad + w = -14 \end{cases}$$
Add -1 times the first equation to the third and fourth equations:
$$\begin{cases} x + y + z + w = 10 \\ y + 3z - w = 7 \\ z - w = 1 \\ -4y \; - z \quad = -24 \end{cases}$$
Switch the positions of equations 3 and 4:
$$\begin{cases} x + y + z + w = 10 \\ y + 3z - w = 7 \\ -4y \; - z \quad = -24 \\ z - w = 1 \end{cases}$$
Add 4 times the second equation to the third:
$$\begin{cases} x + y + z \; + w = 10 \\ y + 3z \; - w = 7 \\ 11z - 4w = 4 \\ z \; - w = 1 \end{cases}$$
Switch the positions of equations 3 and 4:
$$\begin{cases} x + y + z \; + w = 10 \\ y + 3z \; - w = 7 \\ z \; - w = 1 \\ 11z - 4w = 4 \end{cases}$$
Add -11 times the third equation to the fourth:
$$\begin{cases} x + y + z + w = 10 \\ y + 3z - w = 7 \\ z - w = 1 \\ 7w = -7 \\ w = -1 \end{cases}$$
Use back substitution:
$$z - (-1) = 1$$
$$z + 1 = 1$$
$$z = 0$$
$$y + 3(0) - (-1) = 7$$
$$y + 1 = 7$$
$$y = 6$$
$$x + 6 + 0 - 1 = 10$$
$$x + 5 = 10$$
$$x = 5$$
Solution: $(x, y, z, w) = (5, 6, 0, -1)$

17.
$$\begin{cases} x \; - 4y \; + z = 18 \\ 2x \; - 7y - 2z = 4 \\ 3x - 11y \; - z = 22 \end{cases}$$
Add -2 times the first equation to the second, and -3 times the first equation to the third:
$$\begin{cases} x - 4y \; + z = 18 \\ y - 4z = -32 \\ y - 4z = -32 \end{cases}$$
Add -1 times the second equation to the third:

$$\begin{cases} x - 4y + z = 18 \\ y - 4z = -32 \\ 0z = 0 \end{cases}$$
z can be arbitrary.
Then
$$y - 4z = -32$$
$$y = 4z - 32$$
$$x - 4(4z - 32) + z = 18$$
$$x - 16z + 128 + z = 18$$
$$x = 15z - 110$$
Solution: $(15z - 110, 4z - 32, z)$

19.
$$\begin{cases} x - 2y + 3z = -2 \\ 3x - 6y + 9z = -6 \\ -2x + 4y - 6z = 4 \end{cases}$$
Add -3 times the first equation to the second, and 2 times the first equation to the third:
$$\begin{cases} x - 2y + 3z = -2 \\ 0y + 0z = 0 \\ 0z = 0 \end{cases}$$
In this case, both y and z can be chosen arbitrarily.
Then
$$x = 2y - 3z - 2$$
Solution: $(2y - 3z - 2, y, z)$

21.
$$\begin{cases} 2x \; - y \; + 4z = 0 \\ 3x + 2y \; - z = 0 \\ 9x \; - y + 11z = 0 \end{cases}$$
Divide the first equation through by 2:
$$\begin{cases} x - \frac{1}{2}y \; + 2z = 0 \\ 3x + 2y \; - z = 0 \\ 9x \; - y + 11z = 0 \end{cases}$$
Add -3 times the first equation to the second, and -9 times the first equation to the third:
$$\begin{cases} x - \frac{1}{2}y + 2z = 0 \\ \frac{7}{2}y - 7z = 0 \\ \frac{7}{2}y - 7z = 0 \end{cases}$$
Add -1 times the second equation to the third:
$$\begin{cases} x - \frac{1}{2}y + 2z = 0 \\ \frac{7}{2}y - 7z = 0 \\ 0z = 0 \end{cases}$$
z can be arbitrary.
Then
$$\frac{7}{2}y - 7z = 0$$
$$\frac{7}{2}y = 7z$$
$$y = 2z$$
$$x - \frac{1}{2}(2z) + 2z = 0$$
$$x - z + 2z = 0$$
$$x + z = 0$$
$$x = -z$$
Solution: $(x, y, z) = (-z, 2z, z)$

23. $\begin{cases} x - 4y + z = 18 \\ 2x - 7y - 2z = 4 \\ 3x - 11y - z = 10 \end{cases}$

Add -2 times the first equation to the second, and -3 times the first equation to the third:

$\begin{cases} x - 4y + z = 18 \\ y - 4z = -32 \\ y - 4z = -44 \end{cases}$

The last two equations imply that $-32 = -44$; the system is inconsistent.

25. $\begin{cases} x + 3y - 2z = 10 \\ 2x + y + z = 4 \\ 5y - 5z = 16 \end{cases}$

Add -2 times first to second:

$\begin{cases} x + 3y - 2z = 10 \\ -5y + 5z = 16 \\ 5y - 5z = 16 \end{cases}$

Add the second equation to the third:

$\begin{cases} x + 3y - 2z = 10 \\ -5y + 5z = -16 \\ 0z = 0 \end{cases}$

z can be arbitrary. Then

$$-5y + 5z = -16$$
$$-5y = -5z - 16$$
$$y = z + \tfrac{16}{5}$$
$$x + 3(z + \tfrac{16}{5}) - 2z = 10$$
$$x + 3z + \tfrac{48}{5} - 2z = \tfrac{50}{5}$$
$$x = -z + \tfrac{2}{5}$$

Solution: $(x, y, z) = (-z + \tfrac{2}{5}, z + \tfrac{16}{5}, z)$

27. $\begin{cases} x + 2y = 10 \\ x^2 + y^2 - 10x = 0 \end{cases}$

Solve the first equation for x in terms of y:

$$x = 10 - 2y$$

Substitute in the second equation:

$$(10 - 2y)^2 + y^2 - 10(10 - 2y) = 0$$
$$100 - 40y + 4y^2 + y^2 - 100 + 20y = 0$$
$$5y^2 - 20y = 0$$
$$5y(y - 4) = 0$$
$$y = 0 \text{ or } y = 4$$

When $y = 0$, $x = 10 - 2(0) = 10$.
When $y = 4$, $x = 10 - 2(4) = 2$.
Solutions: $(x, y) = (10, 0)$; $(x, y) = (2, 4)$

29. $\begin{cases} x^2 + y^2 - 4x + 6y = 12 \\ x^2 + y^2 + 10x + 4y = 96 \end{cases}$

Add -1 times the first equation to the second:

$\begin{cases} x^2 + y^2 - 4x + 6y = 12 \\ 14x - 2y = 84 \end{cases}$

Solve the second equation for y in terms of x:

$$-2y = -14x + 84$$
$$y = 7x - 42$$

Substitute $y = 7x - 42$ in the first equation:

$$x^2 + (7x - 42)^2 - 4x + 6(7x - 42) = 12$$
$$x^2 + 49x^2 - 588x + 1764 - 4x + 42x - 252 = 12$$
$$50x^2 - 550x + 1500 = 0$$
$$x^2 - 11x + 30 = 0$$
$$(x - 6)(x - 5) = 0$$
$$x = 6 \text{ or } x = 5$$

When $x = 6$, $y = 7(6) - 42 = 0$.
When $x = 5$, $y = 7(5) - 42 = -7$.
Solutions: $(x, y) = (6, 0)$; $(x, y) = (5, -7)$

31. $\begin{cases} y = 4x^2 - 2 \\ y = x^2 + 1 \end{cases}$

Equating the two expressions for y,

$$4x^2 - 2 = x^2 + 1$$
$$3x^2 = 3$$
$$x^2 = 1$$
$$x = \pm 1$$

When $x = 1$, $y = 4(1)^2 - 2 = 2$.
When $x = -1$, $y = 4(-1)^2 - 2 = 2$.
Solutions: $(x, y) = (1, 2)$; $(x, y) = (-1, 2)$

33. $\begin{cases} 2x - 3y = 12 \\ x + 4y = -5 \end{cases}$

Switch the positions of the two equations:

$\begin{cases} x + 4y = -5 \\ 2x - 3y = 12 \end{cases}$

Add -2 times the first equation to the second:

$\begin{cases} x + 4y = -5 \\ -11y = 22 \end{cases}$
$$y = -2$$

Use back substitution:

$$x + 4(-2) = -5$$
$$x - 8 = -5$$
$$x = 3$$

Solution: $(x, y) = (3, -2)$

35. $\begin{cases} 2x - 3y = 6 \\ x + 2y = -4 \end{cases}$

Switch the positions of the two equations:

$\begin{cases} x + 2y = -4 \\ 2x - 3y = 6 \end{cases}$

Add -2 times the first equation to the second:

$\begin{cases} x + 2y = -4 \\ -7y = 14 \end{cases}$
$$y = -2$$

Use back substitution:

$$x + 2(-2) = -4$$
$$x - 4 = -4$$
$$x = 0$$

Solution: $(x, y) = (0, -2)$

37.
$$\begin{cases} x + y + z = 6 \\ -x + y + z = 18 \\ 4x - y + z = 12 \end{cases}$$
Add the first equation to the second, and -4 times the first equation to the third:
$$\begin{cases} x + y \ \ + z = 6 \\ 2y + 2z = 24 \\ -5y - 3z = -12 \end{cases}$$
Divide the second equation through by 2, then add 5 times the second equation to the third:
$$\begin{cases} x + y + z = 6 \\ y + z = 12 \\ 2z = 48 \\ z = 24 \end{cases}$$
Use back substitution:
$$y + 24 = 12$$
$$y = -12$$
$$x - 12 + 24 = 6$$
$$x + 12 = 6$$
$$x = -6$$
Solution: $(x, y, z) = (-6, -12, 24)$

39.
$$\begin{cases} x^2 + y^2 = 4 \\ x + 2y = 2\sqrt{5} \end{cases}$$
Solve the second equation for x in terms of y:
$$x = 2\sqrt{5} - 2y$$
Substitute $x = 2\sqrt{5} - 2y$ in the first equation:
$$(2\sqrt{5} - 2y)^2 + y^2 = 4$$
$$20 - 8\sqrt{5}y + 4y^2 + y^2 = 4$$
$$5y^2 - 8\sqrt{5}y + 16 = 0$$
$$(\sqrt{5}y - 4)^2 = 0$$
$$\sqrt{5}y - 4 = 0$$
$$y = \frac{4}{\sqrt{5}} = \frac{4\sqrt{5}}{5}$$
$$x = 2\sqrt{5} - 2\left(\frac{4\sqrt{5}}{5}\right)$$
$$= 2\sqrt{5} - \frac{8\sqrt{5}}{5}$$
$$= \frac{2\sqrt{5}}{5}$$
Solution: $(x, y) = \left(\frac{2\sqrt{5}}{5}, \frac{4\sqrt{5}}{5}\right)$

41.
$$\begin{cases} x + 2y = 4 \\ ax + 3y = b \end{cases}$$
Add $-a$ times the first equation to the second:
$$\begin{cases} x + 2y = 4 \\ (3 - 2a)y = b - 4a \end{cases}$$
If the system is to have infinitely many solutions,

$$\begin{cases} b - 4a = 0 \\ 3 - 2a = 0 \\ \quad -2a = -3 \\ \quad a = \frac{3}{2} \end{cases}$$
$$b - 4\left(\frac{3}{2}\right) = 0$$
$$b - 6 = 0$$
$$b = 6$$

43. Let $h =$ hundreds digit.
Let $t =$ tens digit.
Let $u =$ units digit.
$$\begin{cases} 100h + 10t + u = 19(h + t + u) \\ 100u + 10t + h - (100h + 10t + u) = 297 \\ t = u + 3 \end{cases}$$
Collecting and arranging:
$$\begin{cases} 81h - 9t - 18u = 0 \\ -99h \quad + 99u = 297 \\ t \ \ - u = 3 \end{cases}$$
Divide the first equation by 9 and the second by -99:
$$\begin{cases} 9h - t - 2u = 0 \\ -h \quad + u = 3 \\ t \ \ - u = 3 \end{cases}$$
Exchange the positions of the first and second equations:
$$\begin{cases} -h \quad + u = 3 \\ 9h - t - 2u = 0 \\ t \ \ - u = 3 \end{cases}$$
Add 9 times the first equation to the second:
$$\begin{cases} -h \quad + u = 3 \\ -t + 7u = 27 \\ t \ \ - u = 3 \end{cases}$$
Add the second equation to the third:
$$\begin{cases} -h \quad + u = 3 \\ -t + 7u = 27 \\ 6u = 30 \\ u = 5 \end{cases}$$
Use back substitution:
$$-t + 7(5) = 27$$
$$-t + 35 = 27$$
$$-t = -8$$
$$t = 8$$
$$-h + 5 = 3$$
$$-h = -2$$
$$h = 2$$
The number is 285.

45. $y = ax^3 + bx^2 + cx + d$

Since each given point must satisfy the equation of the curve, we can generate a system of 4 equations in a, b, c, and d by substituting each ordered pair (x, y) in turn.

$$\begin{cases} -6 = -8a + 4b - 2c + d \\ 5 = -a + b - c + d \\ 3 = a + b + c + d \\ 14 = 8a + 4b + 2c + d \end{cases}$$

Reposition the equations:

$$\begin{cases} a + b + c + d = 3 \\ -a + b - c + d = 5 \\ 8a + 4b + 2c + d = 14 \\ -8a + 4b - 2c + d = -6 \end{cases}$$

Add the first equation to the second, -8 times the first equation to the third, and 8 times the first equation to the fourth:

$$\begin{cases} a + b + c + d = 3 \\ 2b + 2d = 8 \\ -4b - 6c - 7d = -10 \\ 12b + 6c + 9d = 18 \end{cases}$$

Add twice the second equation to the third, and -6 times the second equation to the fourth.

$$\begin{cases} a + b + c + d = 3 \\ 2b + 2d = 8 \\ -6c - 3d = 6 \\ 6c - 3d = -30 \end{cases}$$

Add the third equation to the fourth:

$$\begin{cases} a + b + c + d = 3 \\ 2b + 2d = 8 \\ -6c - 3d = 6 \\ -6d = -24 \\ d = 4 \end{cases}$$

Use back substitution:

$$-6c - 3(4) = 6$$
$$-6c - 12 = 6$$
$$-6c = 18$$
$$c = -3$$
$$2b + 2(4) = 8$$
$$2b = 0$$
$$b = 0$$
$$a + 0 - 3 + 4 = 3$$
$$a + 1 = 3$$
$$a = 2$$

The curve is $y = 2x^3 - 3x + 4$.

47. An equation of a circle can have the form

$$x^2 + y^2 + Ax + By + C = 0$$

Since the three given point satisfy the equation of the circle, we can generate 3 equations in the unknowns A, B, and C by substituting each point (x, y) in turn.

Substitute $(0, 0)$: $C = 0$

Substitute $(4, 0)$: $16 + 4A + C = 0$

Substitute $(\frac{72}{25}, \frac{96}{25})$:

$$\left(\frac{72}{25}\right)^2 + \left(\frac{96}{25}\right)^2 + \frac{72}{25}A + \frac{96}{25}B + C = 0$$

Since we know that $C = 0$, we can immediately use back substitution:

$$16 + 4A + 0 = 0$$
$$4A = -16$$
$$A = -4$$
$$\left(\frac{72}{25}\right)^2 + \left(\frac{96}{25}\right)^2 + \frac{72}{25}(-4) + \frac{96}{25}B = 0$$
$$\frac{5184 + 9216}{25} - 288 + 96B = 0$$
$$576 - 288 + 96B = 0$$
$$96B = -288$$
$$B = -3$$

The equation of the circle is

$$x^2 + y^2 - 4x - 3y = 0$$
$$x^2 - 4x + 4 + y^2 - 3y + \frac{9}{4} = \frac{25}{4}$$
$$(x - 2)^2 + (y - \frac{3}{2})^2 = \frac{25}{4}$$

The radius of the circle is $\frac{5}{2}$.

49. Use synthetic division progressively three times. Set each remainder equal to 0, and solve for a, b, and c.

$$\begin{array}{r|ccccc}
1 & 1 & 0 & a & b & c \\
  &   & 1 & 1 & a+1 & a+b+1 \\
\hline
1 & 1 & 1 & a+1 & a+b+1 & \boxed{a+b+c+1} \\
  &   & 1 & 2 & a+3 & \\
\hline
1 & 1 & 2 & a+3 & \boxed{2a+b+4} & \\
  &   & 1 & 3 & & \\
\hline
  & 1 & 3 & \boxed{a+6} & & \\
\end{array}$$

$$\begin{cases} a + b + c + 1 = 0 \\ 2a + b + 4 = 0 \\ a + 6 = 0 \\ a = -6 \end{cases}$$

Use back substitution:

$$2(-6) + b + 4 = 0$$
$$b - 8 = 0$$
$$b = 8$$
$$-6 + 8 + c + 1 = 0$$
$$c + 3 = 0$$
$$c = -3$$

The polynomial is $x^4 - 6x^2 + 8x - 3$.

51. Let x = original length.
Let y = original width.
$$\begin{cases} xy = 120 \\ (x-3)(y+4) = 144 \end{cases}$$
Solve the first equation for y in terms of x:
$$y = \frac{120}{x}$$
Substitute in the second equation:
$$(x-3)\left(\frac{120}{x} + 4\right) = 144$$
$$(x-3)\frac{(120 + 4x)}{x} = 144$$
$$(x-3)(4)(30 + x) = 144x$$
$$(x-3)(x+30) = 36x$$
$$x^2 + 27x - 90 - 36x = 0$$
$$x^2 - 9x - 90 = 0$$
$$(x-15)(x+6) = 0$$
$$x = 15 \text{ or } x = -6$$
-6 does not satisfy the original condition of the problem. $x = 15$ inches = original length
$$y = \frac{120}{15} = 8 \text{ inches = original width}$$

53. Set standard range values.
Press $\boxed{Y=}$
Enter $\boxed{(}\,500 - X\,\hat{}\,4\,\boxed{)}\,\hat{}\,4\,\boxed{x^{-1}}$ as Y_1
Enter $-\boxed{(}\,500 - X\,\hat{}\,4\,\boxed{)}\,\hat{}\,4\,\boxed{x^{-1}}$ as Y_2
Enter $1 - x$ as Y_3
Press $\boxed{\text{GRAPH}}$
Zooming in on the two intersection points, we find that they are $(4.383, -3.383)$ and $(-3.383, 4.383)$.

Problem Set 11.2

1. $2x - y = 4$
$x - 3y = -2$
$$\begin{bmatrix} 2 & -1 & 4 \\ 1 & -3 & -2 \end{bmatrix}$$

3. $x - 2y + z = 3$
$2x + y = 5$
$x + y + 3z = -4$
$$\begin{bmatrix} 1 & -2 & 1 & 3 \\ 2 & 1 & 0 & 5 \\ 1 & 1 & 3 & -4 \end{bmatrix}$$

5. Rearrange the equations:
$2x - 3y = -4$
$3x + y = -2$
$$\begin{bmatrix} 2 & -3 & -4 \\ 3 & 1 & -2 \end{bmatrix}$$

7. We rearrange some equations:
$x = 5$
$x + 2y - z = 4$
$3x - y - 5z = -13$
$$\begin{bmatrix} 1 & 0 & 0 & 5 \\ 1 & 2 & -1 & 4 \\ 3 & -1 & -5 & -13 \end{bmatrix}$$

9.
$$\begin{bmatrix} 1 & -2 & 3 \\ 0 & 1 & -4 \end{bmatrix}$$
The matrix is in a triangular form which indicates that the system has a unique solution.

11.
$$\begin{bmatrix} 1 & -3 & 5 \\ 2 & -6 & -10 \end{bmatrix}$$
Add -2 times the first row to the second:
$$\begin{bmatrix} 1 & -3 & 5 \\ 0 & 0 & -20 \end{bmatrix}$$
The triangular form indicates that the system has no solution.

13.
$$\begin{bmatrix} 1 & -2 & 4 & -2 \\ 0 & 3 & 1 & 4 \\ 0 & 0 & 1 & -3 \end{bmatrix}$$
The triangular form of the matrix indicates that the system has a unique solution.

15.
$$\begin{bmatrix} 2 & 1 & 5 & 4 \\ 0 & 3 & -2 & 10 \\ 0 & 3 & -2 & 10 \end{bmatrix}$$
Add -1 times the second row to the third:
$$\begin{bmatrix} 2 & 1 & 5 & 4 \\ 0 & 3 & -2 & 10 \\ 0 & 0 & 0 & 0 \end{bmatrix}$$
The triangular form of the matrix indicates that the system has infinitely many solutions.

17.
$$\begin{bmatrix} 3 & 2 & -1 & 0 \\ 0 & 1 & 0 & -4 \\ 0 & 1 & 0 & 5 \end{bmatrix}$$

Add -1 times the second row to the third.

$$\begin{bmatrix} 3 & 2 & -1 & 0 \\ 0 & 1 & 0 & -4 \\ 0 & 0 & 0 & 9 \end{bmatrix}$$

The triangular form of the matrix indicates that the system has no solution.

19. $x + 2y = 5$
$2x - 5y = -8$

$$\begin{bmatrix} 1 & 2 & 5 \\ 2 & -5 & -8 \end{bmatrix}$$

Add -2 times the first row to the second:

$$\begin{bmatrix} 1 & 2 & 5 \\ 0 & -9 & -18 \end{bmatrix}$$

$x + 2y = 5$
$-9y = -18$
$y = 2$

Use back substitution:
$x + 2(2) = 5$
$x = 1$
Solution: $(1, 2)$

21. $3x - 2y = 1$
$-6x + 4y = -2$

$$\begin{bmatrix} 3 & -2 & 1 \\ -6 & 4 & -2 \end{bmatrix}$$

Add twice the first row to the second:

$$\begin{bmatrix} 3 & -2 & 1 \\ 0 & 0 & 0 \end{bmatrix}$$

There are infinitely many solutions.
Let y be arbitrary.
$3x - 2y = 1$
$x = \dfrac{1 + 2y}{3}$

Solution: $\left(\dfrac{1 + 2y}{3}, y \right)$

23. $3x - 2y + 5z = -10$
$y - 4z = 8$
$2y + z = 7$

$$\begin{bmatrix} 3 & -2 & 5 & -10 \\ 0 & 1 & -4 & 8 \\ 0 & 2 & 1 & 7 \end{bmatrix}$$

Add -2 times the second row to the third:

$$\begin{bmatrix} 3 & -2 & 5 & -10 \\ 0 & 1 & -4 & 8 \\ 0 & 0 & 9 & -9 \end{bmatrix}$$

$3x - 2y + 5z = -10$
$y - 4z = 8$
$9z = -9$
$z = -1$

By back substitution:
$y - 4(-1) = 8$
$y = 4$
$3x - 2(4) + 5(-1) = -10$
$3x - 13 = -10$
$3x = 3$
$x = 1$
Solution: $(1, 4, -1)$

25. $x + y - 3z = 10$
$2x + 5y + z = 18$
$5x + 8y - 8z = 48$

$$\begin{bmatrix} 1 & 1 & -3 & 10 \\ 2 & 5 & 1 & 18 \\ 5 & 8 & -8 & 48 \end{bmatrix}$$

Add -2 times row 1 to row 2, and -5 times row 1 to row 3:

$$\begin{bmatrix} 1 & 1 & -3 & 10 \\ 0 & 3 & 7 & -2 \\ 0 & 3 & 7 & -2 \end{bmatrix}$$

Add -1 times row 2 to row 3:

$$\begin{bmatrix} 1 & 1 & -3 & 10 \\ 0 & 3 & 7 & -2 \\ 0 & 0 & 0 & 0 \end{bmatrix}$$

The system has infinitely many solutions. Let z be arbitrary.

$$3y + 7z = -2$$
$$3y = -7z - 2$$
$$y = \frac{-7z}{3} - \frac{2}{3}$$
$$x + y - 3z = 10$$
$$x + \left(\frac{-7z}{3} - \frac{2}{3}\right) - 3z = 10$$
$$x - \frac{16z}{3} - \frac{2}{3} = 10$$
$$x = \frac{16z}{3} + \frac{32}{3}$$

Solution: $\left(\frac{16z}{3} + \frac{32}{3}, \frac{-7z}{3} - \frac{2}{3}, z\right)$

27.
$$2x + 5y + 2z = 6$$
$$x + 2y - z = 3$$
$$3x - y + 2z = 9$$

$$\begin{bmatrix} 2 & 5 & 2 & 6 \\ 1 & 2 & -1 & 3 \\ 3 & -1 & 2 & 9 \end{bmatrix}$$

Exchange rows 1 and 2:

$$\begin{bmatrix} 1 & 2 & -1 & 3 \\ 2 & 5 & 2 & 6 \\ 3 & -1 & 2 & 9 \end{bmatrix}$$

Add -2 times row 1 to row 2 and -3 times row 1 to row 3:

$$\begin{bmatrix} 1 & 2 & -1 & 3 \\ 0 & 1 & 4 & 0 \\ 0 & -7 & 5 & 0 \end{bmatrix}$$

Add 7 times row 2 to row 3:

$$\begin{bmatrix} 1 & 2 & -1 & 3 \\ 0 & 1 & 4 & 0 \\ 0 & 0 & 33 & 0 \end{bmatrix}$$

$$x + 2y - z = 3$$
$$y + 4z = 0$$
$$33z = 0$$
$$z = 0$$

By back substitution:
$$y + 4(0) = 0$$
$$y = 0$$
$$x + 2(0) - 0 = 3$$
$$x = 3$$

Solution: $(3, 0, 0)$

29.
$$x + 1.2y - 2.3z = 8.1$$
$$1.3x + .7y + .4z = 6.2$$
$$.5x + 1.2y + .5z = 3.2$$

$$\begin{bmatrix} 1 & 1.2 & -2.3 & 8.1 \\ 1.3 & 0.7 & 0.4 & 6.2 \\ 0.5 & 1.2 & 0.5 & 3.2 \end{bmatrix}$$

Add -1.3 times the first row to the second, and $-.5$ times the first row to the third:

$$\begin{bmatrix} 1 & 1.2 & -2.3 & 8.1 \\ 0 & -0.86 & 3.39 & -4.33 \\ 0 & 0.6 & 1.65 & -0.85 \end{bmatrix}$$

Multiply the second row by 3 and the third by 4.3:

$$\begin{bmatrix} 1 & 1.2 & -2.3 & 8.1 \\ 0 & -2.58 & 10.17 & -12.99 \\ 0 & 2.58 & 7.095 & -3.655 \end{bmatrix}$$

Add the second row to the third:

$$\begin{bmatrix} 1 & 1.2 & -2.3 & 8.1 \\ 0 & -2.58 & 10.17 & -12.99 \\ 0 & 0 & 17.265 & -16.645 \end{bmatrix}$$

$$x + 1.2y - 2.3z = 8.1$$
$$-2.58y + 10.17z = -12.99$$
$$17.265z = -16.645$$
$$z \approx -.96$$

By back substitution:
$$-2.58y + 10.17(-.96) = -12.99$$
$$-.258y = -3.2268$$
$$y \approx 1.25$$
$$x + 1.2(1.25) - 2.3(-.96) = 8.1$$
$$x \approx 4.4$$

Solution: $(4.4, 1.25, -.96)$

31.
$$\begin{bmatrix} 3 & -2 & 5 \\ 0 & 1 & -3 \end{bmatrix}$$

The matrix is already in triangular form. It shows that the system has a unique solution.

33.
$$\begin{bmatrix} 2 & -1 & 4 & 6 \\ 0 & 4 & -1 & 5 \\ 0 & 0 & 2 & 1 \end{bmatrix}$$

The matrix is already in triangular form. It shows that the system has a unique solution.

35.
$$\begin{bmatrix} 1 & 2 & 3 & 4 & 5 \\ 0 & 3 & 2 & 1 & 0 \\ 0 & 0 & 0 & 3 & -4 \\ 0 & 0 & 0 & -9 & 15 \end{bmatrix}$$

Add 3 times row 3 to row 4:

$$\begin{bmatrix} 1 & 2 & 3 & 4 & 5 \\ 0 & 3 & 2 & 1 & 0 \\ 0 & 0 & 0 & 3 & -4 \\ 0 & 0 & 0 & 0 & 3 \end{bmatrix}$$

The triangular form of the matrix shows that the system has no solution.

37.
$$3x - 2y + 4z = 0$$
$$x - y + 3z = 1$$
$$4x + 2y - z = 3$$

$$\begin{bmatrix} 3 & -2 & 4 & 0 \\ 1 & -1 & 3 & 1 \\ 4 & 2 & -1 & 3 \end{bmatrix}$$

Exchange rows 1 and 2:

$$\begin{bmatrix} 1 & -1 & 3 & 1 \\ 3 & -2 & 4 & 0 \\ 4 & 2 & -1 & 3 \end{bmatrix}$$

Add -3 times row 1 to row 2, and -4 times row 1 to row 3:

$$\begin{bmatrix} 1 & -1 & 3 & 1 \\ 0 & 1 & -5 & -3 \\ 0 & 6 & -13 & -1 \end{bmatrix}$$

Add -6 times row 2 to row 3:

$$\begin{bmatrix} 1 & -1 & 3 & 1 \\ 0 & 1 & -5 & -3 \\ 0 & 0 & 17 & 17 \end{bmatrix}$$

$$x - y + 3z = 1$$
$$y - 5z = -3$$
$$17z = 17, \; z = 1$$

By back substitution:
$$y - 5(1) = -3$$
$$y = 2$$
$$x - 2 + 3(1) = 1, \; x = 0$$
Solution: $(0, 2, 1)$

39.
$$2x + 4y - z = 8$$
$$4x + 9y + 3z = 42$$
$$8x + 17y + z = 58$$

$$\begin{bmatrix} 2 & 4 & -1 & 8 \\ 4 & 9 & 3 & 42 \\ 8 & 17 & 1 & 58 \end{bmatrix}$$

Divide row 1 by 2:

$$\begin{bmatrix} 1 & 2 & -\frac{1}{2} & 4 \\ 4 & 9 & 3 & 42 \\ 8 & 17 & 1 & 58 \end{bmatrix}$$

Add -4 times row 1 to row 2 and -8 times row 1 to row 3:

$$\begin{bmatrix} 1 & 2 & -\frac{1}{2} & 4 \\ 0 & 1 & 5 & 26 \\ 0 & 1 & 5 & 26 \end{bmatrix}$$

Add -1 times row 2 to row 3:

$$\begin{bmatrix} 1 & 2 & -\frac{1}{2} & 4 \\ 0 & 1 & 5 & 26 \\ 0 & 0 & 0 & 0 \end{bmatrix}$$

The system will have infinitely many solutions. Let z be arbitrary.
$$y + 5z = 26$$
$$y = -5z + 26$$
$$x + 2y - \tfrac{1}{2}z = 4$$
$$x + 2(26 - 5z) - \tfrac{1}{2}z = 4$$
$$x = \frac{21z}{2} - 48$$
Solution: $\left(\frac{21z}{2} - 48, -5z + 26, z \right)$

41. $y = ax^2 + bx + c$

Since the given points satisfy the equation of the parabola, we can generate 3 equations in the 3 unknowns a, b, and c by substituting each point (x, y) in turn:
$$-32 = 4a - 2b + c$$
$$4 = a + b + c$$
$$-12 = 9a + 3b + c$$

$$\begin{bmatrix} 4 & -2 & 1 & -32 \\ 1 & 1 & 1 & 4 \\ 9 & 3 & 1 & -12 \end{bmatrix}$$

Exchange rows 1 and 2:

$$\begin{bmatrix} 1 & 1 & 1 & 4 \\ 4 & -2 & 1 & -32 \\ 9 & 3 & 1 & -12 \end{bmatrix}$$

Add -4 times row 1 to row 2, and -9 times row 1 to row 3:

$$\begin{bmatrix} 1 & 1 & 1 & 4 \\ 0 & -6 & -3 & -48 \\ 0 & -6 & -8 & -48 \end{bmatrix}$$

Add -1 times row 2 to row 3:

$$\begin{bmatrix} 1 & 1 & 1 & 4 \\ 0 & -6 & -3 & -48 \\ 0 & 0 & -5 & 0 \end{bmatrix}$$

$$\begin{aligned} a + b + c &= 4 \\ -6b - 3c &= -48 \\ -5c &= 0 \\ c &= 0 \end{aligned}$$

By back substitution:

$$\begin{aligned} -6b - 3(0) &= -48 \\ -6b &= -48 \\ b &= 8 \\ a + 8 + 0 &= 4 \\ a &= -4 \end{aligned}$$

The equation of the parabola is $y = -4x^2 + 8x$.

43. Given

$$\alpha - \beta + \gamma - \delta = 110$$

From the figure we can generate 3 additional equations:

$$\begin{aligned} \gamma + \delta + 20 &= 180 \\ \alpha + \beta + 70 &= 180 \\ \alpha + \delta + 50 &= 180 \end{aligned}$$

Simplifying and aligning the equations gives us the system

$$\begin{aligned} \alpha - \beta + \gamma - \delta &= 110 \\ \gamma + \delta &= 160 \\ \alpha + \beta &= 110 \\ \alpha \qquad + \delta &= 130 \end{aligned}$$

$$\begin{bmatrix} 1 & -1 & 1 & -1 & 110 \\ 0 & 0 & 1 & 1 & 160 \\ 1 & 1 & 0 & 0 & 110 \\ 1 & 0 & 0 & 1 & 130 \end{bmatrix}$$

Add -1 times row 1 to rows 3 and 4:

$$\begin{bmatrix} 1 & -1 & 1 & -1 & 110 \\ 0 & 0 & 1 & 1 & 160 \\ 0 & 2 & -1 & 1 & 0 \\ 0 & 1 & -1 & 2 & 20 \end{bmatrix}$$

Exchange rows 2 and 4:

$$\begin{bmatrix} 1 & -1 & 1 & -1 & 110 \\ 0 & 1 & -1 & 2 & 20 \\ 0 & 2 & -1 & 1 & 0 \\ 0 & 0 & 1 & 1 & 160 \end{bmatrix}$$

Add -2 times row 2 to row 3:

$$\begin{bmatrix} 1 & -1 & 1 & -1 & 110 \\ 0 & 1 & -1 & 2 & 20 \\ 0 & 0 & 1 & -3 & -40 \\ 0 & 0 & 1 & 1 & 160 \end{bmatrix}$$

Add -1 times row 3 to row 4:

$$\begin{bmatrix} 1 & -1 & 1 & -1 & 110 \\ 0 & 1 & -1 & 2 & 20 \\ 0 & 0 & 1 & -3 & -40 \\ 0 & 0 & 0 & 4 & 200 \end{bmatrix}$$

$$\begin{aligned} \alpha - \beta + \gamma - \delta &= 110 \\ \beta - \gamma + 2\delta &= 20 \\ \gamma - 3\delta &= -40 \\ 4\delta &= 200 \\ \delta &= 50 \end{aligned}$$

By back substitution:

$$\begin{aligned} \gamma - 3(50) &= -40 \\ \gamma &= 110 \\ \beta - 110 + 2(50) &= 20 \\ \beta - 10 &= 20 \\ \beta &= 30 \\ \alpha - 30 + 110 - 50 &= 110 \\ \alpha + 30 &= 110 \\ \alpha &= 80 \end{aligned}$$

The angles are: $\alpha = 80°$, $\beta = 30°$, $\gamma = 110°$, $\delta = 50°$.

45. $\dfrac{2x^2 - 21x + 44}{(x-2)^2(x+3)} = \dfrac{a}{x-2} + \dfrac{b}{(x-2)^2} + \dfrac{c}{x+3}$

Clearing fractions, we get

$2x^2 - 21x + 44$
$= a(x-2)(x+3) + b(x+3) + c(x-2)^2$
$= a(x^2 + x - 6) + b(x+3) + c(x^2 - 4x + 4)$
$= x^2(a+c) + x(a+b-4c) + (-6a+3b+4c)$

Equating the coefficients of x^2 and x, and the constant terms, produces 3 linear equations:

$$\begin{aligned} a \quad\quad\, + c &= 2 \\ a + b - 4c &= -21 \\ -6a + 3b + 4c &= 44 \end{aligned}$$

$$\begin{bmatrix} 1 & 0 & 1 & 2 \\ 1 & 1 & -4 & -21 \\ 6 & 3 & 4 & 44 \end{bmatrix}$$

Add -1 times row 1 to row 2, and 6 times row 1 to row 3:

$$\begin{bmatrix} 1 & 0 & 1 & 2 \\ 0 & 1 & -5 & -23 \\ 0 & 3 & 10 & 56 \end{bmatrix}$$

Add -3 times row 2 to row 3:

$$\begin{bmatrix} 1 & 0 & 1 & 2 \\ 0 & 1 & -5 & -23 \\ 0 & 0 & 25 & 125 \end{bmatrix}$$

$$\begin{aligned} a \quad\;\; + c &= 2 \\ b - 5c &= -23 \\ 25c &= 125 \\ c &= 5 \end{aligned}$$

By back substitution:

$$\begin{aligned} b - 5(5) &= -23 \\ b &= 2 \\ a + 5 &= 2 \\ a &= -3 \end{aligned}$$

Solution: $a = -3$, $b = 2$, $c = 5$

47. Let A represent the number of pounds of brand A Wanda needs, B represent the number of pounds of brand B Wanda needs, and C represent the number of pounds of brand C Wanda needs.

Wanda wants 100 pounds of fertilizer that will be 19% phosphate, 34% potash, and 47% nitrogen. That means she will have 19 pounds of phosphate, 34 pounds of potash, and 47 pounds of nitrogen in her mixture. Each brand contributes a portion of each element; the contributions can be described by the following system of equations:

$$\begin{aligned} .10A + .20B + .20C &= 19 \\ .30A + .40B + .30C &= 34 \\ .60A + .40B + .50C &= 47 \end{aligned}$$

Multiplying each equation by 10 to clear decimals, the matrix for the system becomes

$$\begin{bmatrix} 1 & 2 & 2 & 190 \\ 3 & 4 & 3 & 340 \\ 6 & 4 & 5 & 470 \end{bmatrix}$$

Add -3 times row 1 to row 2 and -6 times row 1 to row 3:

$$\begin{bmatrix} 1 & 2 & 2 & 190 \\ 0 & -2 & -3 & -230 \\ 0 & -8 & -7 & -670 \end{bmatrix}$$

Add -4 times row 2 to row 3:

$$\begin{bmatrix} 1 & 2 & 2 & 190 \\ 0 & -2 & -3 & -230 \\ 0 & 0 & 5 & 250 \end{bmatrix}$$

$$\begin{aligned} A + 2B + 2C &= 190 \\ -2B - 3C &= -230 \\ 5C &= 250 \\ C &= 50 \end{aligned}$$

By back substitution:

$$\begin{aligned} -2B - 3(50) &= -230 \\ -2B &= -80 \\ B &= 40 \\ A + 2(40) + 2(50) &= 190 \\ A &= 10 \end{aligned}$$

Wanda needs 10 pounds of A, 40 pounds of B, and 50 pounds of C.

49. Press MATRX ▷ to select the MATRX EDT menu. Press 1 to create matrix A.

Press 4 to create a matrix with 4 rows; press ENTER. Press 5 to create a matrix with 5 columns; press ENTER.

Enter the elements of the matrix one by one: 1, 2, 3, 4, 5, 6, 7, 8, 9, 0, 2, 1, 3, 2, −1, 4, 5, −3, 6, 2. Press 2nd QUIT to return to the home screen.

Begin on a blank line.

Press MATRX.

Use the *ROW+(instruction to multiply row 1 by −6 and add it to row 2. The resulting matrix is displayed and stored in ANS:

    *ROW+(−6, [A], 1, 2)

Press MATRX.

Multiply row 1 of ANS by −2 and add it to row 3:

    *ROW+(−2, ANS, 1, 3)

Press MATRX.
Multiply row 1 of ANS by -4 and add it to row 4:
   *ROW+(-4, ANS, 1, 4)
Press MATRX.
Multiply row 2 of ANS by 3 and add it to row 3.
   *ROW+(3, ANS, 2, 3)
Press MATRX.
Multiply row 2 of ANS by 3 and add it to row 4:
   *ROW+(3, ANS, 2, 4)
Press MATRX.
Multiply row 3 of ANS by 3 and add it to row 4:
   *ROW+(3, ANS, 3, 4)
The triangular form of the matrix is

$$\begin{bmatrix} 1 & 2 & 3 & 4 & 5 \\ 0 & 1 & 2 & 3 & 6 \\ 0 & 0 & 3 & 3 & 7 \\ 0 & 0 & 0 & 8 & 21 \end{bmatrix}$$

The matrix corresponds to the system of equations
$$x + 2y + 3z + 4w = 5$$
$$y + 2z + 3w = 6$$
$$3z + 3w = 7$$
$$8w = 21$$
$$w = \frac{21}{8} = 2.625$$
By back substitution:
$$3z + 3(2.625) = 7$$
$$z = -0.2917$$
$$y + 2(-0.2917) + 3(2.625) = 6$$
$$y = -1.2916$$
$$x + 2(-1.2916) + 3(-0.2917) + 4(2.625) = 5$$
$$x = -2.0417$$

Problem Set 11.3

1.
$$A = \begin{bmatrix} 2 & -1 \\ 3 & 7 \end{bmatrix}, B = \begin{bmatrix} 6 & 5 \\ -2 & 3 \end{bmatrix}$$

$$A + B = \begin{bmatrix} 8 & 4 \\ 1 & 10 \end{bmatrix}$$

$$A - B = \begin{bmatrix} -4 & -6 \\ 5 & 4 \end{bmatrix}$$

$$3A = \begin{bmatrix} 6 & -3 \\ 9 & 21 \end{bmatrix}$$

3. $A = \begin{bmatrix} 3 & -2 & 5 \\ 4 & 0 & -3 \end{bmatrix}, B = \begin{bmatrix} 2 & 6 & -1 \\ 4 & 3 & -3 \end{bmatrix}$

$$A + B = \begin{bmatrix} 5 & 4 & 4 \\ 8 & 3 & -6 \end{bmatrix}$$

$$A - B = \begin{bmatrix} 1 & -8 & 6 \\ 0 & -3 & 0 \end{bmatrix}$$

$$3A = \begin{bmatrix} 9 & -6 & 15 \\ 12 & 0 & -9 \end{bmatrix}$$

5. $AB = \begin{bmatrix} 2 & -1 \\ 3 & 7 \end{bmatrix}\begin{bmatrix} 6 & 5 \\ -2 & 3 \end{bmatrix}$

$$= \begin{bmatrix} 12+2 & 10-3 \\ 18-14 & 15+21 \end{bmatrix}$$

$$= \begin{bmatrix} 14 & 7 \\ 4 & 36 \end{bmatrix}$$

$$BA = \begin{bmatrix} 6 & 5 \\ -2 & 3 \end{bmatrix}\begin{bmatrix} 2 & -1 \\ 3 & 7 \end{bmatrix}$$

$$= \begin{bmatrix} 12+15 & -6+35 \\ -4+9 & 2+21 \end{bmatrix}$$

$$= \begin{bmatrix} 27 & 29 \\ 5 & 23 \end{bmatrix}$$

7. $AB = \begin{bmatrix} 1 & -1 & 2 \\ 3 & 4 & -4 \\ 2 & 1 & 3 \end{bmatrix}\begin{bmatrix} 0 & 2 & -3 \\ 1 & 2 & 3 \\ -1 & -2 & 4 \end{bmatrix}$

$$= \begin{bmatrix} 0-1-2 & 2-2-4 & -3-3+8 \\ 0+4+4 & 6+8+8 & -9+12-16 \\ 0+1-3 & 4+2-6 & -6+3+12 \end{bmatrix}$$

$$= \begin{bmatrix} -3 & -4 & 2 \\ 8 & 22 & -13 \\ -2 & 0 & 9 \end{bmatrix}$$

$$\mathbf{BA} = \begin{bmatrix} 0 & 2 & -3 \\ 1 & 2 & 3 \\ -1 & -2 & 4 \end{bmatrix} \begin{bmatrix} 1 & -1 & 2 \\ 3 & 4 & -4 \\ 2 & 1 & 3 \end{bmatrix}$$

$$= \begin{bmatrix} 0+6-6 & 0+8-3 & 0-8-9 \\ 1+6+6 & -1+8+3 & 2-8+9 \\ -1-6+8 & 1-8+4 & -2+8+12 \end{bmatrix}$$

$$= \begin{bmatrix} 0 & 5 & -17 \\ 13 & 10 & 3 \\ 1 & -3 & 18 \end{bmatrix}$$

9. $\mathbf{AB} = \begin{bmatrix} 1 & -2 & 3 & 4 \\ 3 & 2 & -5 & 1 \end{bmatrix} \begin{bmatrix} 1 & 2 \\ 3 & 4 \end{bmatrix}$

Incompatible

$$\mathbf{BA} = \begin{bmatrix} 1 & 2 \\ 3 & 4 \end{bmatrix} \begin{bmatrix} 1 & -2 & 3 & 4 \\ 3 & 2 & -5 & 1 \end{bmatrix}$$

$$= \begin{bmatrix} 1+6 & -2+4 & 3-10 & 4+2 \\ 3+12 & -6+8 & 9-20 & 12+4 \end{bmatrix}$$

$$= \begin{bmatrix} 7 & 2 & -7 & 6 \\ 15 & 2 & -11 & 16 \end{bmatrix}$$

11. $\mathbf{AB} = \begin{bmatrix} 3 & 1 & -1 \\ 2 & 4 & 2 \\ -3 & 2 & -1 \end{bmatrix} \begin{bmatrix} 1 \\ 2 \\ 3 \end{bmatrix}$

$$= \begin{bmatrix} 3+2-3 \\ 2+8+6 \\ -3+4-3 \end{bmatrix}$$

$$= \begin{bmatrix} 2 \\ 16 \\ -2 \end{bmatrix}$$

$$\mathbf{BA} = \begin{bmatrix} 1 \\ 2 \\ 3 \end{bmatrix} \begin{bmatrix} 3 & 1 & -1 \\ 2 & 4 & 2 \\ -3 & 2 & -1 \end{bmatrix}$$

Incompatible

13. $\mathbf{AB} = \begin{bmatrix} 0 & 0 \\ 0 & 0 \end{bmatrix} \begin{bmatrix} 2 & -1 \\ 3 & 4 \end{bmatrix}$

$$= \begin{bmatrix} 0+0 & 0+0 \\ 0+0 & 0+0 \end{bmatrix}$$

$$= \begin{bmatrix} 0 & 0 \\ 0 & 0 \end{bmatrix}$$

$$\mathbf{BA} = \begin{bmatrix} 2 & -1 \\ 3 & 4 \end{bmatrix} \begin{bmatrix} 0 & 0 \\ 0 & 0 \end{bmatrix}$$

$$= \begin{bmatrix} 0+0 & 0+0 \\ 0+0 & 0+0 \end{bmatrix}$$

$$= \begin{bmatrix} 0 & 0 \\ 0 & 0 \end{bmatrix}$$

15. $\mathbf{X} = 2\begin{bmatrix} -1 & 4 & 3 \\ -2 & 0 & 4 \end{bmatrix} - \begin{bmatrix} 2 & 1 & -3 \\ 1 & 5 & 0 \end{bmatrix}$

$$= \begin{bmatrix} -2 & 8 & 6 \\ -4 & 0 & 8 \end{bmatrix} - \begin{bmatrix} 2 & 1 & -3 \\ 1 & 5 & 0 \end{bmatrix}$$

$$= \begin{bmatrix} -4 & 7 & 9 \\ -5 & -5 & 8 \end{bmatrix}$$

17. $\mathbf{B} + \mathbf{C} = \begin{bmatrix} 2 & 4 \\ 6 & 1 \end{bmatrix} + \begin{bmatrix} -1 & -2 \\ 3 & 6 \end{bmatrix}$

$$= \begin{bmatrix} 1 & 2 \\ 9 & 7 \end{bmatrix}$$

$$\mathbf{A}(\mathbf{B} + \mathbf{C}) = \begin{bmatrix} 2 & -1 \\ 3 & 4 \end{bmatrix} \begin{bmatrix} 1 & 2 \\ 9 & 7 \end{bmatrix}$$

$$= \begin{bmatrix} 2-9 & 4-7 \\ 3+36 & 6+28 \end{bmatrix}$$

$$= \begin{bmatrix} -7 & -3 \\ 39 & 34 \end{bmatrix}$$

234

$$\mathbf{AB} = \begin{bmatrix} 2 & -1 \\ 3 & 4 \end{bmatrix} \begin{bmatrix} 2 & 4 \\ 6 & 1 \end{bmatrix}$$

$$= \begin{bmatrix} 4-6 & 8-1 \\ 6+24 & 12+4 \end{bmatrix}$$

$$= \begin{bmatrix} -2 & 7 \\ 30 & 16 \end{bmatrix}$$

$$\mathbf{AC} = \begin{bmatrix} 2 & -1 \\ 3 & 4 \end{bmatrix} \begin{bmatrix} -1 & -2 \\ 3 & 6 \end{bmatrix}$$

$$= \begin{bmatrix} -2-3 & -4-6 \\ -3+12 & -6+24 \end{bmatrix}$$

$$= \begin{bmatrix} -5 & -10 \\ 9 & 18 \end{bmatrix}$$

$$\mathbf{AB} + \mathbf{AC} = \begin{bmatrix} -2 & 7 \\ 30 & 16 \end{bmatrix} + \begin{bmatrix} -5 & -10 \\ 9 & 18 \end{bmatrix}$$

$$= \begin{bmatrix} -7 & -3 \\ 39 & 34 \end{bmatrix}$$

$\mathbf{A(B+C)} = \mathbf{AB} + \mathbf{AC}$, illustrating the distributive law.

19. Element in third row and second column is:
$$8.09(6.31) - 6.73(-5.32) + 5.03(1.34) = 93.5917$$

21.
$$2\mathbf{B} = \begin{bmatrix} 2 & -6 & 4 \\ 10 & 0 & 6 \\ -10 & 4 & 2 \end{bmatrix}$$

$$\mathbf{A} - 2\mathbf{B} = \begin{bmatrix} 4 & -1 & 3 \\ 2 & 5 & 3 \\ 6 & 2 & 1 \end{bmatrix} - \begin{bmatrix} 2 & -6 & 4 \\ 10 & 0 & 6 \\ -10 & 4 & 2 \end{bmatrix}$$

$$= \begin{bmatrix} 2 & 5 & -1 \\ -8 & 5 & -3 \\ 16 & -2 & -1 \end{bmatrix}$$

$$\mathbf{AB} = \begin{bmatrix} 4 & -1 & 3 \\ 2 & 5 & 3 \\ 6 & 2 & 1 \end{bmatrix} \begin{bmatrix} 1 & -3 & 2 \\ 5 & 0 & 3 \\ -5 & 2 & 1 \end{bmatrix}$$

$$= \begin{bmatrix} 4-5-15 & -12+0+6 & 8-3+3 \\ 2+25-15 & -6+0+6 & 4+15+3 \\ 6+10-5 & -18+0+2 & 12+6+1 \end{bmatrix}$$

$$= \begin{bmatrix} -16 & -6 & 8 \\ 12 & 0 & 22 \\ 11 & -16 & 19 \end{bmatrix}$$

$$\mathbf{A}^2 = \begin{bmatrix} 4 & -1 & 3 \\ 2 & 5 & 3 \\ 6 & 2 & 1 \end{bmatrix} \begin{bmatrix} 4 & -1 & 3 \\ 2 & 5 & 3 \\ 6 & 2 & 1 \end{bmatrix}$$

$$= \begin{bmatrix} 16-2+18 & -4-5+6 & 12-3+3 \\ 8+10+18 & -2+25+6 & 6+15+3 \\ 24+4+6 & -6+10+2 & 18+6+1 \end{bmatrix}$$

$$= \begin{bmatrix} 32 & -3 & 12 \\ 36 & 29 & 24 \\ 34 & 6 & 25 \end{bmatrix}$$

23.
$$\mathbf{AB} = \begin{bmatrix} 1 & 2 & 3 & 4 \end{bmatrix} \begin{bmatrix} 2 \\ 1 \\ -1 \\ -2 \end{bmatrix}$$

$$= \begin{bmatrix} 2+2-3-8 \end{bmatrix} = \begin{bmatrix} -7 \end{bmatrix}$$

$$\mathbf{BA} = \begin{bmatrix} 2 \\ 1 \\ -1 \\ -2 \end{bmatrix} \begin{bmatrix} 1 & 2 & 3 & 4 \end{bmatrix}$$

$$= \begin{bmatrix} 2 & 4 & 6 & 8 \\ 1 & 2 & 3 & 4 \\ -1 & -2 & -3 & -4 \\ -2 & -4 & -6 & -8 \end{bmatrix}$$

25. $(\mathbf{A}+\mathbf{B})^2 = (\mathbf{A}+\mathbf{B})(\mathbf{A}+\mathbf{B})$
$$= \mathbf{A}(\mathbf{A}+\mathbf{B}) + \mathbf{B}(\mathbf{A}+\mathbf{B})$$
$$= \mathbf{A}^2 + \mathbf{A}\mathbf{B} + \mathbf{B}\mathbf{A} + \mathbf{B}^2$$

We know that $\mathbf{A}$ and $\mathbf{B}$ must be the same size in order to be added. Furthermore, for $\mathbf{A}\mathbf{B}+\mathbf{B}\mathbf{A}$ to equal $2\mathbf{A}\mathbf{B}$, the multiplication must be commutative. This will only occur if $\mathbf{A}$ and $\mathbf{B}$ are square matrices.

27. $\mathbf{A}^2 = \begin{bmatrix} 0 & a \\ 0 & 0 \end{bmatrix}\begin{bmatrix} 0 & a \\ 0 & 0 \end{bmatrix}$

$= \begin{bmatrix} 0 & 0 \\ 0 & 0 \end{bmatrix}$

$\mathbf{B}^3 = \begin{bmatrix} 0 & a & b \\ 0 & 0 & c \\ 0 & 0 & 0 \end{bmatrix}\begin{bmatrix} 0 & a & b \\ 0 & 0 & c \\ 0 & 0 & 0 \end{bmatrix}\begin{bmatrix} 0 & a & b \\ 0 & 0 & c \\ 0 & 0 & 0 \end{bmatrix}$

$= \begin{bmatrix} 0 & 0 & ac \\ 0 & 0 & 0 \\ 0 & 0 & 0 \end{bmatrix}\begin{bmatrix} 0 & a & b \\ 0 & 0 & c \\ 0 & 0 & 0 \end{bmatrix}$

$= \begin{bmatrix} 0 & 0 & 0 \\ 0 & 0 & 0 \\ 0 & 0 & 0 \end{bmatrix}$

The nth power of a strictly upper triangular $n \times n$ matrix is the zero matrix.

29. $\mathbf{A}^2 = \begin{bmatrix} 1 & 2 & 0 \\ 0 & 1 & 0 \\ 0 & 0 & 1 \end{bmatrix}\begin{bmatrix} 1 & 2 & 0 \\ 0 & 1 & 0 \\ 0 & 0 & 1 \end{bmatrix}$

$= \begin{bmatrix} 1+0+0 & 2+2+0 & 0+0+0 \\ 0+0+0 & 0+1+0 & 0+0+0 \\ 0+0+0 & 0+0+0 & 0+0+1 \end{bmatrix}$

$= \begin{bmatrix} 1 & 4 & 0 \\ 0 & 1 & 0 \\ 0 & 0 & 1 \end{bmatrix}$

$\mathbf{A}^3 = \begin{bmatrix} 1 & 2 & 0 \\ 0 & 1 & 0 \\ 0 & 0 & 1 \end{bmatrix}\begin{bmatrix} 1 & 4 & 0 \\ 0 & 1 & 0 \\ 0 & 0 & 1 \end{bmatrix}$

$= \begin{bmatrix} 1+0+0 & 4+2+0 & 0+0+0 \\ 0+0+0 & 0+1+0 & 0+0+0 \\ 0+0+0 & 0+0+0 & 0+0+1 \end{bmatrix}$

$= \begin{bmatrix} 1 & 6 & 0 \\ 0 & 1 & 0 \\ 0 & 0 & 1 \end{bmatrix}$

$\mathbf{A}^4 = \begin{bmatrix} 1 & 2 & 0 \\ 0 & 1 & 0 \\ 0 & 0 & 1 \end{bmatrix}\begin{bmatrix} 1 & 6 & 0 \\ 0 & 1 & 0 \\ 0 & 0 & 1 \end{bmatrix}$

$= \begin{bmatrix} 1+0+0 & 6+2+0 & 0+0+0 \\ 0+0+0 & 0+1+0 & 0+0+0 \\ 0+0+0 & 0+0+0 & 0+0+1 \end{bmatrix}$

$= \begin{bmatrix} 1 & 8 & 0 \\ 0 & 1 & 0 \\ 0 & 0 & 1 \end{bmatrix}$

$\mathbf{A}^n = \begin{bmatrix} 1 & 2n & 0 \\ 0 & 1 & 0 \\ 0 & 0 & 1 \end{bmatrix}$

31. Let $\mathbf{B} = \begin{bmatrix} a & b & c \\ d & e & f \\ g & h & i \end{bmatrix}$

$\mathbf{A}\mathbf{B} = \begin{bmatrix} 3 & 0 & 0 \\ 0 & -4 & 0 \\ 0 & 0 & 5 \end{bmatrix}\begin{bmatrix} a & b & c \\ d & e & f \\ g & h & i \end{bmatrix}$

$= \begin{bmatrix} 3a & 3b & 3c \\ -4d & -4e & -4f \\ 5g & 5h & 5i \end{bmatrix}$

Multiplication on the left by $\mathbf{A}$ multiplies the three rows of $\mathbf{B}$ by 3, -4, and 5 respectively.

$\mathbf{B}\mathbf{A} = \begin{bmatrix} a & b & c \\ d & e & f \\ g & h & i \end{bmatrix}\begin{bmatrix} 3 & 0 & 0 \\ 0 & -4 & 0 \\ 0 & 0 & 5 \end{bmatrix}$

$$= \begin{bmatrix} 3a & -4b & 5c \\ 3d & -4e & 5f \\ 3g & -4h & 5i \end{bmatrix}$$

Multiplication on the right by $\mathbf{A}$ multiplies the three columns of $\mathbf{B}$ by 3, -4, and 5 respectively.

33. a) $\mathbf{UX} = \begin{bmatrix} 4 & 3 & 2 \\ 5 & 1 & 2 \\ 3 & 4 & 1 \end{bmatrix} \begin{bmatrix} 1 \\ 2 \\ 3 \end{bmatrix} = \begin{bmatrix} 16 \\ 13 \\ 14 \end{bmatrix}$

$\mathbf{UX}$ represents the wages paid A, B, and C for their work on Monday.

b) $\mathbf{VX} = \begin{bmatrix} 3 & 6 & 1 \\ 4 & 2 & 2 \\ 5 & 1 & 3 \end{bmatrix} \begin{bmatrix} 1 \\ 2 \\ 3 \end{bmatrix} = \begin{bmatrix} 18 \\ 14 \\ 16 \end{bmatrix}$

$\mathbf{VX}$ represents the wages paid A, B, and C for their work on Tuesday.

c) $\mathbf{U} + \mathbf{V} = \begin{bmatrix} 4 & 3 & 2 \\ 5 & 1 & 2 \\ 3 & 4 & 1 \end{bmatrix} + \begin{bmatrix} 3 & 6 & 1 \\ 4 & 2 & 2 \\ 5 & 1 & 3 \end{bmatrix}$

$$= \begin{bmatrix} 7 & 9 & 3 \\ 9 & 3 & 4 \\ 8 & 5 & 4 \end{bmatrix}$$

$\mathbf{U} + \mathbf{V}$ represents the combined output for Monday and Tuesday.

d) $(\mathbf{U} + \mathbf{V})\mathbf{X} = \begin{bmatrix} 7 & 9 & 3 \\ 9 & 3 & 4 \\ 8 & 5 & 4 \end{bmatrix} \begin{bmatrix} 1 \\ 2 \\ 3 \end{bmatrix} = \begin{bmatrix} 34 \\ 27 \\ 30 \end{bmatrix}$

$(\mathbf{U} + \mathbf{V})\mathbf{X} =$ represents the wages paid A, B, and C for Monday and Tuesday together.

35. a) $\mathbf{U} + \mathbf{V} = \begin{bmatrix} u_1 & u_2 \\ -u_2 & u_1 \end{bmatrix} + \begin{bmatrix} v_1 & v_2 \\ -v_2 & v_1 \end{bmatrix}$

$$= \begin{bmatrix} u_1 + v_1 & u_2 + v_2 \\ -(u_2 + v_2) & u_1 + v_1 \end{bmatrix}$$

$$\mathbf{UV} = \begin{bmatrix} u_1 & u_2 \\ -u_2 & u_1 \end{bmatrix} \begin{bmatrix} v_1 & v_2 \\ -v_2 & v_1 \end{bmatrix}$$

$$= \begin{bmatrix} u_1 v_1 - u_2 v_2 & u_1 v_2 + u_2 v_1 \\ -(u_2 v_1 + u_1 v_2) & -u_2 v_2 + u_1 v_1 \end{bmatrix}$$

b) $\mathbf{I}^2 = \begin{bmatrix} 1 & 0 \\ 0 & 1 \end{bmatrix} \begin{bmatrix} 1 & 0 \\ 0 & 1 \end{bmatrix}$

$$= \begin{bmatrix} 1 & 0 \\ 0 & 1 \end{bmatrix} = \mathbf{I}$$

$$\mathbf{J}^2 = \begin{bmatrix} 0 & 1 \\ -1 & 0 \end{bmatrix} \begin{bmatrix} 0 & 1 \\ -1 & 0 \end{bmatrix}$$

$$= \begin{bmatrix} -1 & 0 \\ 0 & -1 \end{bmatrix} = -\mathbf{I}$$

c) $\mathbf{U} + \mathbf{V} = (u_1 + v_1)\mathbf{I} + (u_2 + v_2)\mathbf{J}$;
$\mathbf{UV} = (u_1 v_1 - u_2 v_2)\mathbf{I} + (u_1 v_2 + u_2 v_1)\mathbf{J}$

d) If we identify $\mathbf{I}$ with 1 and $\mathbf{J}$ with i, the system of matrices behaves like the complex numbers.

37. Press $\boxed{\text{MATRX}}$ $\boxed{\triangleright}$ to edit a matrix.
Press 1 to enter the dimensions and elements of matrix A. Press each number followed by $\boxed{\text{ENTER}}$:
4,4; 1, 2, 0, 0; 0, 1, 0, 0; 0, 0, 1, 3; 0, 0, 0, 1
Press $\boxed{\text{MATRX}}$ $\boxed{\triangleright}$ again; press 2 to enter the dimensions and elements of matrix B. Press each number followed by $\boxed{\text{ENTER}}$:
4, 4; 1, 0, 0, 3; 0, 1, 0, 0; 0, 0, 1, 0; 0, 0, 0, 1
Press $\boxed{\text{2nd}}$ $\boxed{\text{QUIT}}$ to return to the home screen.
Press $\boxed{\text{2nd}}$ $\boxed{\text{[A]}}$ $\boxed{\text{ENTER}}$ to make sure you entered the values properly; do the same for $\boxed{\text{B}}$.
Press $\boxed{\text{2nd}}$ $\boxed{\text{[A]}}$ $\boxed{\text{2nd}}$ $\boxed{\text{[A]}}$ $\boxed{\text{2nd}}$ $\boxed{\text{[A]}}$ $\boxed{\text{2nd}}$ $\boxed{\text{[A]}}$ $\boxed{\text{2nd}}$ $\boxed{\text{[A]}}$

$\boxed{\text{2nd}}$ $\boxed{\text{[B]}}$ $\boxed{\text{2nd}}$ $\boxed{\text{[B]}}$ $\boxed{\text{ENTER}}$

Press $\boxed{+}$ $\boxed{\text{2nd}}$ $\boxed{\text{[B]}}$ $\boxed{\text{2nd}}$ $\boxed{\text{[A]}}$ $\boxed{\text{ENTER}}$

$$\mathbf{A}^5\mathbf{B}^2 + \mathbf{BA} = \begin{bmatrix} 2 & 12 & 0 & 9 \\ 0 & 2 & 0 & 0 \\ 0 & 0 & 2 & 18 \\ 0 & 0 & 0 & 2 \end{bmatrix}$$

237

1.

$$\text{Let } \mathbf{M} = \begin{bmatrix} 2 & 3 \\ -1 & -1 \end{bmatrix}. \quad D = 2(-1) - (3)(-1) = 1$$

$$\mathbf{M}^{-1} = \begin{bmatrix} \frac{-1}{1} & \frac{-3}{1} \\ \frac{-1}{1} & \frac{2}{1} \end{bmatrix} = \begin{bmatrix} -1 & -3 \\ 1 & 2 \end{bmatrix}$$

3. $\text{Let } \mathbf{M} = \begin{bmatrix} 6 & -14 \\ 0 & 2 \end{bmatrix}. \quad D = 6(2) - 0(-14) = 12$

$$\mathbf{M}^{-1} = \begin{bmatrix} \frac{2}{12} & \frac{-14}{12} \\ -\frac{0}{12} & \frac{6}{12} \end{bmatrix} = \begin{bmatrix} \frac{1}{6} & \frac{7}{6} \\ 0 & \frac{1}{2} \end{bmatrix}$$

5. $\text{Let } \mathbf{M} = \begin{bmatrix} 1 & 0 \\ 0 & 1 \end{bmatrix}. \quad D = 1(1) - 0(0) = 1$

$$\mathbf{M}^{-1} = \begin{bmatrix} \frac{1}{1} & -\frac{0}{1} \\ -\frac{0}{1} & \frac{1}{1} \end{bmatrix} = \begin{bmatrix} 1 & 0 \\ 0 & 1 \end{bmatrix}$$

7. $\text{Let } \mathbf{M} = \begin{bmatrix} a & 0 \\ 0 & b \end{bmatrix}. \quad D = a(b) - 0(0) = ab$

$$\mathbf{M}^{-1} = \begin{bmatrix} \frac{b}{ab} & -\frac{0}{ab} \\ -\frac{0}{ab} & \frac{a}{ab} \end{bmatrix} = \begin{bmatrix} \frac{1}{a} & 0 \\ 0 & \frac{1}{b} \end{bmatrix}$$

9.

	M		**I**

$$\begin{bmatrix} 1 & 3 \\ 2 & 4 \end{bmatrix} \begin{bmatrix} 1 & 0 \\ 0 & 1 \end{bmatrix}$$

$$\begin{bmatrix} 1 & 3 \\ 0 & -2 \end{bmatrix} \begin{bmatrix} 1 & 0 \\ -2 & 1 \end{bmatrix}$$ Add -2 times row 1 to row 2

$$\begin{bmatrix} 1 & 3 \\ 0 & 1 \end{bmatrix} \begin{bmatrix} 1 & 0 \\ 1 & -\frac{1}{2} \end{bmatrix}$$ Divide row 2 by -2

$$\begin{bmatrix} 1 & 0 \\ 0 & 1 \end{bmatrix} \begin{bmatrix} -2 & \frac{3}{2} \\ 1 & -\frac{1}{2} \end{bmatrix}$$ Add -3 times row 2 to row 1

$$\mathbf{M}^{-1} = \begin{bmatrix} -2 & \frac{3}{2} \\ 1 & -\frac{1}{2} \end{bmatrix}$$

11.

M			**I**		

$$\begin{bmatrix} 1 & 1 & 1 \\ 1 & -1 & 2 \\ 3 & 2 & 0 \end{bmatrix} \begin{bmatrix} 1 & 0 & 0 \\ 0 & 1 & 0 \\ 0 & 0 & 1 \end{bmatrix}$$

Add -1 times row 1 to row 2 and -3 times row 1 to row 3:

$$\begin{bmatrix} 1 & 1 & 1 \\ 0 & -2 & 1 \\ 0 & -1 & -3 \end{bmatrix} \begin{bmatrix} 1 & 0 & 0 \\ -1 & 1 & 0 \\ -3 & 0 & 1 \end{bmatrix}$$

Exchange row 2 and 3:

$$\begin{bmatrix} 1 & 1 & 1 \\ 0 & -1 & -3 \\ 0 & -2 & 1 \end{bmatrix} \begin{bmatrix} 1 & 0 & 0 \\ -3 & 0 & 1 \\ -1 & 1 & 0 \end{bmatrix}$$

Add -2 times row 2 to row 3:

$$\begin{bmatrix} 1 & 1 & 1 \\ 0 & -1 & -3 \\ 0 & 0 & 7 \end{bmatrix} \begin{bmatrix} 1 & 0 & 0 \\ -3 & 0 & 1 \\ 5 & 1 & -2 \end{bmatrix}$$

Divide row 3 by 7:

$$\begin{bmatrix} 1 & 1 & 1 \\ 0 & -1 & -3 \\ 0 & 0 & 1 \end{bmatrix} \begin{bmatrix} 1 & 0 & 0 \\ -3 & 0 & 1 \\ \frac{5}{7} & \frac{1}{7} & -\frac{2}{7} \end{bmatrix}$$

Add 3 times row 3 to row 2:

$$\begin{bmatrix} 1 & 1 & 1 \\ 0 & -1 & 0 \\ 0 & 0 & 1 \end{bmatrix} \begin{bmatrix} 1 & 0 & 0 \\ -\frac{6}{7} & \frac{3}{7} & \frac{1}{7} \\ \frac{5}{7} & \frac{1}{7} & -\frac{2}{7} \end{bmatrix}$$

Add -1 times row 3 to row 1:

$$\begin{bmatrix} 1 & 1 & 0 \\ 0 & -1 & 0 \\ 0 & 0 & 1 \end{bmatrix} \begin{bmatrix} \frac{2}{7} & -\frac{1}{7} & \frac{2}{7} \\ -\frac{6}{7} & \frac{3}{7} & \frac{1}{7} \\ \frac{5}{7} & \frac{1}{7} & -\frac{2}{7} \end{bmatrix}$$

Add row 2 to row 1:

$$\begin{bmatrix} 1 & 0 & 0 \\ 0 & -1 & 0 \\ 0 & 0 & 1 \end{bmatrix} \begin{bmatrix} -\frac{4}{7} & \frac{2}{7} & \frac{3}{7} \\ -\frac{6}{7} & \frac{3}{7} & \frac{1}{7} \\ \frac{5}{7} & \frac{1}{7} & -\frac{2}{7} \end{bmatrix}$$

Multiply row 2 by -1:

$$\begin{bmatrix} 1 & 0 & 0 \\ 0 & 1 & 0 \\ 0 & 0 & 1 \end{bmatrix} \begin{bmatrix} -\frac{4}{7} & \frac{2}{7} & \frac{3}{7} \\ \frac{6}{7} & -\frac{3}{7} & -\frac{1}{7} \\ \frac{5}{7} & \frac{1}{7} & -\frac{2}{7} \end{bmatrix}$$

$$\mathbf{M}^{-1} = \begin{bmatrix} -\frac{4}{7} & \frac{2}{7} & \frac{3}{7} \\ \frac{6}{7} & -\frac{3}{7} & -\frac{1}{7} \\ \frac{5}{7} & \frac{1}{7} & -\frac{2}{7} \end{bmatrix}$$

13.

$$\begin{matrix} \mathbf{M} & & & \mathbf{I} \end{matrix}$$

$$\begin{bmatrix} 3 & 1 & 2 \\ 4 & 1 & -6 \\ 1 & 0 & 1 \end{bmatrix} \begin{bmatrix} 1 & 0 & 0 \\ 0 & 1 & 0 \\ 0 & 0 & 1 \end{bmatrix}$$

Interchange row 1 and row 3:

$$\begin{bmatrix} 1 & 0 & 1 \\ 4 & 1 & -6 \\ 3 & 1 & 2 \end{bmatrix} \begin{bmatrix} 0 & 0 & 1 \\ 0 & 1 & 0 \\ 1 & 0 & 0 \end{bmatrix}$$

Add -4 times row 1 to row 2,
and -3 times row 1 to row 3:

$$\begin{bmatrix} 1 & 0 & 1 \\ 0 & 1 & -10 \\ 0 & 1 & -1 \end{bmatrix} \begin{bmatrix} 0 & 0 & 1 \\ 0 & 1 & -4 \\ 1 & 0 & -3 \end{bmatrix}$$

Add -1 times row 2 to row 3:

$$\begin{bmatrix} 1 & 0 & 1 \\ 0 & 1 & -10 \\ 0 & 0 & 9 \end{bmatrix} \begin{bmatrix} 0 & 0 & 1 \\ 0 & 1 & -4 \\ 1 & -1 & 1 \end{bmatrix}$$

Divide row 3 by 9:

$$\begin{bmatrix} 1 & 0 & 1 \\ 0 & 1 & -10 \\ 0 & 0 & 1 \end{bmatrix} \begin{bmatrix} 0 & 0 & 1 \\ 0 & 1 & -4 \\ \frac{1}{9} & -\frac{1}{9} & \frac{1}{9} \end{bmatrix}$$

Add 10 times row 3 to row 2,
and -1 times row 3 to row 1:

$$\begin{bmatrix} 1 & 0 & 0 \\ 0 & 1 & 0 \\ 0 & 0 & 1 \end{bmatrix} \begin{bmatrix} -\frac{1}{9} & \frac{1}{9} & \frac{8}{9} \\ \frac{10}{9} & -\frac{1}{9} & -\frac{26}{9} \\ \frac{1}{9} & -\frac{1}{9} & \frac{1}{9} \end{bmatrix}$$

$$\mathbf{M}^{-1} = \begin{bmatrix} -\frac{1}{9} & \frac{1}{9} & \frac{8}{9} \\ \frac{10}{9} & -\frac{1}{9} & -\frac{26}{9} \\ \frac{1}{9} & -\frac{1}{9} & \frac{1}{9} \end{bmatrix}$$

15.

$$\begin{matrix} \mathbf{M} & & & & \mathbf{I} \end{matrix}$$

$$\begin{bmatrix} 1 & 2 & 1 & 1 \\ 0 & 2 & 3 & 2 \\ 0 & 0 & 1 & 3 \\ 0 & 0 & 0 & 4 \end{bmatrix} \begin{bmatrix} 1 & 0 & 0 & 0 \\ 0 & 1 & 0 & 0 \\ 0 & 0 & 1 & 0 \\ 0 & 0 & 0 & 1 \end{bmatrix}$$

Divide row 4 by 4:

$$\begin{bmatrix} 1 & 2 & 1 & 1 \\ 0 & 2 & 3 & 2 \\ 0 & 0 & 1 & 3 \\ 0 & 0 & 0 & 1 \end{bmatrix} \begin{bmatrix} 1 & 0 & 0 & 0 \\ 0 & 1 & 0 & 0 \\ 0 & 0 & 1 & 0 \\ 0 & 0 & 0 & \frac{1}{4} \end{bmatrix}$$

Add -3 times row 4 to row 3, -2 times row 4 to row 2, and -1 times row 4 to row 1:

$$\begin{bmatrix} 1 & 2 & 1 & 0 \\ 0 & 2 & 3 & 0 \\ 0 & 0 & 1 & 0 \\ 0 & 0 & 0 & 1 \end{bmatrix} \begin{bmatrix} 1 & 0 & 0 & -\frac{1}{4} \\ 0 & 1 & 0 & -\frac{1}{2} \\ 0 & 0 & 1 & -\frac{3}{4} \\ 0 & 0 & 0 & \frac{1}{4} \end{bmatrix}$$

Add -1 times row 3 to row 1
and -3 times row 3 to row 2:

$$\begin{bmatrix} 1 & 2 & 0 & 0 \\ 0 & 2 & 0 & 0 \\ 0 & 0 & 1 & 0 \\ 0 & 0 & 0 & 1 \end{bmatrix} \begin{bmatrix} 1 & 0 & -1 & \frac{1}{2} \\ 0 & 1 & -3 & \frac{7}{4} \\ 0 & 0 & 1 & -\frac{3}{4} \\ 0 & 0 & 0 & \frac{1}{4} \end{bmatrix}$$

Add -1 times row 2 to row 1:

$$\begin{bmatrix} 1 & 0 & 0 & 0 \\ 0 & 2 & 0 & 0 \\ 0 & 0 & 1 & 0 \\ 0 & 0 & 0 & 1 \end{bmatrix} \begin{bmatrix} 1 & -1 & 2 & -\frac{5}{4} \\ 0 & 1 & -3 & \frac{7}{4} \\ 0 & 0 & 1 & -\frac{3}{4} \\ 0 & 0 & 0 & \frac{1}{4} \end{bmatrix}$$

Divide row 2 by 2:

$$\begin{bmatrix} 1 & 0 & 0 & 0 \\ 0 & 1 & 0 & 0 \\ 0 & 0 & 1 & 0 \\ 0 & 0 & 0 & 1 \end{bmatrix} \begin{bmatrix} 1 & -1 & 2 & -\frac{5}{4} \\ 0 & \frac{1}{2} & -\frac{3}{2} & \frac{7}{8} \\ 0 & 0 & 1 & -\frac{3}{4} \\ 0 & 0 & 0 & \frac{1}{4} \end{bmatrix}$$

$$\mathbf{M}^{-1} = \begin{bmatrix} 1 & -1 & 2 & -\frac{5}{4} \\ 0 & \frac{1}{2} & -\frac{3}{2} & \frac{7}{8} \\ 0 & 0 & 1 & -\frac{3}{4} \\ 0 & 0 & 0 & \frac{1}{4} \end{bmatrix}$$

17.
$$\begin{bmatrix} 1 & 3 & 4 \\ 2 & 1 & -1 \\ 4 & 7 & 7 \end{bmatrix}$$

$$\begin{bmatrix} 1 & 3 & 4 \\ 0 & -5 & -9 \\ 0 & -5 & -9 \end{bmatrix}$$ Add -2 times row 1 to row 2 and -4 times row 1 to row 3

$$\begin{bmatrix} 1 & 3 & 4 \\ 0 & -5 & -9 \\ 0 & 0 & 0 \end{bmatrix}$$ Add -1 times row 2 to row 3

Since the third row of the matrix is all zeros, the matrix can never be transformed into **I**. Therefore, the matrix has no inverse.

19. Let $\mathbf{A} = \begin{bmatrix} 1 & 1 & 1 \\ 1 & -1 & 2 \\ 3 & 2 & 0 \end{bmatrix}$, $\mathbf{X} = \begin{bmatrix} x \\ y \\ z \end{bmatrix}$, $\mathbf{B} = \begin{bmatrix} 2 \\ -1 \\ 5 \end{bmatrix}$.

Then the system can be represented as
$$\mathbf{AX} = \mathbf{B}$$
and the solution is
$$\mathbf{X} = \mathbf{A}^{-1}\mathbf{B}$$

$$\begin{bmatrix} x \\ y \\ z \end{bmatrix} = \begin{bmatrix} -\frac{4}{7} & \frac{2}{7} & \frac{3}{7} \\ \frac{6}{7} & -\frac{3}{7} & -\frac{1}{7} \\ \frac{5}{7} & \frac{1}{7} & -\frac{2}{7} \end{bmatrix} \begin{bmatrix} 2 \\ -1 \\ 5 \end{bmatrix}$$

$$= \frac{1}{7} \begin{bmatrix} -4 & 2 & 3 \\ 6 & -3 & -1 \\ 5 & 1 & -2 \end{bmatrix} \begin{bmatrix} 2 \\ -1 \\ 5 \end{bmatrix}$$

$$= \frac{1}{7} \begin{bmatrix} -8-2+15 \\ 12+3-5 \\ 10-1-10 \end{bmatrix} = \frac{1}{7} \begin{bmatrix} 5 \\ 10 \\ -1 \end{bmatrix}$$

$$\begin{bmatrix} x \\ y \\ z \end{bmatrix} = \begin{bmatrix} \frac{5}{7} \\ \frac{10}{7} \\ -\frac{1}{7} \end{bmatrix} \quad x = \frac{5}{7}, \ y = \frac{10}{7}, \ z = -\frac{1}{7}$$

21. Let $\mathbf{A} = \begin{bmatrix} 3 & 1 & 2 \\ 4 & 1 & -6 \\ 1 & 0 & 1 \end{bmatrix}$, $\mathbf{B} = \begin{bmatrix} 3 \\ 2 \\ 6 \end{bmatrix}$, $\mathbf{X} = \begin{bmatrix} x \\ y \\ z \end{bmatrix}$.

Then the system can be represented as
$$\mathbf{AX} = \mathbf{B}$$
and the solution is
$$\mathbf{X} = \mathbf{A}^{-1}\mathbf{B}$$

$$\begin{bmatrix} x \\ y \\ z \end{bmatrix} = \begin{bmatrix} -\frac{1}{9} & \frac{1}{9} & \frac{8}{9} \\ \frac{10}{9} & -\frac{1}{9} & -\frac{26}{9} \\ \frac{1}{9} & -\frac{1}{9} & \frac{1}{9} \end{bmatrix} \begin{bmatrix} 3 \\ 2 \\ 6 \end{bmatrix}$$

$$\begin{bmatrix} x \\ y \\ z \end{bmatrix} = \begin{bmatrix} -\frac{3}{9}+\frac{2}{9}+\frac{48}{9} \\ \frac{30}{9}-\frac{2}{9}-\frac{156}{9} \\ \frac{3}{9}-\frac{2}{9}+\frac{6}{9} \end{bmatrix} = \begin{bmatrix} \frac{47}{9} \\ -\frac{128}{9} \\ \frac{7}{9} \end{bmatrix}$$

$$x = \frac{47}{9}, \ y = -\frac{128}{9}, \ z = \frac{7}{9}$$

23. Let $\mathbf{M} = \begin{bmatrix} 4 & -3 \\ 5 & -\frac{15}{4} \end{bmatrix}$. $D = 4(-\frac{15}{4}) - 5(-3) = 0$

Because $D = 0$, $\mathbf{M}^{-1}$ does not exist.

25.

$$\begin{array}{cc} \mathbf{M} & \mathbf{I} \end{array}$$

$$\begin{bmatrix} 1 & -2 & 1 \\ 3 & 0 & 2 \\ 1 & 2 & \frac{1}{2} \end{bmatrix} \begin{bmatrix} 1 & 0 & 0 \\ 0 & 1 & 0 \\ 0 & 0 & 1 \end{bmatrix}$$

Interchange rows 2 and 3:

$$\begin{bmatrix} 1 & -2 & 1 \\ 1 & 2 & \frac{1}{2} \\ 3 & 0 & 2 \end{bmatrix} \begin{bmatrix} 1 & 0 & 0 \\ 0 & 0 & 1 \\ 0 & 1 & 0 \end{bmatrix}$$

Add -1 times row 1 to row 2 and -3 times row 1 to row 3:

$$\begin{bmatrix} 1 & -2 & 1 \\ 0 & 4 & -\frac{1}{2} \\ 0 & 6 & -1 \end{bmatrix} \begin{bmatrix} 1 & 0 & 0 \\ -1 & 0 & 1 \\ -3 & 1 & 0 \end{bmatrix}$$

Divide row 2 by 4:

$$\begin{bmatrix} 1 & -2 & 1 \\ 0 & 1 & -\frac{1}{8} \\ 0 & 6 & -1 \end{bmatrix} \begin{bmatrix} 1 & 0 & 0 \\ -\frac{1}{4} & 0 & \frac{1}{4} \\ -3 & 1 & 0 \end{bmatrix}$$

Add -6 times row 2 to row 3:

$$\begin{bmatrix} 1 & -2 & 1 \\ 0 & 1 & -\frac{1}{8} \\ 0 & 0 & -\frac{1}{4} \end{bmatrix} \begin{bmatrix} 1 & 0 & 0 \\ -\frac{1}{4} & 0 & \frac{1}{4} \\ -\frac{6}{4} & 1 & -\frac{6}{4} \end{bmatrix}$$

Multiply row 3 by -4:

$$\begin{bmatrix} 1 & -2 & 1 \\ 0 & 1 & -\frac{1}{8} \\ 0 & 0 & 1 \end{bmatrix} \begin{bmatrix} 1 & 0 & 0 \\ -\frac{1}{4} & 0 & \frac{1}{4} \\ 6 & -4 & 6 \end{bmatrix}$$

Add $\frac{1}{8}$ times row 3 to row 2,
and -1 times row 3 to row 1.

$$\begin{bmatrix} 1 & -2 & 0 \\ 0 & 1 & 0 \\ 0 & 0 & 1 \end{bmatrix} \begin{bmatrix} -5 & 4 & -6 \\ \frac{1}{2} & -\frac{1}{2} & 1 \\ 6 & -4 & 6 \end{bmatrix}$$

Add 2 times row 2 to row 1:

$$\begin{bmatrix} 1 & 0 & 0 \\ 0 & 1 & 0 \\ 0 & 0 & 1 \end{bmatrix} \begin{bmatrix} -4 & 3 & -4 \\ \frac{1}{2} & -\frac{1}{2} & 1 \\ 6 & -4 & 6 \end{bmatrix}$$

$$\mathbf{M}^{-1} = \begin{bmatrix} -4 & 3 & -4 \\ \frac{1}{2} & -\frac{1}{2} & 1 \\ 6 & -4 & 6 \end{bmatrix}$$

27.

$$\begin{matrix} \mathbf{M} & \qquad & \mathbf{I} \end{matrix}$$

$$\begin{bmatrix} 2 & 0 & 0 \\ 0 & 3 & 0 \\ 0 & 0 & -4 \end{bmatrix} \begin{bmatrix} 1 & 0 & 0 \\ 0 & 1 & 0 \\ 0 & 0 & 1 \end{bmatrix}$$

Divide row 1 by 2, row 2 by 3, and row 3 by -4:

$$\begin{bmatrix} 1 & 0 & 0 \\ 0 & 1 & 0 \\ 0 & 0 & 1 \end{bmatrix} \begin{bmatrix} \frac{1}{2} & 0 & 0 \\ 0 & \frac{1}{3} & 0 \\ 0 & 0 & -\frac{1}{4} \end{bmatrix}$$

$$\mathbf{M}^{-1} = \begin{bmatrix} \frac{1}{2} & 0 & 0 \\ 0 & \frac{1}{3} & 0 \\ 0 & 0 & -\frac{1}{4} \end{bmatrix}$$

29. Let $\mathbf{A} = \begin{bmatrix} 1 & -2 & 1 \\ 3 & 0 & 2 \\ 1 & 2 & \frac{1}{2} \end{bmatrix}$, $\mathbf{B} = \begin{bmatrix} a \\ b \\ c \end{bmatrix}$, $\mathbf{X} = \begin{bmatrix} x \\ y \\ z \end{bmatrix}$.

Then the system can be represented as
$$\mathbf{AX} = \mathbf{B}$$
and the solution is
$$\mathbf{X} = \mathbf{A}^{-1}\mathbf{B}$$

$$\begin{bmatrix} x \\ y \\ z \end{bmatrix} = \begin{bmatrix} -4 & 3 & -4 \\ \frac{1}{2} & -\frac{1}{2} & 1 \\ 6 & -4 & 6 \end{bmatrix} \begin{bmatrix} a \\ b \\ c \end{bmatrix}$$

$$= \begin{bmatrix} -4a + 3b - 4c \\ \frac{a}{2} - \frac{b}{2} + c \\ 6a - 4b + 6c \end{bmatrix}$$

$x = -4a + 3b - 4c$

$y = \frac{a}{2} - \frac{b}{2} + c$

$z = 6a - 4b + 6c$

31. $\begin{bmatrix} 1 & -1 \\ 3 & -3 \end{bmatrix} \begin{bmatrix} 2 & -4 \\ 2 & -4 \end{bmatrix} = \begin{bmatrix} 2-2 & -4+4 \\ 6-6 & -12+12 \end{bmatrix}$

$$= \begin{bmatrix} 0 & 0 \\ 0 & 0 \end{bmatrix}$$

33. a) Let $\mathbf{C} = \begin{bmatrix} a & b & c \\ d & e & f \\ g & h & i \end{bmatrix}$

$$\mathbf{AC} = \begin{bmatrix} 1 & 0 & 0 \\ 0 & 1 & 2 \\ 0 & 0 & 1 \end{bmatrix} \begin{bmatrix} a & b & c \\ d & e & f \\ g & h & i \end{bmatrix}$$

$$= \begin{bmatrix} a & b & c \\ d+2g & e+2h & f+2i \\ g & h & i \end{bmatrix}$$

2 times the third row of $\mathbf{C}$ gets added to the second row of $\mathbf{C}$.

$$\mathbf{BC} = \begin{bmatrix} 1 & 0 & 0 \\ 0 & 1 & -2 \\ 0 & 0 & 1 \end{bmatrix} \begin{bmatrix} a & b & c \\ d & e & f \\ g & h & i \end{bmatrix}$$

$$= \begin{bmatrix} a & b & c \\ d-2g & e-2h & f-2i \\ g & h & i \end{bmatrix}$$

-2 times the third row is added to the second row.

b) $\mathbf{A}$ and $\mathbf{B}$ are multiplicative inverses.

35.

$$\begin{array}{cc} \mathbf{M} & \mathbf{I} \end{array}$$

$$\begin{bmatrix} 1 & 1 & 1 & 1 \\ 1 & 2 & 2 & 2 \\ 1 & 2 & 1 & 1 \\ 1 & 2 & 1 & 2 \end{bmatrix} \begin{bmatrix} 1 & 0 & 0 & 0 \\ 0 & 1 & 0 & 0 \\ 0 & 0 & 1 & 0 \\ 0 & 0 & 0 & 1 \end{bmatrix}$$

Add -1 times row 1 to rows 2, 3, and 4:

$$\begin{bmatrix} 1 & 1 & 1 & 1 \\ 0 & 1 & 1 & 1 \\ 0 & 1 & 0 & 0 \\ 0 & 1 & 0 & 1 \end{bmatrix} \begin{bmatrix} 1 & 0 & 0 & 0 \\ -1 & 1 & 0 & 0 \\ -1 & 0 & 1 & 0 \\ -1 & 0 & 0 & 1 \end{bmatrix}$$

Add -1 times row 2 to row 3 and 4:

$$\begin{bmatrix} 1 & 1 & 1 & 1 \\ 0 & 1 & 1 & 1 \\ 0 & 0 & -1 & -1 \\ 0 & 0 & -1 & 0 \end{bmatrix} \begin{bmatrix} 1 & 0 & 0 & 0 \\ -1 & 1 & 0 & 0 \\ 0 & -1 & 1 & 0 \\ 0 & -1 & 0 & 1 \end{bmatrix}$$

Add -1 times row 3 to row 4:

$$\begin{bmatrix} 1 & 1 & 1 & 1 \\ 0 & 1 & 1 & 1 \\ 0 & 0 & -1 & -1 \\ 0 & 0 & 0 & 1 \end{bmatrix} \begin{bmatrix} 1 & 0 & 0 & 0 \\ -1 & 1 & 0 & 0 \\ 0 & -1 & 1 & 0 \\ 0 & 0 & -1 & 1 \end{bmatrix}$$

Add row 4 to row 3, -1 times row 4 to row 2, and -1 times row 4 to row 1:

$$\begin{bmatrix} 1 & 1 & 1 & 0 \\ 0 & 1 & 1 & 0 \\ 0 & 0 & -1 & 0 \\ 0 & 0 & 0 & 1 \end{bmatrix} \begin{bmatrix} 1 & 0 & 1 & -1 \\ -1 & 1 & 1 & -1 \\ 0 & -1 & 0 & 1 \\ 0 & 0 & -1 & 1 \end{bmatrix}$$

Add row 3 to rows 1 and 2:

$$\begin{bmatrix} 1 & 1 & 0 & 0 \\ 0 & 1 & 0 & 0 \\ 0 & 0 & -1 & 0 \\ 0 & 0 & 0 & 1 \end{bmatrix} \begin{bmatrix} 1 & -1 & 1 & 0 \\ -1 & 0 & 1 & 0 \\ 0 & -1 & 0 & 1 \\ 0 & 0 & -1 & 1 \end{bmatrix}$$

Add -1 times row 2 to row 1, and multiply row 3 by -1:

$$\begin{bmatrix} 1 & 0 & 0 & 0 \\ 0 & 1 & 0 & 0 \\ 0 & 0 & 1 & 0 \\ 0 & 0 & 0 & 1 \end{bmatrix} \begin{bmatrix} 2 & -1 & 0 & 0 \\ -1 & 0 & 1 & 0 \\ 0 & 1 & 0 & -1 \\ 0 & 0 & -1 & 1 \end{bmatrix}$$

$$\mathbf{M}^{-1} = \begin{bmatrix} 2 & -1 & 0 & 0 \\ -1 & 0 & 1 & 0 \\ 0 & 1 & 0 & -1 \\ 0 & 0 & -1 & 1 \end{bmatrix}$$

37. Let $\mathbf{M} = \begin{bmatrix} 1 & a & b \\ 0 & 1 & c \\ 0 & 0 & 1 \end{bmatrix}$ be any Heisenberg matrix.

Compute $\mathbf{M}^{-1}$:

$$\begin{array}{cc} \mathbf{M} & \mathbf{I} \end{array}$$

$$\begin{bmatrix} 1 & a & b \\ 0 & 1 & c \\ 0 & 0 & 1 \end{bmatrix} \begin{bmatrix} 1 & 0 & 0 \\ 0 & 1 & 0 \\ 0 & 0 & 1 \end{bmatrix}$$

Add $-c$ times row 3 to row 2, and $-b$ times row 3 to row 1:

$$\begin{bmatrix} 1 & a & 0 \\ 0 & 1 & 0 \\ 0 & 0 & 1 \end{bmatrix} \begin{bmatrix} 1 & 0 & -b \\ 0 & 1 & -c \\ 0 & 0 & 1 \end{bmatrix}$$

Add $-a$ times row 2 to row 1:

$$\begin{bmatrix} 1 & 0 & 0 \\ 0 & 1 & 0 \\ 0 & 0 & 1 \end{bmatrix} \begin{bmatrix} 1 & -a & -b+ac \\ 0 & 1 & -c \\ 0 & 0 & 1 \end{bmatrix}$$

$$\mathbf{M}^{-1} = \begin{bmatrix} 1 & -a & -b+ac \\ 0 & 1 & -c \\ 0 & 0 & 1 \end{bmatrix},$$ a Heisenberg matrix.

39. $X = A^{-1}B$

Press $\boxed{\text{MATRX}}\,\boxed{\triangleright}\,\boxed{1}$ to edit matrix A.

Enter the dimension 4×4 for matrix A; enter the elements 1, 1, 1, 1; 1, 2, 3, 4; 1, 3, 6, 10; 1, 4, 10, 20

Press $\boxed{\text{2nd}}\,\boxed{\text{QUIT}}$ to return to the home screen.

Press $\boxed{\text{2nd}}\,\boxed{[A]}\,\boxed{\text{ENTER}}$ to make sure you entered the values properly.

a) Press $\boxed{\text{MATRX}}\,\boxed{\triangleright}\,\boxed{2}$ to edit matrix B.

Enter the dimensions 4×1 for matrix B; enter the elements 4, 3, 2, 1.

Press $\boxed{\text{2nd}}\,\boxed{\text{QUIT}}$ to return to the home screen.

Press $\boxed{\text{2nd}}\,\boxed{[A]}\,\boxed{x^{-1}}\,\boxed{\text{2nd}}\,\boxed{[B]}\,\boxed{\text{ENTER}}$

$$X = A^{-1}B = \begin{bmatrix} 5 \\ -1 \\ 0 \\ 0 \end{bmatrix}$$

b) Press $\boxed{\text{MATRX}}\,\boxed{\triangleright}\,\boxed{2}$ to edit matrix B.

Enter new elements 1, 0, 1, 0

Press $\boxed{\text{2nd}}\,\boxed{\text{QUIT}}$ to return to the home screen.

Press $\boxed{\triangle}\,\boxed{\text{ENTER}}$

$$X = A^{-1}B = \begin{bmatrix} 8 \\ -17 \\ 14 \\ -4 \end{bmatrix}$$

c) Press $\boxed{\text{MATRX}}\,\boxed{\triangleright}\,\boxed{2}$ to edit matrix B.

Enter new elements 1, −2, 3, −4

Press $\boxed{\text{2nd}}\,\boxed{\text{QUIT}}$ to return to the home screen.

Press $\boxed{\triangle}\,\boxed{\text{ENTER}}$

$$X = A^{-1}B = \begin{bmatrix} 32 \\ -79 \\ 68 \\ -20 \end{bmatrix}$$

Problem Set 11.5

1. $\begin{vmatrix} 4 & 0 \\ 0 & -2 \end{vmatrix} = 4(-2) - 0 = -8$

3. $\begin{vmatrix} 11 & 4 \\ 0 & 2 \end{vmatrix} = 11(2) - 0 = 22$

5. $\begin{vmatrix} -1 & -7 & 9 \\ 0 & 5 & 4 \\ 0 & 0 & 10 \end{vmatrix}$

Since this is a triangular determinant, its value is the product of the numbers in the main diagonal:

$(-1)(5)(10) = -50$

7. $\begin{vmatrix} 3 & 0 & 8 \\ 10 & 0 & 2 \\ -1 & 0 & -9 \end{vmatrix}$

Since an entire column contains zeros, the value of the determinant is 0.

9. a) Interchanging the rows and columns of the determinant leaves its value unchanged at 12.

b) Interchanging two rows of the determinant negates its value, to −12.

c) Multiplying a row of the determinant by 3 increases its value by a factor of 3, to 36.

d) Adding a multiple of one row of the determinant to another row leaves its value unchanged at 12.

11. $\begin{vmatrix} 5 & 3 \\ 5 & -3 \end{vmatrix} = 5(-3) - (5)(3) = -30$

13. $\begin{vmatrix} 4 & 8 & -2 \\ 1 & -2 & 0 \\ 2 & 4 & 0 \end{vmatrix} \begin{matrix} 4 & 8 \\ 1 & -2 \\ 2 & 4 \end{matrix}$

$= 4(-2)(0) + 8(0)(2) + (-2)(1)(4)$
$\quad - (2)(-2)(-2) - (4)(0)(4) - (0)(1)(8)$
$= -8 - 8$
$= -16$

15. $\begin{vmatrix} 2 & 4 & 1 \\ 1 & 3 & 6 \\ 2 & 3 & -1 \end{vmatrix} \begin{matrix} 2 & 4 \\ 1 & 3 \\ 2 & 3 \end{matrix}$

$= (2)(3)(-1) + (4)(6)(2) + (1)(1)(3)$
$\quad - (2)(3)(1) - (3)(6)(2) - (-1)(1)(4)$
$= -6 + 48 + 3 - 6 - 36 + 4$
$= 7$

17.
$$\begin{vmatrix} 2.03 & 5.41 & -3.14 \\ 0 & 6.22 & 0 \\ -1.93 & 7.13 & 6.34 \end{vmatrix} \begin{matrix} 2.03 & 5.41 \\ 0 & 6.22 \\ -1.93 & 7.13 \end{matrix}$$

$= (2.03)(6.22)(6.34) + (5.41)(0)(-1.93)$
$\quad + (-3.14)(0)(7.13) - (-1.93)(6.22)(-3.14)$
$\quad - (7.13)(0)(2.03) - (6.34)(0)(5.41)$
$= 80.052644 - 37.694444$
$= 42.3582$

19. $\begin{cases} 5x + y = 7 \\ 3x - 4y = 18 \end{cases}$

$$x = \frac{\begin{vmatrix} 7 & 1 \\ 18 & -4 \end{vmatrix}}{\begin{vmatrix} 5 & 1 \\ 3 & -4 \end{vmatrix}} = \frac{-28 - 18}{-20 - 3} = \frac{-46}{-23} = 2$$

$$y = \frac{\begin{vmatrix} 5 & 7 \\ 3 & 18 \end{vmatrix}}{\begin{vmatrix} 5 & 1 \\ 3 & -4 \end{vmatrix}} = \frac{90 - 21}{-23} = \frac{69}{-23} = -3$$

21. $\begin{cases} 5x - 3y + 2z = 18 \\ x + 4y + 2z = -4 \\ 3x - 2y + z = 11 \end{cases}$

$$D = \begin{vmatrix} 5 & -3 & 2 \\ 1 & 4 & 2 \\ 3 & -2 & 1 \end{vmatrix} = \begin{vmatrix} 5 & -3 & 2 \\ 1 & 4 & 2 \\ 3 & -2 & 1 \end{vmatrix} \begin{matrix} 5 & -3 \\ 1 & 4 \\ 3 & -2 \end{matrix}$$

$= 5(4)(1) + (-3)(2)(3) + (2)(1)(-2)$
$\quad - (3)(4)(2) - (-2)(2)(5) - (1)(1)(-3)$
$= 20 - 18 - 4 - 24 + 20 + 3$
$= -3$

$$x = -\frac{1}{3}\begin{vmatrix} 18 & -3 & 2 \\ -4 & 4 & 2 \\ 11 & -2 & 1 \end{vmatrix} = -\frac{1}{3}\begin{vmatrix} 18 & -3 & 2 \\ -4 & 4 & 2 \\ 11 & -2 & 1 \end{vmatrix} \begin{matrix} 18 & -3 \\ -4 & 4 \\ 11 & -2 \end{matrix}$$

$= -\frac{1}{3}[(18)(4)(1) + (-3)(2)(11) + (2)(-4)(-2)$
$\quad - (11)(4)(2) - (-2)(2)(18) - (1)(-4)(-3)]$
$= -\frac{1}{3}(72 - 66 + 16 - 88 + 72 - 12) = -\frac{1}{3}(-6)$
$= 2$

$$y = -\frac{1}{3}\begin{vmatrix} 5 & 18 & 2 \\ 1 & -4 & 2 \\ 3 & 11 & 1 \end{vmatrix} = -\frac{1}{3}\begin{vmatrix} 5 & 18 & 2 \\ 1 & -4 & 2 \\ 3 & 11 & 1 \end{vmatrix} \begin{matrix} 5 & 18 \\ 1 & -4 \\ 3 & 11 \end{matrix}$$

$= -\frac{1}{3}[(5)(-4)(1) + (18)(2)(3) + (2)(1)(11)$
$\quad - 3(-4)(2) - (11)(2)(5) - (1)(1)(18)]$
$= -\frac{1}{3}(-20 + 108 + 22 + 24 - 110 - 18) = -\frac{1}{3}(6)$
$= -2$

$$z = -\frac{1}{3}\begin{vmatrix} 5 & -3 & 18 \\ 1 & 4 & -4 \\ 3 & -2 & 11 \end{vmatrix} = -\frac{1}{3}\begin{vmatrix} 5 & -3 & 18 \\ 1 & 4 & -4 \\ 3 & -2 & 11 \end{vmatrix} \begin{matrix} 5 & -3 \\ 1 & 4 \\ 3 & -2 \end{matrix}$$

$= -\frac{1}{3}[(5)(4)(11) + (-3)(-4)(3) + (18)(1)(-2)$
$\quad - (3)(4)(18) - (-2)(-4)(5) - (11)(1)(-3)]$
$= -\frac{1}{3}(220 + 36 - 36 - 216 - 40 + 33) = -\frac{1}{3}(-3)$
$= 1$

Solution: $(x, y, z) = (2, -2, 1)$

23. a) $\begin{vmatrix} 1 & 2 & 3 \\ 0 & 0 & 0 \\ 1.9 & 2.9 & 3.9 \end{vmatrix} = 0,$

since an entire row is zeros.

b) $\begin{vmatrix} 1 & 2 & 3 \\ 1.1 & 2.2 & 3.3 \\ 1.9 & 2.9 & 3.9 \end{vmatrix} = 1.1 \begin{vmatrix} 1 & 2 & 3 \\ 1 & 2 & 3 \\ 1.9 & 2.9 & 3.9 \end{vmatrix}$

$= 1.1 \begin{vmatrix} 1 & 2 & 3 \\ 0 & 0 & 0 \\ 1.9 & 2.9 & 3.9 \end{vmatrix}$

$= 0,$

since an entire row is zeros.

c) $\begin{vmatrix} 1.1 & 2.2 & 3.3 \\ 4.4 & 5.5 & 6.6 \\ 5.5 & 7.7 & 9.9 \end{vmatrix} = (1.1)^3 \begin{vmatrix} 1 & 2 & 3 \\ 4 & 5 & 6 \\ 5 & 7 & 9 \end{vmatrix}$

$= (1.1)^3 \begin{vmatrix} 1 & 2 & 3 \\ 0 & -3 & -6 \\ 0 & -3 & -6 \end{vmatrix}$

$= (1.1)^3 \begin{vmatrix} 1 & 2 & 3 \\ 0 & -3 & -6 \\ 0 & 0 & 0 \end{vmatrix}$

$= 0,$

since an entire row is zeros.

d) $\begin{vmatrix} 1 & 2 & 3 \\ 0 & 2 & 3 \\ 0 & 0 & 3 \end{vmatrix} = (1)(2)(3) = 6$

Since this is a triangular matrix, its value is the product of the entries in the main diagonal.

e) $\begin{vmatrix} 1 & 1 & 1 \\ 1 & 2 & 3 \\ 1 & 3 & 6 \end{vmatrix} = \begin{vmatrix} 1 & 1 & 1 \\ 0 & 1 & 2 \\ 0 & 2 & 5 \end{vmatrix} = \begin{vmatrix} 1 & 1 & 1 \\ 0 & 1 & 2 \\ 0 & 0 & 1 \end{vmatrix}$
$$= (1)(1)(1)$$
$$= 1$$

f) $\begin{vmatrix} 1 & 1 & 1 \\ 1 & 2 & 4 \\ 1 & 3 & 9 \end{vmatrix} = \begin{vmatrix} 1 & 1 & 1 \\ 0 & 1 & 3 \\ 0 & 2 & 8 \end{vmatrix} = \begin{vmatrix} 1 & 1 & 1 \\ 0 & 1 & 3 \\ 0 & 0 & 2 \end{vmatrix}$
$$= (1)(1)(2)$$
$$= 2$$

25. $\begin{vmatrix} x^2 & x & 1 \\ 1 & 2 & 3 \\ 4 & 2 & 1 \end{vmatrix} = \begin{vmatrix} x^2 & x & 1 \\ 1 & 2 & 3 \\ 4 & 2 & 1 \end{vmatrix}\begin{matrix} x^2 & x \\ 1 & 2 \\ 4 & 2 \end{matrix}$

$$= (x^2)(2)(1) + (x)(3)(4) + (1)(1)(2)$$
$$\quad - (4)(2)(1) - (2)(3)(x^2) - (1)(1)(x)$$
$$= 2x^2 + 12x + 2 - 8 - 6x^2 - x$$
$$= -4x^2 + 11x - 6$$

Solve
$$-4x^2 + 11x - 6 = 0$$
$$4x^2 - 11x + 6 = 0$$
$$(4x - 3)(x - 2) = 0$$
$$4x - 3 = 0 \text{ or } x - 2 = 0$$
$$x = \tfrac{3}{4} \text{ or } \quad x = 2$$

27. $\begin{cases} kx + y = k^2 \\ x + ky = 1 \end{cases}$

$$D = \begin{vmatrix} k & 1 \\ 1 & k \end{vmatrix} = k^2 - 1 = (k-1)(k+1)$$

a) If $D \neq 0$, then the system has a unique solution.
$$(k-1)(k+1) \neq 0$$
$$k \neq 1 \text{ and } k \neq -1$$

If the system doesn't have a unique solution, $k = 1$ or $k = -1$.

b) If $k = 1$ then both equations become $x + y = 1$, and the system has infinitely many solutions.

c) If $k = -1$ then
$$\begin{cases} -x + y = 1 \\ x - y = 1 \end{cases}$$
Adding these two, we get $0 = 2$, so the system has no solution.

29. a) From Problem 28, we know that if $\mathbf{A}$ and $\mathbf{B}$ are square matrices, then $|\mathbf{AB}| = |\mathbf{A}\|\mathbf{B}|$. Therefore,
$$|\mathbf{A}^2| = |\mathbf{A}\|\mathbf{A}| = 12 \cdot 12 = 144$$

b)
$$|2\mathbf{A}| = \begin{vmatrix} 2a_1 & 2b_1 & 2c_1 \\ 2a_2 & 2b_2 & 2c_2 \\ 2a_3 & 2b_3 & 2c_3 \end{vmatrix} = 2 \cdot 2 \cdot 2 |\mathbf{A}| = 8 \cdot 12 = 96$$

c) Since $\mathbf{AA}^{-1} = \mathbf{I}$,
$$|\mathbf{AA}^{-1}| = |\mathbf{I}|$$
$$|\mathbf{A}\|\mathbf{A}^{-1}| = 1$$
$$|\mathbf{A}^{-1}| = \frac{1}{|\mathbf{A}|} = \frac{1}{12}$$

d) $\begin{vmatrix} 2a_1 & 2b_1 & 2c_1 \\ a_2 & b_2 & c_2 \\ -3a_3 & -3b_3 & -3c_3 \end{vmatrix} = 2(-3)|\mathbf{A}| = -6 \cdot 12 = -72$

e) $\begin{vmatrix} a_1 & b_1 & c_1 \\ a_2 - a_1 & b_2 - b_1 & c_2 - c_1 \\ 2a_1 & 2b_1 & 2c_2 \end{vmatrix} = 0$,

since by adding -2 times row 1 to row 3 we can produce a row containing all zeros.

31. a) $|\mathbf{A}^5| = |\mathbf{A}|^5 = (-2)^5 = -32$

b) From Problem 29(c), $|\mathbf{A}^{-1}| = \frac{1}{|\mathbf{A}|} = \frac{1}{-2} = -\frac{1}{2}$

c) $|\mathbf{B}^{-1}\mathbf{AB}| = |\mathbf{B}^{-1}\|\mathbf{A}\|\mathbf{B}|$
Because $|\mathbf{B}^{-1}|$ and $|\mathbf{B}|$ are numbers, multiplication is commutative, so $|\mathbf{B}^{-1}\|\mathbf{A}\|\mathbf{B}| = |\mathbf{B}^{-1}\|\mathbf{B}\|\mathbf{A}|$.
Since $|\mathbf{B}^{-1}\|\mathbf{B}| = 1$, $|\mathbf{B}^{-1}\|\mathbf{B}\|\mathbf{A}| = 1 \cdot |\mathbf{A}| = -2$.

d) Multiplying the matrix $\mathbf{A}^3$ by 3 multiplies every entry in the matrix by 3, causing the value of the determinant of $\mathbf{A}^3$ to increase by a factor of $3^3 = 27$.
$$|3\mathbf{A}^3| = 27|\mathbf{A}^3| = 27|\mathbf{A}|^3 = 27(-2)^3 = -216$$

33.

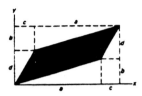

The area of the parallelogram is the area of the rectangle with length $(a+c)$ and width $(b+d)$ minus the areas of the six geometric figures outlined in dashed lines.

$$A = (a+c)(b+d) - bc - bc - \tfrac{1}{2}cd - \tfrac{1}{2}cd - \tfrac{1}{2}ab - \tfrac{1}{2}ab$$
$$= ab + ad + cb + cd - 2bc - cd - ab$$
$$= ad - bc$$
$$= \begin{vmatrix} a & b \\ c & d \end{vmatrix}$$

35. The problem says in effect that $(a+b+c)$ is to be a factor of the expanded determinant. We rewrite the determinant so that it has $(a+b+c)$ in one position on the main diagonal.

$$D = \begin{vmatrix} a & b & c \\ b & c & a \\ c & a & b \end{vmatrix}$$

Add the second row and then the third row to the first:

$$D = \begin{vmatrix} a+b+c & a+b+c & a+b+c \\ b & c & a \\ c & a & b \end{vmatrix}$$

$$= (a+b+c) \begin{vmatrix} 1 & 1 & 1 \\ b & c & a \\ c & a & b \end{vmatrix}$$

We obtain zeros by adding $-b$ times row 1 to row 2 and $-c$ times row 1 to row 3:

$$D = (a+b+c) \begin{vmatrix} 1 & 1 & 1 \\ 0 & c-b & a-b \\ 0 & a-c & b-c \end{vmatrix}$$

$$= (a+b+c) \begin{vmatrix} 1 & 1 & 1 \\ 0 & c-b & a-b \\ 0 & a-c & b-c \end{vmatrix} \begin{matrix} 1 & 1 \\ 0 & c-b \\ 0 & a-c \end{matrix}$$

$$= (a+b+c)(c-b)(b-c) - (a-b)(a-c)$$
$$= (a+b+c)[cb - b^2 - c^2 + bc - (a^2 - ba - ca + bc)]$$
$$= (a+b+c)(-b^2 - c^2 - a^2 + bc + ba + ca)$$
$$P(a,b,c) = -b^2 - c^2 - a^2 + bc + ba + ca$$

37. $x\mathbf{I}$ is the matrix composed of x's along the main diagonal.

$$|\mathbf{A} - x\mathbf{I}| = \begin{vmatrix} 2-x & 5 & 3 \\ -1 & -4-x & -1 \\ 1 & 0 & 1-x \end{vmatrix}$$

$$= (2-x)(-4-x)(1-x) + (5)(-1)(1) + (3)(-1)(0)$$
$$\quad - (1)(-4-x)(3) - (0)-1)(2-x) - (1-x)(-1)(5)$$
$$= -x^3 - x^2 + 10x - 8 - 5 + 12 + 3x + 5 - 5x$$
$$= -x^3 - x^2 + 8x + 4$$

To find the characteristic values, we must find the zeros of $P(x) = -x^3 - x^2 + 8x + 4$.
Set standard range values.
Press $\boxed{Y=}$.
Enter $\boxed{(-)}X \wedge 3 - X \boxed{x^2} + 8X + 4$ as Y_1
Press $\boxed{GRAPH}$.
Zooming in on the three x-intercepts, we find that $x = -3.14$, $x = -.48$, and $x = 2.63$.

Problem Set 11.6

1. Expand according to the third column:

$$\begin{vmatrix} 3 & -2 & 4 \\ 1 & 5 & 0 \\ 3 & 10 & 0 \end{vmatrix} = 4 \begin{vmatrix} 1 & 5 \\ 3 & 10 \end{vmatrix} = 4(10-15) = -20$$

3. Expand according to the first column, after introducing another 0 by adding -1 times row 1 to row 3:

$$\begin{vmatrix} 1 & 2 & 3 \\ 0 & 2 & 3 \\ 1 & 3 & 4 \end{vmatrix} = \begin{vmatrix} 1 & 2 & 3 \\ 0 & 2 & 3 \\ 0 & 1 & 1 \end{vmatrix} = 1 \begin{vmatrix} 2 & 3 \\ 1 & 1 \end{vmatrix}$$
$$= 2 - 3 = -1$$

5. Expand according to the first row:

$$\begin{vmatrix} 3 & 0 & 0 & 0 \\ -1 & 1 & 4 & 2 \\ 2 & 0 & 2 & -3 \\ -4 & 0 & 1 & 5 \end{vmatrix} = 3 \begin{vmatrix} 1 & 4 & 2 \\ 0 & 2 & -3 \\ 0 & 1 & 5 \end{vmatrix}$$

Expand the 3×3 determinant according to the first column:

$$3 \begin{vmatrix} 1 & 4 & 2 \\ 0 & 2 & -3 \\ 0 & 1 & 5 \end{vmatrix} = 3 \cdot 1 \begin{vmatrix} 2 & -3 \\ 1 & 5 \end{vmatrix} = 3(10+3) = 39$$

246

7. Add row 3 to row 1 to get some zeros, then expand according to the first row:

$$
\begin{vmatrix} 3 & 5 & -10 \\ 2 & 4 & 6 \\ -3 & -5 & 12 \end{vmatrix} = \begin{vmatrix} 0 & 0 & 2 \\ 2 & 4 & 6 \\ -3 & -5 & 12 \end{vmatrix}
$$

$$
= 2 \begin{vmatrix} 2 & 4 \\ -3 & -5 \end{vmatrix}
$$
$$
= 2(-10 + 12)
$$
$$
= 4
$$

9. Add twice row 1 to row 2, and -3 times row 1 to row 4 to get more zeros, then expand according to the first column:

$$
\begin{vmatrix} 1 & -2 & 1 & 4 \\ -2 & 5 & -3 & 1 \\ 0 & 7 & -4 & 2 \\ 3 & -2 & 2 & 6 \end{vmatrix} = \begin{vmatrix} 1 & -2 & 1 & 4 \\ 0 & 1 & -1 & 9 \\ 0 & 7 & -4 & 2 \\ 0 & 4 & -1 & -6 \end{vmatrix}
$$

$$
= 1 \begin{vmatrix} 1 & -1 & 9 \\ 7 & -4 & 2 \\ 4 & -1 & -6 \end{vmatrix}
$$

Add the first column to the second, and -9 times the first column to the third to get zeros, then expand according to the first row:

$$
1 \begin{vmatrix} 1 & -1 & 9 \\ 7 & -4 & 2 \\ 4 & -1 & -6 \end{vmatrix} = \begin{vmatrix} 1 & 0 & 0 \\ 7 & 3 & -61 \\ 4 & 3 & -42 \end{vmatrix} = 1 \begin{vmatrix} 3 & -61 \\ 3 & -42 \end{vmatrix}
$$
$$
= 3(-42) - 3(-61)
$$
$$
= 57
$$

11. Expand according to the second column:

$$
\begin{vmatrix} 2 & -3 & 2 \\ 1 & 0 & -4 \\ -1 & 0 & 6 \end{vmatrix} = -(-3) \begin{vmatrix} 1 & -4 \\ -1 & 6 \end{vmatrix} = 3(6 - 4) = 6
$$

13. Add -1 times row 2 to row 1 to get zeros, then expand according to the first row:

$$
\begin{vmatrix} 2 & -3 & 4 & 5 \\ 2 & -3 & 4 & 7 \\ 1 & 6 & 4 & 5 \\ 2 & 6 & 4 & -8 \end{vmatrix} = \begin{vmatrix} 0 & 0 & 0 & -2 \\ 2 & -3 & 4 & 7 \\ 1 & 6 & 4 & 5 \\ 2 & 6 & 4 & -8 \end{vmatrix}
$$

$$
= -(-2) \begin{vmatrix} 2 & -3 & 4 \\ 1 & 6 & 4 \\ 2 & 6 & 4 \end{vmatrix}
$$

Add -1 times row 2 to row 3, then expand according to the third row:

$$
2 \begin{vmatrix} 2 & -3 & 4 \\ 1 & 6 & 4 \\ 2 & 6 & 4 \end{vmatrix} = 2 \begin{vmatrix} 2 & -3 & 4 \\ 1 & 6 & 4 \\ 1 & 0 & 0 \end{vmatrix} = 2 \cdot 1 \begin{vmatrix} -3 & 4 \\ 6 & 4 \end{vmatrix}
$$
$$
= 2(-12 - 24)
$$
$$
= -72
$$

15. Expand according to the second column:

$$
\begin{vmatrix} -1 & 2 & -3 & -1 & -2 \\ -1 & 0 & 2 & 5 & -3 \\ 5 & 0 & 0 & -2 & 4 \\ 0 & 0 & 0 & 6 & 3 \\ 0 & 0 & 0 & 2 & -7 \end{vmatrix} = -2 \begin{vmatrix} -1 & 2 & 5 & -3 \\ 5 & 0 & -2 & 4 \\ 0 & 0 & 6 & 3 \\ 0 & 0 & 2 & -7 \end{vmatrix}
$$

Expand according to the second column:

$$
-2 \begin{vmatrix} -1 & 2 & 5 & -3 \\ 5 & 0 & -2 & 4 \\ 0 & 0 & 6 & 3 \\ 0 & 0 & 2 & -7 \end{vmatrix} = (-2)(-2) \begin{vmatrix} 5 & -2 & 4 \\ 0 & 6 & 3 \\ 0 & 2 & -7 \end{vmatrix}
$$

Expand according to the first column:

$$
4 \begin{vmatrix} 5 & -2 & 4 \\ 0 & 6 & 3 \\ 0 & 2 & -7 \end{vmatrix} = (4)(5) \begin{vmatrix} 6 & 3 \\ 2 & -7 \end{vmatrix}
$$
$$
= 20(-42 - 6)
$$
$$
= -960
$$

17. $D = \begin{vmatrix} 1 & -2 & 1 & 4 \\ -2 & 5 & -3 & 1 \\ 0 & 7 & -4 & 2 \\ 3 & -2 & 2 & 6 \end{vmatrix} = 57$ (by problem 9)

$$x = \frac{1}{57} \begin{vmatrix} 1 & -2 & 1 & 4 \\ -2 & 5 & -3 & 1 \\ 3 & 7 & -4 & 2 \\ 6 & -2 & 2 & 6 \end{vmatrix}$$

Introduce zeros by adding twice row 1 to row 2, -3 times row 1 to row 3, and -6 times row 1 to row 4. Then expand according to the first column.

$$x = \frac{1}{57} \begin{vmatrix} 1 & -2 & 1 & 4 \\ 0 & 1 & -1 & 9 \\ 0 & 13 & -7 & -10 \\ 0 & 10 & -4 & -18 \end{vmatrix} = \frac{1}{57} \begin{vmatrix} 1 & -1 & 9 \\ 13 & -7 & -10 \\ 10 & -4 & -18 \end{vmatrix}$$

Add the first column to the second, and -9 times the first column to the third, then expand according to the first row.

$$x = \frac{1}{57} \begin{vmatrix} 1 & 0 & 0 \\ 13 & 6 & -127 \\ 10 & 6 & -108 \end{vmatrix} = \frac{1}{57} \cdot 1 \begin{vmatrix} 6 & -127 \\ 6 & -108 \end{vmatrix}$$
$$= \frac{1}{57}(-648 + 762)$$
$$= \frac{114}{57}$$
$$= 2$$

19.
$$\begin{vmatrix} a & b & c & d \\ 0 & e & f & g \\ 0 & 0 & h & i \\ 0 & 0 & 0 & j \end{vmatrix} = a \begin{vmatrix} e & f & g \\ 0 & h & i \\ 0 & 0 & j \end{vmatrix} = ae \begin{vmatrix} h & i \\ 0 & j \end{vmatrix}$$
$$= ae(hj - 0)$$
$$= aehj$$

The determinant of a triangular matrix is the product of the entries in the main diagonal.

21.
$$\begin{vmatrix} 1 & 2 & 2.6 & 1.5 \\ 2.3 & 5.6 & -1.3 & 9.8 \\ 2.7 & 1.3 & 4.2 & -1.9 \\ 5.5 & 6.2 & 3.0 & 1.4 \end{vmatrix}$$

Add -2.3 times row 1 to row 2, -2.7 times row 1 to row 3, and -5.5 times row 1 to row 4:

$$\begin{vmatrix} 1 & 2 & 2.6 & 1.5 \\ 0 & 1 & -7.28 & 6.35 \\ 0 & -4.1 & -2.82 & -5.95 \\ 0 & -4.8 & -11.30 & -6.85 \end{vmatrix}$$

Add 4.1 times row 2 to row 3, and 4.8 times row 2 to row 4:

$$\begin{vmatrix} 1 & 2 & 2.6 & 1.5 \\ 0 & 1 & -7.28 & 6.35 \\ 0 & 0 & -32.688 & 20.085 \\ 0 & 0 & -46.244 & 23.63 \end{vmatrix}$$

Add $-\frac{46.244}{32.688}$ times row 3 to row 4:

$$\begin{vmatrix} 1 & 2 & 2.6 & 1.5 \\ 0 & 1 & -7.28 & 6.35 \\ 0 & 0 & -32.688 & 20.085 \\ 0 & 0 & 0 & -4.8018 \end{vmatrix}$$
$$= (1)(1)(-32.668)(-4.8018)$$
$$= 156.8659$$

23. a) Add -2 times column 1 to column 4, then expand according to the first row.

$$\begin{vmatrix} 1 & 0 & 0 & 2 \\ 2.7 & 5 & 0 & 8.9 \\ 3.4 & 0 & 6 & 9.1 \\ 3 & 0 & 0 & 4 \end{vmatrix} = \begin{vmatrix} 1 & 0 & 0 & 0 \\ 2.7 & 5 & 0 & 3.5 \\ 3.4 & 0 & 6 & 2.3 \\ 3 & 0 & 0 & -2 \end{vmatrix}$$
$$= 1 \begin{vmatrix} 5 & 0 & 3.5 \\ 0 & 6 & 2.3 \\ 0 & 0 & -2 \end{vmatrix}$$

(Note the triangular determinant.)
$$= (1)(5)(6)(-2)$$
$$= -60$$

b) Add $-\frac{b}{a}$ times column 1 to column 4, then expand according to the first row.

$$\begin{vmatrix} a & 0 & 0 & b \\ ? & e & 0 & ? \\ ? & 0 & f & ? \\ c & 0 & 0 & d \end{vmatrix} = \begin{vmatrix} a & 0 & 0 & 0 \\ ? & e & 0 & ? \\ ? & 0 & f & ? \\ c & 0 & 0 & d - \frac{bc}{a} \end{vmatrix}$$

$$= a \begin{vmatrix} e & 0 & ? \\ 0 & f & ? \\ 0 & 0 & d - \frac{bc}{a} \end{vmatrix}$$

(Note the triangular determinant.)
$$= (a)(e)(f)\left(d - \frac{bc}{a}\right)$$
$$= aef\left(\frac{ad - bc}{a}\right)$$
$$= ef(ad - bc)$$

25. Subtract row 1 from rows 2 and 3:

$$\begin{vmatrix} n+1 & n+2 & n+3 \\ n+4 & n+5 & n+6 \\ n+7 & n+8 & n+9 \end{vmatrix} = \begin{vmatrix} n+1 & n+2 & n+3 \\ 3 & 3 & 3 \\ 6 & 6 & 6 \end{vmatrix}$$

$$= 3 \cdot 6 \begin{vmatrix} n+1 & n+2 & n+3 \\ 1 & 1 & 1 \\ 1 & 1 & 1 \end{vmatrix}$$

Subtract row 2 from row 3:

$$= 18 \begin{vmatrix} n+1 & n+2 & n+3 \\ 1 & 1 & 1 \\ 0 & 0 & 0 \end{vmatrix}$$

$$= 18(0)$$
$$= 0$$

27.
$$\begin{vmatrix} a_1 & b_1 & 0 & 0 \\ a_2 & b_2 & 0 & 0 \\ 0 & 0 & c_1 & d_1 \\ 0 & 0 & c_2 & d_2 \end{vmatrix}$$

$$= a_1 \begin{vmatrix} b_2 & 0 & 0 \\ 0 & c_1 & d_1 \\ 0 & c_2 & d_2 \end{vmatrix} - a_2 \begin{vmatrix} b_1 & 0 & 0 \\ 0 & c_1 & d_1 \\ 0 & c_2 & d_2 \end{vmatrix}$$

$$= a_1 b_2 \begin{vmatrix} c_1 & d_1 \\ c_2 & d_2 \end{vmatrix} - a_2 b_1 \begin{vmatrix} c_1 & d_1 \\ c_2 & d_2 \end{vmatrix}$$

$$= (a_1 b_2 - a_2 b_1) \begin{vmatrix} c_1 & d_1 \\ c_2 & d_2 \end{vmatrix}$$

$$= \begin{vmatrix} a_1 & b_1 \\ a_2 & b_2 \end{vmatrix} \begin{vmatrix} c_1 & d_1 \\ c_2 & d_2 \end{vmatrix}$$

29. $D_2 = \begin{vmatrix} 1 & 1 \\ 1 & 2 \end{vmatrix} = 2 - 1 = 1$

$$D_3 = \begin{vmatrix} 1 & 1 & 1 \\ 1 & 2 & 3 \\ 1 & 3 & 6 \end{vmatrix}$$

Subtract row 1 from rows 2 and 3:

$$= \begin{vmatrix} 1 & 1 & 1 \\ 0 & 1 & 2 \\ 0 & 2 & 5 \end{vmatrix} = 1 \cdot \begin{vmatrix} 1 & 2 \\ 2 & 5 \end{vmatrix} = 5 - 4 = 1$$

$$D_4 = \begin{vmatrix} 1 & 1 & 1 & 1 \\ 1 & 2 & 3 & 4 \\ 1 & 3 & 6 & 10 \\ 1 & 4 & 10 & 20 \end{vmatrix}$$

$$= \begin{vmatrix} 1 & 1 & 1 & 1 \\ 0 & 1 & 2 & 3 \\ 0 & 1 & 3 & 6 \\ 0 & 1 & 4 & 10 \end{vmatrix}$$ Subtract row 1 from row 2, row 2 from row 3, and row 3 from row 4

$$= 1 \cdot \begin{vmatrix} 1 & 2 & 3 \\ 1 & 3 & 6 \\ 1 & 4 & 10 \end{vmatrix}$$

$$= \begin{vmatrix} 1 & 2 & 3 \\ 0 & 1 & 3 \\ 0 & 1 & 4 \end{vmatrix}$$ Subtract row 1 from row 2 and row 2 from row 3

$$= 1 \cdot \begin{vmatrix} 1 & 3 \\ 1 & 4 \end{vmatrix} = 4 - 3 = 1$$

31. $D = \begin{vmatrix} 1 & 1 & 1 & 1 \\ 1 & 2 & 3 & 4 \\ 1 & 3 & 6 & 10 \\ 1 & 4 & 10 & 20 \end{vmatrix} = 1$ (by Problem 29)

$$x = 1 \begin{vmatrix} 0 & 1 & 1 & 1 \\ 0 & 2 & 3 & 4 \\ 0 & 3 & 6 & 10 \\ 1 & 4 & 10 & 20 \end{vmatrix} = (-1) \begin{vmatrix} 1 & 1 & 1 \\ 2 & 3 & 4 \\ 3 & 6 & 10 \end{vmatrix}$$

Subtract column 2 from column 3 and column 1 from column 2 $= (-1) \begin{vmatrix} 1 & 0 & 0 \\ 2 & 1 & 1 \\ 3 & 3 & 4 \end{vmatrix}$

$$= (-1) \begin{vmatrix} 1 & 1 \\ 3 & 4 \end{vmatrix} = -1$$

$$y = 1 \cdot \begin{vmatrix} 1 & 0 & 1 & 1 \\ 1 & 0 & 3 & 4 \\ 1 & 0 & 6 & 10 \\ 1 & 1 & 10 & 20 \end{vmatrix} = 1 \begin{vmatrix} 1 & 1 & 1 \\ 1 & 3 & 4 \\ 1 & 6 & 10 \end{vmatrix}$$

Subtract
row 2 from row 3 and
row 1 from row 2

$$= \begin{vmatrix} 1 & 1 & 1 \\ 0 & 2 & 3 \\ 0 & 3 & 6 \end{vmatrix}$$

$$= 1 \begin{vmatrix} 2 & 3 \\ 3 & 6 \end{vmatrix}$$

$$= 12 - 9$$

$$= 3$$

$$z = 1 \begin{vmatrix} 1 & 1 & 0 & 1 \\ 1 & 2 & 0 & 4 \\ 1 & 3 & 0 & 10 \\ 1 & 4 & 1 & 20 \end{vmatrix} = (-1) \begin{vmatrix} 1 & 1 & 1 \\ 1 & 2 & 4 \\ 1 & 3 & 10 \end{vmatrix}$$

Subtract
row 2 from row 3 and
row 1 from row 2

$$= (-1) \begin{vmatrix} 1 & 1 & 1 \\ 0 & 1 & 3 \\ 0 & 1 & 6 \end{vmatrix}$$

$$= (-1) \begin{vmatrix} 1 & 3 \\ 1 & 6 \end{vmatrix}$$

$$= (-1)(6 - 3) = -3$$

$$w = 1 \begin{vmatrix} 1 & 1 & 1 & 0 \\ 1 & 2 & 3 & 0 \\ 1 & 3 & 6 & 0 \\ 1 & 4 & 10 & 1 \end{vmatrix} = 1 \begin{vmatrix} 1 & 1 & 1 \\ 1 & 2 & 3 \\ 1 & 3 & 6 \end{vmatrix}$$

$$= 1 \cdot 1 \text{ (from Problem 29)}$$

$$= 1$$

Solution: $(x, y, z, w) = (-1, 3, -3, 1)$

33. If two rows (or two columns) of a determinant, D, are identical, then $D = 0$, since subtracting one of the identical rows from the other gives a row of 0's.

a) $|C_2| = \begin{vmatrix} a_1 b_1 & a_1 b_2 \\ a_1 b_1 & a_2 b_2 \end{vmatrix} = a_1 a_2 \begin{vmatrix} b_1 & b_2 \\ b_1 & b_2 \end{vmatrix} = 0$

$|C_3| = \begin{vmatrix} a_1 b_1 & a_1 b_2 & a_1 b_3 \\ a_2 b_1 & a_2 b_2 & a_2 b_3 \\ a_3 b_1 & a_3 b_2 & a_3 b_3 \end{vmatrix} = a_1 a_2 \begin{vmatrix} b_1 & b_2 & b_3 \\ b_1 & b_2 & b_3 \\ a_3 b_1 & a_3 b_2 & a_3 b_3 \end{vmatrix}$

$$= 0$$

$|C_4| = \begin{vmatrix} a_1 b_1 & a_1 b_2 & a_1 b_3 & a_1 b_4 \\ a_2 b_1 & a_2 b_2 & a_2 b_3 & a_2 b_4 \\ a_3 b_1 & a_3 b_2 & a_3 b_3 & a_3 b_4 \\ a_4 b_1 & a_4 b_2 & a_4 b_3 & a_4 b_4 \end{vmatrix}$

As with $|C_3|$, we factor a_1 from the first row and a_2 from the second, leaving two identical rows.
$|C_4| = 0$
The conjecture would be that $|C_n| = 0$ for $n \geq 2$. The proof is an immediate extension of the evaluation of $|C_4|$ above.

b) $|C_2| = \begin{vmatrix} a_1 - b_1 & a_1 - b_2 \\ a_2 - b_1 & a_2 - b_2 \end{vmatrix}$

Subtract the first row from the second row:

$$= \begin{vmatrix} a_1 - b_1 & a_1 - b_2 \\ a_2 - a_1 & a_2 - a_1 \end{vmatrix}$$

$$= (a_2 - a_1) \begin{vmatrix} a_1 - b_1 & a_1 - b_2 \\ 1 & 1 \end{vmatrix}$$

$$= (a_2 - a_1)[(a_1 - b_1 - (a_1 - b_2)]$$

$$= (a_2 - a_1)(b_2 - b_1)$$

$|C_3| = \begin{vmatrix} a_1 - b_1 & a_1 - b_2 & a_1 - b_3 \\ a_2 - b_1 & a_2 - b_2 & a_2 - b_3 \\ a_3 - b_1 & a_3 - b_2 & a_3 - b_3 \end{vmatrix}$

Subtract the first row from the second row, and subtract the first row from the third. It is clear that all the b terms drop out, leaving

$|C_3| = \begin{vmatrix} a_1 - b_1 & a_1 - b_2 & a_1 - b_3 \\ a_2 - a_1 & a_2 - a_1 & a_2 - a_1 \\ a_3 - a_1 & a_3 - a_1 & a_3 - a_1 \end{vmatrix}$

Factoring $a_2 - a_1$ from the second row and $a_3 - a_1$ from the third row, we are left with two identical rows of 1's. Hence $|C_3| = 0$.
$|C_4|$ is exactly the same as $|C_3|$, with the addition of a column and a row, and the process is identical. Indeed, the general conjecture that

$$|C_n| = 0 \text{ for all } n \geq 3$$

goes through in the same way.

35. $A_{3,4,5} = \begin{bmatrix} \begin{bmatrix} \begin{bmatrix} 1 & \frac{1}{2} & \frac{1}{3} \end{bmatrix} & \frac{1}{4} \\ \frac{1}{2} & \frac{1}{3} & \frac{1}{4} & \frac{1}{5} \\ \frac{1}{3} & \frac{1}{4} & \frac{1}{5} & \frac{1}{6} \\ \frac{1}{4} & \frac{1}{5} & \frac{1}{6} & \frac{1}{7} \end{bmatrix} & \begin{matrix} \frac{1}{5} \\ \frac{1}{6} \\ \frac{1}{7} \\ \frac{1}{8} \end{matrix} \\ \frac{1}{5} & \frac{1}{6} & \frac{1}{7} & \frac{1}{8} & \frac{1}{9} \end{bmatrix}$

a) Clear matrix A: Press $\boxed{0}\boxed{\text{STO}}\boxed{\text{2nd}}\boxed{[A]}\boxed{\text{ENTER}}$
Press $\boxed{\text{MATRX}}\boxed{\triangleright}\boxed{1}$ to edit matrix A. Enter the dimensions (3×3) and whichever elements of A_3 that can be entered exactly:
$(1,1) = 1$; $(1,2) = .5$;
$(2,1) = .5$; $(2,3) = .25$;
$(3,2) = .25$; $(3,3) = .2$
Press $\boxed{\text{2nd}}\boxed{\text{QUIT}}$ to return to the home screen.
Press $3\boxed{x^{-1}}\boxed{\text{STO}}\boxed{\text{2nd}}\boxed{[A]}\boxed{(}1\boxed{\text{ALPHA}}\boxed{,}3\boxed{)}\boxed{\text{ENTER}}$
Press $\boxed{\triangle}$ and edit the expression for $(2,2)$ and press $\boxed{\text{ENTER}}$. Repeat for $(3,1)$.
Press $\boxed{\text{MATRX}}\boxed{5}\boxed{\text{2nd}}\boxed{[A]}\boxed{\text{ENTER}}$

$|A_3| = 4.62962963 \times 10^{-4}$

b) Press $\boxed{\text{MATRX}}\boxed{\triangleright}\boxed{1}$ to edit matrix A. Enter new dimensions for A: 4×4. Enter whichever new elements of A_4 that can be entered exactly:
$(1,4) = .25$;
$(2,4) = .2$;
$(4,1) = .25$; $(4,2) = .2$;
Press $\boxed{\text{2nd}}\boxed{\text{QUIT}}$ to return to the home screen.
Press $6\boxed{x^{-1}}\boxed{\text{STO}}\boxed{\text{2nd}}\boxed{[A]}\boxed{(}3\boxed{\text{ALPHA}}\boxed{,}4\boxed{)}\boxed{\text{ENTER}}$
Press $\boxed{\triangle}$ and edit the expression for $(4,3)$ and press $\boxed{\text{ENTER}}$. Change the 6 to a 7 for $(4,4)$.
Press $\boxed{\text{MATRX}}\boxed{5}\boxed{\text{2nd}}\boxed{[A]}\boxed{\text{ENTER}}$

$|A_4| = 1.653439 \times 10^{-7}$

c) Press $\boxed{\text{MATRX}}\boxed{\triangleright}\boxed{1}$ to edit matrix A. Enter new dimensions for A: 5×5. Enter whichever new elements of A_5 that can be entered exactly:
$(1,5) = .2$;
$(4,5) = .125$;
$(5,1) = .2$; $(5,4) = .125$
Press $\boxed{\text{2nd}}\boxed{\text{QUIT}}$ to return to the home screen.
Press $6\boxed{x^{-1}}\boxed{\text{STO}}\boxed{\text{2nd}}\boxed{[A]}\boxed{(}2\boxed{\text{ALPHA}}\boxed{,}5\boxed{)}\boxed{\text{ENTER}}$
Press $\boxed{\triangle}$ and edit the expression for a 7 for $(3,5)$ and press $\boxed{\text{ENTER}}$. Change to a 6 for $(5,2)$; a 7 for $(5,3)$; a 9 for $(5,5)$
Press $\boxed{\text{MATRX}}\boxed{5}\boxed{\text{2nd}}\boxed{[A]}\boxed{\text{ENTER}}$

$|A_5| = 3.749295 \times 10^{-12}$

As n grows large, $|A_n|$ approaches 0.

Problem Set 11.7

1.

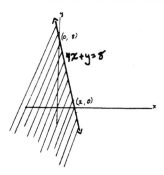

Since the origin $(0,0)$ makes $4x + y \leq 8$ a true statement, the origin and all other points on that same side of the line are part of the solution. The line itself is also part of the solution since the equality sign is included in the inequality.

3.

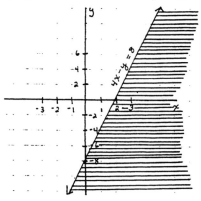

Since the origin $(0,0)$ makes $x \leq 3$ a true statement, the origin and all other points on that side of the line $x = 3$ are part of the solution. The line itself is also part of the solution since the equality sign is included in the inequality.

5. Since the origin $(0,0)$ makes $4x - y \geq 8$ a false statement, the points on the side of the line that does not include the origin are part of the solution. The line itself is also part of the solution since the equality sign is included in the inequality.

7. $4x + y \le 8$
$\quad\quad y \le -4x + 8$
$\quad 2x + 3y \le 14$
$\quad\quad y \le -\frac{2}{3}x + \frac{14}{3}$
$\quad x \ge 0,\ y \ge 0$

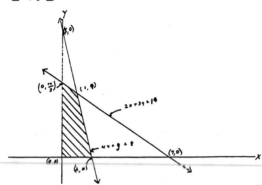

The region lies below the line $4x + y = 8$, below the line $2x + 3y = 14$, to the right of the x-axis, and above the y-axis. The region will include its boundary lines.
Find the point of intersection of
$\quad\quad 4x + y = 8$
$\quad\quad 2x + 3y = 14$
$\begin{cases} 4x + y = 8 \\ 4x + 6y = 28 \end{cases}$
$\quad\quad\overline{\quad\quad 5y = 20}$
$\quad\quad\quad\quad y = 4$
$\quad\quad 4x + 4 = 8$
$\quad\quad\quad\quad x = 1$
The point of intersection of the two lines is $(1, 4)$.

9. $4x + y \le 8$
$\quad\quad y \le -4x + 8$
$\quad x - y \le -2$
$\quad\quad -y \le -x - 2$
$\quad\quad y \ge x + 2$
$\quad\quad x \ge 0$

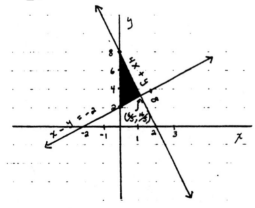

The region lies below the line $4x + y = 8$, above the line $x - y = -2$, and to the right of the y-axis. The region will include its boundary lines.
Find the point of intersection of
$\begin{cases} 4x + y = 8 \\ x - y = -2 \end{cases}$
$\quad\quad\overline{\quad\quad 5x = 6}$
$\quad\quad\quad\quad x = \frac{6}{5}$
$\quad\quad \frac{6}{5} - y = -2$
$\quad\quad\quad -y = -\frac{16}{5}$
$\quad\quad\quad\quad y = \frac{16}{5}$

11.

Vertex	$P = 2x + y$
$(0,0)$	0
$(0, \frac{14}{3})$	$\frac{14}{3}$
$(1, 4)$	6
$(2, 0)$	4

The maximum value of P is 6, occurring at $(1, 4)$.
The minimum value of P is 0, occurring at $(0, 0)$.

13.

Vertex	$P = 2x - y$
$(0, 2)$	-2
$(0, 8)$	-8
$(\frac{6}{5}, \frac{16}{5})$	$-\frac{4}{5}$

The maximum value of P is $-\frac{4}{5}$, occurring at $(\frac{6}{5}, \frac{16}{5})$.
The minimum value of P is -8, occurring at $(0, 8)$.

15. $x + y \ge 4$
$\quad\quad y \ge -x + 4$
$\quad\quad x \ge 0$
$\quad\quad y \ge 0$

The region lies above the line $x + y = 4$, to the right of the line $x = 2$, and above the x-axis. The region will include its boundary lines.

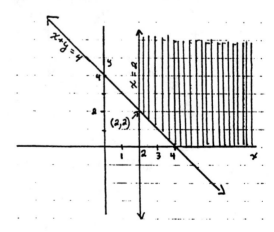

Vertex	$5x + 2y$
$(2,2)$	14
$(4,0)$	20

The minimum value of $5x + 2y$ is 14, occurring at $(2,2)$.

17.
$$4x + y \geq 7$$
$$y \geq -4x + 7$$
$$2x + 3y \geq 6$$
$$y \geq -\tfrac{2}{3}x + 2$$
$$x \geq 1$$
$$y \geq 0$$

The region will lie above the line $4x + y = 7$, above the line $2x + 3y = 6$, to the right of the line $x = 1$, and above the x-axis. The region will include its boundary lines.

Find the point of intersection of
$$\begin{cases} 4x + y = 7 \\ 2x + 3y = 6 \end{cases}$$
$$\begin{array}{r} 4x + y = 7 \\ -4x - 6y = -12 \\ \hline -5y = -5 \\ y = 1 \end{array}$$
$$4x + 1 = 7$$
$$4x = 6$$
$$x = \tfrac{3}{2}$$

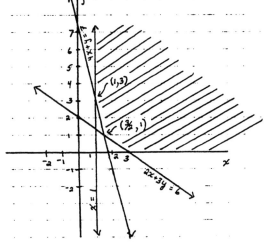

Vertex	$2x + y$
$(1,3)$	5
$(\tfrac{3}{2}, 1)$	4
$(3,0)$	6

The minimum value of $2x + y$ is 4, occurring at $(\tfrac{3}{2}, 1)$.

19.
$$y \leq 4x - x^2$$
$$y \leq x$$
$$x \geq 0$$
$$y \geq 0$$

The region lies below the parabola $y = 4x - x^2$, below the line $y = x$, above the x-axis and to the right of the y-axis. The region will include its boundary lines.
$$y = 4x - x^2$$
$$y = -(x^2 - 4x)$$
$$y - 4 = -(x^2 - 4x + 4)$$
$$y - 4 = -(x - 2)^2$$

This is a parabola with vertex $(2,4)$, opening downward. The intersections of the line $y = x$ and the parabola are
$$x = 4x - x^2$$
$$x^2 - 3x = 0$$
$$x(x - 3) = 0$$
$$x = 0, \ y = 0 \ \text{or} \ x = 3, \ y = 3$$

21. $y \leq \log_2 x$
$$x \leq 8$$
$$y \geq 0$$

The region will lie below the curve $y = \log_2 x$, to the left of the line $x = 8$, and above the x-axis. The region will include its boundary lines.

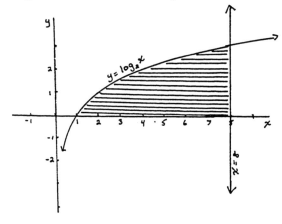

23.
$$4x + y \le 8$$
$$y \le -4x + 8$$
$$x - y \ge -2$$
$$-y \ge -x - 2$$
$$y \le x + 2$$
$$x \ge 0,\ y \ge 0$$

The region lies below the line $4x + y = 8$, below the line $x - y = -2$, to the right of the y-axis, and above the x-axis. The region will include its boundary lines.

From Problem 9, we know that the lines $4x + y = 8$ and $x - y = -2$ intersect at $\left(\frac{6}{5}, \frac{16}{5}\right)$.

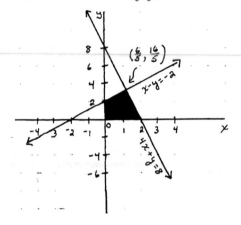

25. From Problem 23, we can obtain the following vertices.

Vertex	$P = 2x - y$
$(0,0)$	0
$(0,2)$	-2
$\left(\frac{6}{5}, \frac{16}{5}\right)$	$-\frac{4}{5}$
$(2,0)$	4

The maximum value of P is 4, occurring at $(2,0)$.
The minimum value of P is -2, occurring at $(0,2)$.

27.
$$x - y \ge -1$$
$$-y \ge -x - 1$$
$$y \le x + 1$$
$$x - 2y \le 5$$
$$-2y \le -x + 5$$
$$y \ge \frac{x}{2} - \frac{5}{2}$$
$$3x + y \le 10$$
$$y \le -3x + 10$$
$$x \ge 0,\ y \ge 0$$

The region lies below the line $x - y = -1$, above the line $x - 2y \le 5$, below the line $3x + y = 10$, above the x-axis, and to the right of the y-axis.

The region will include its boundary lines.

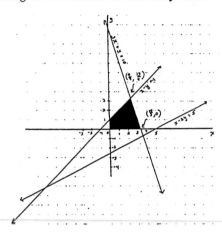

The graph reveals that one vertex is the point of intersection of $3x + y = 10$ and $x - y = -1$.
$$\begin{cases} 3x + y = 10 \\ \underline{x - y = -1} \\ 4x = 9 \\ x = \frac{9}{4} \end{cases}$$
$$\frac{9}{4} - y = -1$$
$$-y = -\frac{13}{4}$$
$$y = \frac{13}{4}$$

Vertex	$x + y$
$(0,0)$	0
$(0,1)$	1
$\left(\frac{9}{4}, \frac{13}{4}\right)$	$\frac{11}{2}$
$\left(\frac{10}{3}, 0\right)$	$\frac{10}{3}$

The minimum value of $x + y$ is 0, occurring at $(0,0)$. The maximum value of $x + y$ is $\frac{11}{2}$, occurring at $\left(\frac{9}{4}, \frac{13}{4}\right)$.

29.

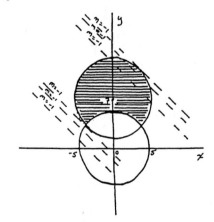

$$x^2 + y^2 \geq 25$$
$$x^2 + (y-7)^2 \leq 32$$

The region lies inside the circle $x^2 + (y-7)^2 = 32$ and outside the circle $x^2 + y^2 = 25$. The region will include its boundary lines.

$$P = 2x + 2y$$
$$2y = -2x + P$$
$$y = -x + \frac{P}{2}$$

No matter what value P has, the line $P = 2x + 2y$ will have slope $m = -1$. We want to find the points in the region at which a line with slope $m = -1$ first enters and last leaves the region. The line will be tangent to the large circle at those points. That means the slope of the radii to those points, call them (x,y), will be 1, because they are perpendicular to the tangent lines. Since the center of the circle is $(0,7)$, we have the equation

$$\frac{y-7}{x-0} = = 1$$
$$y - 7 = x$$

Let $x = y - 7$ in the equation of the large circle:
$$(y-7)^2 + (y-7)^2 = 32$$
$$2(y-7)^2 = 32$$
$$(y-7)^2 = 16$$
$$y - 7 = \pm 4$$
$$y = 11 \text{ or } y = 3$$

When $y = 11$, $x = 11 - 7 = 4$.
When $y = 3$, $x = 3 - 7 = -4$.
P will first satisfy the inequality of the larger circle at $(-4,3)$, but we must check that it also satisfies the other inequality: $(-4)^2 + (3)^2 = 25 \geq 25$. So, the minimum value of P is $2(-4) + 2(3) = -2$.
The maximum value of P occurs at $(4,11)$, and is $2(4) + 2(11) = 30$.

31. Let x represent the number of items that can be produced on production line A in one week.
Let y represent the number of items that can be produced on production line B in one week.
Then $T = x + y$ represents the total number of items that can be produced in one week. We want to maximize T.
Restriction on time: $4x + 3y \leq 900$
$$y \leq -\frac{4}{3}x + 300$$
Restriction on cost: $5x + 6y \leq 1500$
$$y \leq -\frac{5}{6}x + 300$$
And, of course, $x \geq 0$, $y \geq 0$

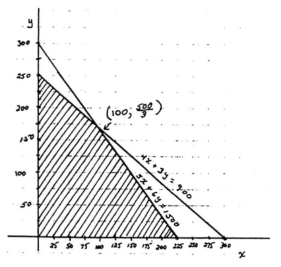

$\left(100, \frac{500}{3}\right)$

$4x + 3y = 900$
$5x + 6y = 1500$

Find the point of intersection:
$$\begin{cases} 4x + 3y = 900 \\ 5x + 6y = 1500 \end{cases}$$
$$-8x - 6y = -1800$$
$$\underline{5x + 6y = 1500}$$
$$-3x = -300$$
$$x = 100$$
$$4(100) + 3y = 900$$
$$3y = 500$$
$$y = \frac{500}{3} = 166\frac{2}{3}$$

Vertex	$T = x + y$
$(0,0)$	0
$(0,250)$	250
$(100, 166\frac{2}{3})$	$266\frac{2}{3}$
$(225, 0)$	225

The maximum number of items that can be produced is $266\frac{2}{3}$, if we are to consider $\frac{2}{3}$ of an item as being produced (e.g., if production of that item can be completed the following week). If we don't consider $\frac{2}{3}$ of an item as being produced, then the nearest point to the intersection is $(100, 166)$, with $T = 266$. However, there are 7 other integral points in the region with $T = 266$: $(96, 170)$, $(97, 169)$, $(98, 168)$, $(99, 167)$, $(100, 166)$, $(101, 165)$, $(102, 164)$.

33. Let x represent the number of campers produced.
Let y represent the number of trailers produced.
Then $P = 600x + 800y$ represents the manufacturer's profit. We want to maximize P.

255

Restriction on wood: $3x + 6y \leq 42$
$$x + 3y \leq 14$$
$$y \leq -\frac{x}{3} + \frac{14}{3}$$

Restriction on workers: $7x + 7y \leq 56$
$$x + y \leq 8$$
$$y \leq -x + 8$$

Aluminum restriction: $3x + y \leq 16$
$$y \leq -3x + 16$$

And, of course, $x \geq 0$, $y \geq 0$.

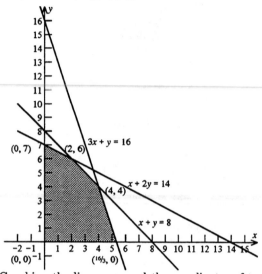

Graphing the lines, we need the coordinates of two intersections:

$x + 2y = 14$		$3x + y = 16$
$x + y = 8$		$x + y = 8$
$y = 6$		$2x = 8$
$x = 2$		$x = 4$
		$y = 4$

Vertex	$P = 600x + 800y$
$(0,0)$	0
$(0,7)$	5600
$(2,6)$	6000
$(4,4)$	5600
$(\frac{16}{3},0)$	3200

The maximum profit of \$6000 will be achieved when $x = 2$ and $y = 6$; that is, when 2 campers and 6 trailers are produced.

35. Let x represent the number of pounds of food A a person should buy.
Let y represent the number of pounds of food B a person should buy.
Then $C = 3x + 1.40y$ represents the person's cost.

Restriction on carbohydrates: $5x + 2y \geq 60$
$$y \geq -\frac{5}{2}x + 30$$

Restriction on protein: $3x + 2y \geq 40$
$$y \geq -\frac{3}{2}x + 20$$

Restriction on fat: $5x + y \geq 35$
$$y \geq -5x + 35$$

And, of course, $x \geq 0$, $y \geq 0$

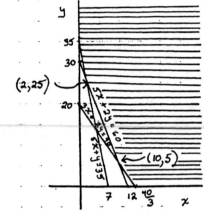

Looking at the graph, we need two intersections:

$5x + 2y = 60$		$5x + 2y = 60$
$5x + y = 35$		$3x + 2y = 40$
$y = 25$		$2x = 20$
$x = 2$		$x = 10$
		$y = 5$

Vertex	$C = 3x + 1.40y$
$(0,35)$	49
$(2,25)$	41
$(10,5)$	37
$(\frac{40}{3},0)$	40

The minimum cost is \$37, occurring when $x = 10$, and $y = 5$; that is, when you buy 10 pound of A, and 5 pounds of B.

37. Divide the polygon into two regions by the line $y - 2x = 0$. Then in region A, $y - 2x \geq 0$ and $|y - 2x| + y + x = y - 2x + y + x = 2y - x$. In region B, $y - 2x < 0$ and $|y - 2x| = -y + 2x + y + x = 3x$.

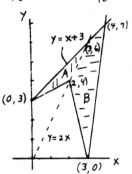

The equation of the line between $(0,3)$ and $(4,7)$ is $y = x + 3$. Find where $y = 2x$ intersects:

$y = 2x$
$2y = 2x + 6$
$y = 6$
$x = 3$

They intersect at $(3,6)$. Note also that $y = 2x$ passes through $(2,4)$.

256

Vertex	$2y - x$
$(0,3)$	6
$(2,4)$	6
$(3,6)$	9

On region A the maximum is 9, occurring at $(3,6)$.
For region B,

Vertex	$3x$
$(3,0)$	9
$(4,7)$	12

On region B the maximum is 12, occurring at $(4,7)$.
Thus, the maximum value of $|y - 2x| + y + x$ on the entire polygon is 12.

39. $x^4 + y^4 = 6561$
$$y = \pm(6561 - x^4)^{1/4}$$
Press $\boxed{Y=}$.
Enter $(6561 - X \wedge 4) \wedge 4 \boxed{x^{-1}}$ as Y_1.
Enter $\boxed{(-)}(6561 - X \wedge 4) \wedge 4 \boxed{x^{-1}}$ as Y_2.
The maximum value of $P = 2x + y$ will occur when the line $y = -2x + P$ is tangent to the curve $x^4 + y^4 = 6561$ in quadrant 1.
Experiment by entering $-2x + P$ as Y_3, for different values of P. We find that when we enter $Y_3 = -2x + 23.13$, and press $\boxed{\text{GRAPH}}$, the line is tangent to the curve. Thus, $P = 23.13$.

Chapter 11 Review Problem Set

1. True.

2. False. Such a system can have either no solution, exactly one solution, or infinitely many solutions.

3. True.

4. False. The determinant would be multiplied 3^2.

5. True. You can produce a row of zeros.

6. True.

7. True.

8. False. This changes the sign of the determinant.

9. False. Consider $P = x + y$ on a set that consists of the entire first quadrant.

10. True.

11. $\begin{cases} 3x + 2y = 7 \\ 2x - y = 7 \end{cases}$
Add 2 times the second equation to the first:
$$7x = 21, \; x = 3$$
$$3(3) + 2y = 7$$
$$2y = -2, \; y = -1$$
Solution: $(3, -1)$

12. $\begin{cases} y = 3x - 2 \\ 6x - 2y = 5 \end{cases}$
$$6x - 2(3x - 2) = 5$$
$$6x - 6x + 4 = 5$$
$$4 = 5$$
The system has no solution.

13. $\begin{cases} \dfrac{3}{x} + \dfrac{1}{y} = 9 \\[2mm] \dfrac{2}{x} - \dfrac{3}{y} = -5 \end{cases}$
Let $u = \dfrac{1}{x}$ and $v = \dfrac{1}{y}$:
$$\begin{cases} 3u + v = 9 \\ 2u - 3v = -5 \end{cases}$$
Add 3 times the first equation to the second:
$$11u = 22$$
$$u = 2$$
$$3(2) + v = 9$$
$$v = 3$$
$$2 = \frac{1}{x} \qquad 3 = \frac{1}{y}$$
$$x = \frac{1}{2} \qquad y = \frac{1}{3}$$
Solution: $\left(\frac{1}{2}, \frac{1}{3}\right)$

14. $\begin{cases} x^2 + y^2 = 13 \\ 2x^2 - 3y^2 = -19 \end{cases}$
Add 3 times the first equation to the second:
$$5x^2 = 20$$
$$x^2 = 4$$
$$x = \pm 2$$
When $x^2 = 4$,
$$4 + y^2 = 13$$
$$y^2 = 9$$
$$y = \pm 3$$
Solution: $(2, 3), (2, -3), (-2, 3), (-2, -3)$

15. $\begin{cases} 2x + y - 4z = 3 \\ 3y + z = 7 \\ z = -2 \end{cases}$
By back substitution,
$$3y - 2 = 7$$
$$3y = 9, \; y = 3$$
$$2x + 3 - 4(-2) = 3$$
$$2x + 11 = 3$$
$$2x = -8, \; x = -4$$
Solution: $(-4, 3, -2)$

16. $\begin{cases} 2x - 3y + z = 4 \\ 3y - 5z = 6 \\ -9y + 15z = 18 \end{cases}$
Adding 3 times the second equation to the third,
$$0 = 36$$
Therefore, there are no solutions.

17.
$$\begin{cases} x - 2y + 4z = 16 \\ 2x - 3y - z = 4 \\ x + 3y + 2z = 5 \end{cases}$$
Add -2 times the first equation to the second, and -1 times the first equation to the third:
$$\begin{aligned} x - 2y + 4z &= 16 \\ y - 9z &= -28 \\ 5y - 2z &= -11 \end{aligned}$$
Add -5 times the second equation to the third:
$$\begin{aligned} x - 2y + 4z &= 16 \\ y - 9z &= -28 \\ 43z &= 129 \\ z &= 3 \end{aligned}$$
By back substitution,
$$\begin{aligned} y - 9(3) &= -28 \\ y - 27 &= -28 \\ y &= -1 \\ x - 2(-1) + 4(3) &= 16 \\ x + 14 &= 16 \\ x &= 2 \end{aligned}$$
Solution: $(2, -1, 3)$

18.
$$\begin{cases} x - 3y + z = 6 \\ 2x + y - 2z = 5 \\ -4x - 2y + 4z = -10 \end{cases}$$
If we multiply the second equation by -2, we see that the last two equations are equivalent.
The system reduces to
$$\begin{cases} x - 3y + z = 6 \\ 2x + y - 2z = 5 \end{cases}$$
Adding the first equation to 3 times the second,
$$\begin{aligned} 7x - 5z &= 21 \\ 7x &= 5z + 21 \\ x &= \tfrac{5}{7}z + 3 \end{aligned}$$
Subtracting the second equation from 2 times the first equation:
$$\begin{aligned} -7y + 4z &= 7 \\ -7y &= -4z + 7 \\ y &= \tfrac{4}{7}z - 1 \end{aligned}$$
Solution: $(x, y, z) = \left(\tfrac{5}{7}z + 3, \tfrac{4}{7}z - 1, z\right)$

19.
$$\begin{cases} x^2 + y^2 = 13 \\ 2x^2 - 3y^2 = -14 \end{cases}$$
Add -2 times the first equation to the second:
$$\begin{aligned} -5y^2 &= -40 \\ y^2 &= 8 \\ y &= \pm\sqrt{8} = \pm 2\sqrt{2} \end{aligned}$$
When $y^2 = 8$,
$$\begin{aligned} x^2 + 8 &= 13 \\ x^2 &= 5 \\ x &= \pm\sqrt{5} \end{aligned}$$
Solution:
$$(\sqrt{5}, 2\sqrt{2}), \qquad (\sqrt{5}, -2\sqrt{2}), \qquad (-\sqrt{5}, 2\sqrt{2}),$$

$$(-\sqrt{5}, -2\sqrt{2})$$

20.
$$\begin{cases} 5 \cdot 2^x - 3^y = 1 \\ 2^{x+2} + 3^{y+2} = 40 \end{cases}$$
$2^{x+2} = 2^x \cdot 2^2$ and $3^{y+2} = 3^y \cdot 3^2$
Let $a = 2^x$ and let $b = 3^y$:
$$\begin{cases} 5a - b = 1 \\ 4a + 9b = 40 \end{cases}$$
Add 9 times the first equation to the second:
$$\begin{aligned} 49a &= 49 \\ a &= 1 \\ 5 - b &= 1 \\ b &= 4 \end{aligned}$$
So, $2^x = 1 \Rightarrow x = 0$, and
$$\begin{aligned} 3^y &= 4 \\ y \ln 3 &= \ln 4 \\ y &= \frac{\ln 4}{\ln 3} \approx 1.26186 \end{aligned}$$

21.
$$\begin{bmatrix} 2 & -1 & 4 \\ 0 & 1 & 3 \end{bmatrix} + 2 \begin{bmatrix} 1 & 0 & 3 \\ 4 & -1 & 5 \end{bmatrix}$$
$$= \begin{bmatrix} 2 & -1 & 4 \\ 0 & 1 & 3 \end{bmatrix} + \begin{bmatrix} 2 & 0 & 6 \\ 8 & -2 & 10 \end{bmatrix}$$
$$= \begin{bmatrix} 4 & -1 & 10 \\ 8 & -1 & 13 \end{bmatrix}$$

22.
$$\begin{bmatrix} 5 & -4 \\ 3 & 1 \end{bmatrix} \begin{bmatrix} 2 & 5 \\ 3 & -1 \end{bmatrix} = \begin{bmatrix} -2 & 29 \\ 9 & 14 \end{bmatrix}$$

23.
$$4\begin{bmatrix} 2 \\ -1 \\ 0 \end{bmatrix} - 3\begin{bmatrix} 8 \\ -5 \\ 9 \end{bmatrix} = \begin{bmatrix} 8 \\ -4 \\ 0 \end{bmatrix} + \begin{bmatrix} -24 \\ 15 \\ -27 \end{bmatrix}$$
$$= \begin{bmatrix} -16 \\ 11 \\ -27 \end{bmatrix}$$

24.
$$\begin{bmatrix} 2 & -3 \\ 4 & 1 \\ 0 & 5 \end{bmatrix} \begin{bmatrix} 4 & 0 & 1 \\ 2 & -3 & 6 \end{bmatrix} = \begin{bmatrix} 2 & 9 & -16 \\ 18 & -3 & 10 \\ 10 & -15 & 30 \end{bmatrix}$$

25. Note that $\begin{bmatrix} 1 & 0 \\ 0 & 1 \end{bmatrix}$ is the identity matrix.

$$\begin{bmatrix} 1 & 0 \\ 0 & 1 \end{bmatrix}\begin{bmatrix} 5 & 2 \\ -3 & 4 \end{bmatrix}\begin{bmatrix} b & 0 \\ 0 & b \end{bmatrix}$$

$$= \begin{bmatrix} 5 & 2 \\ -3 & 4 \end{bmatrix}\begin{bmatrix} b & 0 \\ 0 & b \end{bmatrix}$$

$$= \begin{bmatrix} 5b & 2b \\ -3b & 4b \end{bmatrix}$$

26. $\begin{bmatrix} 6 & 11 \\ 0 & 0 \end{bmatrix}\begin{bmatrix} 11 & 0 \\ -6 & 0 \end{bmatrix} = \begin{bmatrix} 0 & 0 \\ 0 & 0 \end{bmatrix}$

27. $\begin{bmatrix} 4 & 3 \\ 2 & 3 \end{bmatrix}\left\{\begin{bmatrix} 4 & 3 \\ 2 & 3 \end{bmatrix}^{-1} + \begin{bmatrix} 1 & 0 \\ 0 & 1 \end{bmatrix}\right\}$

$$= \begin{bmatrix} 4 & 3 \\ 2 & 3 \end{bmatrix}\begin{bmatrix} 4 & 3 \\ 2 & 3 \end{bmatrix}^{-1} + \begin{bmatrix} 4 & 3 \\ 2 & 3 \end{bmatrix}\begin{bmatrix} 1 & 0 \\ 0 & 1 \end{bmatrix}$$

$$= \begin{bmatrix} 1 & 0 \\ 0 & 1 \end{bmatrix} + \begin{bmatrix} 4 & 3 \\ 2 & 3 \end{bmatrix}$$

$$= \begin{bmatrix} 5 & 3 \\ 2 & 4 \end{bmatrix}$$

28. Note that the inverse of a diagonal matrix is found by taking the reciprocals of the diagonal elements. (See Problem Set 8.4, No. 28.)

$$\begin{bmatrix} 1 & -2 & 3 \\ 0 & 2 & 1 \\ 1 & -2 & 4 \end{bmatrix}\begin{bmatrix} 1 & 0 & 0 \\ 0 & \frac{1}{2} & 0 \\ 0 & 0 & \frac{1}{3} \end{bmatrix}^{-1}$$

$$= \begin{bmatrix} 1 & -2 & 3 \\ 0 & 2 & 1 \\ 1 & -2 & 4 \end{bmatrix}\begin{bmatrix} 1 & 0 & 0 \\ 0 & 2 & 0 \\ 0 & 0 & 3 \end{bmatrix}$$

$$= \begin{bmatrix} 1 & -4 & 9 \\ 0 & 4 & 3 \\ 1 & -4 & 12 \end{bmatrix}$$

29.
$$\begin{array}{cc} \mathbf{M} & \mathbf{I} \end{array}$$
$$\begin{bmatrix} 1 & -2 & 3 \\ 0 & 2 & 1 \\ 1 & -2 & 4 \end{bmatrix}\begin{bmatrix} 1 & 0 & 0 \\ 0 & 1 & 0 \\ 0 & 0 & 1 \end{bmatrix}$$

Subtract row 1 from row 3:
$$\begin{bmatrix} 1 & -2 & 3 \\ 0 & 2 & 1 \\ 0 & 0 & 1 \end{bmatrix}\begin{bmatrix} 1 & 0 & 0 \\ 0 & 1 & 0 \\ -1 & 0 & 1 \end{bmatrix}$$

Subtract row 3 from row 2, and 3 times row 3 from row 1:
$$\begin{bmatrix} 1 & -2 & 0 \\ 0 & 2 & 0 \\ 0 & 0 & 1 \end{bmatrix}\begin{bmatrix} 4 & 0 & -3 \\ 1 & 1 & -1 \\ -1 & 0 & 1 \end{bmatrix}$$

Add row 2 to row 1:
$$\begin{bmatrix} 1 & 0 & 0 \\ 0 & 2 & 0 \\ 0 & 0 & 1 \end{bmatrix}\begin{bmatrix} 5 & 1 & -4 \\ 1 & 1 & -1 \\ -1 & 0 & 1 \end{bmatrix}$$

Divide row 2 by 2:
$$\begin{bmatrix} 1 & 0 & 0 \\ 0 & 1 & 0 \\ 0 & 0 & 1 \end{bmatrix}\begin{bmatrix} 5 & 1 & -4 \\ \frac{1}{2} & \frac{1}{2} & -\frac{1}{2} \\ -1 & 0 & 1 \end{bmatrix}$$

$$\mathbf{M}^{-1} = \begin{bmatrix} 5 & 1 & -4 \\ \frac{1}{2} & \frac{1}{2} & -\frac{1}{2} \\ -1 & 0 & 1 \end{bmatrix}$$

30. In matrix form our system of equations becomes
$$\begin{bmatrix} 1 & -2 & 3 \\ 0 & 2 & 1 \\ 1 & -2 & 4 \end{bmatrix}\begin{bmatrix} x \\ y \\ z \end{bmatrix} = \begin{bmatrix} 2 \\ -4 \\ 0 \end{bmatrix}$$

By Problem 29, the inverse of the coefficient matrix is
$$\begin{bmatrix} 5 & 1 & -4 \\ \frac{1}{2} & \frac{1}{2} & -\frac{1}{2} \\ -1 & 0 & 1 \end{bmatrix}$$

Therefore,

$$\begin{bmatrix} x \\ y \\ z \end{bmatrix} = \begin{bmatrix} 5 & 1 & -4 \\ \frac{1}{2} & \frac{1}{2} & -\frac{1}{2} \\ -1 & 0 & 1 \end{bmatrix} \begin{bmatrix} 2 \\ -4 \\ 0 \end{bmatrix} = \begin{bmatrix} 6 \\ -1 \\ -2 \end{bmatrix}$$

Solution: $(x, y, z) = (6, -1, -2)$

31. $\begin{vmatrix} 3 & -4 \\ -5 & 6 \end{vmatrix} = (3)(6) - (-5)(-4) = 18 - 20 = -2$

32. $\begin{vmatrix} 5 & -2 \\ -10 & 4 \end{vmatrix}^2 = \begin{vmatrix} 5 & -2 \\ -10 & 4 \end{vmatrix}^2 = (20 - 20)^2 = 0^2 = 0$

33. $\begin{vmatrix} 3 & 0 & -2 \\ -3 & 1 & 0 \\ 1 & 5 & 4 \end{vmatrix} = \begin{vmatrix} 3 & 0 & -2 \\ -3 & 1 & 0 \\ 7 & 5 & 0 \end{vmatrix}$ Add twice row 1 to row 3

$= -2 \begin{vmatrix} -3 & 1 \\ 7 & 5 \end{vmatrix}$ Expand according to column 3

$= -2(-15 - 7)$

$= 44$

34. Since the determinant is triangular, it is equal to the product of the entries on the main diagonal.

$\begin{vmatrix} 5 & 1 & 19 \\ 0 & -\frac{1}{2} & 1 \\ 0 & 0 & 4 \end{vmatrix} = (5)(-\frac{1}{2})(4) = -10$

35. $\begin{vmatrix} 2 & -1 & 4 \\ 1 & 3 & -2 \\ 4 & 5 & 0 \end{vmatrix} = \begin{vmatrix} 4 & 5 & 0 \\ 1 & 3 & -2 \\ 4 & 5 & 0 \end{vmatrix}$ Add twice row 2 to row 1

$= -(-2) \begin{vmatrix} 4 & 5 \\ 4 & 5 \end{vmatrix}$ Expand according to column 3

$= 2(0)$

$= 0$

36. $\begin{vmatrix} 2 & -1 & 4 \\ 1 & 3 & -2 \\ 4 & 5 & 8 \end{vmatrix} = 2 \begin{vmatrix} 2 & -1 & 2 \\ 1 & 3 & -1 \\ 4 & 5 & 4 \end{vmatrix}$ Factor a 2 from column 3

$= 2 \begin{vmatrix} 2 & -1 & 0 \\ 1 & 3 & -2 \\ 4 & 5 & 0 \end{vmatrix}$ Subtract column 1 from column 3

$= 2 \cdot -(-2) \begin{vmatrix} 2 & -1 \\ 4 & 5 \end{vmatrix}$ Expand by column 3

$= 4(10 - (-4)) = 4 \cdot 14$

$= 56$

37. $\begin{vmatrix} 3 & 1 & 0 & 0 \\ 5 & 2 & 0 & 0 \\ 0 & 0 & 4 & -1 \\ 0 & 0 & -6 & 2 \end{vmatrix}$

$= 3 \begin{vmatrix} 2 & 0 & 0 \\ 0 & 4 & -1 \\ 0 & -6 & 2 \end{vmatrix} - 1 \begin{vmatrix} 5 & 0 & 0 \\ 0 & 4 & -1 \\ 0 & -6 & 2 \end{vmatrix}$

$= 3 \cdot 2 \begin{vmatrix} 4 & -1 \\ -6 & 2 \end{vmatrix} - 1 \cdot 5 \begin{vmatrix} 4 & -1 \\ -6 & 2 \end{vmatrix}$

$= (6 - 5) \begin{vmatrix} 4 & -1 \\ -6 & 2 \end{vmatrix} = 1(8 - 6) = 2$

38. $\begin{vmatrix} 3 & 0 & 2 & 0 & -1 \\ 0 & 2 & 0 & 3 & 2 \\ -3 & 3 & 1 & -3 & 0 \\ 0 & 0 & 3 & 0 & 1 \\ -1 & 1 & 0 & -1 & 0 \end{vmatrix}$

Subtract 3 times column 5 from column 3:

$= \begin{vmatrix} 3 & 0 & 5 & 0 & -1 \\ 0 & 2 & -6 & 3 & 2 \\ -3 & 3 & 1 & -3 & 0 \\ 0 & 0 & 0 & 0 & 1 \\ -1 & 1 & 0 & -1 & 0 \end{vmatrix}$

Expand according to the fourth row:

$= (-1) \begin{vmatrix} 3 & 0 & 5 & 0 \\ 0 & 2 & -6 & 3 \\ -3 & 3 & 1 & -3 \\ -1 & 1 & 0 & -1 \end{vmatrix}$

Add column 1 to column 2, and -1 times column 1 to column 4:

$$= (-1) \begin{vmatrix} 3 & 3 & 5 & -3 \\ 0 & 2 & -6 & 3 \\ -3 & 0 & 1 & 0 \\ -1 & 0 & 0 & 0 \end{vmatrix}$$

Expand according to the fourth row:

$$= (1)(-1) \begin{vmatrix} 3 & 5 & -3 \\ 2 & -6 & 3 \\ 0 & 1 & 0 \end{vmatrix}$$

Expand according to the third row:

$$= (-1)(-1) \begin{vmatrix} 3 & -3 \\ 2 & 3 \end{vmatrix} = (3)(3) - (-3)(2) = 15$$

39. In Problem 33, we computed D, the determinant of the system. Since $D \neq 0$, the system has a unique solution. So the one solution must be $(0,0,0)$.

40. $D = \begin{vmatrix} 2 & -1 & 4 \\ 1 & 3 & -2 \\ 4 & 5 & 8 \end{vmatrix} = 56$, by Problem 36

$$y = \frac{1}{56} \begin{vmatrix} 2 & 0 & 4 \\ 1 & -2 & -2 \\ 4 & 1 & 8 \end{vmatrix}$$

Add 2 times row 3 to row 2:

$$y = \frac{1}{56} \begin{vmatrix} 2 & 0 & 4 \\ 9 & 0 & 14 \\ 4 & 1 & 8 \end{vmatrix} = \frac{1}{56}(-1) \begin{vmatrix} 2 & 4 \\ 9 & 14 \end{vmatrix}$$

$$= -\frac{1}{56}[(2)(14) - (4)(9)] = \frac{8}{56} = \frac{1}{7}$$

41.
$$x + y \leq 7$$
$$y \leq -x + 7$$
$$-3x + 4y \geq 0$$
$$y \geq \frac{3x}{4}$$

$x \geq 0$, $y \geq 0$

The region lies below the line $x + y = 7$, above the line $-3x + 4y = 0$, to the right of the y-axis, and above the x-axis. The region will include its boundary lines.

Find the point of intersection of
$$\begin{cases} x + y = 7 \\ -3x + 4y = 0 \end{cases}$$
$$7y = 21$$
$$y = 3$$
$$x + 3 = 7$$
$$x = 4$$

42.
$$2x + y \leq 8$$
$$x + y \leq 6$$
$$3x + 2y \geq 12$$

$2x + y = 8$ has axis-intercepts $(0,8)$ and $(4,0)$.
$x + y = 6$ has axis-intercepts $(0,6)$ and $(6,0)$.
$3x + 2y = 12$ has axis-intercepts $(0,6)$ and $(4,0)$.

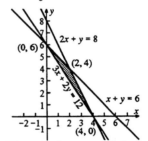

The solution set will be below the line $2x + y = 8$, below the line $x + y = 6$, above the line $3x + 2y = 12$, and will include its boundary lines.
Find the point of intersection of
$$\begin{cases} 2x + y = 8 \\ x + y = 6 \end{cases}$$
$$x = 2$$
$$y = 4$$

43.
$$3x + y \leq 15$$
$$y \leq -3x + 15$$
$$x + y \leq 7$$
$$y \leq -x + 7$$
$$x \geq 0, \ y \geq 0$$

The region lies below the line $3x + y = 15$, below the line $x + y = 7$, to the right of the y-axis, and above the x-axis. The region will include its boundary lines.

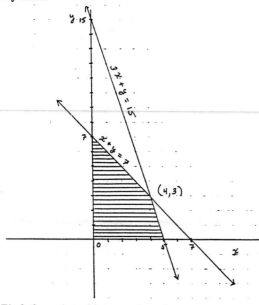

Find the point of intersection of
$$\begin{cases} 3x + y = 15 \\ x + y = 7 \end{cases}$$
$$2x = 8$$
$$x = 4$$
$$4 + y = 7$$
$$y = 3$$

44. $x \geq 0, \ y \geq 0$
$$x - 2y + 4 \geq 0$$
$$x - 2y \geq -4$$
$$2y - x \leq 4$$

$$x + y - 11 \geq 0$$
$$x + y \geq 11$$
$2y - x = 4$ has axis-intercepts $(0, 2)$ and $(-4, 0)$.
$x + y = 11$ has axis-intercepts $(0, 11)$ and $(11, 0)$.

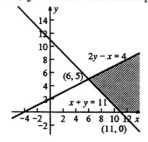

The solution set will be to the right of the y-axis, above the x-axis, below the line $2y - x = 4$, above the line $x + y = 11$, and will include its boundary lines.
Find the point of intersection:
$$\begin{cases} x + y = 11 \\ x - 2y = -4 \end{cases}$$
$$3y = 15$$
$$y = 5$$
$$x = 6$$

45.

Vertex	$P = 2x + y$
$(0,0)$	0
$(0,7)$	7
$(4,3)$	11

The maximum value of P is 11, occurring at $(4, 3)$.

46.

Vertex	$P = x + 2y$
$(0,6)$	12
$(2,4)$	10
$(4,0)$	4

The maximum value of P is 12, occurring at $(0, 6)$.

47.

Vertex	$P = x + 2y$
$(0,0)$	0
$(0,7)$	14
$(4,3)$	10
$(5,0)$	5

The minimum value of P is 0, occurring at $(0, 0)$.
The maximum value of P is 14, occurring at $(0, 7)$.

48.

Vertex	$P = 2x + 3y$
$(6,5)$	27
$(11,0)$	22

The minimum value of P is 22, occurring at $(11, 0)$. We also notice that since x and y can each (independently) get arbitrarily large, $P(x, y)$ will not attain a largest value. P has no maximum.

49. Let x represent the number of grams of 30% alloy that should be melted, y represent the number of grams of 40% alloy, and z represent the number of grams of 60% alloy.
There will be 100 grams of combined alloy:
$$x + y + z = 100$$
There will be 10 grams more of 60% alloy used than the 30% alloy:
$$z = x + 10$$
The new alloy will contain 45% nickel:
$$.30x + .40y + .60z = .45(100)$$
$$\begin{cases} 3x + 4y + 6z = 450 \\ x + y + z = 100 \\ -x + z = 10 \end{cases}$$

Subtract 3 times the second equation from the first, and add the second equation to the third:

$$\begin{cases} y + 3z = 150 \\ x + y + z = 100 \\ y + 2z = 110 \end{cases}$$

Subtracting the third equation from the first,

$$z = 40$$

Substituting into the original equations,

$$40 = x + 10$$
$$x = 30$$
$$30 + y + 40 = 100$$
$$y = 30$$

The metallurgist must combine 30 grams of the 30% alloy, 30 grams of the 40% alloy, and 40 grams of the 60% alloy.

50. The manufacturer should produce x units of model A and y units of model B.

$P = 100x + 120y$ is his profit, which we want to maximize.

$x \geq 0$, $y \geq 0$

$x \geq 50$ and $x \leq 110$

$y \geq 75$ and $y \leq 100$

Total daily production cannot exceed 180 units:

$x + y \leq 180$

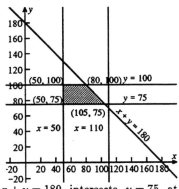

$x + y = 180$ intersects $y = 75$ at $x = 105$, and it intersects $y = 100$ at $x = 80$.

The solution set will be to the right of $x = 50$, above $y = 75$, below $y = 100$, below $x + y = 180$, and will include its boundary lines. (The other constraints are redundant.)

Vertex	$P = 100x + 120y$
$(50, 75)$	$\$14,000$
$(50, 100)$	$17,000$
$(100, 80)$	$19,600$ (Maximum)
$(105, 75)$	$19,500$

100 units of model A and 80 units of model B should be produced for a profit of $19,600.

Chapter 12: Sequences, Counting Problems, and Probability

Problem Set 12.1

1. a) 1, 3, 5, 7, $\boxed{9}$, $\boxed{11}$

   b) 17, 14, 11, 8, $\boxed{5}$, $\boxed{2}$

   c) $1, \frac{1}{2}, \frac{1}{4}, \frac{1}{8}, \boxed{\frac{1}{16}}, \boxed{\frac{1}{32}}$

   d) 1, 9, 25, 49, $\boxed{81}$, $\boxed{121}$

3. a) $a_n = 2n + 3$
   $a_4 = 2(4) + 3 = 11$
   $a_{20} = 2(20) + 3 = 43$

   b) $a_n = \frac{n}{n+1}$

   $a_5 = \frac{5}{5+1} = \frac{5}{6}$

   $a_9 = \frac{9}{9+1} = \frac{9}{10}$

   c) $a_n = (2n - 1)^2$
   $a_4 = (2 \cdot 4 - 1)^2 = (7)^2 = 49$
   $a_5 = (2 \cdot 5 - 1)^2 = (9)^2 = 81$

   d) $a_n = (-3)^n$
   $a_3 = (-3)^3 = -27$
   $a_4 = (-3)^4 = 81$

5. a) $a_n = 2n - 1$
   b) $a_n = 17 - 3(n - 1)$
   c) $a_n = (\frac{1}{2})^{n-1}$
   d) $a_n = (2n - 1)^2$

7. a) $a_1 = 2; a_n = a_{n-1} + 3$
   $a_2 = 2 + 3 = 5$
   $a_3 = 5 + 3 = 8$
   $a_4 = 8 + 3 = 11$
   $a_5 = 11 + 3 = 14$

   b) $a_1 = 2; a_n = 3a_{n-1}$
   $a_2 = 3(2) = 6$
   $a_3 = 3(6) = 18$
   $a_4 = 3(18) = 54$
   $a_5 = 3(54) = 162$

   c) $a_1 = 8; a_n = \frac{1}{2}a_{n-1}$

   $a_2 = \frac{1}{2}(8) = 4$

   $a_3 = \frac{1}{2}(4) = 2$

   $a_4 = \frac{1}{2}(2) = 1$

   $a_5 = \frac{1}{2}(1) = \frac{1}{2}$

   d) $a_1 = 1; a_n = a_{n-1} + 8(n - 1)$
   $a_2 = 1 + 8(2 - 1) = 1 + 8 = 9$
   $a_3 = 9 + 8(3 - 1) = 9 + 16 = 25$
   $a_4 = 25 + 8(4 - 1) = 25 + 24 = 49$
   $a_5 = 49 + 8(5 - 1) = 49 + 32 = 81$

9. a) 1, 3, 5, 7, ...
   $a_1 = 1; a_n = a_{n-1} + 2$

   b) 17, 14, 11, ,8, ...
   $a_1 = 17; a_n = a_{n-1} - 3$

   c) $1, \frac{1}{2}, \frac{1}{4}, \frac{1}{8}, ...$
   $a_1 = 1; a_n = \frac{1}{2}(a_{n-1})$

   d) 1, 9, 25, 49, ...
   From Problem 7, part (d),
   $a_1 = 1; a_n = a_{n-1} + 8(n - 1)$

11. $a_n = 2n + 1$
    $a_1 = 2(1) + 1 = 3$
    $a_2 = 2(2) + 1 = 5$
    $a_3 = 2(3) + 1 = 7$
    $a_4 = 2(4) + 1 = 9$
    $a_5 = 2(5) + 1 = 11$
    $a_6 = 2(6) + 1 = 13$
    $A_6 = a_1 + a_2 + a_3 + a_4 + a_5 + a_6$
    $= 3 + 5 + 7 + 9 + 11 + 13 = 48$

13. $a_n = (-2)^n$
    $a_1 = (-2)^1 = -2$
    $a_2 = (-2)^2 = 4$
    $a_3 = (-2)^3 = -8$
    $a_4 = (-2)^4 = 16$
    $a_5 = (-2)^5 = -32$
    $a_6 = (-2)^6 = 64$
    $A_6 = a_1 + a_2 + a_3 + a_4 + a_5 + a_6$
    $= -2 + 4 - 8 + 16 - 32 + 64 = 42$

15. $a_n = n^2 - 2$
    $a_1 = (1)^2 - 2 = -1$
    $a_2 = (2)^2 - 2 = 2$
    $a_3 = (3)^2 - 2 = 7$
    $a_4 = (4)^2 - 2 = 14$
    $a_5 = (5)^2 - 2 = 23$
    $a_6 = (6)^2 - 2 = 34$
    $A_6 = a_1 + a_2 + a_3 + a_4 + a_5 + a_6$
    $= -1 + 2 + 7 + 14 + 23 + 34 = 79$

17. $a_1 = 4; a_n = a_{n-1} + 3$
    $a_1 = 4$
    $a_2 = 4 + 3 = 7$
    $a_3 = 7 + 3 = 10$
    $a_4 = 10 + 3 = 13$
    $a_5 = 13 + 3 = 16$
    $a_6 = 16 + 3 = 19$
    $A_6 = a_1 + a_2 + a_3 + a_4 + a_5 + a_6$
    $= 4 + 7 + 10 + 13 + 16 + 19 = 69$

19. a) $b_3 + b_4 + \cdots + b_{20} = \sum_{i=3}^{20} b_i$

b) $1^2 + 2^2 + \cdots + 19^2 = \sum_{i=1}^{19} i^2$

c) $1 + \frac{1}{2} + \frac{1}{3} + \cdots + \frac{1}{n} = \sum_{i=1}^{n} \frac{1}{i}$

21. a) $\sum_{i=1}^{5} (i^2 - 1)$

$= (1^2 - 1) + (2^2 - 1) + (3^2 - 1) + (4^2 - 1) + (5^2 - 1)$
$= 0 + 3 + 8 + 15 + 24 = 50$

b) $\sum_{i=1}^{5} (2^i - i^2)$

$= (2^1 - 1^2) + (2^2 - 2^2) + (2^3 - 3^2) + (2^4 - 4^2)$
$\qquad\qquad\qquad\qquad + (2^5 - 5^2)$
$= 1 + 0 + (-1) + 0 + 7 = 7$

c) $\sum_{i=2}^{5} \frac{1}{i-1} = \frac{1}{2-1} + \frac{1}{3-1} + \frac{1}{4-1} + \frac{1}{5-1}$

$\qquad = 1 + \frac{1}{2} + \frac{1}{3} + \frac{1}{4}$

$\qquad = \frac{12}{12} + \frac{6}{12} + \frac{4}{12} + \frac{3}{12} = \frac{25}{12}$

d) $\sum_{i=2}^{7} (2i + 5)$

$= [2(2) + 5] + [2(3) + 5] + [2(4) + 5] + [2(5) + 5]$
$\qquad\qquad\qquad\qquad + [2(6) + 5] + [2(7) + 5]$

$= (2)(2 + 3 + 4 + 5 + 6 + 7) + (5)(6)$

$= 2(27) + 30 = 84$

23. a) $a_n = 5n + 2$
$a_5 = 5(5) + 2 = 27$
$a_{16} = 5(16) + 2 = 82$

b) $a_n = n(n-1)$
$a_5 = 5(4) = 20$
$a_{16} = 16(15) = 240$

25. $a_n = 3a_{n-1} - 1$; $b_n = n^2 - 3n$; $c_n = a_n - b_n$

a) $a_1 = 2$; $b_1 = 1^2 - 3(1) = -2$; $c_1 = 2 - (-2) = 4$

b) $a_2 = 3(2) - 1 = 5$; $b_2 = 2^2 - 3(2) = -2$;
$c_2 = 5 - (-2) = 7$

c) $a_3 = 3(5) - 1 = 14$; $b_3 = 3^2 - 3(3) = 0$;
$c_3 = 14 - 0 = 14$

d) $a_4 = 3(14) - 1 = 41$; $b_4 = 4^2 - 3(4) = 4$;
$c_4 = 41 - 4 = 37$

e) $a_5 = 3(41) - 1 = 122$; $b_5 = 5^2 - 3(5) = 10$;
$c_5 = 122 - 10 = 112$

27. a) 2, 6, 18, 54, ...
$a_1 = 2$; $a_n = 3a_{n-1}$

b) 2, 6, 10, 14, ...
$a_1 = 2$; $a_n = a_{n-1} + 4$

c) 2, 4, 8, 14, ...
$a_1 = 2$; $a_n = a_{n-1} + 2(n-1)$

d) 2, 4, 6, 10, 16, ...
$a_1 = 2$; $a_2 = 2$; $a_n = a_{n-1} + a_{n-2}$

e) 2, 1, $\frac{1}{2}$, $\frac{1}{4}$, ...

$a_1 = 2$; $a_n = \frac{1}{2}a_{n-1}$

f) 2, 5, 10, 17, ...
$a_1 = 2$; $a_n = a_{n-1} + 2n - 1$

29. $\frac{1}{7} = .\overline{142857}$ Let a_n be the nth digit of the decimal expansion of $\frac{1}{7}$.

$a_1 = 1$
$a_2 = 4$
$a_3 = 2$
$a_4 = 8$
$a_5 = 5$
$a_6 = 7$

In general, $a_n = a_m$ where m is the remainder on dividing by 6 and where a_0 is defined to be equal to a_6. Since $\frac{8}{6} = 1$ remainder 2, $a_8 = a_2 = 4$.

Since $\frac{27}{6} = 4$ remainder 3, $a_{27} = a_3 = 2$.

Since $\frac{53}{6} = 8$ remainder 5, $a_{53} = a_5 = 5$.

31. a) $a_n = 2n - 1$
$a_1 = 2(1) - 1 = 1$
$a_2 = 2(2) - 1 = 3$
$a_3 = 2(3) - 1 = 5$
$a_4 = 2(4) - 1 = 7$
$a_5 = 2(5) - 1 = 9$

b) $A_1 = a_1 = 1$
$A_2 = a_1 + a_2 = 1 + 3 = 4$
$A_3 = a_1 + a_2 + a_3 = 1 + 3 + 5 = 9$
$A_4 = a_1 + a_2 + a_3 + a_4 = 1 + 3 + 5 + 7 = 16$
$A_5 = a_1 + a_2 + a_3 + a_4 + a_5$
$\qquad = 1 + 3 + 5 + 7 + 9 = 25$

c) $A_n = n^2$

33. a) $c_3 + c_4 + \cdots + c_{112} = \sum_{i=3}^{112} c_i$

b) $4^2 + 5^2 + 6^2 + \cdots 104^2 = \sum_{i=4}^{104} i^2$

c) $12 + 18 + 24 + \cdots + 60 = 6 \cdot 2 + 6 \cdot 3 + \cdots + 6 \cdot 10$
$\qquad\qquad\qquad\qquad = \sum_{i=2}^{10} 6i$

35. If the oranges are piled in a pyramid, we can assume that the second level of the pyramid is a square 14×14, the third level a square 13×13, and so on up to 1.

$$15^2 + 14^2 + 13^2 + \cdots + 1^2 = \sum_{i=1}^{15} i^2$$

37. a) $1, 4, 9, 16, \ldots; s_n = n^2$

b) $s_1 = 1; s_n = s_{n-1} + 2n - 1$

c) $1, 3, 6, 10, 15, \ldots; t_1 = 1; t_n = t_{n-1} + n$

d)

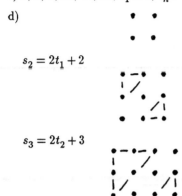

$s_2 = 2t_1 + 2$

$s_3 = 2t_2 + 3$

$s_4 = 2t_3 + 4$
$s_{n+1} = 2t_n + (n+1)$

e) $\quad 2t_n + (n+1) = s_{n+1}$
$2t_n + n + 1 = (n+1)^2$
$2t_n = (n+1)^2 - (n+1)$
$2t_n = (n+1)(n+1-1)$
$2t_n = (n+1)(n)$

$$t_n = \frac{(n+1)(n)}{2}$$

39. a) On the last day of the first month there will still be only 1 pair of rabbits, since the first pair has not started to reproduce.

b) On the last day of the second month there will still be 1 pair; the first litter to the original pair will be be born the next day.

c) At the end of the third month there will be 2 pairs.

d) At the end of the fourth month there will be 3 pairs. The original pair will have had a second litter at the beginning of that month, but the first of their litters will not have its first litter until the next day.

e) At the end of the fifth month, there will be 5 pairs. The original pair and the first of their litters will each have had a litter at the beginning of the month.

f) The sequence representing the number of pairs at the end of each month is $1, 1, 2, 3, 5, \ldots$.
Let f_n represent the number of pairs at the end of month n. Then,
$f_1 = 1; f_2 = 1; f_n = f_{n-1} + f_{n-2}$

41. The rectangle in Figure 3 has dimensions 8×13, that is, $f_6 \times f_7$. Its area is

$$f_1^2 + f_2^2 + f_3^2 + f_4^2 + f_5^2 + f_6^2 = f_6 f_7.$$

To build the next larger rectangle, affix a square 13×13 to the right side of the existing rectangle. The new dimensions are 13×21, i.e., $f_7 \times f_8$. Thus,

$$f_1^2 + f_2^2 + f_3^2 + f_4^2 + f_5^2 + f_6^2 + f_7^2 = f_7 f_8.$$

In general, $f_1^2 + f_2^2 + f_3^2 + \cdots f_n^2 = f_n f_{n+1}$.

Problem Set 12.2

1. a) $1, 4, 7, 10, \boxed{13}, \boxed{16}$

b) $2, 2.3, 2.6,, 2.9, \boxed{3.2}, \boxed{3.5}$

c) $28, 24, 20, 16, \boxed{12}, \boxed{8}$

Each is an arithmetic sequence with common differences of $3, .3,$ and -4, respectively.

3. a) $d = 4 - 1 = 3; a_{30} = a_1 + 29d = 1 + 29(3) = 88$

b) $d = 2.3 - 2 = .3$
$a_{30} = a_1 + 29d = 2 + 29(.3) = 10.7$

c) $d = 24 - 28 = -4$
$a_{30} = a_1 + 29d = 29 + 28(-4) = -88$

5. a) $A_{30} = \frac{30}{2}(1 + 88) = 15(89) = 1335$

b) $B_{30} = \frac{30}{2}(2 + 10.7) = 15(12.7) = 190.5$

c) $C_{30} = \frac{30}{2}[28 + (-88)] = 15(-60) = -900$

7. Given the formula
$$a_n = a_1 + (n - 1)d,$$
we can solve for d:
$$d = \frac{a_n - a_1}{n - 1}$$
If $n = 40$, $a_{40} = 24.5$, and $a_1 = 5$,
$$d = \frac{24.5 - 5}{39} = .5$$

9. a) $2 + 4 + 6 + \cdots + 200$
Notice that each term is of the form $2n$.
When $2n = 200$, $n = 100$.
$$A_{100} = \tfrac{100}{2}(2 + 200) = 50(202) = 10{,}100$$

b) $1 + 3 + 5 + \cdots + 199$
Notice that each term is of the form $2n - 1$.
When $2n - 1 = 199$,
$$2n = 200$$
$$n = 100$$
$$A_{100} = \tfrac{100}{2}(1 + 199) = 50(200) = 10{,}000$$

c) $3 + 6 + 9 + \cdots + 198$
Notice that each term is of the form $3n$.
When $3n = 198$, $n = 66$.
$$A_{66} = \tfrac{66}{2}(3 + 198) = 33(201) = 6{,}633$$

11. To compute the amount of material needed, add the terms of the arithmetic sequence with $n = 17$, $a_1 = 30$, and $a_{17} = 15$.
$$A_{17} = (\tfrac{17}{2})(30 + 15) = \tfrac{17}{2}(45) = 382.5 \text{ cm}$$

13. $3, a, b, c, d, 7$ is an arithmetic sequence. $n = 6$, $a_1 = 3$, and $a_6 = 7$. Let $D =$ common difference.
$$a_6 = a_1 + (n-1)D$$
$$7 - 3 = 5D$$
$$4 = 5D$$
$$D = \tfrac{4}{5}$$
So the sequence is $3, 3\tfrac{4}{5}, 4\tfrac{3}{5}, 5\tfrac{2}{5}, 6\tfrac{1}{5}, 7$.

15. The first multiple of 9 past 200 is 207. The last multiple of 9 before 300 is 297.
Consider the sequence $207, 216, \ldots, 297$ with $a_1 = 207$, $a_n = 297$, and $d = 9$. To find n, solve $a_n = a_1 + (n-1)d$ for n.
$$n = 1 + \frac{a_n - a_1}{d}$$
$$= 1 + \frac{297 - 207}{9} = 1 + \frac{90}{9} = 11$$
So, there are 11 numbers between 200 and 300 that are multiples of 9. The sum of these numbers is
$$A_{11} = \tfrac{11}{2}(207 + 297) = \tfrac{11}{2}(504) = 2772$$

17. a) $\sum\limits_{i=2}^{60} (i - 5) = -3 + (-2) + \cdots + 55$
$n = 59$; $a_1 = -3$; $a_{59} = 55$
$$A_{59} = \tfrac{59}{2}(-3 + 55) = \tfrac{59}{2}(52) = 1534$$

b) $\sum\limits_{i=1}^{40} (4 - 3i) = 1 + (-2) + \cdots + (-116)$
$n = 40$; $a_1 = 1$; $a_{40} = -116$
$$A_{40} = \tfrac{40}{2}[1 + (-116)] = 20(-115) = -2300$$

c) $\sum\limits_{i=1}^{100} (3i + 2) = 5 + 8 + \cdots + 302$
$n = 100$; $a_1 = 5$; $a_{100} = 302$
$$A_{100} = \tfrac{100}{2}(5 + 302) = 50(307) = 15{,}350$$

d) $\sum\limits_{2}^{100} (2i - 3) = 1 + 3 + \cdots + 197$
$n = 99$; $a_1 = 1$; $a_{99} = 197$
$$A_{99} = \tfrac{99}{2}(1 + 197) = \tfrac{99}{2}(198) = 9{,}801$$

19. a) $3 \cdot 4 + 4 \cdot 5 + 5 \cdot 6 + \cdots + 99 \cdot 100 = \sum\limits_{i=3}^{99} i(i+1)$

b) $8 + 12 + 16 + \cdots + 80 = \sum\limits_{i=1}^{19}[8 + 4(i-1)]$
$$= \sum\limits_{i=1}^{19}(4 + 4i)$$
or, if you start with $i = 2$,
$$= \sum\limits_{2}^{20} 4i$$

c) $7 + 10 + 13 + \cdots + 91 = \sum\limits_{i=1}^{29}[7 + 3(i-1)]$
$$= \sum\limits_{i=1}^{29}(4 + 3i)$$
or, if you start with $i = 2$,
$$= \sum\limits_{i=2}^{30}(3i + 1)$$

21. $20, 19.25, 18.5, 17.75, \ldots$
a) $d = 19.25 - 20 = -.75$

b) $a_{51} = 20 + 50(-.75) = 20 - 37.5 = -17.5$

c) $A_{51} = \tfrac{51}{2}[20 + (-17.5)] = \tfrac{51}{2}(2.5) = 63.75$

23. $6 + 6.8 + 7.6 + \cdots + 37.2 + 38$
$d = 6.8 - 6 = .8$; $a_1 = 6$; $a_n = 38$
$$38 = 6 + (n-1)(.8)$$
$$32 = .8(n-1)$$
$$40 = n - 1$$
$$41 = n$$
$$A_{41} = \tfrac{41}{2}(6 + 38) = \tfrac{41}{2}(44) = 902$$

25. The terms of the sequence can be written as
$$a_1, a_1 + d, a_1 + 2d, a_1 + 3d, \ldots a_1 + (n-1)$$
$$A_5 = a_1 + (a_1 + d) + (a_1 + 2d) + (a_1 + 3d)$$
$$+ (a_1 + 5d)$$
$$= 5a_1 + 10d$$
$$A_{20} = a_1 + (a_1 + d) + (a_1 + 2d) + \cdots + (a_1 + 19d)$$
$$= 20a_1 + d(1 + 2 + \cdots + 19)$$
$$= 20a_1 + d \cdot \tfrac{19}{2}(1 + 19)$$
$$= 20a_1 + 190d$$

Since $A_5 = 50$ and $A_{20} = 650$,

$$\begin{cases} 5a_1 + 10d = 50 \\ 20a_1 + 190d = 650 \end{cases}$$
$$\begin{cases} a_1 + 2d = 10 \\ 2a_1 + 19d = 65 \end{cases}$$

Subtract twice the first equation from the second:
$$15d = 45, \ d = 3$$
$$a_1 + 2(3) = 10$$
$$a_1 = 4$$

With this information,
$$a_{15} = 4 + 14(3) = 46$$
$$A_{15} = \tfrac{15}{2}(4 + 46) = \tfrac{15}{2}(50) = 375$$

27. a) $\displaystyle\sum_{i=1}^{100}(2i+1) = 3 + 5 + \cdots + 201$

$n = 100; \ a_1 = 3; \ a_{100} = 201$

$$A_{100} = \tfrac{100}{2}(3 + 201) = 50(204) = 10{,}200$$

b) $\displaystyle\sum_{i=1}^{100}(-4i+2) = -2 + (-6) + \cdots + (-398)$

$n = 100; \ a_1 = -2, \ a_{100} = 398$

$$A_{100} = \tfrac{100}{2}[-2 + (-398)] = 50(-400) = -20{,}000$$

29. Person 1 shook hands with 299 people; person 2 shook hands with 298 additional people; person 3 shook hands with 297 additional people. The pattern continues until person 299 shook hands with only 1 remaining person.

The total number of handshakes is
$$A_n = 299 + 298 + 297 + \cdots + 1$$
$n = 299; \ a_1 = 299; \ a_{299} = 1$
$$A_n = \tfrac{299}{2}(299 + 1) = \tfrac{299}{2}(300) = 44{,}850 \text{ handshakes}$$

31. $A_n = 70 + 69 + 68 + \cdots + 10$

$n = 70 - 10 + 1 = 61; \ a_1 = 70; \ a_{61} = 10$
$$A_{61} = \tfrac{61}{2}(70 + 10) = \tfrac{61}{2}(80) = 2440 \text{ logs}$$

33. Let A_n represent the amount of money Jose will have after n years.
$$A_1 = 1000 + 1000(.095)$$
$$A_2 = A_1 + 1000(.095) = 1000 + 2(1000)(.095)$$
$$A_3 = A_2 + 1000(.095) = 1000 + 3(1000)(.095)$$
$$A_n = 1000 + n(1000)(.095)$$
$$A_{10} = 1000 + 10(1000)(.095) = \$1950$$

35. Let a_i represent the amount of money Ikeda pays each month, and A_n represent the sum of the payments for n months.
$$a_1 = 6000(.01) + 200$$
After the first payment, the loan balance is reduced to \$5800.
$$a_2 = 5800(.01) + 200$$
For the last payment, the loan balance has been reduced to \$200:
Let a_n represent the last payment.

$$a_n = .01(200) + 200$$
$$A_n = .01(6000 + 5800 + \cdots + 200) + 200n$$
Since $200n = 6000,$
$$n = 30$$
$$6000 + 5800 + \cdots + 200 = \tfrac{30}{2}(6000 + 2000) = 93{,}000$$
The sum of all the payments, then, is
$$A_{30} = .01(93{,}000) + 200(30) = 930 + 6000 = \$6930$$

37. Let $a_1 = \tfrac{1}{2}, \ a_2 = \tfrac{1}{3} + \tfrac{2}{3}, \ a_3 = \tfrac{1}{4} + \tfrac{2}{4} + \tfrac{3}{4}$, and so on.

Then $\quad a_n = \dfrac{1}{n+1} + \dfrac{2}{n+1} + \dfrac{3}{n+1} + \cdots + \dfrac{n}{n+1}$

$$= \tfrac{n}{2}\left(\dfrac{1}{n+1} + \dfrac{n}{n+1}\right) = \dfrac{n}{2}$$

$$a_n - a_{n-1} = \tfrac{n}{2} - \tfrac{n-1}{2} = \tfrac{1}{2}$$

Therefore, $a_1, \ a_2, \ a_3, \ \ldots, \ a_{99}$ is an arithmetic sequence with $d = \tfrac{1}{2}$. Therefore,

$$a_1 + a_2 + \cdots + a_{99} = \tfrac{99}{2}\left(\tfrac{1}{2} + \tfrac{99}{2}\right) = 99(25) = 2475$$

39. The number of revolutions in 18 minutes is $(33\tfrac{1}{3})(18) = 600$. The circumferences of the consecutive circles form (approximately) an arithmetic sequence starting with $2\pi \cdot 3$ inches and ending with $2\pi \cdot 6$ inches. Therefore, the total length of the playing groove is

$$\tfrac{600}{2}(6\pi + 12\pi) = (5400\pi \text{ inches}) \div 12 \approx 1414 \text{ feet}.$$

Problem Set 12.3

1. a) $\tfrac{1}{2}$, 1, 2, 4, $\boxed{8}$, $\boxed{16}$

b) 8, 4, 2, 1, $\boxed{\tfrac{1}{2}}$, $\boxed{\tfrac{1}{4}}$

c) .3, .03, .003, .0003, $\boxed{.00003}$, $\boxed{.000003}$

3. $r =$ the ratio of any 2 terms, $\dfrac{a_n}{a_{n-1}}$.

a) $r = \tfrac{2}{1} = 2; \ a_n = a_1 r^{n-1} = (\tfrac{1}{2})2^{n-1} = 2^{n-2}$

b) $r = \tfrac{4}{8} = \tfrac{1}{2}$

$$a_n = a_1 r^{n-1}$$
$$= (8)(\tfrac{1}{2})^{n-1}$$
$$= 2^3(\tfrac{1}{2})^{n-1}$$
$$= (\tfrac{1}{2})^{-3}(\tfrac{1}{2})^{n-1}$$
$$= (\tfrac{1}{2})^{n-4}$$

c) $r = \tfrac{.03}{.3} = .1$

$$a_n = a_1 r^{n-1}$$
$$= .3(.1)^{n-1}$$
$$= 3(.1)(.1)^{n-1}$$
$$= 3(.1)^n$$
$$= 3 \times 10^{-n}$$

5. a) $a_{30} = 2^{30-2} = 2^{28} = 268,435,456$

b) $a_{30} = (\frac{1}{2})^{30-4} = (\frac{1}{2})^{26} = 1.4 \times 10^{-8}$

c) $a_{30} = 3 \times 10^{-30}$

7. Use the formula $A_n = \dfrac{a_1(1 - r^n)}{1 - r}$

a) $A_5 = \dfrac{\frac{1}{2}(1 - 2^5)}{1 - 2} = \dfrac{\frac{1}{2}(1 - 32)}{-1} = \dfrac{-31}{2(-1)} = \dfrac{31}{2}$

b) $A_5 = \dfrac{8(1 - (\frac{1}{2})^5)}{1 - \frac{1}{2}} = \dfrac{8(1 - \frac{1}{32})}{\frac{1}{2}}$

$= \dfrac{8 \cdot \frac{31}{32}}{\frac{1}{2}} = \dfrac{8 \cdot 31}{32} \cdot \dfrac{2}{1} = \dfrac{31}{2}$

c) $A_5 = \dfrac{.3(1 - (\frac{1}{10})^5)}{1 - \frac{1}{10}} = \dfrac{.3(1 - .00001)}{.9}$

$= \dfrac{.3(.99999)}{.9} = .3(1.1111) = .33333$

9. Use the formula $A_n = \dfrac{a_1(1 - r^n)}{1 - r}$

a) $A_{30} = \dfrac{\frac{1}{2}(1 - 2^{30})}{1 - 2} = \dfrac{\frac{1}{2}(1 - 2^{30})}{-1}$

$= \dfrac{2^{30} - 1}{2} = 536,870,911.5$

b) $A_{30} = \dfrac{8(1 - (\frac{1}{2})^{30})}{1 - \frac{1}{2}} = \dfrac{8(1 - (\frac{1}{2})^{30})}{\frac{1}{2}}$

$= 16(1 - (\frac{1}{2})^{30}) = 16 - 2^4 \cdot \dfrac{1}{2^{30}}$

$= 16 - (\frac{1}{2})^{26}$

c) $A_{30} = \dfrac{.3(1 - (\frac{1}{10})^{30})}{1 - \frac{1}{10}}$

$= \dfrac{.3(1 - (\frac{1}{10})^{30})}{.9}$

$= \frac{1}{3}[1 - (.1)^{30}]$

11. $a_1 = 100 \cdot 2 =$ the number of bacteria present at the end of week 1. $a_2 = 100 \cdot 2^2 =$ the number of bacteria at the end of week 2.

The number of bacteria present at the end of each week form a geometric sequence with $a_1 = 100 \cdot 2$ and $r = 2$. We want to find the number of bacteria at the end of week 10.

$a_{10} = (100 \cdot 2)2^9 = 100 \cdot 2^{10} = 102,400$

13. Let a_i represent the amount Johnny earns on day i.

$a_1 = 1$
$a_2 = 1 \cdot 2$
$a_3 = 1 \cdot 2^2$
$\vdots$
$a_{31} = 1 \cdot 2^{30}$

Johnny's daily earnings form a geometric sequence with $a_1 = 1$ and $r = 2$.

$$A_{31} = \dfrac{1(1 - 2^{31})}{1 - 2} = 2^{31} - 1 = \$2,147,483,647$$

15. a) $\sum\limits_{i=1}^{\infty} (\frac{1}{3})^i = \frac{1}{3} + (\frac{1}{3})^2 + (\frac{1}{3})^3 + \cdots$

$a_1 = \frac{1}{3}, \; r = \frac{1}{3}. \; \sum\limits_{i=1}^{\infty}(\frac{1}{3})^i = \dfrac{\frac{1}{3}}{1 - \frac{1}{3}} = \dfrac{\frac{1}{3}}{\frac{2}{3}} = \dfrac{1}{2}$

b) $a_1 = (\frac{2}{5})^2 = \frac{4}{25}; \; r = \frac{2}{5}$

$\sum\limits_{i=2}^{\infty}(\frac{2}{5})^i = \dfrac{\frac{4}{25}}{1 - \frac{2}{5}} = \dfrac{\frac{4}{25}}{\frac{3}{5}} = \dfrac{4}{25} \cdot \dfrac{5}{3} = \dfrac{4}{15}$

17. Let D represent the total distance the ball travels before coming to rest.

$D = 10 + 5 + \frac{5}{2} + \frac{5}{4} + \cdots$

$\quad\quad + 5 + \frac{5}{2} + \frac{5}{4} + \cdots$

$\quad 10 + 2(5 + \frac{5}{2} + \frac{5}{4} + \cdots)$

$5 + \frac{5}{2} + \frac{5}{4} + \cdots$ is a geometric series with $a_1 = 5$ and $r = \frac{1}{2}$. $5 + \frac{5}{2} + \frac{5}{4} + \cdots = \dfrac{5}{1 - \frac{1}{2}} = \dfrac{5}{\frac{1}{2}} = 10$. Therefore,

$D = 10 + 2(10) = 30$ feet.

19. $.11\overline{1} = \frac{1}{10} + \frac{1}{100} + \frac{1}{1000} + \cdots = \frac{1}{10} + (\frac{1}{10})^2 + (\frac{1}{10})^3 + \cdots$

$a_1 = \frac{1}{10}; \; r = \frac{1}{10}$

$.11\overline{1} = \dfrac{\frac{1}{10}}{1 - \frac{1}{10}} = \dfrac{\frac{1}{10}}{\frac{9}{10}} = \dfrac{1}{9}$

21. $.2525\overline{25} = \frac{25}{100} + \frac{25}{10000} + \frac{25}{1000000} + \cdots$

$= 25\left(\frac{1}{100}\right) + 25\left(\frac{1}{100}\right)^2 + 25\left(\frac{1}{100}\right)^3$

$a_1 = \frac{25}{100}; \; r = \frac{1}{100}$

$.2525\overline{25} = \dfrac{\frac{25}{100}}{1 - \frac{1}{100}} = \dfrac{\frac{25}{100}}{\frac{99}{100}} = \dfrac{25}{99}$

23. $1.23\overline{34} = \frac{12}{10} + \frac{34}{1000} + \frac{34}{10000} + \cdots$

$$= \frac{12}{10} + \frac{34}{1000} + \frac{34}{1000}\left(\frac{1}{10}\right)^2 + \cdots$$

With $a_1 = \frac{34}{1000}$ and $r = \left(\frac{1}{10}\right)^2$

$$1.23\overline{34} = \frac{12}{10} + \frac{\frac{34}{1000}}{1 - \frac{1}{100}} = \frac{12}{10} + \frac{\frac{34}{1000}}{\frac{99}{100}}$$

$$= \frac{12}{10} + \frac{34}{1000} \cdot \frac{100}{99} = \frac{12}{10} + \frac{34}{990}$$

$$= \frac{12(99) + 34}{990} = \frac{1222}{990}$$

$$= \frac{611}{495}$$

25. $a_n = 625(0.2)^{n-1}$
$a_1 = 625(0.2)^0 = 625 \cdot 1 = 625$
$a_2 = 625(0.2)^1 = 625(0.2) = 125$
$a_3 = 625(0.2)^2 = 625(.04) = 25$
$a_4 = 625(0.2)^3 = 625(.008) = 5$
$a_5 = 625(0.2)^4 = 625(.0016) = 1$
More easily, if one realizes that $(0.2) = \frac{1}{5}$, one can obtain each number in the sequence by multiplying the previous number by $\frac{1}{5}$.

27. a) A geometric sequence with $a_1 = 130$ and $r = \frac{1}{2}$.

$a_n = 130(\frac{1}{2})^{n-1} = 65 \cdot 2(\frac{1}{2})^{n-1} = 65(\frac{1}{2})^{n-2}$.

b) Neither an arithmetic nor a geometric sequence.

c) An arithmetic sequence with $a_1 = 105$ and $d = 2$.
$a_n = 105 + 2(n-1) = 103 + 2n$

d) A geometric sequence with $a_1 = 100(1.05)$ and $r = 1.05$.
$a_n = 100(1.05)(1.05)^{n-1} = 100(1.05)^n$

e) Neither an arithmetic nor a geometric sequence.

f) A geometric sequence with $a_1 = 3$ and $r = -2$.
$a_n = 3(-2)^{n-1}$

29. After 10 years, $100 will be worth $100(1.08)^{10}$
$= \$215.89$. We want to know the smallest value of n for which
$$100(1.08)^n \geq 250$$
$$(1.08)^n \geq 2.5$$
Using logarithms,
$$n \log 1.08 \geq \log 2.5$$
$$n \geq \frac{\log 2.5}{\log 1.08}$$
$$n \geq 11.91$$
The amount will first exceed $250 after 12 years.

31. $A_{10} = 100(1.08)^{10} + 100(1.08)^9 + \cdots + 100(1.08)^1$
For simplicity, we add in reverse order and use the formula for the sum of a geometric series with $a_1 = 100(1.08)$ and $r = 1.08$:

$$A_{10} = \frac{100(1.08)[1 - (1.08)^{10}]}{1 - 1.08}$$

$$= \frac{100(1.08)[1 - 2.1589]}{-.08}$$

$$= \frac{108(-1.1589)}{-.08}$$

$$= \$1564.52$$

33. Total increase in spending (in billions)
$= 1(.75) + 1(.75)^2 + 1(.75)^3 + \cdots$. This is an infinite geometric series with $a_1 = .75$ and $r = .75$.

Total increase in spending $= \frac{.75}{1 - .75} = \frac{.75}{.25}$
$= 3$ billion dollars

35.

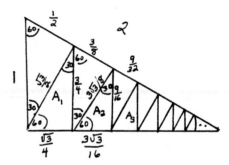

If the side opposite the 30° angle of a 30°-60°-90° triangle is of length x, the hypotenuse is of length $2x$, and the side opposite the 60° angle is of length $\sqrt{3}x$. We can calculate the lengths of the sides of the first few triangles, noticing the pattern. Each leg of triangle A_n is $\frac{3}{4}$ the length of the corresponding leg of triangle A_{n-1}. Since the area of a right triangle is $\frac{1}{2}$ the product of the two legs, the area of triangle A_n is $\frac{3}{4} \cdot \frac{3}{4}$(Area of A_{n-1}), or $\frac{9}{16}$ of the area of the preceding triangle. The sum of all the areas is

$$A_1 + A_2 + A_3 + \cdots$$

$$= \frac{3\sqrt{3}}{32} + \frac{9}{16} \cdot \frac{3\sqrt{3}}{32} + \left(\frac{9}{16}\right)^2 \frac{3\sqrt{3}}{32} + \cdots$$

This is a geometric series with $a_1 = \frac{3\sqrt{3}}{2}$ and $r = \frac{9}{16}$.

The sum of all the areas is

$$\frac{\frac{3\sqrt{3}}{32}}{1 - \frac{9}{16}} = \frac{\frac{3\sqrt{3}}{32}}{\frac{7}{16}} = \frac{3\sqrt{3}}{32} \cdot \frac{16}{7} = \frac{3\sqrt{3}}{14}$$

37. $P = (1+3)(1+3^2)(1+3^4)(1+3^8)(1+3^{16})$

$= (1+3+3^2+3^3)(1+3^4+3^8+3^{12})(1+3^{16})$

$= (1+3+3^2+3^3+3^4+3^5+3^6+3^7+3^8+3^9$
$\quad + 3^{10}+3^{11}+3^{12}+3^{13}+3^{14}+3^{15})(1+3^{16})$

$= 1+3+3^2+3^3+3^4+3^5+\cdots+3^{29}+3^{30}+3^{31}$

With $a_1 = 1$, $r = 3$, and $n = 32$,

$P = \dfrac{1(1-3^{32})}{1-3} = \dfrac{3^{32}-1}{2} \approx 9.2651 \times 10^{14}$

39. Since $\quad a + ar = 5$,
$\qquad a(1+r) = 5$

$\qquad\qquad a = \dfrac{5}{1+r}$

Since the sum of the first 6 terms is 65,

$65 = a + ar + ar^2 + \cdots + ar^5$

$\quad = \dfrac{a(1-r^6)}{1-r}$

$\quad = \dfrac{5}{1+r} \cdot \dfrac{(1-r^6)}{1-r}$

$\quad = \dfrac{5(1-r^6)}{1-r^2}$

$\quad = \dfrac{5(r^6-1)}{r^2-1}$

$\quad = \dfrac{5(r^2-1)(r^4+r^2+1)}{r^2-1}$

$\quad = 5(r^4+r^2+1)$

So, $\quad 5(r^4+r^2+1) = 65$
$\qquad\qquad r^4+r^2+1 = 13$
$\qquad\qquad r^4+r^2-12 = 0$
$\qquad (r^2+4)(r^2-3) = 0$
$\qquad\qquad r^2 = -4 \text{ or } r^2 = 3$

Since we are not interested in imaginary numbers, we will say $r^2 = 3$.

$A_4 = a + ar + ar^2 + ar^3$

$\quad = a(1+r+r^2+r^3)$

$\quad = \dfrac{5}{1+r}[(1+r)+r^2(1+r)]$

$\quad = 5 + 5r^2$

$\quad = 5 + 5 \cdot 3 = 20$

41. Tom and Joel are closing the distance between them at 20 miles per hour, so it will take 5 hours of cycling. This means that Corky ran for 5 hours, covering $5 \times 25 = 125$ miles.

Problem Set 12.4

1. P_n: $1+2+3+\cdots+n = \dfrac{n(n+1)}{2}$

P_1: $\dfrac{1(1+1)}{2} = \dfrac{1 \cdot 2}{2} = 1$; P_1 is true

P_k: $1+2+3+\cdots+k = \dfrac{k(k+1)}{2}$

P_{k+1}: $1+2+3+\cdots+k+(k+1)$
$\qquad = \dfrac{k(k+1)}{2}+k+1$

$\qquad = \dfrac{k(k+1)}{2}+\dfrac{2(k+1)}{2}$

$\qquad = \dfrac{k+1}{2}(k+2)$

So $P_k \Rightarrow P_{k+1}$. Hence by mathematical induction the statement is true for all positive integers n.

3. P_n: $3+7+11+\cdots+(4n-1) = n(2n+1)$
P_1: $1(2 \cdot 1+1) = 1(3) = 3$; P_1 is true
P_k: $3+7+11+\cdots+(4k-1) = k(2k+1)$
P_{k+1}: $3+7+11+\cdots+(4k-1)+[4(k+1)-1]$
$\qquad = k(2k+1)+[4(k+1)-1]$
$\qquad = 2k^2+k+4k+3$
$\qquad = 2k^2+5k+3$
$\qquad = (k+1)(2k+3)$
$\qquad = (k+1)[2(k+1)+1]$
So $P_k \Rightarrow P_{k+1}$. Hence by mathematical induction the statement is true for all positive integers n.

5. P_n:
$1 \cdot 2 + 2 \cdot 3 + 3 \cdot 4 + \cdots + n(n+1) = \frac{1}{3}n(n+1)(n+2)$

P_1: $\frac{1}{3} \cdot 1(1+1)(1+2) = \frac{1}{3}(2)(3) = 2$; P_1 is true

P_k:
$1 \cdot 2 + 2 \cdot 3 + 3 \cdot 4 + \cdots + k(k+1) = \frac{1}{3}k(k+1)(k+2)$

P_{k+1}
$= 1 \cdot 2 + 2 \cdot 3 + 3 \cdot 4 + \cdots + k(k+1) + (k+1)(k+2)$
$= \frac{1}{3}k(k+1)(k+2)+(k+1)(k+2)$

$= (k+1)(k+2)(\frac{1}{3}k+1)$

$= (k+1)(k+2)\frac{1}{3}(k+3)$

$= \frac{1}{3}(k+1)(k+2)(k+3)$
So $P_k \Rightarrow P_{k+1}$. Hence by mathematical induction the statement is true for all positive integers n.

7. P_n: $2+2^2+2^3+\cdots+2^n = 2(2^n-1)$
P_1: $2(2^1-1) = 2(1) = 2$; P_1 is true
P_k: $2+2^2+2^3+\cdots+2^k = 2(2^k-1)$
P_{k+1}: $2+2^2+2^3+\cdots+2^k+2^{k+1}$
$\qquad = 2^{k+1}-2+2^{k+1} = 2 \cdot 2^{k+1}-2$
$\qquad = 2(2^{k+1}-1)$
So $P_k \Rightarrow P_{k+1}$. Hence by mathematical induction the statement is true for all positive integers n.

9. P_n is true for every integer $n \geq 8$.

11. We can only say that P_1 is true.

13. P_1, P_3, P_5, and in general P_n is true, for n an odd integer.

15. P_n is true for all positive integers n.

17. P_1, P_4, P_{16}, and in general P_n is true for $n = 4^k$, $k \geq 0$.

19. $n + 5 < 2^n$ is not true for $n = 1$, 2, or 3. It is true for $n = 4$.
P_k: $k + 5 < 2^k$
P_{k+1}: $\quad (k+1) + 5 = (k+5) + 1$
$$< 2^k + 1$$
$$< 2^k + 2^k$$
$$= 2^k(2)$$
$$= 2^{k+1}$$
So $P_k \Rightarrow P_{k+1}$. Hence by mathematical induction, P_n is true for $n \geq 4$.

21. $\log_{10} n < n$ is true for $n = 1$
P_k: $\log_{10} k < k$
P_{k+1}: $\quad \log_{10}(k+1) < \log_{10} 10k$
$$= \log_{10} k + \log_{10} 10$$
$$= \log_{10} k + 1$$
$$< k + 1$$
So $P_k \Rightarrow P_{k+1}$. Hence by mathematical induction, P_n is true for $n \geq 1$.

23. $(1 + x)^n \geq 1 + nx$ for $n = 1$
P_k: $(1 + x)^k \geq 1 + kx$
P_{k+1}: $\quad (1 + x)^{k+1} = (1 + x)^k (1 + x)$
$$\geq (1 + kx)(1 + x)$$
$$= 1 + kx + x + kx^2$$
$$= 1 + (k+1)x + kx^2$$
$$\geq 1 + (k+1)x$$
So $P_k \Rightarrow P_{k+1}$. Hence by mathematical induction, P_n is true for $n \geq 1$.

25. $x + y$ is a factor of $x^2 - y^2$, so the statement is true for $n = 1$. Assume the statement is true for k. That is, there is some polynomial $Q(x, y)$ such that
$$x^{2k} - y^{2k} = (x + y)Q(x, y)$$
For $n = k + 1$,
$$x^{2(k+1)} - y^{2(k+1)} = x^{2k+2} - y^{2k+2}$$
$$= x^{2k+2} - x^{2k}y^2 + x^{2k}y^2 - y^{2k+2}$$
$$= x^{2k}(x^2 - y^2) + y^2(x^{2k} - y^{2k})$$
$$= x^{2k}(x + y)(x - y) + y^2(x + y)Q(x, y)$$
$$= (x + y)[x^{2k}(x - y) + y^2 Q(x, y)]$$
Let $P(x, y)$ be the polynomial
$x^{2k}(x - y) + y^2 Q(x, y)$. Then
$$x^{2(k+1)} - y^{2(k+1)} = (x + y)P(x, y)$$

That is, $x + y$ is a factor of $x^{2(k+1)} - y^{2(k+1)}$ if it is a factor of $x^{2k} - y^{2k}$. Hence, by mathematical induction, $x + y$ is a factor of $x^{2n} - y^{2n}$ for $n \geq 1$.

27. The statement is true for $n = 1$, since $(1)^2 - 1 = 0$, which is even. Assume the statement is true for $n = k$. That is, for $n = k$, $k^2 - k = 2q$ for some integer q.
$$(k+1)^2 - (k+1) = k^2 + 2k + 1 - k - 1$$
$$= k^2 - k + 2k$$
$$= 2q + 2k$$
$$= 2(q + k)$$
Since $q + k$ is an integer, $(k+1)^2 - (k+1)$ is even. Hence by mathematical induction, $n^2 - n$ is even for $n \geq 1$.

29. a) The formula is true for $n = 1$, since $\frac{1}{2}(1)(1+1) = 1 = 1^2$. Assume the formula is true for $n = k$; that is $1 + 2 + 3 + \cdots + k = \frac{1}{2}k(k+1)$.
For $n = k + 1$,
$$1 + 2 + 3 + \cdots + k + (k+1) = \frac{1}{2}k(k+1) + (k+1)$$
$$= (k+1)(\tfrac{1}{2}k + 1)$$
$$= (k+1)\tfrac{1}{2}(k+2)$$
$$= \tfrac{1}{2}(k+1)(k+2)$$
Thus if the formula is true for $n = k$, it is also true for $n = k + 1$. By mathematical induction, it is true for all $n \geq 1$.

b) See proof in textbook.

c) The formula is true for $n = 1$ since
$\frac{1}{4}(1)^2(1+1)^2 = \frac{1}{4} \cdot 4 = 1 = 1^3$.
Assume the formula is true for $n = k$; that is,
$1^3 + 2^3 + 3^3 + \cdots + k^3 = \frac{1}{4}k^2(k+1)^2$.

For $n = k + 1$,
$$1^3 + 2^3 + \cdots + k^3 + (k+1)^3 = \tfrac{1}{4}k^2(k+1)^2 + (k+1)^3$$
$$= \tfrac{1}{4}(k+1)^2[k^2 + 4(k+1)]$$
$$= \tfrac{1}{4}(k+1)^2(k^2 + 4k + 4)$$
$$= \tfrac{1}{4}(k+1)^2(k+2)^2$$
Thus the formula is true for $n = k + 1$ if it is true for $n = k$. Hence by mathematical induction, the formula is true for all $n \geq 1$.

d) The formula is true for $n = 1$ since
$$\tfrac{1}{30}(1)(1+1)(6 \cdot 1 + 9 \cdot 1 + 1 - 1) = \tfrac{1}{30}(2)(15) = 1 = 1^4$$
Assume the formula is true for $n = k$; that is,
$$1^4 + 2^4 + 3^4 + \cdots + k^4$$
$$= \tfrac{1}{30}k(k+1)(6k^3 + 9k^2 + k - 1)$$

reamginfineream

For $n = k+1$,
$$1^4 + 2^4 + 3^4 + \cdots + k^4 + (k+1)^4$$
$$= \tfrac{1}{30}k(k+1)(6k^3 + 9k^2 + k - 1) + (k+1)^4$$
$$= \tfrac{1}{30}(k+1)[k(6k^3 + 9k^2 + k - 1) + 30(k+1)^3]$$
$$= \tfrac{1}{30}(k+1)[6k^4 + 9k^3 + k^2 - k + 30k^3 + 90k^2$$
$$+ 90k + 30]$$
$$= \tfrac{1}{30}(k+1)[6k^4 + 39k^3 + 91k^2 + 89k + 30]$$

Use synthetic division to show that $k+2$ is a factor of the 4th degree polynomial:

$$\begin{array}{r|rrrrr} -2 & 6 & 39 & 91 & 89 & 30 \\ & & -12 & -54 & -74 & -30 \\ \hline & 6 & 27 & 37 & 15 & \underline{}\,0 \end{array}$$

So, $1^4 + 2^4 + \cdots + k^4 + (k+1)^4$
$$= \tfrac{1}{30}(k+1)(k+2)(6k^3 + 27k^2 + 37k + 15)$$
$$= \tfrac{1}{30}(k+1)(k+2)[(6k^3 + 18k^2 + 18k + 6)$$
$$+ (9k^2 + 18k + 9) + (k+1) - 1]$$
$$= \tfrac{1}{30}(k+1)(k+2)[6(k+1)^3 + 9(k+1)^2$$
$$+ (k+1) - 1]$$

Thus the formula is true for $n = k+1$ if it is true for $n = k$. Hence by mathematical induction, the formula is true for all $n \geq 1$.

31. a) $\displaystyle\sum_{k=1}^{100}(3k+1)$
$$= (3 \cdot 1 + 1) + (3 \cdot 2 + 1) + (3 \cdot 3 + 1) + \cdots$$
$$+ (3 \cdot 100 + 1)$$
$$= 3 \cdot 1 + 3 \cdot 2 + 3 \cdot 3 + \cdots + 3 \cdot 100 + 1(100)$$
$$= 3(1 + 2 + \cdots + 100) + 1(100)$$
$$= 3(\tfrac{1}{2})(100)(101) + 100$$
$$= 3(50)(101) + 100$$
$$= 15{,}250$$

b) $\displaystyle\sum_{k=1}^{10}(k^2 - 3k)$
$$= [1^2 - 3(1)] + [2^2 - 3(2)] + \cdots + [10^2 - 3(10)]$$
$$= 1^2 + 2^2 + \cdots + 10^2 - 3(1 + 2 + \cdots + 10)$$
$$= \tfrac{1}{6}(10)(11)(21) - 3(\tfrac{1}{2})(10)(11)$$
$$= 5(11)(7) - 3(5)(11)$$
$$= 5(11)(7 - 3)$$
$$= 5(44)$$
$$= 220$$

c) $\displaystyle\sum_{k=1}^{10}(k^3 + 3k^2 + 3k + 1)$
$$= \sum_{k=1}^{10}k^3 + 3\sum_{k=1}^{10}k^2 + 3\sum_{k=1}^{10}k + \sum_{k=1}^{10}1$$
$$= \tfrac{1}{4}(10)^2(11)^2 + 3(\tfrac{1}{6})(10)(11)(21)$$
$$+ 3(\tfrac{1}{2})(10)(11) + 10$$
$$= 3025 + 1155 + 165 + 10$$
$$= 4355$$

d) $\displaystyle\sum_{k=1}^{n}(6k^2 + 2k) = 6\sum_{k=1}^{n}k^2 + 2\sum_{k=1}^{n}k$
$$= 6(\tfrac{1}{6})(n)(n+1)(2n+1) + 2(\tfrac{1}{2})(n)(n+1)$$
$$= n(n+1)(2n+1+1)$$
$$= n(n+1)(2n+2)$$
$$= 2n(n+1)^2$$

33. The formula is true for $n = 2$ since
$$\frac{2+1}{2(2)} = \frac{3}{4} = 1 - \frac{1}{4} = 1 - \frac{1}{2^2}$$
Assume the formula is true for $n = k$; that is,
$$\left(1 - \tfrac{1}{4}\right)\left(1 - \tfrac{1}{9}\right)\left(1 - \tfrac{1}{16}\right)\cdots\left(1 - \tfrac{1}{k^2}\right) = \frac{k+1}{2k}$$
For $n = k+1$,
$$\left(1 - \tfrac{1}{4}\right)\left(1 - \tfrac{1}{9}\right)\left(1 - \tfrac{1}{16}\right)\cdots\left(1 - \tfrac{1}{k^2}\right)\left(1 - \frac{1}{(k+1)^2}\right)$$
$$= \left(\frac{k+1}{2k}\right)\left(1 - \frac{1}{(k+1)^2}\right)$$
$$= \frac{k+1}{2k} \cdot \frac{(k+1)^2 - 1}{(k+1)^2}$$
$$= \frac{(k+1)^2 - 1}{2k(k+1)} = \frac{k^2 + 2k + 1 - 1}{2k(k+1)}$$
$$= \frac{k^2 + 2k}{2k(k+1)} = \frac{k(k+2)}{2k(k+1)}$$
$$= \frac{k+2}{2(k+1)} = \frac{(k+1)+1}{2(k+1)}$$
Thus, the formula is true for $n = k+1$ if it is true for $n = k$. Hence, by mathematical induction, the formula is true for $n \geq 2$.

35. The inequality is true for $n = 3$, since
$$\frac{1}{3+1} + \frac{1}{3+2} + \frac{1}{3(2)} = \frac{1}{4} + \frac{1}{5} + \frac{1}{6} = \frac{15 + 12 + 10}{60}$$
$$= \frac{37}{60} > \frac{3}{5}.$$ Assume the inequality is true for $n = k$;
that is, $\dfrac{1}{k+1} + \dfrac{1}{k+2} + \cdots + \dfrac{1}{2k} > \dfrac{3}{5}$.
For $n = k+1$, the left side of the inequality is
$$\frac{1}{k+2} + \frac{1}{k+3} + \cdots + \frac{1}{2k} + \frac{1}{2k+1} + \frac{1}{2k+2}$$
$$= \left(\frac{1}{k+1} + \frac{1}{k+2} + \frac{1}{k+3} + \cdots + \frac{1}{2k}\right)$$
$$- \frac{1}{k+1} + \frac{1}{2k+1} + \frac{1}{2(k+1)}$$
$$= \left(\frac{1}{k+1} + \frac{1}{k+2} + \frac{1}{k+3} + \cdots + \frac{1}{2k}\right)$$
$$+ \frac{-2(2k+1) + (2k+2) + (2k+1)}{2(k+1)(2k+1)}$$
$$= \left(\frac{1}{k+1} + \frac{1}{k+2} + \frac{1}{k+3} + \cdots + \frac{1}{2k}\right)$$
$$+ \frac{1}{2(k+1)(2k+1)}$$
$$> \frac{1}{k+1} + \frac{1}{k+2} + \frac{1}{k+3} + \cdots + \frac{1}{2k} > \frac{3}{5}$$

Thus the inequality is true for $n = k + 1$ if it is true for $n = k$. By mathematical induction, the inequality is true for all $n \geq 3$.

37. The statement is true when $n = 3$ since it then says the sum of the interior angles of a triangle is $180°$. Assume that the statement is true for any k-sided convex polygon. Any $(k+1)$-sided convex polygon can be dissected into a k-sided polygon and a triangle. Its angles add up to
$$(k-2)(180°) + 180° = [(k+1) - 2]180$$
Thus if the statement is true for a k-sided polygon it is true for a $(k+1)$-sided polygon. By mathematical induction it is true for any n-sided polygon, $n \geq 3$.
Each exterior angle of a polygon is ($180°$ – the interior angle). The sum of the exterior angles is ($n \cdot 180$ – the sum of the interior angles)
$$= (n)180° - (n-2)180° = 180°[n - n - (-2)]$$
$$= (180°)(2) = 360°$$

39. Note that adding a line to n lines creates n new intersection points, and thus $n + 1$ additional regions, so $R_{n+1} = R_n + n + 1$. Now use induction. The statement is true for for $n = 1$. With n lines, there will be 2 regions, and $\frac{1^2 + 1 + 2}{2} = \frac{4}{2} = 2$.

Assume the statement is true for $n = k$ lines; that is, $R_k = \frac{k^2 + k + 2}{2}$. For $k + 1$ lines we have
$$R_{k+1} = R_k + k + 1 = \frac{k^2 + k + 2}{2} + k + 1$$
$$= \frac{k^2 + k + 2 + 2k + 2}{2}$$
$$= \frac{(k^2 + 2k + 1) + (k + 1) + 2}{2}$$
$$= \frac{(k+1)^2 + (k+1) + 2}{2}$$

Thus if the formula is true for $n = k$, it is true for $n = k + 1$. By induction, it is true for all $n \geq 1$.

41. For $n = 1$, the proposition read $f_1^2 = f_1 f_2$. This is true because $f_1 = f_2 = 1$. For $n = 2$, it reads $f_1^2 + f_2^2 = f_2 \cdot f_3$. This is true because $1^2 + 1^2 = 1 \cdot 2$. Now suppose that the proposition is true for $n = k$; that is $f_1^2 + f_2^2 + \cdots + f_k^2 = f_k \cdot f_{k+1}$. We must show that it follows that
$$f_1^2 + f_2^2 + \cdots + f_k^2 + f_{k+1}^2 = f_{k+1} f_{k+2}.$$
$$(f_1^2 + f_2^2 + \cdots + f_k^2) + f_{k+1}^2 = f_k \cdot f_{k+1} + f_{k+1}^2$$
$$= f_{k+1}(f_k + f_{k+1})$$
$$= f_{k+1} f_{k+2}$$

43. Let P_n be the proposition $a_n = \frac{2}{3}[1 - (-\frac{1}{2})^n]$, $n \geq 0$. For $n = 0, 1, k, k+1$, and $k+2$ we have

P_0: $a_0 = \frac{2}{3}[1 - (-\frac{1}{2})^0] = \frac{2}{3}(1 - 1) = 0$ This is true.

P_1: $a_1 = \frac{2}{3}[1 - (-\frac{1}{2})^1] = \frac{2}{3}(\frac{3}{2}) = 1$ This is true.

P_k: $a_k = \frac{2}{3}[1 - (-\frac{1}{2})^k]$; P_{k+1}: $a_k = \frac{2}{3}[1 - (-\frac{1}{2})^{k+1}]$

P_{k+2}: $a_{k+2} = \frac{2}{3}[1 - (-\frac{1}{2})^{k+2}]$
Assume that P_k and P_{k+1} are both true. We must show that P_{k+2} is then true.
$$a_{k+2} = \frac{a_{k+1} + a_k}{2} = \frac{\frac{2}{3}[2 - (-\frac{1}{2})^{k+1} - (-\frac{1}{2})^k]}{2}$$
$$= \frac{2}{3}[1 + (-\frac{1}{2})^{k+2} + (-\frac{1}{2})^{k+1}]$$
$$= \frac{2}{3}[1 + (-\frac{1}{2})^{k+1}(-\frac{1}{2} + 1)]$$
$$= \frac{2}{3}[1 - (-\frac{1}{2})^{k+1}(-\frac{1}{2})]$$
$$= \frac{2}{3}(1 - (-\frac{1}{2})^{k+2})$$

Therefore, P_k and $P_{k+1} \Rightarrow P_{k+2}$.

Problem Set 12.5

1. a) $3! = 3 \cdot 2 \cdot 1 = 6$

    b) $(3!)(2!) = 3 \cdot 2 \cdot 1 \cdot 2 \cdot 1 = 12$

    c) $\frac{10!}{8!} = \frac{10 \cdot 9 \cdot 8 \cdot 7 \cdot 6 \cdot 5 \cdot 4 \cdot 3 \cdot 2 \cdot 1}{8 \cdot 7 \cdot 6 \cdot 5 \cdot 4 \cdot 3 \cdot 2 \cdot 1} = 90$

3. a) $_5P_2 = 5 \cdot 4 = 20$ **b)** $_9P_4 = 9 \cdot 8 \cdot 7 \cdot 6 = 3024$

    c) $_{10}P_3 = 10 \cdot 9 \cdot 8 = 720$

5. The president can be chosen in 6 ways; the secretary can be chosen in 5 ways. $6 \cdot 5 = 30$.

7. a) There are 6 even numbers from 1 to 12 inclusive, so there 6 ways of choosing just one of them.

    b) There are 3 numbers greater than 9 (10, 11 or 12) and 2 numbers less than 3 (1 and 2), so there are 5 ways of picking just one of those.

9. a) Since there are 6 even numbers in the box, there are 6 ways to draw an even number the first time; if the card is replaced, there will be 6 ways to draw an even number the second time. There are $6 \cdot 6 = 36$ ways in which both numbers drawn will be even.

    b) There are 36 ways for both numbers drawn to be even, and by the same argument, 36 ways for both numbers to be odd. There are $36 + 36 = 72$ ways for both numbers to be even or odd.

    c) There are 3 ways in which the first number can be greater than 9; there are 2 ways in which the second number can be less than 3. There are $3 \cdot 2 = 6$ ways in which the first number will be greater than 9 and the second one less than 3.

11. $_8P_4 = 8 \cdot 7 \cdot 6 \cdot 5 = 1680$ words

13. a) There are 5 ways of choosing the road for the going trip, and 5 ways of choosing the return route. So, there are $5 \cdot 5 = 25$ different round trips.

b) There are 5 ways of making the first part of the trip, but only 4 ways left to return if he is to avoid returning on the same road, or $5 \cdot 4 = 20$ routes altogether.

15. There are 3 choices for salad, 20 choices for pizza, and 4 choices for dessert. So there are $3 \cdot 20 \cdot 4 = 240$ different meals.

17. There are $_5P_3 = 5 \cdot 4 \cdot 3 = 60$ 3-letter words, $_5P_4 = 5 \cdot 4 \cdot 3 \cdot 2 = 120$ 4-letter words, and $_5P_5 = 5 \cdot 4 \cdot 3 \cdot 2 \cdot 1 = 120$ 5-letter words. Thus, the number of 3 or 4 or 5-letter code words that can be formed is $60 + 120 + 120 = 300$.

19. There are 10 ways to make a one-dip cone; there are $10 \cdot 10 = 100$ ways to make a 2-dip cone; and there are $10 \cdot 10 \cdot 10 = 1000$ ways to make a 3-dip cone. Thus there are $10 + 100 + 1000 = 1110$ ways to make different-looking cones.

21. a) Letters T and N are fixed; the only choice is which of the other 6 is picked for the second place, and which of the 5 remaining is selected for the third place. So there are $6 \cdot 5 = 30$ ways of obtaining the 4-letter code word.

b) There are 3 ways of selecting the first letter, a consonant, and 2 ways of selecting the last letter consonant. Then there are 6 ways of selecting the second letter and 5 ways of selecting the third letter; these may be either vowels or consonants. Thus there are $3 \cdot 2 \cdot 6 \cdot 5 = 180$ ways of forming the code word.

c) If the code word is to have vowels only, there are 5 ways of selecting the first, 4 ways for the second, and 3 and 2 ways of selecting the last two letters, or $5 \cdot 4 \cdot 3 \cdot 2 = 120$ code words.

d) Since there are only three consonants, they must all be present in each code word. The fourth letter must be a vowel and there are 5 of them. So there are 5 different sets of letters. Each set of 4 letters may be arranged in 4! ways, so the total number of code words is $5 \cdot 4! = 120$ words.

e) The vowels must be in the 4th position, the third and fourth, the second, third, and fourth, or all four positions. Since these situations are independent, the number of code words will be the sum of the individual code words in each case. Since there are only 3 consonants, no words can be formed with no vowels.

(1) Vowel in the fourth position:
$N_1 = 5 \cdot 3 \cdot 2 \cdot 1 = 30$
(2) Vowels in third and fourth positions:
$N_2 = 5 \cdot 4 \cdot 3 \cdot 2 = 120$
(3) Vowels in second, third and fourth positions:
$N_3 = 5 \cdot 4 \cdot 3 \cdot 3 = 180$
(4) Vowels in all four positions:
$N_4 = 5 \cdot 4 \cdot 3 \cdot 2 = 120$

Total number of code words
$= N_1 + N_2 + N_3 + N_4$
$= 30 + 120 + 180 + 120 = 450$

23. a) $_{12}P_9 = 12 \cdot 11 \cdot 10 \cdot 9 \cdot 8 \cdot 7 \cdot 6 \cdot 5 \cdot 4 = 79{,}833{,}600$

b) There are 2 choices for pitcher; the remaining 8 positions on the team can be filled by any of 10 remaining people. The number of ways a team can be formed is $2 \cdot 10 \cdot 9 \cdot 8 \cdot 7 \cdot 6 \cdot 5 \cdot 4 \cdot 3 = 3{,}628{,}800$.

c) There are 2 choices for pitcher; then there will still be 11 people who can fill the remaining 8 positions. The number of ways a team can be formed is $2 \cdot 11 \cdot 10 \cdot 9 \cdot 8 \cdot 7 \cdot 6 \cdot 5 \cdot 4 = 13{,}305{,}600$.

25. $\dfrac{7!}{3!\,2!\,2!} = \dfrac{7 \cdot 6 \cdot 5 \cdot 4 \cdot 3 \cdot 2}{3 \cdot 2 \cdot 2 \cdot 2} = 210$

27. MISSISSIPPI has 4 I's, 4 S's, and 2 P's. Hence the number of 11-letter code words is
$\dfrac{11!}{4!\,4!\,2!} = \dfrac{11 \cdot 10 \cdot 9 \cdot 8 \cdot 7 \cdot 6 \cdot 5 \cdot 4 \cdot 3 \cdot 2 \cdot 1}{4 \cdot 3 \cdot 2 \cdot 2 \cdot 4 \cdot 3 \cdot 2 \cdot 1 \cdot 2 \cdot 1}$
$= 11 \cdot 10 \cdot 9 \cdot 7 \cdot 5 = 34{,}650$

29. $a^3 b c^6$ will expand to have 3 a's, 1 b, and 6 c's. Those letters can be arranged in
$\dfrac{10!}{3!\,6!} = \dfrac{10 \cdot 9 \cdot 8 \cdot 7}{3 \cdot 2} = 840$ ways

31. Each shortest route from A to B must include 4 steps East and 5 steps North. The problem is the same as asking how many unique arrangements there are of the letters NNNNNEEEE. There are
$\dfrac{9!}{5!\,4!} = \dfrac{9 \cdot 8 \cdot 7 \cdot 6}{4 \cdot 3 \cdot 2} = 126$ arrangements
Thus there are 126 shortest routes.

33. a) $\dfrac{11!}{8!} = \dfrac{11 \cdot 10 \cdot 9 \cdot 8!}{8!} = 11 \cdot 10 \cdot 9 = 990$

b) $\dfrac{11!}{8!\,3!} = \dfrac{11 \cdot 10 \cdot 9 \cdot 8!}{8!\,3 \cdot 2 \cdot 1} = 165$

c) $11! - 8! = 11 \cdot 10 \cdot 9 \cdot 8! - 8!$
$= 8!(11 \cdot 10 \cdot 9 - 1)$
$= 40{,}320(990 - 1)$
$= 40{,}320(989)$
$= 39{,}876{,}480$

275

d) $\dfrac{8!}{2^6} = \dfrac{8 \cdot 7 \cdot 6 \cdot 5 \cdot 4 \cdot 3 \cdot 2 \cdot 1}{2 \cdot 2 \cdot 2 \cdot 2 \cdot 2 \cdot 2} = 7 \cdot 6 \cdot 5 \cdot 3 = 630$

e) $\dfrac{(n+1)! - n!}{n!} = \dfrac{n!(n+1-1)}{n!} = n$

f) $\dfrac{(n+1)! + n!}{(n+1)! - n!} = \dfrac{n!(n+1+1)}{n!(n+1-1)} = \dfrac{n+2}{n}$

35. a) $_{10}P_{10} = 10! = 3{,}628{,}800$

b) Any of 10 horses can come in first; any of the remaining 9 can come in second; any of the remaining 8 can come in third. There are $10 \cdot 9 \cdot 8 = 720$ possibilities for the first three places.

37. a) We must count the number of ways in which we can arrange the 6 letters. Since there are 3 C's, the number of 6 letter words is $\dfrac{6!}{3!} = \dfrac{6 \cdot 5 \cdot 4 \cdot 3!}{3!} = 120$.

b) Consider the three C's as an inseparable block. They can start in positions 1, 2, 3, or 4 of the word. Once they are inserted, the remaining 3 letters can fill the remaining 3 positions in 3! ways. There are thus $4 \cdot 3! = 4 \cdot 3 \cdot 2 = 24$ such words.

39.
$$\frac{1}{k!} - \frac{1}{(k+1)!} = \frac{(k+1)! - k!}{k!(k+1)!}$$
$$= \frac{k!(k+1-1)}{k!(k+1)!}$$
$$= \frac{k}{(k+1)!}$$

Having shown this,
$$\frac{1}{2!} + \frac{2}{3!} + \frac{3}{4!} + \cdots + \frac{(n-1)}{n!} + \frac{n}{(n+1)!}$$
$$= \left(\frac{1}{1!} - \frac{1}{2!}\right) + \left(\frac{1}{2!} - \frac{1}{3!}\right) + \left(\frac{1}{3!} - \frac{1}{4!}\right) + \cdots$$
$$\cdots + \left(\frac{1}{(n-1)!} - \frac{1}{n!}\right) + \left(\frac{1}{n!} - \frac{1}{(n+1)!}\right)$$
$$= \frac{1}{1!} - \frac{1}{(n+1)!}$$
$$= 1 - \frac{1}{(n+1)!}$$

41. For the area code, there are 8 choices for the first digit, 2 choices for the second, and 10 choices for the third. So there are $8 \cdot 2 \cdot 10 = 160$ possible area codes.

For the exchange, there are 8 choices for the first digit, 8 choices for the second, and 10 choices for the third. So there are $8 \cdot 8 \cdot 10 = 640$ possible exchanges.

We can choose any of 10 numbers for each of the four digits of the line number. However, we must not count the one arrangement that gives us 4 zeros. So there are $10 \cdot 10 \cdot 10 \cdot 10 - 1 = 9999$ pos-

sible line numbers.

Putting all this together, there are
$(8 \cdot 2 \cdot 10)(8 \cdot 8 \cdot 10)(10^4 - 1) = 160 \cdot 640 \cdot 9999$
$= 1{,}023{,}897{,}600$ possible telephone numbers.

43. We can create numbers having from one to five digits. There are 5 choices for each digit that we fill.
Number of 1-digit numbers: 5
Number of 2-digit numbers: $5 \cdot 5 = 25$
Number of 3-digit numbers: $5 \cdot 5 \cdot 5 = 125$
Number of 4-digit numbers: $5^4 = 625$
Number of 5-digit numbers: $5^5 = 3125$
Adding these results, there are 3905 different numbers between 0 and 60,000 that use only the digits 1, 2, 3, 4, or 5.

45. Seat the first person anywhere at the round table. The number of different arrangements is then the number of ways the remaining 5 people can be seated, that is, $5! = 120$ ways.

47. Since there have to be $(n-1)$ losers and each game produces one loser, there must be $(n-1)$ games.

Problem Set 12.6

1. a) $_{10}P_3 = 10 \cdot 9 \cdot 8 = 720$

b) $_{10}C_3 = \dfrac{10 \cdot 9 \cdot 8}{3 \cdot 2 \cdot 1} = 120$

c) $_5P_5 = 5 \cdot 4 \cdot 3 \cdot 2 \cdot 1 = 120$

d) $_5C_5 = \dfrac{5!}{5!} = 1$

e) $_6P_1 = 6$

f) $_6C_1 = \dfrac{6}{1} = 6$

3. a) $_{20}C_{17} = {_{20}C_3} = \dfrac{20 \cdot 19 \cdot 18}{3 \cdot 2 \cdot 1} = 1140$

b) $_{100}C_{97} = {_{100}C_3} = \dfrac{100 \cdot 99 \cdot 98}{3 \cdot 2 \cdot 1} = 161{,}700$

5. $_8C_3 = \dfrac{8 \cdot 7 \cdot 6}{3 \cdot 2 \cdot 1} = 56$

7. $_{12}C_4 = \dfrac{12 \cdot 11 \cdot 10 \cdot 9}{4 \cdot 3 \cdot 2 \cdot 1} = 495$

9. This is an arrangement problem:
$_{10}P_3 = 10 \cdot 9 \cdot 8 = 720$

11. a) $_{30}C_4 = \dfrac{30 \cdot 29 \cdot 28 \cdot 27}{4 \cdot 3 \cdot 2 \cdot 1} = 27{,}405$

b) $_{25}C_3 = \dfrac{25 \cdot 24 \cdot 23}{3 \cdot 2 \cdot 1} = 2300$

13. $_{10}C_2$ tells us how many games there will be if each team plays every other team once. So $2 \cdot {}_{10}C_2$ will tell us how many games there are when each team plays every other team twice.

$$2 \cdot {}_{10}C_2 = 2 \cdot \frac{10 \cdot 9}{2 \cdot 1} = 90$$

15. $_{12}C_{10} = {}_{12}C_2 = \frac{12 \cdot 11}{2 \cdot 1} = 66$

17.
$$\begin{aligned}
&{}_6C_1 + {}_6C_2 + {}_6C_3 + {}_6C_4 + {}_6C_5 + {}_6C_6 \\
&= {}_6C_1 + {}_6C_2 + {}_6C_3 + {}_6C_2 + {}_6C_1 + {}_6C_6 \\
&= 2 \cdot {}_6C_1 + 2 \cdot {}_6C_2 + {}_6C_3 + {}_6C_6 \\
&= 2 \cdot \frac{6}{1} + 2 \cdot \frac{6 \cdot 5}{2 \cdot 1} + \frac{6 \cdot 5 \cdot 4}{3 \cdot 2 \cdot 1} + 1 \\
&= 12 + 30 + 20 + 1 \\
&= 63
\end{aligned}$$

19. a) Two women can be chosen in $_4C_2$ ways; 1 man can be chosen in $_6C_1$ ways. The committee can be chosen in $_4C_2 \cdot {}_6C_1 = \frac{4 \cdot 3}{2 \cdot 1} \cdot \frac{6}{1} = 36$ ways.

b) There are $_{10}C_3$ ways to choose a committee. There are $_6C_3$ ways to choose a committee with only men, no women. Thus there are $_{10}C_3 - {}_6C_3$ committees with at least one woman.

$$_{10}C_3 - {}_6C_3 = \frac{10 \cdot 9 \cdot 8}{3 \cdot 2 \cdot 1} - \frac{6 \cdot 5 \cdot 4}{3 \cdot 2 \cdot 1} = 120 - 20 = 100$$

c) There are $_4C_3$ ways to choose a committee that is all women; there are $_6C_3$ ways to choose a committee that is all men. Therefore, there are $_4C_3 + {}_6C_3$ committees where all 3 members are of the same sex.

$$_4C_3 + {}_6C_3 = \frac{4 \cdot 3 \cdot 2}{3 \cdot 2 \cdot 1} + \frac{6 \cdot 5 \cdot 4}{3 \cdot 2 \cdot 1} = 4 + 20 = 24$$

21. a) 1 black ball can be selected in $_4C_1$ ways; 2 white balls can be selected in $_7C_2$ ways. By the multiplication principle, 1 black and 2 white balls can be selected in $_4C_1 \cdot {}_7C_2 = \frac{4}{1} \cdot \frac{7 \cdot 6}{2 \cdot 1} = 84$ ways.

b) 3 black balls can be selected in $_4C_3$ ways; 3 white balls can be selected in $_7C_3$ ways. By the addition principle, 3 balls of one color can be selected in
$$_4C_3 + {}_7C_3 = \frac{4 \cdot 3 \cdot 2}{3 \cdot 2 \cdot 1} + \frac{7 \cdot 6 \cdot 5}{3 \cdot 2 \cdot 1} = 4 + 35 = 39 \text{ ways.}$$

c) 3 balls can be selected in $_{11}C_3$ ways. 3 white balls, (that is, no black balls) can be selected in $_7C_3$ ways. At least one black ball can be selected in
$$_{10}C_3 - {}_7C_3 = \frac{11 \cdot 10 \cdot 9}{3 \cdot 2 \cdot 1} - \frac{7 \cdot 6 \cdot 5}{3 \cdot 2 \cdot 1} = 165 - 35 = 130$$
ways.

23. There are $_{26}C_{13}$ ways of selecting a hand with only red cards.

25. There are 4 ways of choosing the ace, 4 ways of choosing the king, and so on. Hence by the multiplication principle, there are 4^{13} different hands.

27. A hand with two kings can be chosen in $_4C_2 \cdot {}_{48}C_{11}$ ways. A hand with 3 kings can be chosen in $_4C_3 \cdot {}_{48}C_{10}$ ways. A hand with 4 kings can be chosen in $_4C_4 \cdot {}_{48}C_9$ ways. By the addition principle there are $_4C_2 \cdot {}_{48}C_{11} + {}_4C_3 \cdot {}_{48}C_{10} + {}_4C_4 \cdot {}_{48}C_9$ hands with 2, 3, or 4 kings.

29. A poker hand consists of any combination of 5 cards out of the 52-card deck: $_{52}C_5$

31. There are $_4C_2$ ways of selecting 2 cards of a particular kind, and $_{13}C_2$ ways of selecting the two kinds. There remains one of the 5 cards to be selected, out of the 44 remaining. Hence there are $_4C_2 \cdot {}_4C_2 \cdot {}_{13}C_2 \cdot 44 = 123{,}552$ hands altogether.

33. Each coin can fall in either of 2 ways: heads or tails. Thus the four coins can fall in $2 \cdot 2 \cdot 2 \cdot 2 = 16$ ways.

35. 2 people can be chosen from labor in $_5C_2$ ways; 2 people can be chosen from business in $_4C_2$ ways; 2 people can be chosen from the public in $_3C_2$ ways. By the multiplication principle, a committee can be chosen in
$$_5C_2 \cdot {}_4C_2 \cdot {}_3C_2 = \frac{5 \cdot 4}{3 \cdot 1} \cdot \frac{4 \cdot 3}{2 \cdot 1} = \frac{3 \cdot 2}{2 \cdot 1} = 180 \text{ ways.}$$

37. The president, secretary, and treasurer can be chosen in $_{12}P_3$ ways. A committee of 3 can then be chosen from the remaining 9 people in $_9C_3$ ways. By the multiplication principle, the positions can be filled in
$$_{12}P_3 \cdot {}_9C_3 = 12 \cdot 11 \cdot 10 \cdot \frac{9 \cdot 8 \cdot 7}{3 \cdot 2 \cdot 1} = 110{,}880 \text{ ways.}$$

39. a) Each question can be answered in 2 ways; there are thus $2^{10} = 1024$ sets of answers.

b) Out of the 10 questions, exactly four can have the correct answers in $_{10}C_4 = \frac{10 \cdot 9 \cdot 8 \cdot 7}{4 \cdot 3 \cdot 2 \cdot 1} = 210$ ways.

41. Since Mary can choose to give her daughter any 1 coin, any 2 coins, or any 3, 4, 5, or 6 coins, the number of different sums of money she can give her daughter is $_6C_1 + {}_6C_2 + {}_6C_3 + {}_6C_4 + {}_6C_5 + {}_6C_6$
$$\begin{aligned}
&= {}_6C_1 + {}_6C_2 + {}_6C_3 + {}_6C_2 + {}_6C_1 + {}_6C_6 \\
&= \frac{6}{1} + \frac{6 \cdot 5}{2 \cdot 1} + \frac{6 \cdot 5 \cdot 4}{3 \cdot 2 \cdot 1} + \frac{6 \cdot 5}{2 \cdot 1} + \frac{6}{1} + 1 \\
&= 6 + 15 + 20 + 15 + 6 + 1 \\
&= 63
\end{aligned}$$

43. One must count the number of triangles, quadrilaterals, ..., decagons possible and add the totals. This would be

$$_{10}C_3 + {}_{10}C_4 + {}_{10}C_5 + {}_{10}C_6 + {}_{10}C_7 + {}_{10}C_8 + {}_{10}C_9 \\ + {}_{10}C_{10}$$
$$= 120 + 210 + 252 + 210 + 120 + 45 + 10 + 1$$
$$= 968$$

45. Consider splitting n presents between 2 people.

One way to count the number of ways the presents can be distributed is to say that the first present has 2 choices of where it should go; the second present has 2 choices; the third has 2 choices, and so on. By the multiplication principle, there are 2^n different ways to distribute the n presents.

A second approach is to say that the first person can get $_nC_0$ presents, or he can get $_nC_1$ presents, or $_nC_2$ presents, and so on. That is, there are

$$_nC_0 + {}_nC_1 + {}_nC_2 + \cdots + {}_nC_n$$

different combinations of presents possible.

Thus, $_nC_0 + {}_nC_1 + {}_nC_2 + \cdots + {}_nC_n = 2^n$.

47. $\displaystyle\sum_{j=0}^{n} ({}_nC_j)^2 = ({}_nC_0)^2 + ({}_nC_1)^2 + ({}_nC_2)^2 + \cdots + ({}_nC_n)^2$

$$= {}_nC_0\,{}_nC_n + {}_nC_1\,{}_nC_{n-1} + {}_nC_2\,{}_nC_{n-2} + \cdots \\ \cdots + {}_nC_n\,{}_nC_0$$

$$= {}_{2n}C_n \text{ by problem 46}$$

49. $S = {}_{n+1}C_1 + {}_{n+2}C_2 + {}_{n+3}C_3 + \cdots + {}_{n+k}C_k$

$_{n+1}C_0 + S = \underbrace{{}_{n+1}C_0 + {}_{n+1}C_1}_{} + {}_{n+2}C_2 + {}_{n+3}C_3 + \\ \cdots + {}_{n+k}C_k$

$$= \underbrace{{}_{n+2}C_1 + {}_{n+2}C_2}_{} + {}_{n+3}C_3 + \cdots \\ \cdots + {}_{n+k}C_k$$

$$= {}_{n+3}C_2 + {}_{n+3}C_3 + \cdots + {}_{n+k}C_k$$
$$= \cdots$$
$$= {}_{n+k}C_{k-1} + {}_{n+k}C_k$$
$$= {}_{n+k+1}C_k$$

Therefore,

$$S = {}_{n+k+1}C_k - {}_{n+1}C_0$$
$$= {}_{n+k+1}C_k - 1$$

51. A rectangle is determined by choosing 2 horizontal lines and 2 vertical lines from the $n \times n$ grid. The 2 horizontal lines can be chosen from the total of $n+1$ lines in $_{n+1}C_2$ ways. Similarly the 2 vertical lines can be chosen in $_{n+1}C_2$ ways. Therefore, the total number of rectangles is

$$\left({}_{n+1}C_2\right)^2 = \left(\frac{(n+1)n}{2}\right)^2$$

1. $(x+y)^3 = {}_3C_0 x^3 y^0 + {}_3C_1 x^2 y^1 + {}_3C_2 x^1 y^2 + {}_3C_3 x^0 y^3$
$$= x^3 + 3x^2 y + 3xy^2 + y^3$$

3. $(x - 2y)^3$
$$= {}_3C_0 x^3 (-2y)^0 + {}_3C_1 x^2 (-2y)^1 + {}_3C_2 x(-2y)^2 \\ + {}_3C_3 x^0 (-2y)^3$$
$$= x^3 + 3x^2(-2y) + 3x(4y^2) + (-8y^3)$$
$$= x^3 - 6x^2 y + 12xy^2 - 8y^3$$

5. $(c^2 - 3d^3)^4$
$$= {}_4C_0 (c^2)^4 (-3d^3)^0 + {}_4C_1 (c^2)^3 (-3d^3)^1 \\ + {}_4C_2 (c^2)^2 (-3d^3)^2 + {}_4C_3 (c^2)^1 (-3d^3)^3 \\ + {}_4C_4 (c^2)^0 (-3d^3)^4$$
$$= c^8 + 4c^6(-3d^3) + 6c^4(9d^6) + 4c^2(-27d^9) + (81d^{12})$$
$$= c^8 - 12c^6 d^3 + 54c^4 d^6 - 108c^2 d^9 + 81d^{12}$$

7. $(ab^2 - bc)^5$
$$= {}_5C_0 (ab^2)^5 (-bc)^0 + {}_5C_1 (ab^2)^4 (-bc)^1 \\ + {}_5C_2 (ab^2)^3 (-bc)^2 + {}_5C_3 (ab^2)^2 (-bc)^3 \\ + {}_5C_4 (ab^2)^1 (-bc)^4 + {}_5C_5 (ab^2)^0 (-bc)^5$$
$$= a^5 b^{10} + 5a^4 b^8 (-bc) + 10a^3 b^6 (b^2 c^2) \\ + 10a^2 b^4 (-b^3 c^3) + 5ab^2 (b^4 c^4) + (-b^5 c^5)$$
$$= a^5 b^{10} - 5a^4 b^9 c + 10a^3 b^8 c^2 - 10a^2 b^7 c^3 \\ + 5ab^6 c^4 - b^5 c^5$$

9. $(x + y)^{20}$
$$= {}_{20}C_0 x^{20} y^0 + {}_{20}C_1 x^{19} y^1 + {}_{20}C_2 x^{18} y^2 + \cdots$$
$$= x^{20} + \frac{20}{1} x^{19} y + \frac{20 \cdot 19}{2 \cdot 1} x^{18} y^2 + \cdots$$
$$= x^{20} + 20x^{19} y + 190x^{18} y^2 + \cdots$$

11. $\left(x + \dfrac{1}{x^5}\right)^{20} = {}_{20}C_0 x^{20}\left(\dfrac{1}{x^5}\right)^0 + {}_{20}C_1 x^{19}\left(\dfrac{1}{x^5}\right)^1 \\ + {}_{20}C_2 x^{18}\left(\dfrac{1}{x^5}\right)^2 + \cdots$

$$= x^{20} + 20\frac{x^{19}}{x^5} + 190\frac{x^{18}}{x^{10}} + \cdots$$
$$= x^{20} + 20x^{14} + 190x^8 + \cdots$$

13. A term involving z^9 will arise when we raise z^3 to the third power. The term we want is

$$_{10}C_3 (y^2)^7 (-z^3)^3 = \frac{10 \cdot 9 \cdot 8}{3 \cdot 2 \cdot 1} y^{14} (-z)^9 = -120 y^{14} z^9$$

15. The term that involves a^3 is

$$_{12}C_9 (2a)^3 (-b)^9 = -{}_{12}C_3 (8a^3) b^9$$
$$= -\frac{12 \cdot 11 \cdot 10}{3 \cdot 2 \cdot 1} \cdot 8a^3 b^9$$
$$= -1760 a^3 b^9$$

278

17. $20(1.02)^8 = 20(1 + .02)^8$
$$= 20[_8C_0(1)^8(.02)^0 + {}_8C_1(1)^7(.02)^1$$
$$+ {}_8C_2(1)^6(.02)^2 + \cdots]$$
$$= 20[1 + 8(.02) + \frac{8 \cdot 7}{2}(.0004) + \cdots]$$
$$\approx 20[1 + .16 + .0112] = 20(1.1712)$$
$$= 23.424$$

19. $500(1.005)^{20}$
$$= 500(1 + .005)^{20}$$
$$= 500[_{20}C_0(1)^{20}(.005)^0 + {}_{20}C_1(1)^{19}(.005)^1$$
$$+ {}_{20}C_2(1)^{18}(.005)^2 + \cdots]$$
$$\approx 500[1 + 20(.005) + \frac{20 \cdot 19}{2 \cdot 1}(.000025)]$$
$$= 500[1 + .10 + .00475] = 500(1.10475)$$
$$= 552.375$$

21. $100(1.02)^{20}$
$$= 100(1.02)^{20}$$
$$= 100[_{20}C_0(1)^{20}(.02)^0 + {}_{20}C_1(1)^{19}(.02)^1$$
$$+ {}_{20}C_2(1)^{18}(.02)^2 + {}_{20}C_3(1)^{17}(.02)^3$$
$$+ {}_{20}C_4(1)^{16}(.02)^4 + \cdots]$$
$$= 100[1 + 20(.02) + \frac{20 \cdot 19}{2 \cdot 1}(.0004)$$
$$+ \frac{20 \cdot 19 \cdot 18}{3 \cdot 2 \cdot 1}(.000008) + \frac{20 \cdot 19 \cdot 18 \cdot 17}{4 \cdot 3 \cdot 2 \cdot 1}(.00000016)$$
$$+ \cdots]$$
$$= 100[1 + .4 + .076 + .00912 + .0007752 + \cdots]$$
$$\approx 100(1.4858952) \approx 148.5$$
$$\approx 149$$

23. a) $2^4 = 16$ subsets

b) $2^5 = 32$ subsets

c) $2^6 = 64$ sunsets

25. The number of committees with at least 3 members is the same as the number of subsets with at least 3 members and this equals
$$2^9 - {}_9C_0 - {}_9C_1 - {}_9C_2 = 512 - 1 - 9 - 36 = 466$$

27. To produce any given term in the expansion of $(\mathbf{A} + \mathbf{B})^2$, each of the two factors $\mathbf{A} + \mathbf{B}$ contributes either an $\mathbf{A}$ or a $\mathbf{B}$. There are $2 \cdot 2 = 4$ ways in which they can make this contribution, hence there are 4 terms in the expansion. Since the matrix multiplication is not commutative, none of the terms can be combined.

A similar argument shows that there will be $2 \cdot 2 \cdot 2 = 8$ terms in the expansion of $(\mathbf{A} + \mathbf{B})^3$.

29. a) $(2x + \frac{1}{2})^8$
$$= {}_8C_0(2x)^8(\tfrac{1}{2})^0 + {}_8C_1(2x)^7(\tfrac{1}{2})^1 + {}_8C_2(2x)^6(\tfrac{1}{2})^2$$
$$+ {}_8C_3(2x)^5(\tfrac{1}{2})^3 + {}_8C_4(2x)^4(\tfrac{1}{2})^4 + {}_8C_5(2x)^3(\tfrac{1}{2})^5$$
$$+ {}_8C_6(2x)^2(\tfrac{1}{2})^6 + {}_8C_7(2x)^1(\tfrac{1}{2})^7 + {}_8C_8(2x)^0(\tfrac{1}{2})^8$$
$$= 2^8 x^8 + 8 \cdot 2^7 x^7 \cdot \tfrac{1}{2} + \frac{8 \cdot 7}{2} \cdot 2^6 x^6 \cdot \frac{1}{2^2} + \frac{8 \cdot 7 \cdot 6}{3 \cdot 2 \cdot 1} \cdot 2^5 x^5 \cdot \frac{1}{2^3}$$
$$+ \frac{8 \cdot 7 \cdot 6 \cdot 5}{4 \cdot 3 \cdot 2 \cdot 1} \cdot 2^4 x^4 \cdot \frac{1}{2^4} + \frac{8 \cdot 7 \cdot 6}{3 \cdot 2 \cdot 1} \cdot 2^3 x^3 \cdot \frac{1}{2^5}$$
$$+ \frac{8 \cdot 7}{2 \cdot 1} \cdot 2^2 x^2 \cdot \frac{1}{2^6} + 8 \cdot 2x \cdot \frac{1}{2^7} + 1 \cdot \frac{1}{2^8}$$
$$= 256x^8 + 512x^7 + 448x^6 + 224x^5 + 70x^4 + 14x^3$$
$$+ \tfrac{7}{4}x^2 + \tfrac{1}{8}x + \frac{1}{256}$$

b) $(1 + \sqrt{3})^6$
$$= {}_6C_0(1)^6(\sqrt{3})^0 + {}_6C_1(1)^5(\sqrt{3})^1 + {}_6C_2(1)^4(\sqrt{3})^2$$
$$+ {}_6C_3(1)^3(\sqrt{3})^3 + {}_6C_4(1)^2(\sqrt{3})^4 + {}_6C_5(1)^1(\sqrt{3})^5$$
$$+ {}_6C_6(1)^0(\sqrt{3})^6$$
$$= 1 + 6\sqrt{3} + \frac{6 \cdot 5}{2 \cdot 1} \cdot 3 + \frac{6 \cdot 5 \cdot 4}{3 \cdot 2 \cdot 1} \cdot 3\sqrt{3} + \frac{6 \cdot 5}{2 \cdot 1} \cdot 9$$
$$+ 6 \cdot 9\sqrt{3} + 27$$
$$= 1 + 6\sqrt{3} + 45 + 60\sqrt{3} + 135 + 54\sqrt{3} + 27$$
$$= 208 + 120\sqrt{3}$$

31. The term involving z^{15} will be the term involving $(z^3)^5$. That term is
$$_8C_5(x)^3(-2z^3)^5 = \frac{8 \cdot 7 \cdot 6}{3 \cdot 2 \cdot 1} \cdot x^3(-32z^{15}) = -1792x^3z^{15}$$

33. a) $\dfrac{(x + h)^n - x^n}{h}$
$$= \frac{x^n + {}_nC_1 x^{n-1}h + {}_nC_2 x^{n-2}h^2 + \cdots + {}_nC_n h^n - x^n}{h}$$
$$= {}_nC_1 x^{n-1} + {}_nC_2 x^{n-2}h + \cdots + {}_nC_n h^{n-1}$$
The only term that does not involve h is
$$_nC_1 x^{n-1} = nx^{n-1}$$

b) $\dfrac{(x + h)^{10} + 2(x + h)^4 - x^{10} - 2x^4}{h}$
$$= \frac{1}{h}[x^{10} + {}_{10}C_1 x^9 h + \cdots + {}_{10}C_{10} h^{10} + 2x^4$$
$$+ 2 \cdot {}_4C_1 x^3 h + \cdots + 2 \cdot {}_4C_4 h^4 - x^{10} - 2x^4]$$
The only terms that will not involve h are
$$_{10}C_1 x^9 + 2 \cdot {}_4C_1 x^3 = 10x^9 + 8x^3$$

35. We are looking for a term of the form $(x^2)^m\left(\frac{1}{x}\right)^n$ such that $2m = n$. Since $m + n = 12$, we get $3m = 12$, $m = 4$, $n = 8$. The constant term is
$$_{12}C_8(3x^2)^4\left(\frac{1}{3x}\right)^8 = \frac{12 \cdot 11 \cdot 10 \cdot 9}{4 \cdot 3 \cdot 2 \cdot 1} \cdot \frac{3^4 x^8}{3^8 x^8} = \frac{990}{162} = \frac{55}{9}$$

279

37. $(1.01)^{50} = (1 + .01)^{50}$

$= {}_{50}C_0(1)^{50}(.01)^0 + {}_{50}C_1(1)^{49}(.01)^1$
$\qquad\qquad + {}_{50}C_2(1)^{48}(.01)^2 + \cdots$

$= 1 + 50(.01) + \dfrac{50 \cdot 49}{2}(.0001) + \cdots$

$= 1.5 + \text{positive numbers}$

> 1.5

39. 2^{12} committees of any size can be selected. The number of committees consisting of 3 or more numbers is $2^{12} - {}_{12}C_0 - {}_{12}C_1 - {}_{12}C_2$
$\qquad = 4096 - 1 - 12 - 66 = 4017$

41. a) $(x + y + z)^3$

$= \dfrac{3!\,x^3}{3!\,0!\,0!} + \dfrac{3!\,y^3}{3!\,0!\,0!} + \dfrac{3!\,z^3}{3!\,0!\,0!} + \dfrac{3!\,xy^2}{1!\,2!\,0!} + \dfrac{3!\,x^2y}{2!\,1!\,0!} + \dfrac{3!\,xz^2}{1!\,0!\,2!}$

$\qquad + \dfrac{3!\,x^2z}{2!\,0!\,1!} + \dfrac{3!\,yz^2}{0!\,1!\,2!} + \dfrac{3!\,y^2z}{0!\,2!\,1!} + \dfrac{3!\,xyz}{1!\,1!\,1!}$

$= x^3 + y^3 + z^3 + 3xy^2 + 3x^2y + 3xz^2 + 3x^2z$
$\qquad\qquad + 3yz^2 + 3y^2z + 6xyz$

b) The term involving x^2y^4z is

$\dfrac{7!}{2!\,4!\,1!}(2x)^2y^4z = \dfrac{7 \cdot 6 \cdot 5}{2} \cdot 4x^2y^4z = 420x^2y^4z$

The coefficient of the term is 420.

43. $\displaystyle\sum_{k=0}^{n} {}_nC_k \cdot 2^k = {}_nC_0(2)^0 + {}_nC_1(2)^1 + {}_nC_2(2)^2 + \cdots$
$\qquad\qquad\qquad\qquad\qquad \cdots + {}_nC_n(2)^n$

$= {}_nC_0(1)^n(2^0) + {}_nC_1(1)^{n-1}(2)^1$
$\qquad + {}_nC_2(1)^{n-2}(2)^2 + \cdots + {}_nC_n(1)^0(2)^n$

$= (1 + 2)^n$

$= 3^n$

45. a) $P(0) = 1$; $P(1) = 1 + 1 = 2$; $P(2) = 1 + 2 + 1 = 4$
For $k \le n$,
$P(k) = 1 + k + {}_kC_2 + {}_kC_3 + \cdots + {}_kC_k$
$\qquad = (1+1)^k$
$\qquad = 2^k$

b) $P(n+1) = 1 + {}_{n+1}C_1 + {}_{n+1}C_2 + \cdots + {}_{n+1}C_n$
$\qquad = 2^{n+1} - {}_{n+1}C_{n+1}$
$\qquad = 2^{n+1} - 1$

c) $P(n+2) = 1 + {}_{n+2}C_1 + {}_{n+2}C_2 + \cdots + {}_{n+2}C_n$
$\qquad = 2^{n+2} - {}_{n+2}C_{n+1} - {}_{n+2}C_{n+2}$
$\qquad = 2^{n+2} - (n+2) - 1$
$\qquad = 2^{n+2} - n - 3$

1. Let n be the number of spots on the upper face.
 a) $P(n = 3) = \frac{1}{6}$
 b) $P(n > 3) = \frac{3}{6} = \frac{1}{2}$
 c) $P(n < 3) = \frac{2}{6} = \frac{1}{3}$
 d) $P(n \text{ is even}) = \frac{3}{6} = \frac{1}{2}$
 e) $P(n \text{ is odd}) = \frac{3}{6} = \frac{1}{2}$. Or, use the fact that n must be odd or even, so
 $P(n \text{ is odd}) = 1 - P(n \text{ is even}) = 1 - \frac{1}{2} = \frac{1}{2}$

3. HHH, HHT, HTH, HTT, TTT, TTH, THT, THH
 a) $P(3 \text{ heads}) = \frac{1}{8}$
 b) $P(\text{exactly 2 heads}) = \frac{3}{8}$
 c) $P(\text{more than 1 head}) = \frac{4}{8} = \frac{1}{2}$

5. a) $P[(1,1) \text{ or } (2,2) \text{ or } (3,3) \text{ or } (4,4) \text{ or } (5,5) \text{ or } (6,6)]$
 $= \frac{6}{36} = \frac{1}{6}$
 b) $P[(1,2) \text{ or } (2,4) \text{ or } (3,6) \text{ or } (2,1) \text{ or } (4,2) \text{ or } (6,3)]$
 $= \frac{6}{36} = \frac{1}{6}$
 c) $P[(1,3) \text{ or } (1,4) \text{ or } (1,5) \text{ or } (1,6) \text{ or } (2,4) \text{ or } (2,5) \text{ or } (2,6) \text{ or } (3,1) \text{ or } (3,5) \text{ or } (3,6) \text{ or } (4,1) \text{ or } (4,2) \text{ or } (4,6) \text{ or } (5,1) \text{ or } (5,2) \text{ or } (5,3) \text{ or } (6,1) \text{ or } (6,2) \text{ or } (6,3) \text{ or } (6,4)] = \frac{20}{36} = \frac{5}{9}$

7. a) With 4 possible outcomes for each tetrahedron, there are $4 \cdot 4 = 16$ different results possible when rolling 2 tetrahedra.
 b) $P[(3,4) \text{ or } (4,3)] = \frac{2}{16} = \frac{1}{8}$
 c) $P(\text{sum} < 7) = 1 - P(\text{sum} \ge 7)$
 $\qquad = 1 - P[(3,4) \text{ or } (4,3) \text{ or } (4,4)]$
 $\qquad = 1 - \frac{3}{16} = \frac{13}{16}$

9. a) The probability of being born in a certain state is dependent on the number of births in that state, not on the number of states.

 b) Smoking and drinking are not disjoint events; some people smoke and drink.

 c) The probability of winning plus the probability of losing should equal 1.

 d) Ties are possible in football.

11. a) $\frac{16}{50} = \frac{8}{25}$ **b)** $\frac{100}{240} = \frac{5}{12}$
 c) $\frac{4 + 16 + 100}{300} = \frac{120}{300} = \frac{2}{5}$ **d)** $\frac{32}{300} = \frac{8}{75}$

13. $P(\text{drawn in order 1, 2, 3, 4}) = \frac{1}{4} \cdot \frac{1}{3} \cdot \frac{1}{2} \cdot 1 = \frac{1}{24}$

15. If the odds of winning are 2 to 9, the probability of winning is $\frac{2}{11}$. $P(\text{losing}) = 1 - \frac{2}{11} = \frac{9}{11}$

17. $P(2 \text{ heads}) = \frac{1}{4}$. The odds in favor of getting 2 heads are 1 to $(4 - 1)$, or 1 to 3.

19. a) $\frac{26}{52} = \frac{1}{2}$

b) $\frac{13}{52} = \frac{1}{4}$

c) $\frac{4}{52} = \frac{1}{13}$

21. There are $_{52}C_3$ ways to pick 3 cards from a standard deck.

a) There $_{26}C_3$ ways to pick 3 red cards.

$$P(3 \text{ red cards}) = \frac{_{26}C_3}{_{52}C_3}$$

$$= \frac{\frac{26 \cdot 25 \cdot 24}{3 \cdot 2 \cdot 1}}{\frac{52 \cdot 51 \cdot 50}{3 \cdot 2 \cdot 1}} = \frac{26 \cdot 25 \cdot 24}{52 \cdot 51 \cdot 50}$$

$$= \frac{1}{2} \cdot \frac{1}{2} \cdot \frac{8}{17} = \frac{2}{17}$$

$$\approx .118$$

b) There are $_{13}C_3$ ways to choose 3 diamonds.

$$P(3 \text{ diamonds}) = \frac{_{13}C_3}{_{52}C_3}$$

$$= \frac{\frac{13 \cdot 12 \cdot 11}{3 \cdot 2 \cdot 1}}{\frac{52 \cdot 51 \cdot 50}{3 \cdot 2 \cdot 1}} = \frac{13 \cdot 12 \cdot 11}{52 \cdot 51 \cdot 50}$$

$$= \frac{1}{4} \cdot \frac{4}{17} \cdot \frac{11}{50} = \frac{11}{850}$$

$$\approx .013$$

c) There are $_4C_1 \cdot _{48}C_2$ ways in which 1 of the cards will be a queen and the other two cards will not.

$$P(\text{exactly one queen}) = \frac{_4C_1 \cdot _{48}C_2}{_{52}C_3} = \frac{4 \cdot \frac{48 \cdot 47}{2 \cdot 1}}{\frac{52 \cdot 51 \cdot 50}{3 \cdot 2 \cdot 1}}$$

$$= \frac{24 \cdot 47}{13 \cdot 17 \cdot 25} = \frac{1128}{5525}$$

$$\approx .204$$

d) There are $_4C_3$ ways to choose 3 queens.

$$P(3 \text{ queens}) = \frac{_4C_3}{_{52}C_3} = \frac{4}{\frac{52 \cdot 51 \cdot 50}{3 \cdot 2 \cdot 1}}$$

$$= \frac{1}{13 \cdot 17 \cdot 25} = \frac{1}{5525}$$

$$\approx .0002$$

23. The deck contains 8 kings and queens. The number of ways of choosing 5 of those cards is $_8C_5$.

$$P(\text{all kings and queens}) = \frac{_8C_5}{_{52}C_5} = \frac{\frac{8 \cdot 7 \cdot 6}{3 \cdot 2 \cdot 1}}{\frac{52 \cdot 51 \cdot 50 \cdot 49 \cdot 48}{5 \cdot 4 \cdot 3 \cdot 2 \cdot 1}}$$

$$= \frac{8 \cdot 7 \cdot 6 \cdot 5 \cdot 4}{52 \cdot 51 \cdot 50 \cdot 49 \cdot 48} = \frac{1}{46410}$$

$$\approx .00002$$

25. When 3 dice are tossed, there are $6 \cdot 6 \cdot 6 = 216$ possible outcomes.

a) $P(18) = P(6, 6, 6) = \frac{1}{216}$

b) $P(16) = P[(6, 6, 4)$ or $(6, 4, 6)$ or $(6, 5, 5)$ or $(5, 6, 5)$ or $(5, 5, 6)]$

$$= \frac{6}{216} = \frac{1}{36}$$

c) $P(\text{total} > 4) = 1 - P(\text{total} \leq 4)$.

$P(\text{total} \leq 4) = P[(1, 1, 1)$ or $(1, 1, 2)$ or $(2, 1, 1)]$

$= \frac{4}{216} = \frac{1}{54}$. Therefore, $P(\text{total} > 4) = 1 - \frac{1}{54} = \frac{53}{54}$.

27. a) There are 20 honor cards; therefore, there are $_{20}C_1$ ways to select an honor card.

$P(\text{drawing an honor card}) = \frac{_{20}C_1}{_{52}C_1} = \frac{20}{52} = \frac{5}{13}$.

b) The probability of drawing one of the 27 cards we want is $\frac{27}{52}$.

29. There are $_8C_4$ ways of drawing 4 balls.

a) There are $_3C_1 \cdot _5C_3$ ways to draw 1 black and 3 red balls.

$P(1 \text{ black and 3 red balls})$

$$= \frac{_3C_1 \cdot _5C_3}{_8C_4} = \frac{3 \cdot \frac{5 \cdot 4}{2}}{\frac{8 \cdot 7 \cdot 6 \cdot 5}{4 \cdot 3 \cdot 2}} = \frac{3}{7}$$

b) There are $_5C_2 \cdot _3C_2$ ways to draw 2 red and 2 black balls.

$P(2 \text{ red and 2 black balls})$

$$= \frac{_5C_2 \cdot _3C_2}{_8C_4} = \frac{\frac{5 \cdot 4}{2} \cdot 3}{\frac{8 \cdot 7 \cdot 6 \cdot 5}{4 \cdot 3 \cdot 2}} = \frac{3}{7}$$

c) $P(\text{at least 1 black ball}) = 1 - P(\text{no black balls})$.

$P(\text{no black balls}) = P(\text{all red balls})$

$$= \frac{_5C_4}{_8C_4} = \frac{5}{\frac{8 \cdot 7 \cdot 6 \cdot 5}{4 \cdot 3 \cdot 2 \cdot 1}} = \frac{1}{14}.$$

$P(\text{at least one black ball}) = 1 - \frac{1}{14} = \frac{13}{14}$.

31. There are a total of $_{15}C_3$ ways to draw 3 cards.
a) There are 7 even-numbered cards, so there are $_7C_3$ ways to draw an even number.

$$P(3 \text{ even numbers}) = \frac{_7C_3}{_{15}C_3} = \frac{\frac{7 \cdot 6 \cdot 5}{3 \cdot 2 \cdot 1}}{\frac{15 \cdot 14 \cdot 13}{3 \cdot 2 \cdot 1}} = \frac{1}{13}$$

b) $P(\text{at least 1 number is odd}) =$
$$1 - P(\text{no numbers are odd})$$
$P(\text{no numbers are odd}) = P(3 \text{ numbers are even})$
$$= \tfrac{1}{13} \text{ (from part (a))}.$$
$P(\text{at least 1 number is odd}) = 1 - \tfrac{1}{13} = \tfrac{12}{13}.$

c) If the product of 3 numbers is even, at least one of the numbers must be even.
$P(\text{at least 1 number is even})$
$$= 1 - P(\text{no numbers are even})$$
$P(\text{no numbers are even}) = P(3 \text{ numbers are odd})$

$$= \frac{_8C_3}{_{15}C_3} = \frac{\frac{8 \cdot 7 \cdot 6}{3 \cdot 2 \cdot 1}}{\frac{15 \cdot 14 \cdot 13}{3 \cdot 2 \cdot 1}} = \frac{8}{65}$$

$P(\text{at least 1 number is even}) = 1 - \tfrac{8}{65} = \tfrac{57}{65}.$

33. There are a total of 7! ways in which the men and women can be seated.
a) For the men and women to alternate, the women must sit in the first, third, fifth, and seventh seats. They can do that in 4! ways. The men can be arranged in the 3 remaining seats in 3! ways. Thus,

$$P(\text{men and women alternate}) = \frac{3! \, 4!}{7!} = \frac{3 \cdot 2 \cdot 1}{7 \cdot 6 \cdot 5} = \frac{1}{35}.$$

b) Consider the group of 4 women as an inseparable block. They have only 4 ways to choose a block of seats, but then can arrange themselves in the seats in 4! ways. The men can be arranged in 3! ways in the remaining 3 seats. Thus,

$$P(\text{4 women sitting together}) = \frac{4 \cdot 4! \, 3!}{7!} = \frac{4 \cdot 3 \cdot 2 \cdot 1}{7 \cdot 6 \cdot 5}$$
$$= \frac{4}{35}$$

c) Any of 3 men can be chosen to occupy one end seat, and either of the 2 remaining men can be chosen to occupy the other end seat. The remaining 5 people can be arranged in 5! ways. Thus,

$$P(\text{2 men occupy the end seats}) = \frac{3 \cdot 2 \cdot 5!}{7!} = \frac{3 \cdot 2}{7 \cdot 6} = \frac{1}{7}.$$

35. When a single die is tossed 4 times, there are $6^4 = 1296$ possible outcomes.
a) $(1, 2, 3, 4)$, appearing in that order, can happen in only 1 way. $P[(1, 2, 3, 4)] = \tfrac{1}{1296}.$

b) 1, 2, 3, and 4 appearing in any order can happen in $4! = 24$ ways. Thus,
$$P(1, 2, 3, 4 \text{ appearing in any order}) = \frac{24}{1296} = \frac{1}{54}$$

c) $P(\text{at least one 6}) = 1 - P(\text{no sixes})$
$$= 1 - \frac{5^4}{6^4} = 1 - \frac{625}{1296}$$
$$= \frac{671}{1296}$$

d) The first number can be anything, so there are 6 choices for the first number. But once the first number appears, the remaining numbers must match it. This can occur in only one way.

$$P(\text{the same number appears each time}) = \frac{6}{1296} = \frac{1}{216}$$

37. With 2 M's, 2 T's, and 2 A's, there are $\frac{11!}{2! \, 2! \, 2!}$ code words that can be formed from the letters in MATHEMATICS.
$$P(\text{spelling MATHEMATICS}) = \frac{1}{\frac{11!}{2! \, 2! \, 2!}} = \frac{8}{11!}$$

39.

There are 6 ways of choosing the first vertex, and $_3C_2 = 3$ ways of choosing two points on the opposite side, for a total of $6 \cdot 3 = 18$ triangles.
a) There are 2 right triangles with right angle at B, 2 at C, 2 at D, and 2 at E, for a total of 8. Therefore, the answer is $\frac{8}{18} = \frac{4}{9}$.

b) The base and altitude must each measure 2 units. Thus, the base needs to be AC or DF. For each, 3 triangles are possible. Therefore, the answer is $\frac{6}{18} = \frac{1}{3}$.

41. The total number of ways of choosing 3 sticks is $_8C_3 = 56$. Let a, b, and c represent the lengths of the 3 sides of the triangle formed, with $a < b < c$. Because $c < a + b$, we have the following possibilities.

282

a	b	Possible values of c	Number of triangles possible
1	2	none	0
1	3	none	0
2	3	4	1
2	4	5	1
2	5	6	1
2	6	7	1
2	7	8	1
3	4	5, 6	2
3	5	6, 7	2
3	6	7, 8	2
3	7	8	1
4	5	6, 7, 8	3
4	6	7, 8	2
4	7	8	1
5	6	7, 8	2
5	7	8	1
6	7	8	1
			Total = 22

Therefore, the probability is $\frac{22}{56} = \frac{11}{28}$.

Problem Set 12.9

1. $P(\text{all 1's}) = \frac{1}{6} \cdot \frac{1}{6} \cdot \frac{1}{6} = \frac{1}{216}$.

3. a) $P(A \text{ shows red and } B \text{ shows red}) = \frac{1}{4} \cdot \frac{3}{6} = \frac{1}{8}$.

   b) $P(A \text{ shows not red and } B \text{ shows not red})$
   $= \frac{3}{4} \cdot \frac{3}{6} = \frac{3}{8}$

   c) $P(A \text{ shows red and } B \text{ shows not red}) = \frac{1}{4} \cdot \frac{3}{6} = \frac{1}{8}$

   d) $P(A \text{ shows red and } B \text{ shows red or green})$
   $= \frac{1}{4} \cdot \frac{5}{6} = \frac{5}{24}$

   e) $P[(A \text{ shows green and } B \text{ shows not green})$
   $\quad$ or $(A \text{ shows not green and } B \text{ shows green})]$
   $= \frac{2}{4} \cdot \frac{4}{6} + \frac{2}{4} \cdot \frac{2}{6} = \frac{2}{6} + \frac{1}{6} = \frac{3}{6} = \frac{1}{2}$

5. a) No. Doing well in physics probably depends on having good math skills.

   b) Yes, the events seem unrelated.

   c) Yes, the events are independent.

   d) No, the events are not independent. The probability of getting an odd total $= \frac{1}{2}$. There are 11 ways of getting a 5 on one of the dice, 6 of which have an odd total. $\frac{6}{11} \neq \frac{1}{2}$.

   e) No. It is more likely that a doctor is a man.

   f) Yes, despite superstitions, these events are independent.

7. $P(\text{customer will get all good bolts})$
   $= P(\text{first box is perfect})$
   $\quad \cdot P(\text{second box is perfect}) \cdot P(\text{third box is perfect})$
   $= \frac{9}{10} \cdot \frac{9}{10} \cdot \frac{9}{10} = \frac{729}{1000}$

9. $P(\text{red ball})$
   $= P(\text{first urn is chosen and a red ball is drawn})$
   $\quad + P(\text{second urn is chosen and a red ball is drawn})$
   $= \frac{1}{2} \cdot \frac{3}{10} + \frac{1}{2} \cdot \frac{6}{12} = \frac{3}{20} + \frac{1}{4} = \frac{8}{20} = \frac{2}{5}$

11. $P(2 \text{ balls of the same color})$
    $= P(\text{drawing a red from A and then drawing 2 red}$
    $\quad \text{or 2 green from A})$
    $\quad + P(\text{drawing a green from A and then a green}$
    $\quad \text{from B})$
    $= \frac{6}{10}\left(\frac{6}{10} \cdot \frac{5}{9} + \frac{4}{10} \cdot \frac{3}{9}\right) + \frac{4}{10} \cdot \frac{2}{20} = \frac{3}{5}\left(\frac{1}{3} + \frac{2}{15}\right) + \frac{1}{25}$
    $= \frac{3}{5} \cdot \frac{7}{15} + \frac{1}{25} = \frac{21}{75} + \frac{3}{75} = \frac{24}{75} = \frac{8}{25}$

13. a) $P(\text{both balls are black}) = \frac{5}{10} \cdot \frac{5}{10} = \frac{1}{4}$

    b) $P(\text{both balls are black}) = \frac{5}{10} \cdot \frac{4}{9} = \frac{2}{9}$

15. a) $P(3 \text{ green balls}) = \frac{3}{10} \cdot \frac{3}{10} \cdot \frac{3}{10} = \frac{27}{1000} = .027$

    b) $P(3 \text{ red balls}) = \frac{2}{10} \cdot \frac{2}{10} \cdot \frac{2}{10} = \frac{8}{1000} = \frac{1}{125} = .008$

    c) $P(3 \text{ balls of the same color})$
    $= P(3 \text{ red}) + P(3 \text{ green}) + P(3 \text{ black})$
    $= \frac{8}{1000} + \frac{27}{1000} + \frac{5}{10} \cdot \frac{5}{10} \cdot \frac{5}{10} = \frac{160}{1000} = \frac{4}{25} = .16$

    d) The probability of getting a red, then a green, and then a black would be $\frac{2}{10} \cdot \frac{3}{10} \cdot \frac{5}{10}$. But the order is immaterial, so we multiply this by 6 since there are 3! different arrangements of 3 things. Therefore, the probability is $6 \cdot \frac{2}{10} \cdot \frac{3}{10} \cdot \frac{5}{10} = \frac{180}{1000} = .18$.

17. a) $P(\text{second ball is red}) = \frac{2}{10} = \frac{1}{5}$

    b) $P(\text{second ball is red})$
    $= P(\text{first is red and the second ball is red})$
    $\quad + P(\text{first ball is not red and the second ball is red})$
    $= \frac{2}{10} \cdot \frac{1}{9} + \frac{8}{10} \cdot \frac{2}{9} = \frac{18}{90} = \frac{1}{5}$

19. a) $P(A \cup B) = P(A) + P(B) - P(A \cap B)$
    $\qquad\qquad = .8 + .5 - .4 = .9$

b) $P(B \mid A) = \dfrac{P(A \cap B)}{P(A)} = \dfrac{.4}{.8} = .5$

c) $P(A \mid B) = \dfrac{P(A \cap B)}{P(B)} = \dfrac{.4}{.5} = .8$

21. $P(C) = .30$
$P(D) = .55$
$P(D \mid C) = .90$
a) P(both Catholic and Democrat)
$= P(C \cap D) = P(C)P(D \mid C) = (.30)(.90) = .27$

b) P(Catholic or Democrat)
$= P(C \cup D) = P(C) + P(D) - P(C \cap D)$
$= .30 + .55 - .27 = .58$

c) P(Catholic given that he or she is a Democrat)
$= P(C \mid D) = \dfrac{P(C \cap D)}{P(D)} = \dfrac{.27}{.55} \approx .4909$

23. a) P(Mary will decipher the message and John won't)
$= \frac{1}{2} \cdot \frac{1}{3} = \frac{1}{6}$

b) P(Mary will decipher the message or John will decipher it)
$= P$(Mary will decipher it) $+ P$(John will decipher it)
$\quad - P$(they will both decipher it)
$= \frac{1}{2} + \frac{2}{3} - \frac{1}{2} \cdot \frac{2}{3} = \frac{5}{6}$

25. There are $_3C_2$ sets of 2 games that Mary can win. The probability of Mary winning 2 and John losing the third is $\left(\frac{2}{3}\right)^2\left(\frac{1}{3}\right)$.
P(Mary wins any 2 games and John loses the third)
$= {_3C_2}\left(\frac{2}{3}\right)^2\left(\frac{1}{3}\right) = \dfrac{3 \cdot 4}{27} = \dfrac{4}{9}$

27. a) P(2 white) $= \dfrac{5}{16} \cdot \dfrac{5}{16} = \dfrac{25}{256}$

b) P(2 white) $= \dfrac{5}{16} \cdot \dfrac{4}{15} = \dfrac{1}{12}$

29. a) P(both red) $= \dfrac{3}{10} \cdot \dfrac{3}{10} = \dfrac{9}{100}$

b) P(one white and the other black)
$= P$(first is white and second is black)
$\quad + P$(first is black and second is white)
$= \dfrac{2}{10} \cdot \dfrac{5}{10} + \dfrac{5}{10} \cdot \dfrac{2}{10} = \dfrac{1}{5}$

c) P(both balls have the same color)
$= P$(2 white) $+ P$(2 red) $+ P$(2 black)
$= \dfrac{2}{10} \cdot \dfrac{2}{10} + \dfrac{3}{10} \cdot \dfrac{3}{10} + \dfrac{5}{10} \cdot \dfrac{5}{10} = \dfrac{38}{100} = \dfrac{19}{50}$

d) P(the balls have different colors)
$= 1 - P$(both balls have the same color)
$= 1 - \dfrac{19}{50} = \dfrac{31}{50}$

31. P(getting a 7) $= \dfrac{6}{36}$

P(getting an 11) $= \dfrac{2}{36}$
We ignore all tosses until a 7 or an 11 occurs. Thus,

$$P(7 \mid 7 \text{ or } 11) = \dfrac{P(7 \cap (7 \text{ or } 11))}{P(7 \text{ or } 11)}$$

$$= \dfrac{P(7)}{P(7) + P(11)} = \dfrac{\frac{6}{36}}{\frac{6}{36} + \frac{2}{36}} = \dfrac{6}{8} = \dfrac{3}{4}$$

33. The probability if getting at least one head is 1 minus the probability of getting no heads. The probability of no heads on any one coin toss (i.e., of getting a tail) is $1 - .6 = .4$.

P(getting at least one head)
$= 1 - (.4)^8 = 1 - .000655360 = .999344640$

b) P(at least 7 tails) $= P$(7 tails) $+ P$(8 tails)
$= (.6)(.4)^7(8) + (.4)^8$
$= .007864320 + .00655360$
$= .008519680.$

35. The probability that Amy wins on her first toss is $\frac{1}{2}$; on her second toss, $\frac{1}{8} \cdot \frac{1}{2}$; on her third toss, $\left(\frac{1}{8}\right)^2\left(\frac{1}{2}\right)$; and so on. Therefore,

$$P(\text{Amy wins}) = \frac{1}{2} + \frac{1}{8} \cdot \frac{1}{2} + \left(\frac{1}{8}\right)^2\left(\frac{1}{2}\right) + \cdots$$

$$= \dfrac{a}{1 - r} = \dfrac{\frac{1}{2}}{1 - \frac{1}{8}} = \dfrac{4}{7}$$

Similarly,
$$P(\text{Betty wins}) = \frac{1}{2} \cdot \frac{1}{2} + \frac{1}{8} \cdot \frac{1}{4} + \left(\frac{1}{8}\right)^2\left(\frac{1}{4}\right) + \cdots$$

$$= \dfrac{\frac{1}{4}}{1 - \frac{1}{8}} = \dfrac{2}{7}$$

Finally, P(Candy wins) $= 1 - \frac{4}{7} - \frac{2}{7} = \frac{1}{7}.$

37. Let A mean history precedes English.
Let B mean history immediately precedes English. We seek $P(B \mid A)$. We know that $P(A) = \frac{1}{2}$. Also, $P(A \cap B) = P(B)$. There are 5 ways to schedule the history-English pair and $4!$ ways to schedule the remaining 4 courses, so $P(B) = \dfrac{5 \cdot 4!}{6!} = \dfrac{1}{6}$

Therefore, $P(B \mid A) = \dfrac{P(A \cap B)}{P(A)} = \dfrac{\frac{1}{6}}{\frac{1}{2}} = \dfrac{1}{3}.$

Chapter 12 Review Problem Set
-----

1. False; any sequence a, a, a,... is both arithmetic and geometric.

2. False. A recursion formula for a sequence relates the value of a term to values of previous terms.

3. True

4. True

5. True

6. True

7. False; there are 2^{10} ways to respond to the quiz.

8. True

9. True:
$$_nC_6 = {}_nC_7$$
$$\frac{n!}{6!(n-6)!} = \frac{n!}{7!(n-7)!}$$
$$6!(n-6)! = 7!(n-7)!$$
$$\frac{(n-6)!}{(n-7)!} = \frac{7!}{6!}$$
$$n-6 = 7$$
$$n = 13$$

10. False. Disjoint means not simultaneous.

11. (b) is geometric with $r = -\frac{1}{2}$;
    (c) is geometric with $r = \sqrt{2}$

12. (a) is arithmetic with $d = 3$;
    (d) is arithmetic with $d = 2\pi$

13. a) $d_1 = \pi$; $d_n = d_{n-1} + 2\pi$

    b) $e_1 = 1$, $e_2 = 1$, $e_3 = 1$; $e_n = e_{n-1} + e_{n-2} + e_{n-3}$

14. The nth term is given by $(2n-1)\pi$. The 51st term equals 101π.

15. Use the formula $a_n = a_1 + (n-1)d$ with
    $a_1 = 3$, $d = 3$, $n = 100$:
    $a_{100} = 3 + 99(3) = 300$
    Use the formula $A_n = \frac{n}{2}(a_1 + a_n)$:
    $$A_{100} = \frac{100}{2}(3 + 300) = 50(303) = 15{,}150$$

16. This sequence is geometric with first term 1 and common ratio $\sqrt{2}$. Therefore, the sum of the first 10 terms is given by
    $$\frac{1(1-(\sqrt{2})^{10})}{1-\sqrt{2}} = \frac{1-32}{1-\sqrt{2}} = 31(\sqrt{2}+1)$$

17. Use the formula $S = \frac{a_1}{1-r}$ with $a_1 = -1$, $r = -\frac{1}{2}$:

    The sum of all the terms equals $\dfrac{-1}{1-(-\frac{1}{2})} = \dfrac{-1}{\frac{3}{2}} = -\dfrac{2}{3}$

18. His allowance would be the sum of a geometric sequence with first term $0.01 and common ratio 2. His total allowance would be
    $$\frac{.01(1-2^{30})}{1-2} = \$10{,}737{,}418.23 \quad \text{WOW!}$$

19. $2.22\overline{2} = 2 + \frac{2}{10} + \frac{2}{100} + \frac{2}{1000} + \cdots$

    This is a geometric series with $a_1 = 2$ and $r = \frac{1}{10}$.
    The sum of all the terms is $\dfrac{2}{1-\frac{1}{10}} = \dfrac{2}{\frac{9}{10}} = \dfrac{20}{9}$.

    $2.22\overline{2} = \frac{20}{9}$

20. a) $\displaystyle\sum_{i=1}^{50}(3i-2) = \frac{50}{2}(1+148) = 3725$
    (arithmetic sequence)

    b) $\displaystyle\sum_{i=1}^{\infty}4(\frac{1}{3})^i = \dfrac{\frac{4}{3}}{1-\frac{1}{3}} = 2$

21. $a_3 = a_2 + 2(a_1) = 1 + 2(1) = 3$
    $a_4 = a_3 + 2a_2 = 3 + 2(1) = 5$
    $a_5 = a_4 + 2a_3 = 5 + 2(3) = 11$
    $a_6 = a_5 + 2a_4 = 11 + 2(5) = 21$
    $a_7 = a_6 + 2a_5 = 21 + 2(11) = 43$

22. A straight $1800 annual raise would amount to an arithmetic sequence and we are looking for the tenth term. Her salary would equal
    $$\$30{,}000 + 9 \cdot \$1800 = \$46{,}200.$$
    A 5% annual raise would mean a geometric sequence with common ratio 1.05. Her salary in the tenth year would equal
    $$\$30{,}000(1.05)^9 = \$46{,}539.85.$$
    Thus, the 5% raise would yield a better income during her tenth year by $339.85.

23. A straight $1800 raise each year would give Pat a total income of
    $30{,}000 + (30{,}000 + 1800) + \cdots + (30{,}000 + 9 \cdot 1800)$
    $= \frac{10}{2}(30{,}000 + 46{,}200) = \$381{,}000$
    A 5% raise each year would give Pat a total income of $30{,}000 + 30{,}000(1.05) + 30{,}000(1.05)^2 + \cdots$
    $\cdots + 30{,}000(1.05)^9$
    $$= \frac{30000[1-(1.05)^{10}]}{1-1.05}$$
    $$= \frac{30000(1-1.62889)}{-0.05}$$
    $= \$377{,}336.78$
    A straight $1800 a year raise would give Pat more total income by $381{,}000 - \$377{,}366.78 = \$3{,}663.22.$

24. a) $_{13}P_{10} = 13 \cdot 12 \cdot 11 \cdot 10 \cdot 9 \cdot 8 \cdot 7 \cdot 6 \cdot 5 \cdot 4$
$= 1{,}037{,}836{,}800$

b) $_{13}C_{10} = \dfrac{_{13}P_{10}}{10!} = \dfrac{13 \cdot 12 \cdot 11 \cdot 10 \cdot 9 \cdot 8 \cdot 7 \cdot 6 \cdot 5 \cdot 4}{10 \cdot 9 \cdot 8 \cdot 7 \cdot 6 \cdot 5 \cdot 4 \cdot 3 \cdot 2 \cdot 1}$
$= 13 \cdot 2 \cdot 11 = 286$
Alternatively, we could note that
$_{13}C_{10} = {}_{13}C_3 = \dfrac{13 \cdot 12 \cdot 11}{3 \cdot 2 \cdot 1} = 286$

25. $_{10}P_3 = 10 \cdot 9 \cdot 8 = 720$ ways

26. a) This is a "combinations" problem since the order of the scoops doesn't matter. Therefore, there are $_{20}C_2 = \dfrac{20 \cdot 19}{2 \cdot 1} = 190$ possible cones.

b) There are 190 possible cones with different flavors and 20 more cones that have the same flavor on both scoops. Consequently, there is a total of 210 possible cones.

27. a) Since the A is repeated twice,
$\dfrac{7!}{2!} = 7 \cdot 6 \cdot 5 \cdot 4 \cdot 3 = 2520$ code words can be made.

b) S is repeated 3 times; E is repeated twice.
$\dfrac{7!}{3! \, 2!} = \dfrac{7 \cdot 6 \cdot 5 \cdot 4}{2} = 420$ code words can be made.

28. There are 2 choices for the crust, 2 choices for the cheese, $_5C_3$ choices of meats and $_3C_2$ choices of vegetables. Therefore, there are $2 \cdot 2 \cdot {}_5C_3 \cdot {}_3C_2$
$= 2 \cdot 2 \cdot 10 \cdot 3 = 120$ different pizzas.

29. a) Seat any one of the 12 people anywhere. Then the remaining 11 people can be arranged in $11! = 39{,}916{,}800$ ways.

b) Put King Arthur in his seat. Lancelot has a choice of 2 seats. The remaining 10 people can be arranged in $10!$ ways. All together, there are $2 \cdot 10! = 7{,}257{,}600$ arrangements.

30. We have $_6C_2$ ways to choose the guards, $_4C_1$ ways to pick the center, and $_5C_2$ choices for the forwards. Therefore, there are $_6C_2 \cdot {}_4C_1 \cdot {}_5C_2$
$= \dfrac{6 \cdot 5}{2 \cdot 1} \cdot 4 \cdot \dfrac{5 \cdot 4}{2 \cdot 1} = 600$ different starting lineups.

31. The coach can choose 9 people to play field positions in $_{12}P_9$ ways. He can then arrange the 9 people in $9!$ different batting orders. By the multiplication principle, the coach can make his choices in $9! \cdot {}_{12}P_9$ ways.

32. If we start with 2 letters, $26^2 \cdot 10^5$ plates are possible. If we start with 3 letters, $26^3 \cdot 10^4$ plates are possible. If we start with 4 letters, $26^4 \cdot 10^3$ plates are possible. Therefore, we have a total of
$26^2 \cdot 10^5 + 26^3 \cdot 10^4 + 26^4 \cdot 10^3$
$= 26^2 \cdot 10^3(10^2 + 26^1 10^1 + 26^2)$
$= 700{,}336{,}000$ possible license plates.

33. You have 10 choices for the first digit of the combination and 10 choices for the second. But once the second digit is chosen, there are only 8 possibilities for the third digit. By the multiplication principle, the number of legal combinations is $10 \cdot 10 \cdot 8 = 800$.

34. If $n = 1$, then the set has just two subsets, namely itself and $\emptyset$. Thus, there are 2^1 subsets. Hence, the statement is true for $n = 1$.
Assume the statement is true for $n = k$. That is, assume that every set with k elements has 2^k subsets. Suppose the set S has $k + 1$ elements. Let $a \in S$ and consider the set $S - \{a\}$
$= \{x : x \in S, \, x \neq a\}$. This set has k elements. By our induction hypothesis $S - \{a\}$ has 2^k subsets, all of which are subsets that *do not* contain a. If we add a into each of these subsets we obtain all subsets of S that *do* contain a. Therefore, S has a total of $2^k + 2^k = 2^{k+1}$ subsets.
Hence, by mathematical induction, any set with n elements has 2^n subsets.

35. The statement is true for $n = 4$, since $4! = 24 > 16 = 2^4$. Assume the statement is true for $n = k$; that is, $\qquad k! > 2^k$
Then for $n = k + 1$,
$\begin{aligned}(k+1)! = (k+1)(k!) &> (k+1)2^k \\ &> k \cdot 2^k \\ &> 2 \cdot 2^k \text{ (because } k \geq 4) \\ &= 2^{k+1}\end{aligned}$
So the statement is true for $n = k + 1$, provided it is true for $n = k$. Hence by mathematical induction, the statement is true for all $n \geq 4$.

36. This statement is *false*. It can be easily disproven by letting $n = 11$ and obtaining $11^2 - 11 + 11 = 11^2$.

37. P_n is true when n is a positive even integer.

38. $(x + 2y)^6$
$= x^6 + 6x^5(2y) + {}_6C_2 x^4(2y)^2 + {}_6C_3 x^3(2y)^3$
$\qquad\quad + {}_6C_4 x^2(2y)^4 + {}_6C_5 x(2y)^5 + (2y)^6$
$= x^6 + 6x^5(2y) + 15x^4(2y)^2 + 20x^3(2y)^3$
$\qquad\quad + 15x^2(2y)^4 + 6x(2y)^5 + (2y)^6$
$= x^6 + 12x^5 y + 60x^4 y^2 + 160x^3 y^3 + 240x^2 y^4$
$\qquad\qquad\qquad\qquad\qquad + 192xy^5 + 64y^6$

39. $\displaystyle\sum_{r=0}^{6}{}_6C_r2^r = {}_6C_0(2)^0 + {}_6C_1(2)^1 + \cdots + {}_6C_6(2)^6$

$= {}_6C_0(1)^6(2)^0 + {}_6C_1(1)^5(2)^1 + \cdots$

$\cdots + {}_6C_6(1)^0(2)^6$

$= (1+2)^6 = 3^6$

$= 729$

40. The term that involves y^6 is the term involving $(-3y^2)^3$, namely ${}_8C_5(2x)^5(-3y^2)^3$

$= \dfrac{8\cdot7\cdot6\cdot5\cdot4}{5\cdot4\cdot3\cdot2\cdot1}\cdot32x^5(-27)y^6 = -48{,}384x^5y^6$

41. A person must choose to walk any 3 blocks east and any 5 blocks north. The problem is analogous to asking how many different arrangements can be made using the letters EEENNNNN. The answer to that question is $\dfrac{8!}{3!\,5!} = \dfrac{8\cdot7\cdot6}{3\cdot2\cdot1} = 56$.

42. $(1.00001)^8 = (1+.00001)^8$

$= 1 + 8(.00001) + {}_8C_2(.00001)^2 + {}_8C_3(.00001)^3$
$\quad + {}_8C_4(.00001)^4 + {}_8C_5(.00001)^5{}_8C_6(.00001)^6$
$\quad + {}_8C_7(.00001)^7 + (.00001)^8$

$= 1 + 8(.00001) + 28(.00001)^2 + 56(.00001)^3$
$\quad + 70(.00001)^4 + 56(.00001)^5 + 28(.00001)^6\cdots$

$\approx 1 + 8(.00001) + 28(.00001)^2 + 56(.00001)^3$

$= 1 + .00008 + .0000000028 + .00000000000056$

$= 1.00008\,00028\,00056$

43. There are 2^{26} subsets of any kind. Since there are equal numbers of subsets with even and odd elements, (as found in Section 12.7: Binomial Coefficients,) half the total number of subsets will have an even number of elements.

$\dfrac{2^{26}}{2} = 2^{25}$

44. a) $P(\text{odd}) = \dfrac{1}{2}$

b) $P(\text{multiple of 5}) = \dfrac{1}{5}$

c) $P(\text{digit sum} > 16) = P(89 \text{ or } 98 \text{ or } 99) = \dfrac{3}{100}$

d) There are 9 choices for the first digit and 9 for the second digit except that 00, corresponding to 100, is excluded.

$P(\text{no ones digit}) = \dfrac{9\cdot9 - 1}{100} = \dfrac{80}{100} = \dfrac{4}{5}$

45. There are ${}_{52}C_5$ ways to choose a 5-card poker hand. There are only 4 ways to choose a royal flush: one royal flush to each suit.

$P(\text{royal flush}) = \dfrac{4}{{}_{52}C_5} \approx 1.539 \times 10^{-6}$

46. The number of ways to draw 2 balls from 18 is ${}_{18}C_2$.

a) $P(\text{2 red balls}) = \dfrac{{}_5C_2}{{}_{18}C_2}$

b) $P(\text{2 red or 2 green or 2 black}) = \dfrac{{}_5C_2 + {}_6C_2 + {}_7C_2}{{}_{18}C_2}$

47. a) $P(\text{2 red balls}) = \dfrac{5}{18}\cdot\dfrac{5}{18} \approx .0772$

b) $P(\text{2 red or 2 green or 2 black})$

$= \dfrac{5}{18}\cdot\dfrac{5}{18} + \dfrac{6}{18}\cdot\dfrac{6}{18} + \dfrac{7}{18}\cdot\dfrac{7}{18}$

$\approx .3395$

48. $P(\text{marry or graduate})$

$= P(\text{marry}) + P(\text{graduate}) - P(\text{both})$

$= .8 + .6 - .5 = .9$

49. a) $P(\text{fail English or History})$

$= P(\text{fail English}) + P(\text{fail History}) - P(\text{fail both})$

$= .30 + .20 - .08 = .42$

b) $P(\text{failed English}|\text{failed History})$

$= \dfrac{P(\text{failed English and History})}{P(\text{failed History})} = \dfrac{.08}{.20} = .4$

50. $P(\text{2 boys}) = \dfrac{{}_4C_2}{2^4} = \dfrac{6}{16} = \dfrac{3}{8}$

Chapter 13: Analytic Geometry

1. $x^2 = 8y$. The parabola opens upward.

3. $y^2 = 6x$. The parabola opens to the right.

5. $3y^2 = -5x$. The parabola opens to the left.

7. Since the focus is $(0,6)$, $p = 6$. The parabola opens upward, and has an equation of the form
$$x^2 = 4py$$
$$x^2 = 4(6)y$$
$$x^2 = 24y$$

9. If the directrix is $x = 3$, the parabola opens to the left, and the equation has the form $y^2 = -4px$ with $p = 3$.
$$y^2 = -4(3)x$$
$$y^2 = -12x$$

11. $x^2 = -8y = -4(2)y$
The parabola opens downward. $p = 2$. The focus is $(0,-2)$ and the directrix is $y = 2$.

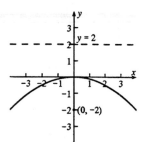

13. $y^2 = \frac{1}{2}x = 4(\frac{1}{8})x$
The parabola opens to the right. $p = \frac{1}{8}$. The focus is $(\frac{1}{8}, 0)$ and the directrix is $x = -\frac{1}{8}$.

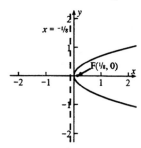

15. $y = \frac{1}{2}x^2$
$$x^2 = 2y$$
$$x^2 = 4(\frac{1}{2})y$$
The parabola opens upward. $p = \frac{1}{2}$. The focus is $(0, \frac{1}{2})$ and the directrix is $y = -\frac{1}{2}$.

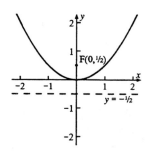

17. $9x = 4y^2$
$$y^2 = \frac{9}{4}x$$
$$y^2 = 4(\frac{9}{16})x$$
The parabola opens to the right. $p = \frac{9}{16}$. The focus is $(\frac{9}{16}, 0)$ and the directrix is $x = -\frac{9}{16}$.

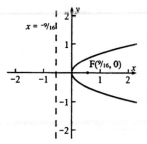

19. Let $y = 1$ in
$$y = 4x^2$$
Then
$$4x^2 = 1$$
$$x^2 = \frac{1}{4}$$
$$x = \pm\sqrt{\frac{1}{4}}$$
$$= \pm\frac{1}{2}$$
Two points on the parabola having a y-coordinate of 1 are $(\frac{1}{2}, 1)$ and $(-\frac{1}{2}, 1)$.

21. The equation has the form
$$x^2 = 4py$$
Since $(2,6)$ lies on the parabola,
$$(2)^2 = 4p(6)$$
$$4 = 24p$$
$$p = \frac{4}{24} = \frac{1}{6}$$
The equation of the parabola is
$$x^2 = 4(\frac{1}{6})y$$
$$x^2 = \frac{2}{3}y$$
$$3x^2 = 2y$$

288

23. Since the parabola opens to the right, the equation is of the form
$$y^2 = 4px$$
Since the focus and the directrix are equidistant from the origin, $p = 3$.
$$y^2 = 4(3)x$$
$$y^2 = 12x$$

25. Since the parabola goes through $(1,2)$ and $(1,-2)$, the x-axis is the axis of symmetry. Since it is also true that the vertex is $(0,0)$, the parabola must open to the right. The equation is of the form
$$y^2 = 4px$$
Since $(1,2)$ lies on the parabola,
$$(2)^2 = 4p(1)$$
$$4 = 4p$$
$$p = 1$$
The equation of the parabola is
$$y^2 = 4x$$

27. The light source should be placed at the focus of the parabola.
$$16y = x^2$$
$$4(4)y = x^2$$
$p = 4$. The light source should be placed at $(0,4)$.

29.

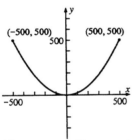

Center the cable on a coordinate system as shown. Then the equation of the parabola that represents the cable is of the form
$$x^2 = 4py$$
Since $(500,500)$ lies on the parabola,
$$(500)^2 = 4p(500)$$
$$500 = 4p$$
$$p = 125$$
Thus the equation of the parabola is
$$x^2 = 4(125)y$$
$$x^2 = 500y$$
We want the y-coordinate when $x = 120$.
$$(120)^2 = 500y$$
$$y = \frac{(120)(120)}{500} = \frac{144}{5} = 28.8$$
120 feet from the center, the strut is 28.8 feet high.

31.

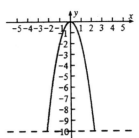

Since this chapter discusses only those parabolas whose vertex is on the origin, position a coordinate system on the arch of the doorway so that the top of the arch is at $(0,0)$ and the base is at $y = -10$. Then the equation of the parabola that represents the doorway is
$$x^2 = -4py$$
The doorway is 2 feet wide 8 feet up; $(1,-2)$ lies on the parabola.
$$(1)^2 = -4p(-2)$$
$$1 = 8p$$
$$p = \frac{1}{8}$$
Thus the equation of the parabola is
$$x^2 = -4(\tfrac{1}{8})y$$
$$x^2 = -\tfrac{1}{2}y$$
The base is 10 feet below the top of the doorway. To find out how wide the doorway is at that point, let $y = -10$:
$$x^2 = -\tfrac{1}{2}(-10)$$
$$x^2 = 5$$
$$x = \pm\sqrt{5}$$
That means that at the base. the doorway is $2\sqrt{5} \approx 4.47$ feet wide.

33.
$$4x = -5y^2$$
$$y^2 = -\tfrac{4}{5}x$$
$$y^2 = -4(\tfrac{1}{5})x$$
The parabola opens leftward. $p = \tfrac{1}{5}$. The focus is $(-\tfrac{1}{5}, 0)$ and the directrix is $x = \tfrac{1}{5}$.

35.

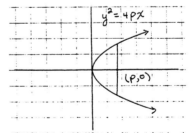

The latus rectum intersects the parabola at two points. The x-coordinate of each of those is p.

When $x = p$,
$$y^2 = 4p^2$$
$$y = \pm 2p$$
Thus, the two intersection points are $(p, 2p)$ and $(p, -2p)$. The length of the latus rectum is the distance between those two points.
$$|(2p) - (-2p)| = 4p$$

37. Place the arch on a coordinate system so that the top of the arch is at $(0,0)$. The y-axis is the axis of symmetry, and the base is at $y = -12$. The equation of the parabola that represents the arch is
$$x^2 = -4py$$
The point $(\frac{5}{2}, -12)$ lies on the parabola.
$$(\tfrac{5}{2})^2 = -4p(-12)$$
$$\tfrac{25}{4} = 48p$$
$$\tfrac{25}{192} = = p$$
The equation of the parabola is
$$x^2 = -4\left(\frac{25}{192}\right)y$$
$$x^2 = -\frac{25}{48}y$$
We want to find the width of the doorway 9 feet above the floor, because that will be the maximum width of the box. Let $y = -3$:
$$x^2 = \frac{-25}{48}(-3)$$
$$x^2 = \frac{25}{16}$$
$$x = \pm\frac{5}{4}$$
The maximum width of the box is $2 \cdot \frac{5}{4} = \frac{5}{2} = 2.5$ ft.

39.

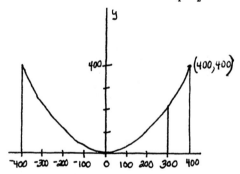

Place a coordinate system on the bridge as shown. Then the equation of the parabola that represents the cable is
$$x^2 = 4py$$
Since $(400, 400)$ lies on the parabola,
$$(400)^2 = 4p(400)$$
$$400 = 4p$$

$p = 100$
Thus the equation of the parabola is
$$x^2 = 4(100)y$$
$$x^2 = 400y$$
100 meters from the tower, $x = 300$.
$$(300)^2 = 400y$$
$$y = \frac{(300)(300)}{400} = \frac{900}{4} = 225$$
The strut is 225 meters high at that point.

41.

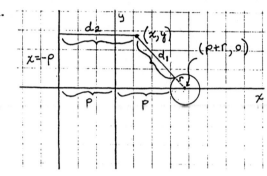

Let (x, y) be any point on the path of the submarine. Then
$$d_1 = d_2$$
$$\sqrt{(r+p-x)^2 + (0-y)^2} - r = p + x$$
$$\sqrt{(r+p-x)^2 + y^2} = p + x + r$$
Squaring both sides and removing parentheses,
$$r^2 + 2pr + p^2 - 2rx - 2px + x^2 + y^2$$
$$= p^2 + 2px + x^2 + 2pr + 2xr + r^2$$
$$-2rx - 2px + y^2 = 2px + 2rx$$
$$y^2 = 4px + 4rx$$
$$y^2 = 4(p+r)x$$

43.

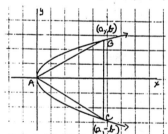

By the symmetry of the figure, if we could find the values of the coordinates (a, b) above, we would know the length of a side of the triangle.
Since $y^2 = 4px$, when $x = a$,
$$y^2 = 4pa$$
$$y = \pm\sqrt{4pa}$$
Therefore, the coordinate $b = \sqrt{4pa} = 2\sqrt{pa}$.
Since the triangle is equilateral,

$$\overline{AB} = \overline{BC}$$
$$\sqrt{a^2 + (2\sqrt{pa})^2} = 2(2\sqrt{pa}) = 4\sqrt{pa}$$
$$a^2 + (2\sqrt{pa})^2 = (4\sqrt{pa})^2$$
$$a^2 + 4pa = 16pa$$
$$a^2 - 12pa = 0$$
$$a(a - 12p) = 0$$
$$a = 0 \quad \text{or} \quad a - 12p = 0$$
$$a = 12p$$

Since we do not want $a = 0$, then $a = 12p$.

$$b = 2\sqrt{pa} = 2\sqrt{p(12p)} = 4p\sqrt{3}$$

Thus $\overline{AB} = \overline{AC} = \overline{BC} = 2(4p\sqrt{3}) = 8p\sqrt{3}$.

45. a) $\overline{FR} = \sqrt{(x - p)^2 + (y - 0)^2}$
$$= \sqrt{x^2 - 2px + p^2 + y^2}$$

Since $y^2 = 4px$,
$$\overline{FR} = \sqrt{x^2 - 2px + p^2 + 4px}$$
$$= \sqrt{x^2 + 2px + p^2}$$
$$= \sqrt{(x + p)^2}$$
$$= x + p$$

$$\overline{RG} = p - x$$

Therefore,
$$\overline{FR} + \overline{RG} = x + p + p - x = 2p$$

b) Let H have coordinates (a, b), where $b^2 = 4pa$.
$$\overline{FH} = \sqrt{(a - p)^2 + (b - 0)^2}$$
$$= \sqrt{a^2 - 2ap + p^2 + b^2}$$
$$= \sqrt{a^2 + 2ap + p^2}$$
$$= \sqrt{(a + p)^2}$$
$$= a + p$$

$$\overline{HG} = \sqrt{(p - a)^2 + (y - b)^2}$$
$$> \sqrt{(p - a)^2}$$
$$= p - a$$

Therefore,
$$\overline{FH} + \overline{HG} > (a + p) + (p - a) = 2p$$

47. Set new range values:
Ymin=0
Ymax=60
Yscl=5
Press $\boxed{Y=}$.
Enter X$\boxed{x^2}$ as Y_1.
Enter $10X - 21$ as Y_2.
Enter $\boxed{(-)}.1X + 9.3$ as Y_3.
Press $\boxed{\text{GRAPH}}$.

Zooming in on the point of intersection in Quadrant II, we find it is $B(-3.10, 9.61)$.

Zooming in on the second of the two intersection points in Quadrant I, we find it is $C(7, 49)$.

Since the slopes of the two lines are negative reciprocals of each other, triangle ABC is a right triangle, with the right angle at A.

The area of the triangle is
$\frac{1}{2}(\overline{AB})(\overline{AC})$
$$= \frac{1}{2}\sqrt{(-3.10 - 3)^2 + (9.61 - 9)^2}\sqrt{(7 - 3)^2 + (49 - 9)^2}$$
$$= \frac{1}{2}\sqrt{37.5821}\sqrt{1616}$$
$$= 123.22$$

Problem Set 13.2

1. $\dfrac{x^2}{7} + \dfrac{y^2}{16} = 1$

The larger number is under y^2, so this is a vertical ellipse.
$a^2 = 16$, $a = 4$. Major diameter $= 2a = 8$.
$b^2 = 7$, $b = \sqrt{7}$. Minor diameter $= 2b = 2\sqrt{7}$.

3. $\dfrac{x^2}{36} + \dfrac{y^2}{20} = 1$

Because the larger number is under x^2, this is a horizontal ellipse.
$a^2 = 36$, $a = 6$. Major diameter $= 2a = 12$.
$b^2 = 20$, $b = \sqrt{20} = 2\sqrt{5}$
Minor diameter $= 2b = 4\sqrt{5}$.

5. $4x^2 + 9y^2 = 4$
$$\frac{4x^2}{4} + \frac{9y^2}{4} = 1$$
$$\frac{x^2}{1} + \frac{y^2}{\frac{4}{9}} = 1$$

Because the larger number is under x^2, this is a horizontal ellipse.
$a^2 = 1$, $a = 1$. Major diameter $= 2a = 2$.
$b^2 = \frac{4}{9}$, $b = \frac{2}{3}$. Minor diameter $= 2b = \frac{4}{3}$.

7. $4k^2x^2 + k^2y^2 = 1$
$$\frac{x^2}{\frac{1}{4k^2}} + \frac{y^2}{\frac{1}{k^2}} = 1$$

Since $k^2 > 0$, $\dfrac{1}{k^2} > \dfrac{1}{4k^2}$.
Therefore, this is a vertical ellipse.

$a^2 = \dfrac{1}{k^2}$, $a = \dfrac{1}{k}$. Major diameter $= 2a = \dfrac{2}{k}$.

Minor diameter $= 2b = \dfrac{1}{k}$

9. $\frac{x^2}{25} + \frac{y^2}{9} = 1$

Because the larger number is under x^2, this is a horizontal ellipse.

$$a^2 = 25 \qquad b^2 = 9 \qquad c^2 = a^2 - b^2$$
$$a = 5 \qquad b = 3 \qquad = 25 - 9$$
$$= 16$$
$$c = 4$$

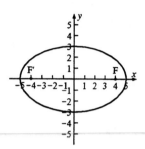

11. $\frac{x^2}{1} + \frac{y^2}{4} = 1$

Since the larger number is under y^2, this is a vertical ellipse.

$$a^2 = 4 \qquad b^2 = 1 \qquad c^2 = a^2 - b^2$$
$$a = 2 \qquad b = 1 \qquad = 4 - 1$$
$$= 3$$
$$c = \sqrt{3}$$

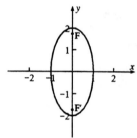

13. This is a vertical ellipse with $a = 5$ and $c = 3$.
$$a^2 = 25 \qquad c^2 = 9 \qquad b^2 = a^2 - c^2$$
$$= 25 - 9$$
$$= 16$$
$$b = 4$$

$\frac{x^2}{16} + \frac{y^2}{25} = 1$. Eccentricity $= \frac{c}{a} = \frac{3}{5}$

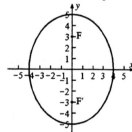

15. Since the vertex is on the same horizontal line as the center, this is a horizontal ellipse.

$$a = 7 \qquad c = 3 \qquad b^2 = a^2 - c^2$$
$$a^2 = 49 \qquad c^2 = 9 \qquad = 49 - 9$$
$$= 40$$
$$b = \sqrt{40} = 2\sqrt{10}$$

Eccentricity $= \frac{c}{a} = \frac{3}{7}$.

$$\frac{x^2}{49} + \frac{y^2}{40} = 1$$

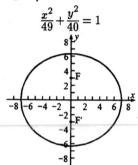

17.
$$2a = 14 \qquad 2b = 4 \qquad c^2 = a^2 - b^2$$
$$a = 7 \qquad b = 2 \qquad = 49 - 4$$
$$a^2 = 49 \qquad b^2 = 4 \qquad = 45$$
$$c = \sqrt{45} = 3\sqrt{5}$$

Eccentricity $= \frac{c}{a} = \frac{3\sqrt{5}}{7}$

Since the ellipse is horizontal, a^2 goes under x^2.

$$\frac{x^2}{49} + \frac{y^2}{4} = 1$$

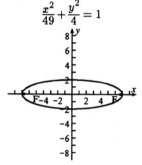

19. Since the vertices are on the x-axis, this is a horizontal ellipse with an equation of the form

$$\frac{x^2}{a^2} + \frac{y^2}{b^2} = 1$$

$a = 9$, $a^2 = 81$.
We can find b by replacing (x, y) with $(3, \sqrt{8})$:

$$\frac{(3)^2}{81} + \frac{(\sqrt{8})^2}{b^2} = 1$$

$$\frac{8}{b^2} = 1 - \frac{9}{81}$$

$$\frac{8}{b^2} = \frac{72}{81}$$

$$b^2 = \frac{8(81)}{72}$$

$b^2 = 9$

$b = 3$

So the equation of the ellipse is $\frac{x^2}{81} + \frac{y^2}{9} = 1$.

$c^2 = a^2 - b^2 = 81 - 9 = 72$, $c = \sqrt{72} = 6\sqrt{2}$.

Eccentricity $= \frac{c}{a} = \frac{6\sqrt{2}}{9} = \frac{2\sqrt{2}}{3}$.

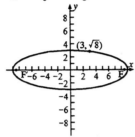

21. The equation of the bottom half of the ellipse is of the form $\frac{x^2}{a^2} + \frac{y^2}{b^2} = 1$.

Since the diameter is 12, $a = 6$, $a^2 = 36$.

Since the depth is 2, $b = 2$, $b^2 = 4$.

$c^2 = a^2 - b^2 = 36 - 4 = 32$, $c = \sqrt{32} = 4\sqrt{2}$.

Eccentricity $= \frac{c}{a} = \frac{4\sqrt{2}}{6} = \frac{2\sqrt{2}}{3}$.

23.

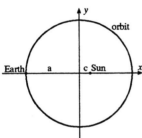

The earth will be farthest from the sun when it is at the far end of the major diameter of the elliptical orbit, a distance of $(a + c)$ kilometers.

$2a = 29{,}914{,}000$

$a = 14{,}957{,}000$

Eccentricity $= \frac{c}{a} = \frac{c}{14{,}957{,}000} = .0167$

$c \approx 249{,}782$

The earth's greatest distance from the sun is

$14{,}957{,}000 + 249{,}782 = 15{,}206{,}781$ km

25.
$$4x^2 + 25y^2 = 100$$
$$\frac{4x^2}{100} + \frac{25y^2}{100} = 1$$
$$\frac{x^2}{25} + \frac{y^2}{4} = 1$$

$a^2 = 25$ $b^2 = 4$ $c^2 = a^2 - b^2$

$a = 5$ $b = 2$ $= 25 - 24$

$= 21$

$c = \sqrt{21}$

$e = \frac{c}{a} = \frac{\sqrt{21}}{5}$

27. Since the foci are on the y-axis, this is a vertical ellipse. Because the foci are $(0, \pm 4)$, $c = 4$.

$e = \frac{c}{a} = \frac{4}{a} = \frac{1}{3}$, $a = 12$

$a^2 = 144$ $c^2 = 16$ $b^2 = a^2 - c^2$

$= 144 - 16$

$= 128$

The equation of the ellipse is $\frac{x^2}{128} + \frac{y^2}{144} = 1$

29. Place the arch on a coordinate system with the major diameter on the x-axis, the center of the diameter at $(0, 0)$. Then the equation of the ellipse that represents the arch is of the form

$$\frac{x^2}{a^2} + \frac{y^2}{b^2} = 1$$

with $a = \frac{10}{2} = 5$, $a^2 = 25$ and $b = 4$, $b^2 = 16$.

Thus the equation of the ellipse is

$$\frac{x^2}{25} + \frac{y^2}{16} = 1$$

When $y = 2$,

$$\frac{x^2}{25} + \frac{4}{16} = 1$$
$$\frac{x^2}{25} = \frac{3}{4}$$
$$x^2 = \frac{75}{4}$$
$$x = \sqrt{\frac{75}{4}} = \frac{5\sqrt{3}}{2}$$

The box can be $2\left(\frac{5\sqrt{3}}{2}\right) = 5\sqrt{3}$ feet wide.

31.

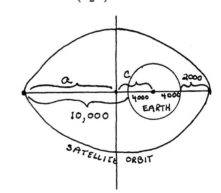

The satellite will be farthest from the earth when it is at one end of the major diameter of the orbit, and closest to the earth when it is at the other end. From the diagram, we can construct the following equations:

$$a + c = 10000 + 4000$$
$$\underline{a - c = 2000 + 4000}$$
$$2a = 20000$$
$$a = 10000$$
$$10000 + c = 14000$$
$$c = 4000$$
$$a^2 = 100{,}000{,}000; \quad c^2 = 16{,}000{,}000$$
$$b^2 = a^2 - c^2 = 84{,}000{,}000$$
$$b = \sqrt{84{,}000{,}000} = 2000\sqrt{21}$$

The major diameter of the orbit is $2a = 20{,}000$ mi. The minor diameter of the orbit is $2b = 4000\sqrt{21}$ mi.

Note: Since c turned out to be 4000 miles, the diagram should show the earth tangent to the y-axis.

33.
$$11x^2 + 7y^2 = 77$$
$$\frac{11x^2}{77} + \frac{7y^2}{77} = 1$$
$$\frac{x^2}{7} + \frac{y^2}{11} = 1$$
$$a^2 = 11 \qquad b^2 = 7$$
$$a = \sqrt{11} \qquad b = \sqrt{7}$$
$$\text{Area} = \pi ab = \pi\sqrt{11}\sqrt{7} = \pi\sqrt{77}$$

35. a) The stakes become the foci of an elliptical path.

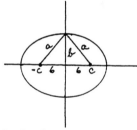

The triangle in the diagram represents one position of the stretched rope.
$$2a + 12 = 32$$
$$2a = 20$$
$$a = 10$$
$$b^2 = a^2 - c^2$$
$$= 100 - 36$$
$$= 64$$
$$b = 8$$

The area of the ellipse, which is the area the dog can cover, is $\pi ab = 80\pi$ square feet.

b) The area would become circular with $r = 16$.
$$A = \pi r^2 = \pi(256) = 256\pi \text{ square feet}$$
The area would increase $256\pi - 80\pi = 176\pi$ square feet.

37. By the definition of an ellipse, $\overline{PR} + \overline{RQ} = 2a$. Since R' is outside the ellipse,
$$\overline{PR'} + \overline{R'Q} > 2a$$
Hence,
$$\overline{PR'} + \overline{R'Q} > \overline{PR} + \overline{RQ}$$

b) Let R' be any point on the line l. From Problem 44 of Section 13.1, we know that $\overline{PR'} + \overline{R'Q}$ is minimized when $\alpha = \beta$. Furthermore, in part (a) of this problem, we showed that the minimum value of $\overline{PR'} + \overline{R'Q}$ occurs when R' is chosen to be R, a point on the ellipse. Therefore, $\alpha = \beta$.

39.
$$\frac{x^2}{64} + \frac{y^2}{49} = 1$$
$$49x^2 + 64y^2 = (64)(69)$$
$$y^2 = 49 - \frac{49}{64}x^2$$
$$y = \pm\sqrt{49 - \frac{49x^2}{64}}$$

Set standard range values.
Press $\boxed{Y=}$.
Enter $\boxed{\sqrt{}}(49 - 49X\boxed{x^2} \div 64)$ as Y_1.
Enter $\boxed{(-)}\boxed{\sqrt{}}(49 - 49X\boxed{x^2} \div 64)$ as Y_2.
Press $\boxed{\text{GRAPH}}$.

Problem Set 13.3

1. $\dfrac{x^2}{16} - \dfrac{y^2}{36} = 1$

Since the x^2 term is positive, this is a horizontal hyperbola.
$$a^2 = 16 \qquad b^2 = 36 \qquad c^2 = a^2 + b^2$$
$$a = 4 \qquad\quad b = 6 \qquad\quad = 52$$
$$c = \sqrt{52} = 2\sqrt{13}$$

3. $\dfrac{x^2}{16} - \dfrac{y^2}{9} = -1$
$$-\frac{x^2}{16} + \frac{y^2}{9} = 1$$

Since the y^2 term is positive, this is a vertical hyperbola.
$$a^2 = 9 \qquad b^2 = 16 \qquad c^2 = a^2 + b^2$$
$$a = 3 \qquad\quad b = 4 \qquad\quad = 25$$
$$c = 5$$

5. $4x^2 - 16y^2 = 1$

$$\frac{x^2}{\frac{1}{4}} - \frac{y^2}{\frac{1}{16}} = 1$$

Since the x^2 term is positive, this is a horizontal hyperbola.

$a^2 = \frac{1}{4}$ $b^2 = \frac{1}{16}$ $c^2 = a^2 + b^2$

$a = \frac{1}{2}$ $b = \frac{1}{4}$ $= \frac{5}{16}$

$$c = \sqrt{\frac{5}{16}} = \frac{1}{4}\sqrt{5}$$

7. $4x^2 - y^2 = 16$

$$\frac{4x^2}{16} - \frac{y^2}{16} = 1$$

$$\frac{x^2}{4} - \frac{y^2}{16} = 1$$

Since the x^2 term is positive, this is a horizontal hyperbola.

$a^2 = 4$ $b^2 = 16$ $c^2 = a^2 + b^2$

$a = 2$ $b = 4$ $= 20$

$$c = \sqrt{20} = 2\sqrt{5}$$

9. $\frac{x^2}{25} - \frac{y^2}{9} = 1$

Since the x^2 term is positive, this is a horizontal hyperbola.

$a^2 = 25$ $b^2 = 9$ $c^2 = a^2 + b^2 = 34$

$a = 5$ $b = 3$ $c = \sqrt{34}$

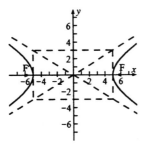

11. $\frac{y^2}{64} - \frac{x^2}{36} = 1$

Since the y^2 term is positive, this is a vertical hyperbola.

$a^2 = 64$ $b^2 = 36$ $c^2 = a^2 + b^2 = 100$

$a = 8$ $b = 6$ $c = \sqrt{100} = 10$

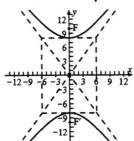

13. Since the vertices are on the y-axis, this is a vertical hyperbola with $a = 3$, $a^2 = 9$.

$$\frac{y^2}{9} - \frac{x^2}{b^2} = 1$$

To find b^2, let $(x, y) = (2, 5)$.

$$\frac{5^2}{9} - \frac{2^2}{b^2} = 1$$

$$\frac{25}{9} - \frac{4}{b^2} = 1$$

$$\frac{-4}{b^2} = \frac{-16}{9}$$

$$b^2 = \frac{4 \cdot 9}{16} = \frac{9}{4}$$

The equation of the hyperbola is

$$\frac{y^2}{9} - \frac{x^2}{\frac{9}{4}} = 1$$

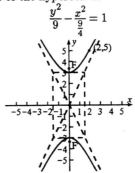

15. Since the foci are on the x-axis, this is a horizontal hyperbola.

$c = 4$ $a = 1$ $b^2 = c^2 - a^2 = 15$

$c^2 = 16$ $a^2 = 1$ $b = \sqrt{15} \approx 3.9$

$$\frac{x^2}{1} - \frac{y^2}{15} = 1$$

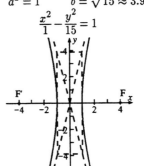

17. Since the vertices are on the x-axis, this is a horizontal hyperbola with $a = 3$, $a^2 = 9$.

Since the asymptote has slope 2,

$$\frac{b}{a} = 1$$

$$\frac{b}{3} = 2$$

$$b = 6$$

$$b^2 = 36$$

$$\frac{x^2}{9} - \frac{y^2}{36} = 1$$

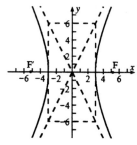

19. Since a focus and vertex are on the y-axis, this is a vertical hyperbola.

$$c = 8 \qquad a = 6 \qquad b^2 = c^2 - a^2$$
$$c^2 = 64 \qquad a^2 = 36 \qquad b^2 = 28$$

The equation of the hyperbola is

$$\frac{y^2}{36} - \frac{x^2}{28} = 1$$

21. Since the eccentricity is greater than 1, the conic is a hyperbola. Since the foci are on the x-axis, it is a horizontal hyperbola, with $c = 12$.

$$e = \frac{c}{a} = \frac{12}{a} = 3, \; a = 4$$
$$a^2 = 16 \qquad c^2 = 144 \quad b^2 = c^2 - a^2 = 128$$

The equation of the hyperbola is

$$\frac{x^2}{16} - \frac{y^2}{128} = 1$$

23. $\frac{x^2}{9} - \frac{y^2}{16} = 1$

$$a^2 = 9 \qquad b^2 = 16 \qquad c^2 = a^2 + b^2 = 25$$
$$c = 5$$

Since this is a horizontal hyperbola, c is the x-coordinate of a focus. The y-coordinate of the point on the hyperbola that lies perpendicularly above the focus is found by letting $x = 5$:

$$\frac{5^2}{9} - \frac{y^2}{16} = 1$$
$$\frac{-y^2}{16} = 1 - \frac{25}{9}$$
$$\frac{-y^2}{16} = -\frac{16}{9}$$
$$y^2 = \frac{16 \cdot 16}{9}$$
$$y = \frac{16}{3}$$

Because of the symmetry of the hyperbola, the length of the focal chord is $2(\frac{16}{3}) = \frac{32}{3}$.

25. Whether this is a vertical or horizontal hyperbola is immaterial, since the slope of the asymptotes is ± 1.

$$\frac{b}{a} = \frac{a}{b} = \pm 1$$
$$a = \pm b$$

$$a^2 = b^2$$
$$c^2 = a^2 + b^2 = 2a^2$$
$$c = a\sqrt{2}$$
$$e = \frac{c}{a} = \frac{a\sqrt{2}}{a} = \sqrt{2}$$

27. For the hyperbola $\frac{x^2}{16} - \frac{y^2}{9} = 1$,

$$a^2 = 16 \qquad b^2 = 9 \qquad c^2 = a^2 + b^2 = 25$$
$$c = 5$$

Therefore, $(-5, 0)$ is a focal point of the hyperbola. An object that strikes a branch of the hyperbola from the opposite focus will be reflected away along the line which passes through the nearby focus, in this case, through $(5, 0)$. The ball strikes the hyperbola at $(8, 3\sqrt{3})$. The equation of the line which passes through $(5, 0)$ and $(8, 3\sqrt{3})$ is

$$y - 0 = \frac{3\sqrt{3} - 0}{8 - 5}(x - 5)$$
$$y = \frac{3\sqrt{3}}{3}(x - 5)$$
$$y = \sqrt{3}(x - 5)$$

When the x-coordinate of the ball is 10, the y-coordinate is $\sqrt{3}(10 - 5) = 5\sqrt{3}$.

29.

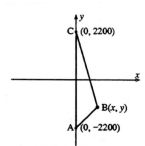

It takes the sound of the shot 6 seconds longer to traverse the distance $\overline{AC} + \overline{CB}$ than it takes for it cover the distance $\overline{AB}$. Since sound travels at the rate of 1100 feet per second,

$$\overline{AC} + \overline{CB} = \overline{AB} + 6(1100)$$
$$4400 + \overline{CB} = \overline{AB} + 6600$$
$$\overline{CB} - \overline{AB} = 2200$$

The curve will be a hyperbola $\frac{y^2}{a^2} - \frac{x^2}{b^2} = 1$, with foci at A and C, where $2a = 2200$ is the constant difference, and $c = \frac{\overline{AC}}{2} = 2200$ is the focal distance.

$$a = 1100 \qquad c = 2200 \qquad b^2 = c^2 - a^2$$
$$= 2200^2 - 1100^2$$
$$= 1100^2(2^2 - 1^2)$$
$$= 3(1100)^2$$

The equation of the hyperbola is

$$\frac{y^2}{1100^2} - \frac{x^2}{3(1100)^2} = 1$$

At the x-axis, the echo arrives $6 - 2 = 4$ seconds after the sound of the shot. Therefore, Brian must be standing below the x-axis, on the negative branch of the hyperbola.

Solve for the negative value of y:

$$y^2 - \frac{x^2}{3} = 1100^2$$

$$y = -\sqrt{1100^2 + \frac{x^2}{3}}$$

$$= -\sqrt{1,210,000 + \frac{x^2}{3}}$$

31. Set new range values:

Xmin=−30
Xmax=30
Xscl=5
Ymin=−30
Ymax=30
Yscl=5

$$\frac{y^2}{144} - \frac{x^2}{25} = 1$$

$$y^2 = 144\left(1 + \frac{x^2}{25}\right)$$

$$y = \pm 12\sqrt{1 + \frac{x^2}{25}}$$

Press $\boxed{Y=}$.

Enter $12 \boxed{\text{2nd}} \boxed{\sqrt{}} (1 + X \boxed{x2} \div 25)$ as Y_1.

Enter $\boxed{(-)} 12 \boxed{\text{2nd}} \boxed{\sqrt{}} (1 + X \boxed{x2} \div 25)$ as Y_2.

Press $\boxed{\text{GRAPH}}$.

Problem Set 13.4

1.
$$u^2 + 2(v+1)^2 - 4(v+1) = 0$$
$$u^2 + 2(v^2 + 2v + 1) - 4v - 4 = 0$$
$$u^2 + 2v^2 + 4v + 2 - 4v - 4 = 0$$
$$u^2 + 2v^2 - 2 = 0$$
$$u^2 + 2v^2 = 2$$
$$\frac{u^2}{2} + \frac{v^2}{1} = 1$$

Ellipse

3.
$$(u+2)^2 + (v-1)^2 - 4(u+2) + 2(v-1) = -4$$
$$u^2 + 4u + 4 + v^2 - 2v + 1 - 4u - 8 + 2v - 2 = -4$$
$$u^2 + v^2 - 5 = -4$$
$$u^2 + v^2 = 1$$

Circle

5.
$$(u+3)^2 - 6(u+3) - 4(v+1) + 13 = 0$$
$$u^2 + 6u + 9 - 6u - 18 - 4v - 4 + 13 = 0$$
$$u^2 - 4v = 0$$
$$u^2 = 4v$$

Parabola

7.
$$x^2 + y^2 + 12x - 2y + 33 = 0$$
Complete the square:
$$x^2 + 12x + 36 + y^2 - 2y + 1 = -33 + 36 + 1$$
$$(x+6)^2 + (y-1)^2 = 4$$
Let $u = x + 6$ and $v = y - 1$. Then
$$u^2 + v^2 = 4; \text{ circle}$$

9.
$$x^2 + y^2 + 12x - 2y + 37 = 0$$
Complete the square:
$$x^2 + 12x + 36 + y^2 - 2y + 1 = -37 + 36 + 1$$
$$(x+6)^2 + (y-1)^2 = 0$$
Let $u = x + 6$ and $v = y - 1$. Then
$$u^2 + v^2 = 0$$
The only pair of numbers that satisfies this equation is $u = v = 0$. The equation describes a point.

11.
$$x^2 - 4y^2 + 12x - 8y + 28 = 0$$
Complete the square:
$$x^2 + 12x + 36 - 4(y^2 + 2y + 1) = -28 + 36 - 4$$
$$(x+6)^2 - 4(y+1)^2 = 4$$
Let $u = y + 6$ and $v = y + 1$. Then
$$u^2 - 4v^2 = 4$$
$$\frac{u^2}{4} - v^2 = 1; \text{ hyperbola}$$

13. $\dfrac{(x+3)^2}{4} + \dfrac{(y+2)^2}{16} = 1$

An ellipse with center at $(-3, -2)$:

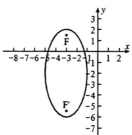

15. $\dfrac{(x+3)^2}{4} - \dfrac{(y+2)^2}{16} = 1$

A horizontal hyperbola with center $(-3, -2)$:

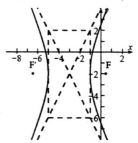

17. $(x+2)^2 = 8(y-1)$

A vertical parabola with vertex at $(-2,1)$:

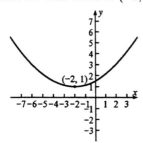

19. $(y-1)^2 = 16$

$y - 1 = \pm\sqrt{16}$

$y = 1 \pm 4$

$y = 5$ or $y = -3$

A pair of horizontal lines.

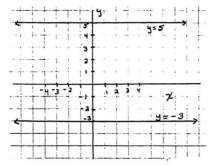

21. $4x^2 + 16x + 4y^2 - 8y = 0$

Complete the square:

$4(x^2 + 4x + 4) + 4(y^2 - 2y + 1) = 0 + 16 + 4$

$4(x+2)^2 + 4(y-1)^2 = 20$

$(x+2)^2 + (y-1)^2 = 5$

A circle of radius $\sqrt{5}$ with center at $(-2,1)$.

23. $4x^2 - 16x + y^2 - 8y = -6$

Complete the square:

$4(x^2 - 4x + 4) + y^2 - 8y + 16 = -6 + 16 + 16$

$4(x-2)^2 + (y-4)^2 = 26$

$\dfrac{4(x-2)^2}{26} + \dfrac{(y-4)^2}{26} = 1$

$\dfrac{(x-2)^2}{\frac{26}{4}} + \dfrac{(y-4)^2}{26} = 1$

A vertical ellipse with center at $(2,4)$.

25. $4x^2 - 16x + y^2 - 8y = -32$

Complete the square:

$4(x^2 - 4x + 4) + y^2 - 8y + 16 = -32 + 16 + 16$

$4(x-2)^2 + (y-4)^2 = 0$

$4(x-2)^2 = -(y-4)^2$

The only pair of numbers that satisfies the

equation is $(2,4)$. The graph of the equation is only that one point.

27. $4x^2 - 16x + y - 8 = 0$

$4(x^2 - 4x + 4) = -y + 8 + 16$

$4(x-2)^2 = -(y - 24)$

$(x-2)^2 = -\frac{1}{4}(y - 24)$

A vertical parabola with vertex at $(2,24)$.

29. $4x^2 - 16x - 9y^2 + 18y + 7 = 0$

$4(x^2 - 4x + 4) - 9(y^2 - 2y + 1) = -7 + 16 - 9$

$4(x-2)^2 - 9(y-1)^2 = 0$

$(y-1)^2 = \frac{4}{9}(x-2)^2$

$y - 1 = \pm\frac{2}{3}(x - 2)$

A pair of straight lines intersecting at $(2,1)$.

31. $9x^2 - 18x + 4y^2 + 16y = 11$

$9(x^2 - 2x + 1) + 4(y^2 + 4y + 4) = 11 + 9 + 16$

$9(x-1)^2 + 4(y+2)^2 = 36$

$\dfrac{(x-1)^2}{4} + \dfrac{(y+2)^2}{9} = 1$

Ellipse centered at $(1,-2)$:

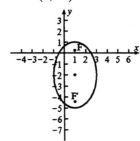

33. $-9x^2 + 18x + 4y^2 + 24y = 9$

$-9(x^2 - 2x + 1) + 4(y^2 + 6y + 9) = 9 - 9 + 36$

$-9(x-1)^2 + 4(y+3)^2 = 36$

$\dfrac{-(x-1)^2}{4} + \dfrac{(y+3)^2}{9} = 1$

This is a vertical hyperbola with $a^2 = 9$, $a = 3$. The distance between the vertices is $2a = 6$.

35. $2y^2 - 4y - 10x = 0$

$2(y^2 - 2y + 1) = 10x + 2$

$2(y-1)^2 = 10(x + \frac{1}{5})$

$(y-1)^2 = 5(x + \frac{1}{5})$

$4p = 5$

$p = \frac{5}{4}$

This is a horizontal parabola opening to the right with vertex at $(-\frac{1}{5}, 1)$. The focus is $\frac{5}{4}$ units to the right of the vertex, at $(-\frac{1}{5} + \frac{5}{4}, 1) = (\frac{21}{20}, 1)$.

The directrix is the vertical line $\frac{5}{4}$ units to the left of the vertex: $x = -\frac{1}{5} - \frac{5}{4} = -\frac{29}{20}$.

37. a) $\dfrac{(x+5)^2}{16} + \dfrac{(y-3)^2}{9} = 1$

Horizontal ellipse centered at $(-5, 3)$. $a = 4$, $b = 3$.

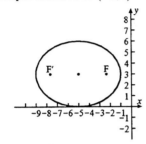

b) $\dfrac{(x+5)^2}{16} - \dfrac{(y-3)^2}{9} = 1$;   horizontal   hyperbola centered at $(-5, 3)$. $a = 4$, $b = 3$.

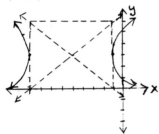

39. If $a < 0$, then $y^2 + ax^2 = x$ is a hyperbola.

If $a = 0$, then $y^2 = x$ is a parabola.

If $a > 0$ but $a \neq 1$, then $y^2 + ax^2 = x$ is an ellipse.

If $a = 1$, then $y^2 + x^2 = x$ is a circle.

41. Since the focus is on the same horizontal line as the vertex, and to the left of the vertex, this is a horizontal parabola opening to the left, with an equation of the form
$$(y-5)^2 = -4p(x-4)$$
p is the distance between the focus and the vertex.
$p = 1$: $(y-5)^2 = -4(x-4)$

43. Since the foci are $(\pm 2, 2)$, the center of the ellipse is $(0, 2)$. The major diameter is parallel to the x-axis. The equation of the ellipse has the form
$$\frac{x^2}{a^2} + \frac{(y-2)^2}{b^2} = 1$$
Let $(x, y) = (0, 0)$.

$$\frac{0}{a^2} + \frac{(-2)^2}{b^2} = 1$$

$$\frac{4}{b^2} = 1$$

$$b^2 = 4$$

We know $c = 2$, $c^2 = 4$. Therefore,
$$a^2 = b^2 + c^2 = 4 + 4 = 8.$$

The equation of the ellipse is
$$\frac{x^2}{8} + \frac{(y-2)^2}{4} = 1$$

45. The asymptotes intersect at the center of the hyperbola.

$$\begin{aligned} y &= 2x - 10 \\ y &= -2x + 2 \\ \hline 2y &= -8 \\ y &= -4 \\ -4 &= 2x - 10 \\ 6 &= 2x \\ 3 &= x \end{aligned}$$

The center of the hyperbola is $(3, -4)$. Since a focus lies on the same vertical line as the center, the equation of the hyperbola is of the form
$$\frac{(y+4)^2}{a^2} - \frac{(x-3)^2}{b^2} = 1$$

The focus is $2 - (-4) = 6$ units above the center. Therefore $c = 6$, $c^2 = 36$. The slope of the asymptotes is $\pm\frac{a}{b}$. Therefore,

$$\frac{a}{b} = 2$$
$$a = 2b$$
$$a^2 = 4b^2$$
$$c^2 = a^2 + b^2$$
$$36 = 4b^2 + b^2$$
$$36 = 5b^2$$
$$\frac{36}{5} = b^2$$
$$a^2 = 4\left(\frac{36}{5}\right) = \frac{144}{5}$$

The equation of the hyperbola is
$$\frac{(y+4)^2}{\frac{144}{5}} - \frac{(x-3)^2}{\frac{36}{5}} = 1$$

$$\frac{5(y+4)^2}{144} - \frac{5(x-3)^2}{36} = 1$$

47. We are given the points $(-1,2)$, $(0,0)$, and $(3,6)$.

a) A vertical parabola has an equation of the form
$$ax^2 + bx + c = y$$
Generate 3 equations by replacing x and y with their given values:

$(-1,2)$: $a - b + c = 2$
$(0,0)$: $c = 0$
$(3,6)$: $9a + 3b + c = 6$

This system reduces to
$$\begin{cases} a - b = 2 & (1) \\ 9a + 3b = 6 & (2) \end{cases}$$
Add 3 times equation (1) to equation (2):
$$12a = 12$$
$$a = 1$$
Substituting into equation (1),
$$1 - b = 2$$
$$b = -1$$
$a = 1$, $b = -1$, $c = 0$. The equation of the parabola is
$$x^2 - x = y$$

b)
A horizontal parabola has an equation of the form
$$ay^2 + by + c = x$$
Generate 3 equations by replacing x and y with their given values:

$(-1,2)$: $4a + 2b + c = -1$
$(0,0)$: $c = 0$
$(3,6)$: $36a + 6b + c = 3$

This system reduces to
$$\begin{cases} 4a + 2b = -1 & (1) \\ 36a + 6b = 3 & (2) \end{cases}$$
Add -3 times equation (1) to equation (2):
$$24a = 6$$
$$a = \tfrac{1}{4}$$
Substituting into equation (1),
$$1 + 2b = -1$$
$$2b = -2$$
$$b = -1$$
$a = \tfrac{1}{4}$, $b = -1$, $c = 0$. The equation of the parabola is
$$\tfrac{1}{4}y^2 - y = x$$

c) A circle will have an equation of the form
$$x^2 + y^2 + Dx + Ey = F$$
Generate 3 equations by replacing x and y with their given values:

$(-1,2)$: $1 + 4 - D + 2E = F$
$(0,0)$: $0 = F$
$(3,6)$: $9 + 36 + 3D + 6E = F$

This system reduces to
$$\begin{cases} -D + 2E = -5 \\ 3D + 6E = -45 \end{cases}$$
Add 3 times equation (1) to equation (2):
$$12E = -60$$
$$E = -5$$

Substituting into equation (1),
$$-D - 10 = -5$$
$$D = -5$$
$D = -5$, $E = -5$, $F = 0$. The equation of the circle is
$$x^2 + y^2 - 5x - 5y = 0$$

49. a) By the definition of an ellipse, the sum of the distances from a point on the ellipse to each of the foci is the constant $2a$. Since $a^2 = 25$, $a = 5$, and the length of $ACB = 2a = 10$.

b) With $a^2 = 25$ and $b^2 = 16$, we know that $c^2 = a^2 - b^2 = 9$, $c = 3$. This represents how far the focus is from the center of the ellipse on the major diameter. Since the center of the ellipse is $(3,0)$, the foci are at $(3-3,0) = (0,0)$ and $(3+3,0) = (6,0)$. Find the point of intersection of $y = x$ and the upper half of the ellipse by letting $y = x$ in the equation of the ellipse:

$$\frac{(x-3)^2}{25} + \frac{x^2}{16} = 1$$
$$16(x-3)^2 + 25x^2 = 400$$
$$16x^2 - 96x + 144 + 25x^2 = 400$$
$$41x^2 - 96x - 256 = 0$$
$$x = \frac{96 + \sqrt{96^2 - 4(41)(-256)}}{2(41)}$$
$$x \approx 3.93$$
$$y = x \approx 3.93$$

CB's length is the distance from $(3.93, 3.93)$ to $(6,0)$.
$$\overline{CB} = \sqrt{(6-3.93)^2 + (0-3.93)^2} = \sqrt{19.7298} \approx 4.44$$

51. $36x^2 + 100y^2 = 3600$
$$\frac{x^2}{100} + \frac{y^2}{36} = 1$$
$(0,6)$ is the upper y-intercept.
$a^2 = 100$, $b^2 = 36$, $c^2 = a^2 - b^2 = 64$.
The foci are $(\pm 8, 0)$.
The equation of the circle will have the form
$$x^2 + y^2 + Dx + Ey = F$$

For $(8,0)$: $64 + 8D = F$
For $(-8,0)$: $64 - 8D = F$
For $(0,6)$: $36 + 6E = F$

Subtract the first two equations:
$$16D = 0$$
$$D = 0$$
By substitution,
$$64 = F$$
$$36 + 6E = 64$$
$$6E = 28$$
$$E = \tfrac{14}{3}$$
$D = 0$, $E = \tfrac{14}{3}$, $F = 64$. The equation of the circle is
$$x^2 + y^2 + \tfrac{14}{3}y = 64$$

53. Set new range values:

Xmin=−5; Xmax=20; Ymin=−5; Ymax=20

$$(y-6)^2 + .5(x-5.5)^2 = 36$$
$$y-6 = \pm\sqrt{36 - .5(x-5.5)^2}$$
$$y = 6 \pm\sqrt{36 - .5(x-5.5)^2}$$

Press $\boxed{Y=}$.

Enter $X \wedge 2$ as Y_1.

Enter $6 + \boxed{\text{2nd}}\,\boxed{\sqrt{}}\,(36 - .5(X - 5.5)\,\boxed{x2})$ as Y_2.

Enter $6 - \boxed{\text{2nd}}\,\boxed{\sqrt{}}\,(36 - .5(X - 5.5)\,\boxed{x2})$ as Y_3.

Press $\boxed{\text{GRAPH}}$.

Zooming in on the four intersection points of the parabola and the ellipse, we find that at those points, $x_1 = -2.731$, $x_2 = -1.671$, $x_3 = .964$, $x_4 = 3.438$.

$$x_1 + x_2 + x_3 + x_4 = 0$$

Problem Set 13.5

1. $x = u\cos 60° - v\sin 60° = \frac{1}{2}u - \frac{\sqrt{3}}{2}v$

$y = u\sin 60° + v\cos 60° = \frac{\sqrt{3}}{2}u + \frac{1}{2}v$

Substitute for x and y in the equation
$$y = \sqrt{3}x$$
$$\frac{\sqrt{3}}{2}u + \frac{1}{2}v = \sqrt{3}\left(\frac{1}{2}u - \frac{\sqrt{3}}{2}v\right)$$
$$\frac{\sqrt{3}}{2}u + \frac{1}{2}v = \frac{\sqrt{3}}{2}u - \frac{3}{2}v$$
$$2v = 0$$
$$v = 0$$

3. $x = u\cos 90° - v\sin 90° = u(0) - v(1) = -v$

$y = u\sin 90° + v\cos 90° = u(1) + v(0) = u$

Substitute for x and y in the equation
$$x^2 + 4y^2 = 16$$
$$(-v)^2 + 4(u)^2 = 16$$
$$v^2 + 4u^2 = 16$$

5. $x = u\cos 45° - v\sin 45°$
$$= u\left(\frac{\sqrt{2}}{2}\right) - v\left(\frac{\sqrt{2}}{2}\right) = \frac{\sqrt{2}}{2}(u-v)$$

$y = u\sin 45° + v\cos 45°$
$$= u\left(\frac{\sqrt{2}}{2}\right) + v\left(\frac{\sqrt{2}}{2}\right) = \frac{\sqrt{2}}{2}(u+v)$$

Substitute for x and y in the equation
$$y^2 = 4\sqrt{2}x$$
$$\left[\frac{\sqrt{2}}{2}(u+v)\right]^2 = 4\sqrt{2}\left[\frac{\sqrt{2}}{2}(u-v)\right]$$
$$\frac{2}{4}(u^2 + 2uv + v^2) = 4(u-v)$$
$$\frac{u^2}{2} + uv + \frac{v^2}{2} = 4u - 4v$$

$$u^2 + 2uv + v^2 = 8u - 8v$$
$$u^2 + 2uv + v^2 - 8u + 8v = 0$$

7. $x = u\cos 45° - v\sin 45° = \frac{\sqrt{2}}{2}u - \frac{\sqrt{2}}{2}v$

$y = u\sin 45° + v\cos 45° = \frac{\sqrt{2}}{2}u + \frac{\sqrt{2}}{2}v$

Substitute for x and y in the equation
$$x^2 - xy + y^2 = 4$$
$$\left(\frac{\sqrt{2}}{2}u - \frac{\sqrt{2}}{2}v\right)^2 - \left(\frac{\sqrt{2}}{2}u - \frac{\sqrt{2}}{2}v\right)\left(\frac{\sqrt{2}}{2}u + \frac{\sqrt{2}}{2}v\right)$$
$$+ \left(\frac{\sqrt{2}}{2}u + \frac{\sqrt{2}}{2}v\right)^2 = 4$$
$$\tfrac{1}{2}u^2 - uv + \tfrac{1}{2}v^2 - (\tfrac{1}{2}u^2 - \tfrac{1}{2}v^2) + \tfrac{1}{2}u^2 + uv + \tfrac{1}{2}v^2 = 4$$
$$\tfrac{1}{2}u^2 + \tfrac{3}{2}v^2 = 4$$
$$u^2 + 3v^2 = 8$$

9. Since $\theta = \cos^{-1}(\frac{3}{5})$,

$\cos\theta = \frac{3}{5}$ and θ is in quadrant I

$\sin\theta = \sqrt{1 - (\frac{3}{5})^2} = \sqrt{\frac{16}{5}} = \frac{4}{5}$

$x = u\cos\theta - v\sin\theta = \frac{3}{5}u - \frac{4}{5}v$

$y = u\sin\theta + v\cos\theta = \frac{4}{5}u + \frac{3}{5}v$

Substitute for x and y in the equation
$$6x^2 - 24xy - y^2 = 30$$
$$6(\tfrac{3}{5}u - \tfrac{4}{5}v)^2 - 24(\tfrac{3}{5}u - \tfrac{4}{5}v)(\tfrac{4}{5}u + \tfrac{3}{5}v) - (\tfrac{4}{5}u + \tfrac{3}{5}v)^2$$
$$= 30$$
$$6(\tfrac{9}{25}u^2 - \tfrac{24}{25}uv + \tfrac{16}{25}v^2) - 24(\tfrac{12}{25}u^2 - \tfrac{7}{25}uv - \tfrac{12}{25}v^2)$$
$$- (\tfrac{16}{25}u^2 + \tfrac{24}{25}uv + \tfrac{9}{25}v^2) = 30$$
$$54u^2 - 144uv + 96v^2 - 288u^2 + 168uv + 288v^2$$
$$- 16u^2 - 24uv - 9v^2 = 750$$
$$-250u^2 + 375v^2 = 750$$
$$-2u^2 + 3v^2 = 6$$

11. $3x^2 + 10xy + 3y^2 + 8 = 0$

$A = 3$, $B = 10$, $C = 3$

$\cot 2\theta = \dfrac{A - C}{B} = \dfrac{0}{10} = 0$

$2\theta = 90°$

$\theta = 45°$

$x = u\cos 45° - v\sin 45°$
$$= u\left(\frac{\sqrt{2}}{2}\right) - v\left(\frac{\sqrt{2}}{2}\right) = \frac{\sqrt{2}}{2}(u-v)$$

$y = u\sin 45° + v\cos 45°$
$$= u\left(\frac{\sqrt{2}}{2}\right) + v\left(\frac{\sqrt{2}}{2}\right) = \frac{\sqrt{2}}{2}(u+v)$$

Substitute for x and y:

$$3\left[\frac{\sqrt{2}}{2}(u-v)\right]^2 + 10\left[\frac{\sqrt{2}}{2}(u-v)\frac{\sqrt{2}}{2}(u+v)\right]$$
$$+ 3\left[\frac{\sqrt{2}}{2}(u+v)\right]^2 + 8 = 0$$

$$\frac{3}{2}(u^2 - 2uv + v^2) + 5(u^2 - v^2)$$
$$+ \frac{3}{2}(u^2 + 2uv + v^2) + 8 = 0$$

$$\frac{3}{2}u^2 - 3uv + \frac{3}{2}v^2 + 5u^2 - 5v^2$$
$$+ \frac{3}{2}u^2 + 3uv + \frac{3}{2}v^2 + 8 = 0$$

$$8u^2 - 2v^2 = -8$$
$$-u^2 + \frac{v^2}{4} = 1$$

The graph is a hyperbola on the uv-axes, with vertices on the v-axis and center at $(0,0)$.

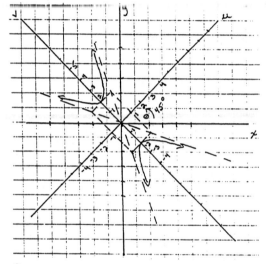

13. $4x^2 - 3xy = 18$
$A = 4$, $B = 3$, $C = 0$

$$\cot 2\theta = \frac{A-C}{B} = \frac{4-0}{-3} = -\frac{4}{3}$$

$$\cos 2\theta = -\frac{4}{5}$$

$$\sin \theta = \sqrt{\frac{1+\frac{4}{5}}{2}} = \sqrt{\frac{9}{5} \cdot \frac{1}{2}} = \frac{3\sqrt{10}}{10}$$

$$\cos \theta = \sqrt{\frac{1-\frac{4}{5}}{2}} = \sqrt{\frac{1}{5} \cdot \frac{1}{2}} = \frac{\sqrt{10}}{10}$$

$$x = \frac{\sqrt{10}}{10}u - \frac{3\sqrt{10}}{10}v$$

$$y = \frac{3\sqrt{10}}{10}u + \frac{\sqrt{10}}{10}v$$

Substitute for x and y:

$$4\left(\frac{\sqrt{10}}{10}u - \frac{3\sqrt{10}}{10}v\right)^2$$
$$- 3\left(\frac{\sqrt{10}}{10}u - \frac{3\sqrt{10}}{10}v\right)\left(\frac{3\sqrt{10}}{10}u + \frac{\sqrt{10}}{10}v\right) = 18$$

$$4\left(\frac{1}{10}u^2 - \frac{3}{5}uv + \frac{9}{10}v^2\right)$$
$$- 3\left(\frac{3}{10}u^2 - \frac{4}{5}uv - \frac{3}{10}v^2\right) = 18$$

$$4u^2 - 24uv + 36v^2 - 9u^2 + 24uv + 9v^2 = 180$$
$$-5u^2 + 45v^2 = 180$$
$$\frac{-u^2}{36} + \frac{v^2}{4} = 1$$

This is the equation of a hyperbola.

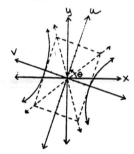

15. $x^2 - 2\sqrt{3}xy + 3y^2 - 12\sqrt{3}x - 12y = 0$
$A = 1$, $B = -2\sqrt{3}$, $C = 3$

$$\cot 2\theta = \frac{A-C}{B} = \frac{-2}{-2\sqrt{3}} = \frac{1}{\sqrt{3}}$$

$$2\theta = 60°$$
$$\theta = 30°$$

$$x = \frac{\sqrt{3}}{2}u - \frac{1}{2}v; \quad y = \frac{1}{2}u + \frac{\sqrt{3}}{2}v$$

Substitute for x and y:

$$\left(\frac{\sqrt{3}}{2}u - \frac{1}{2}v\right)^2 - 2\sqrt{3}\left(\frac{\sqrt{3}}{2}u - \frac{1}{2}v\right)\left(\frac{1}{2}u + \frac{\sqrt{3}}{2}v\right)$$
$$+ 3\left(\frac{1}{2}u + \frac{\sqrt{3}}{2}v\right)^2 - 12\sqrt{3}\left(\frac{\sqrt{3}}{2}u - \frac{1}{2}v\right)$$
$$- 12\left(\frac{1}{2}u + \frac{\sqrt{3}}{2}v\right) = 0$$

$$\frac{3}{4}u^2 - \frac{\sqrt{3}}{2}uv + \frac{1}{4}v^2 - 2\sqrt{3}\left(\frac{\sqrt{3}}{4}u^2 + \frac{1}{2}uv - \frac{\sqrt{3}}{4}v^2\right)$$
$$+ 3\left(\frac{1}{4}u^2 + \frac{\sqrt{3}}{2}uv + \frac{3}{4}v^2\right) - 18u + 6\sqrt{3}v$$
$$- 6u - 6\sqrt{3}v = 0$$

$$\frac{3}{4}u^2 - \frac{\sqrt{3}}{2}uv + \frac{1}{4}v^2 - \frac{3}{2}u^2 - \sqrt{3}uv + \frac{3}{2}v^2$$
$$+ \frac{3}{4}u^2 + \frac{3\sqrt{3}}{2}uv + \frac{9}{4}v^2 - 24u = 0$$

$$4v^2 - 24u = 0$$
$$v^2 = 6u$$

The graph is a parabola on the uv-axes, with vertex at $(0,0)$.

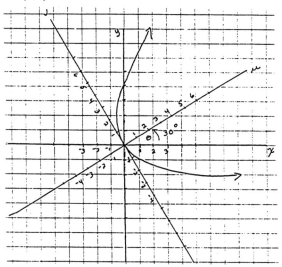

17. $13x^2 + 6\sqrt{3}xy + 7y^2 - 32 = 0$
$A = 13$, $B = 6\sqrt{3}$, $C = 7$

$$\cot 2\theta = \frac{A-C}{B} = \frac{6}{6\sqrt{3}} = \frac{1}{\sqrt{3}}$$
$$2\theta = 60°$$
$$\theta = 30°$$
$$x = \frac{\sqrt{3}}{2}u - \frac{1}{2}v; \; y = \frac{1}{2}u + \frac{\sqrt{3}}{2}v$$

Substitute for x and y:

$$13\left(\frac{\sqrt{3}}{2}u - \frac{1}{2}v\right)^2 + 6\sqrt{3}\left(\frac{\sqrt{3}}{2}u - \frac{1}{2}v\right)\left(\frac{1}{2}u + \frac{\sqrt{3}}{2}v\right)$$
$$+ 7\left(\frac{1}{2}u + \frac{\sqrt{3}}{2}v\right)^2 - 32 = 0$$

$$13\left(\frac{3}{4}u^2 - \frac{\sqrt{3}}{2}uv + \frac{1}{4}v^2\right) + 6\sqrt{3}\left(\frac{\sqrt{3}}{4}u^2 + \frac{1}{2}uv - \frac{\sqrt{3}}{4}v^2\right)$$
$$+ 7\left(\frac{1}{4}u^2 + \frac{\sqrt{3}}{2}uv + \frac{3}{4}v^2\right) - 32 = 0$$

$$\frac{39}{4}u^2 - \frac{13\sqrt{3}}{2}uv + \frac{13}{4}v^2 + \frac{18}{4}u^2 + 3\sqrt{3}uv - \frac{18}{4}v^2$$
$$+ \frac{7}{4}u^2 + \frac{7\sqrt{3}}{2}uv + \frac{21}{4}v^2 - 32 = 0$$
$$16u^2 + 4v^2 = 32$$
$$\frac{u^2}{2} + \frac{v^2}{8} = 1$$

The graph is an ellipse on the uv-axes, with center on the origin.

19. $9x^2 - 24xy + 16y^2 - 60x + 80y + 75 = 0$
$A = 9$, $B = -24$, $C = 16$

$$\cot 2\theta = \frac{A-C}{B} = \frac{-7}{-24} = \frac{7}{24}$$

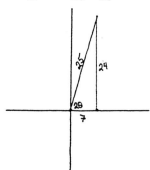

$$\cos 2\theta = \frac{7}{25}$$

$$\sin \theta = \sqrt{\frac{1 - \cos 2\theta}{2}} = \sqrt{\frac{1 - \frac{7}{25}}{2}} = \sqrt{\frac{9}{25}} = \frac{3}{5}$$

$$\cos \theta = \sqrt{\frac{1 + \cos 2\theta}{2}} = \sqrt{\frac{1 + \frac{7}{25}}{2}} = \sqrt{\frac{16}{25}} = \frac{4}{5}$$

$$x = \frac{4}{5}u - \frac{3}{5}v; \; y = \frac{3}{5}u + \frac{4}{5}v$$

Substitute for x and y:

$$9\left(\frac{4}{5}u - \frac{3}{5}v\right)^2 - 24\left(\frac{4}{5}u - \frac{3}{5}v\right)\left(\frac{3}{5}u + \frac{4}{5}v\right)$$
$$+ 16\left(\frac{3}{5}u + \frac{4}{5}v\right)^2 - 60\left(\frac{4}{5}u - \frac{3}{5}v\right)$$
$$+ 80\left(\frac{3}{5}u + \frac{4}{5}v\right) + 75 = 0$$

$$9\left(\frac{16}{25}u^2 - \frac{24}{25}uv + \frac{9}{25}v^2\right) - 24\left(\frac{12}{25}u^2 + \frac{7}{25}uv - \frac{12}{25}v^2\right)$$
$$+ 16\left(\frac{9}{25}u^2 + \frac{24}{25}uv + \frac{16}{25}v^2\right) - 48u + 36v$$
$$+ 48u + 64v + 75 = 0$$

Clear fractions (multiply both sides by 25) and remove parentheses:

$$144u^2 - 216uv + 81v^2 - 288u^2 - 168uv + 288v^2$$
$$+ 144u^2 + 384uv + 256v^2 + 2500v = -1875$$
$$v^2 + 4v + 3 = 0$$
$$(v + 3)(v + 1) = 0$$
$$v = -3 \text{ or } v = -1$$

The graph is 2 straight lines parallel to the u-axis. In fact, the original equation factors as:

$$(3x - 4y)^2 - 20(3x - 4y) + 75 = 0$$
$$(3x - 4y - 5)(3x - 4y - 15) = 0$$

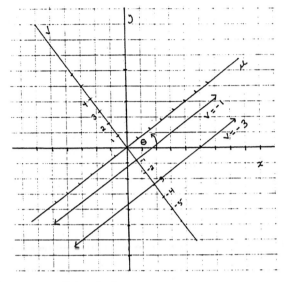

21. $u = x \cos \theta + y \sin \theta$
$= 5 \cos 60° + (-3)\sin 60°$
$= \frac{5}{2} - \frac{3\sqrt{3}}{2}$
$= \frac{5 - 3\sqrt{3}}{2}$

$v = -x \sin \theta + y \cos \theta$
$= -5 \sin 60° + (-3)\cos 60°$
$= \frac{-5\sqrt{3}}{2} - \frac{3}{2}$
$= \frac{-5\sqrt{3} - 3}{2}$

23. $u = x \cos \theta + y \sin \theta$
$= 3\sqrt{2} \cos 45° + \sqrt{2} \sin 45°$
$= 3\sqrt{2}\left(\frac{\sqrt{2}}{2}\right) + \sqrt{2}\left(\frac{\sqrt{2}}{2}\right)$
$= 3 + 1$
$= 4$

$v = -x \sin \theta + y \cos \theta$
$= -3\sqrt{2} \sin 45° + \sqrt{2} \cos 45°$
$= -3\sqrt{2}\left(\frac{\sqrt{2}}{2}\right) + \sqrt{2}\left(\frac{\sqrt{2}}{2}\right)$
$= -3 + 1$
$= -2$

25. $\theta = \tan^{-1}\left(\frac{4}{3}\right)$

$\tan \theta = \frac{4}{3}$

$\cos \theta = \frac{3}{5}$
$\sin \theta = \frac{4}{5}$
$u = x \cos \theta + y \sin \theta = 3\left(\frac{3}{5}\right) + 4\left(\frac{4}{5}\right) = \frac{9}{5} + \frac{16}{5} = 5$
$v = -x \sin \theta + y \cos \theta$
$= -3\left(\frac{4}{5}\right) + 4\left(\frac{3}{5}\right) = \frac{-12}{5} + \frac{12}{5} = 0$

27. $u = x \cos \theta + y \sin \theta$
$= x \cos 60° + y \sin 60°$
$= \frac{1}{2}x + \frac{\sqrt{3}}{2}y$

$v = -x \sin \theta + y \cos \theta$
$= -x \sin 60° + y \cos 60°$
$= -\frac{\sqrt{3}}{2}x + \frac{1}{2}y$

Substitute for u and v in the equation
$$u^2 = 4v$$
$$\left(\frac{1}{2}x + \frac{\sqrt{3}}{2}y\right)^2 = 4\left(-\frac{\sqrt{3}}{2}x + \frac{1}{2}y\right)$$
$$\frac{1}{4}x^2 + \frac{\sqrt{3}}{2}xy + \frac{3}{4}y^2 = -2\sqrt{3}x + 2y$$
$$x^2 + 2\sqrt{3}\,xy + 3y^2 = -8\sqrt{3}x + 8y$$

29. $2x^2 + \sqrt{3}xy + y^2 = 5$

$A = 2,\ B = \sqrt{3},\ C = 1$
$$\cot 2\theta = \frac{A - C}{B} = \frac{1}{\sqrt{3}}$$
$$2\theta = 60°$$
$$\theta = 30°$$
$$x = u \cos 30° - v \sin 30° = \frac{\sqrt{3}}{2}u - \frac{1}{2}v$$
$$y = u \sin 30° + v \sin 30° = \frac{1}{2}u + \frac{\sqrt{3}}{2}v$$

Substitute for x and y:
$$2\left(\frac{\sqrt{3}}{2}u - \frac{1}{2}v\right)^2 + \sqrt{3}\left(\frac{\sqrt{3}}{2}u - \frac{1}{2}v\right)\left(\frac{1}{2}u + \frac{\sqrt{3}}{2}v\right)$$
$$+ \left(\frac{1}{2}u + \frac{\sqrt{3}}{2}v\right)^2 = 5$$

$$2\left(\frac{3}{4}u^2 - \frac{\sqrt{3}}{2}uv + \frac{1}{4}v^2\right) + \sqrt{3}\left(\frac{\sqrt{3}}{4}u^2 + \frac{1}{2}uv - \frac{\sqrt{3}}{4}v^2\right)$$
$$+ \left(\frac{1}{4}u^2 + \frac{\sqrt{3}}{2}uv + \frac{3}{4}v^2\right) = 5$$

$$\frac{3}{2}u^2 - \sqrt{3}uv + \frac{1}{2}v^2 + \frac{3}{4}u^2 + \frac{\sqrt{3}}{2}uv - \frac{3}{4}v^2 + \frac{1}{4}u^2$$
$$+ \frac{\sqrt{3}}{2}uv + \frac{3}{4}v^2 = 5$$

$$\frac{5}{2}u^2 + \frac{1}{2}v^2 = 5$$
$$\frac{u^2}{2} + \frac{v^2}{10} = 1$$

The graph is an ellipse.

31. $(x - 2\sqrt{2})^2 + (y - 2\sqrt{2})^2 = 16$ is the equation of a circle with center $(2\sqrt{2}, 2\sqrt{2})$ and radius 4. When the axes are rotated through $45°$, the u-axis will pass through the center of the circle. The circle will be centered at $(4, 0)$ with regard to the uv-axes. The uv-equation is $(u - 4)^2 + v^2 = 16$.

33. $13x^2 + 24xy + 3y^2 = 105$

$A = 13,\ B = 24,\ C = 3$

$$\cot 2\theta = \frac{A - C}{B} = \frac{10}{24} = \frac{5}{12}$$

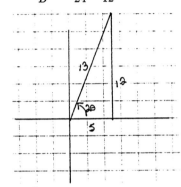

$\cos 2\theta = \frac{5}{13}$

$$\sin\theta = \sqrt{\frac{1 - \cos 2\theta}{2}} = \sqrt{\frac{1 - \frac{5}{13}}{2}} = \sqrt{\frac{8}{26}} = \frac{2}{\sqrt{13}}$$

$$\cos\theta = \sqrt{1 - \sin^2\theta} = \sqrt{1 - \frac{4}{13}} = \sqrt{\frac{9}{13}} = \frac{3}{\sqrt{13}}$$

$x = u\cos\theta - v\sin\theta = \frac{3}{\sqrt{13}}u - \frac{2}{\sqrt{13}}v$

$y = \frac{2}{\sqrt{13}}u + \frac{3}{\sqrt{13}}v$

Substitute for x and y:

$$13\left(\frac{3}{\sqrt{13}}u - \frac{2}{\sqrt{13}}v\right)^2$$
$$+ 24\left(\frac{3}{\sqrt{13}}u - \frac{2}{\sqrt{13}}v\right)\left(\frac{2}{\sqrt{13}}u + \frac{3}{\sqrt{13}}v\right)$$
$$+ 3\left(\frac{2}{\sqrt{13}}u + \frac{3}{\sqrt{13}}v\right)^2 = 105$$

$$13\left(\frac{9}{13}u^2 - \frac{12}{13}uv + \frac{4}{13}v^2\right) + 24\left(\frac{6}{13}u^2 + \frac{5}{13}uv - \frac{6}{13}v^2\right)$$
$$+ 3\left(\frac{4}{13}u^2 + \frac{12}{13}uv + \frac{9}{13}v^2\right) = 105$$

$$9u^2 - 12uv + 4v^2 + \frac{144}{13}u^2 + \frac{120}{13}uv - \frac{144}{13}v^2$$
$$+ \frac{12}{13}u^2 + \frac{36}{13}uv + \frac{27}{13}v^2 = 105$$
$$21u^2 - 5v^2 = 105$$
$$\frac{u^2}{5} - \frac{v^2}{21} = 1$$

The graph is a hyperbola.

35. Let $x = u\cos\alpha - v\sin\alpha$ and $y = u\sin\alpha + v\cos\alpha$. Substitute for x and y in the equation

$$x\cos\alpha + y\sin\alpha = d$$
$$u\cos^2\alpha - v\sin\alpha\cos\alpha + u\sin^2\alpha + v\cos\alpha\sin\alpha = d$$
$$u(\cos^2\alpha + \sin^2\alpha) = d$$
$$u = d$$

The equation $u = d$ represents a line $|d|$ units along the u-axis from the origin.

37. $5x + 12y = 39$

If $ax + by = c$, then $a = 5$, $b = 12$, and $c = 39$. From Problem 36, the perpendicular distance from the origin is

$$\frac{|c|}{\sqrt{a^2 + b^2}} = \frac{39}{\sqrt{25 + 144}} = \frac{39}{\sqrt{169}} = \frac{39}{13} = 3$$

39. Assume that $Ax^2 + Bxy + Cy^2 = 1$ has been transformed by a rotation of axes into $au^2 + cv^2 = 1$ or $\frac{u^2}{\frac{1}{a}} + \frac{v^2}{\frac{1}{c}} = 1$. Since the area of the ellipse is

$\pi(\frac{1}{2}\text{ major diameter})(\frac{1}{2}\text{ minor diameter})$, the area of the given ellipse is $\pi\sqrt{\frac{1}{a}}\sqrt{\frac{1}{c}} = \frac{\pi}{\sqrt{ac}}$.

From Problem 38,
$$b^2 - 4ac = B^2 - 4AC$$

In our transformation, $b = 0$. So,
$$-4ac = B^2 - 4AC$$
$$ac = \frac{4AC - B^2}{4}$$

The area of the ellipse is

$$\frac{\pi}{\sqrt{ac}} = \frac{\pi}{\sqrt{\frac{4AC - B^2}{4}}} = \frac{2\pi}{\sqrt{4AC - B^2}}$$

41. Set standard range values.

Press $\boxed{\text{Y=}}$.

Enter X + $\boxed{\text{2nd}}\boxed{\sqrt{}}$ (32 − 3X$\boxed{x^2}$) as Y_1.

Enter X − $\boxed{\text{2nd}}\boxed{\sqrt{}}$ (32 − 3X$\boxed{x^2}$) as Y_2.

Press $\boxed{\text{GRAPH}}$.

x_0 is the largest value of x on the ellipse, y_0 is the largest value of y on the ellipse, x_1 is the smallest value of x on the ellipse, and y_1 is the smallest value of y on the ellipse. The vertices of the circumscribed rectangle are $A(x_0, y_0)$, $B(x_1, y_0)$, $C(x_1, y_1)$, and $D(x_0, y_1)$. Since replacing x with $-x$ and y with $-y$ in the equation $4x^2 - 2xy + y^2 = 32$ has no effect, the ellipse is symmetrical about the origin.

Notice that at the maximum and minimum x-values, there is a one-pixel gap in the display, because at these points of tangency with the rectangle the graph approximates a vertical line.

Zoom in to find that $x_0 = 3.27$ and $y_0 = 6.53$. By symmetry, $x_1 = -x_0$ and $y_1 = -y_0$.
Thus, the vertices of the rectangle are $A(3.27, 6.53)$, $B(-3.27, 6.53)$, $C(-3.27, -6.53)$ and $D(3.27, -6.53)$.

Problem Set 13.6

1. Both functions are linear.

$x = 3s + 1$, $y = -2s + 5$

$2x = 6s + 2$, $3y = -6s + 15$

$2x + 3y = 17$

3. One function is not linear.
$x = 2t - 1$

$t = \dfrac{x + 1}{2}$

$y = 2t^2 + t$

$\quad = 2\left(\dfrac{x+1}{2}\right)^2 + \dfrac{x+1}{2}$

$\quad = \dfrac{x^2 + 2x + 1}{2} + \dfrac{x+1}{2}$

$\quad = \dfrac{1}{2}x^2 + \dfrac{3}{2}x + 1$

5. Cosine and sine with equal coefficients.

$x = 2 \cos t$, $y = 2 \sin t$

$\quad x^2 + y^2 = (2 \cos t)^2 + (2 \sin t)^2$

$\quad\quad\quad\quad = 4 \cos^2 t + 4 \sin^2 t$

$\quad\quad\quad\quad = 4(\cos^2 t + \sin^2 t)$

$\quad\quad\quad\quad = 4$

7. Cosine and sine with unequal coefficients.

$x = 2 \cos t$, $y = 3 \sin t$

$\dfrac{x}{2} = \cos t$, $\dfrac{y}{3} = \sin t$

$\left(\dfrac{x}{2}\right)^2 + \left(\dfrac{y}{3}\right)^2 = \cos^2 t + \sin^2 t$

$\quad\dfrac{x^2}{4} + \dfrac{y^2}{9} = 1$

9. One function not linear.

$x = 3t + 1$

$t = \dfrac{x - 1}{3}$

$y = t^3$

$y = \left(\dfrac{x-1}{3}\right)^3$

11. $(0, 0)$ and $(2, -3)$

$x = 2t$, $y = -3t$

13. $(1, 2)$ and $(4, -5)$

$\dfrac{b}{a} = \dfrac{-5 - 2}{4 - 1} = \dfrac{-7}{3}$

$x = 1 + 3t$, $y = 2 - 7t$

15. $x = 3 + 2t$, $y = -5 - 4t$

$\quad 2x = 6 + 4t$

$\quad 2x + y = 1$

$\quad\quad y = -2x + 1$

Slope: -2, y-intercept: 1

17. $\dfrac{x^2}{25} + \dfrac{y^2}{3} = 1 = \cos^2 t + \sin^2 t$

$\dfrac{x}{5} = \cos t$, $\dfrac{y}{\sqrt{3}} = \sin t$

$x = 5 \cos t$, $y = \sqrt{3} \sin t$

19. $\dfrac{x^2}{25} + \dfrac{y^2}{64} = 1 = \cos^2 t + \sin^2 t$

$\dfrac{x}{5} = \cos t$, $\dfrac{y}{8} = \sin t$

$x = 5 \cos t$, $y = 8 \sin t$

21. $\quad 9x^2 - 16y^2 = 144$

$\quad\quad \dfrac{x^2}{16} - \dfrac{y^2}{9} = 1 = \sec^2 t - \tan^2 t$

$\dfrac{x}{4} = \sec t$, $\dfrac{y}{3} = \tan t$

$x = 4 \sec t$, $y = 3 \tan t$

23. $x = 2t - 1$, $y = t^2 + 2$; $-2 \le t \le 2$

t	-2	-1	0	1	2
x	-5	-3	-1	1	3
y	6	3	2	3	6

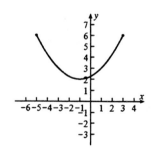

25. $x = t^3$, $y = t^2$; $-2 \le t \le 2$

t	-2	-1	0	1	2
x	-8	-1	0	1	8
y	4	1	0	1	4

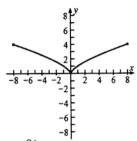

27. $x = \dfrac{1-t^2}{1+t^2}$, $y = \dfrac{2t}{1+t^2}$

$$x^2 + y^2 = \frac{1 - 2t^2 + t^4}{(1+t^2)^2} + \frac{4t^2}{(1+t^2)^2} = \frac{1 + 2t^2 + t^4}{1 + 2t^2 + t^4} = 1$$

The graph is the unit circle with $(-1,0)$ deleted because there is no value of t for which $x = -1$. As t grows very large, x approaches -1 but never reaches it.

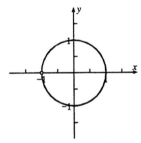

29. $x = 8t - 4 \sin t$, $y = 8 - 4 \cos t$; $0 \le t \le 4\pi$

θ	$0°$	$\frac{\pi}{2}$	π	$\frac{3\pi}{2}$	2π	$\frac{5\pi}{2}$	3π	$\frac{7\pi}{2}$	4π
x	0	$4\pi-4$	8π	$12\pi+4$	16π	$20\pi-4$	24π	$28\pi+4$	32π
y	4	8	12	8	4	8	12	8	4

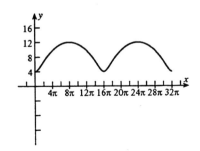

31. a) $x = 64\sqrt{3}t$

$t = \dfrac{x}{64\sqrt{3}}$

$$y = -16t^2 + 64t$$
$$= -16\left(\frac{x}{64\sqrt{3}}\right)^2 + 64\left(\frac{x}{64\sqrt{3}}\right)$$
$$= -\frac{x^2}{768} + \frac{x}{\sqrt{3}}$$

b) Find the values of t for which $y = 0$:

$$0 = -16t^2 + 64t = -16t(t - 4)$$

Thus $y = 0$ when $t = 0$ (firing) and $t = 4$ (landing). The total time of the flight is 4 seconds.

c) The projectile strikes the ground when $t = 4$.

$$x = 64\sqrt{3}(4) = 256\sqrt{3}$$

The range is $256\sqrt{3} \approx 443$ feet.

d) $y = -16(t^2 - 4t)$
$$= -16(t^2 - 4t + 4) + 64$$
$$= -16(t - 2)^2 + 64$$

has a maximum of 64 feet.

33. a) $x = 3 - 2t$, $y = 4 + 3t$

$3x = 9 - 6t$, $2y = 8 + 6t$

$3x + 2y = 17$

b)　　$x = 3t$

　　　$t = \frac{x}{3}$

　　　$y = 4 \cos t$

　　　　$= 4 \cos\left(\frac{x}{3}\right)$

c) $x = 2 \sec t$, $y = 3 \tan t$

　　　$\dfrac{x}{2} = \sec t$, $\dfrac{y}{3} = \tan t$

　　$\left(\dfrac{x}{2}\right)^2 - \left(\dfrac{y}{3}\right)^2 = \sec^2 t - \tan^2 t$

　　　$\dfrac{x^2}{4} - \dfrac{y^2}{9} = 1$

d) One equation is nonlinear; solve the linear for t.

　　　$y = 2t - 1$

　　　$t = \dfrac{y+1}{2}$

　　　$x = 1 - t^3$

　　　　$= 1 - \left(\dfrac{y+1}{2}\right)^3$

　　　　$= 1 - \dfrac{1}{8}(y+1)^3$

e) Both are nonlinear; look for common terms.

$x = t^2 + 2t$

$y = \sqrt[3]{t} - (t^2 + 2t)$

$y = \sqrt[3]{t} - x$

$\sqrt[3]{t} = x + y$

$t = (x + y)^3$

Let $t = (x + y)^3$ in $x = t^2 + 2t$:

$x = [(x + y)^3]^2 + 2(x + y)^3$

$x = (x + y)^6 + 2(x + y)^3$

35. a) $x = 2 \cos t, \ y = 2 \sin t; \ 0 \le t \le \frac{\pi}{2}$

As t increases from 0 to $\frac{\pi}{2}$,

x decreases from 2 to 0 and

y increases from 0 to 2.

$x^2 + y^2 = 4 \cos^2 t + 4 \sin^2 t = 4$

b) $x = \sqrt{t}, \ y = \sqrt{4 - t}; \ 0 \le t \le 4$

As t increases from 0 to 4,

x increases from 0 to 2 and

y decreases from 2 to 0.

$x^2 + y^2 = (\sqrt{t})^2 + (\sqrt{4 - t})^2 = t + 4 - t = 4$

c) $x = t + 1, \ y = \sqrt{3 - 2t - t^2}; \ -1 \le t \le 1$

As t increases from -1 to 1,

x increases from 0 to 2 and

y decreases from 2 to 0.

$x^2 + y^2 = (t + 1)^2 + (\sqrt{3 - 2t - t^2})^2$

$ = t^2 + 2t + 1 + 3 - 2t - t^2$

$ = 4$

d) $x = \frac{2 - 2t}{1 + t}, \ y = \frac{4\sqrt{t}}{1 + t}; \ 0 \le t \le 1$

As t increases from 0 to 1,

x decreases from 2 to 0 and

y increases from 0 to 2.

$x^2 + y^2 = \left(\frac{2 - 2t}{1 + t}\right)^2 + \left(\frac{4\sqrt{t}}{1 + t}\right)^2$

$ = \frac{4 - 8t + 4t^2}{(1 + t)^2} + \frac{16t}{(1 + t)^2}$

$ = \frac{4 + 8t + 4t^2}{1 + 2t + t^2}$

$ = 4$

37. $x = 2 + 3 \cos t, \ y = 1 + 4 \sin t$

t	0	$\frac{\pi}{2}$	π	$\frac{3\pi}{2}$	2π
x	5	2	-1	2	5
y	1	5	1	-3	1

$\frac{x - 2}{3} = \cos t, \ \frac{y - 1}{4} = \sin t$

$\frac{(x - 2)^2}{9} + \frac{(y - 1)^2}{16} = \cos^2 t + \sin^2 t = 1$

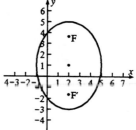

39. $x = (v_0 \cos \alpha)t, \ y = -16t^2 + (v_0 \sin \alpha)t$

a) $t = \frac{x}{v_0 \cos \alpha}$

$y = \frac{-16x^2}{(v_0 \cos \alpha)^2} + \frac{v_0 \sin \alpha}{v_0 \cos \alpha}x$

$ = \frac{-16x^2}{v_0^2 \cos^2 \alpha} + (\tan \alpha)x$

Since v_0 and α are constants, the equation defines a parabola.

b) Find the values of t for which $y = 0$:

$0 = -16t^2 + (v_0 \sin \alpha)t = -t(16t - v_0 \sin \alpha)$

$t = 0$ or $t = \frac{v_0 \sin \alpha}{16}$

The time of flight is $\frac{v_0 \sin \alpha}{16}$.

c) When $t = \frac{v_0 \sin \alpha}{16}$,

$x = \frac{v_0 \cos \alpha(v_0 \sin \alpha)}{16} = \frac{v_0^2 \sin 2\alpha}{32}$

d) The maximum range value will occur when

$\sin 2\alpha = 1$

$2\alpha = 90°$

$\alpha = 45°$

41. $x = 5 \sin^2 t - 4 \cos^2 t$

$y = 5 \sin^2 t + 4 \cos^2 t$

$x + y = 10 \sin^2 t, \ y - x = 8 \cos^2 t$

$\frac{x + y}{10} + \frac{y - x}{8} = \sin^2 t + \cos^2 t = 1$

Multiply by 40 to clear fractions:

$4x + 4y + 5y - 5x = 40$

$$9y - x = 40$$

When $t = 0$, $x = -4$, $y = 4$;

when $t = \frac{\pi}{2}$, $x = 5$, $y = 5$.

The endpoints of the segment are $(-4, 4)$ and $(5, 5)$.

The parameter interval $\frac{\pi}{2} \leq t \leq \pi$ traces the same line segment in the reverse direction.

43.

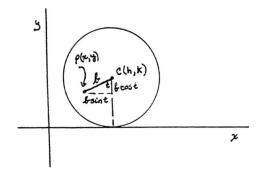

The radius of the circle is $a > b$. Our objective is to express the coordinates of P(x, y) in terms of the wheel's center C(h, k). Since the height of the center is the radius of the circle, $k = a$, and so

$$y = a - b \cos t$$

h, the distance the center has moved when the circle has rotated through an angle of t radians, is equal to the arc, at, and so

$$x = at - b \sin t$$

45.

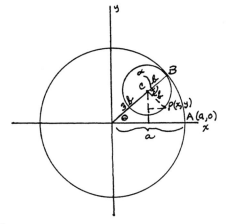

$a = 4b$

The condition of rolling is expressed by the equality of arcs AB and BP:

$$\alpha b = (4b)\theta$$

$$\alpha = 4\theta$$

Using the formula for $\cos 3t$ from Section 8-3, Problem 23,

$$
\begin{aligned}
x &= 3b \cos \theta + b \cos(\alpha - \theta) \\
&= b(3 \cos \theta + \cos 3\theta) \\
&= b[3 \cos \theta + (4 \cos^3 \theta - 3 \cos \theta)] \\
&= 4b \cos^3 \theta \\
&= a \cos^3 \theta
\end{aligned}
$$

Using the formula for $\sin 3t$ from Section 8-3, Example D,

$$
\begin{aligned}
y &= 3b \sin \theta - b \sin(\alpha - \theta) \\
&= b(3 \sin \theta - \sin 3\theta) \\
&= b[3 \sin \theta - (3 \sin \theta - 4 \sin^3 \theta)] \\
&= 4b \cos^3 \theta \\
&= a \cos^3 \theta
\end{aligned}
$$

47. a) Press $\boxed{\text{MODE}}$

Select *Param*, Press $\boxed{\text{ENTER}}$

Press $\boxed{\text{RANGE}}$

Enter TMIN $= -6.28318531$;
XMIN $= -25$; XMAX $= 25$

Press $\boxed{\text{Y=}}$

Enter $4(\boxed{\text{X|T}} - \boxed{\text{SIN}} \, \boxed{\text{X|T}}$ as X_{1T}

Enter $4(1 - \boxed{\text{COS}} \, \boxed{\text{X|T}}$ as Y_{1T}

Press $\boxed{\text{GRAPH}}$

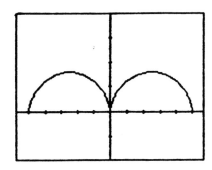

b) Press $\boxed{\text{Y=}}$

Enter $4(\boxed{\text{X|T}} - \boxed{\text{SIN}} \, 2 \, \boxed{\text{X|T}}$ as X_{1T}

Enter $4(1 - \boxed{\text{COS}} \, \boxed{\text{X|T}}$ as Y_{1T}

Press $\boxed{\text{GRAPH}}$

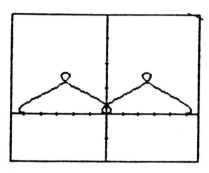

c) Press $\boxed{Y=}$

Enter $4(\boxed{X|T} - \boxed{SIN}\ 3\ \boxed{X|T}$ as X_{1T}

Enter $4(1 - \boxed{COS}\ \boxed{X|T}$ as Y_{1T}

Press $\boxed{GRAPH}$

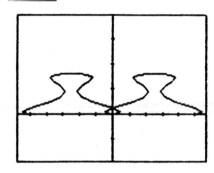

d) Press $\boxed{Y=}$

Enter $4(\boxed{X|T} - \boxed{SIN}\ 3\ \boxed{X|T}$ as X_{1T}

Enter $4(1 - \boxed{COS}\ 2\ \boxed{X|T}$ as Y_{1T}

Press $\boxed{GRAPH}$

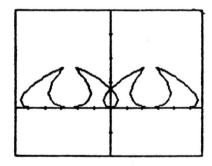

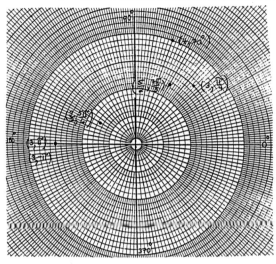

1. $\left(3, \frac{\pi}{4}\right)$ **3.** $\left(\frac{3}{2}, \frac{5\pi}{6}\right)$ **5.** $(3, \pi)$

7. $(3, -\pi)$ **9.** $(4, 70°)$ **11.** $\left(\frac{5}{2}, \frac{7\pi}{3}\right)$

13. $(r, \theta) = \left(4, \frac{\pi}{4}\right)$

$x = r \cos \theta = 4 \cos \frac{\pi}{4} = 4\left(\frac{\sqrt{2}}{2}\right) = 2\sqrt{2}$

$y = r \sin \theta = 4 \sin \frac{\pi}{4} = 4\left(\frac{\sqrt{2}}{2}\right) = 2\sqrt{2}$

$(x, y) = (2\sqrt{2}, 2\sqrt{2})$

15. $(r, \theta) = (3, \pi)$

$x = r \cos \theta = 3 \cos \pi = 3(-1) = -3$

$y = r \sin \theta = 3 \sin \pi = 3(0) = 0$

$(x, y) = (-3, 0)$

17. $(r, \theta) = \left(10, \frac{4\pi}{3}\right)$

$x = r \cos \theta = 10 \cos \frac{4\pi}{3} = 10\left(-\frac{1}{2}\right) = -5$

$y = r \sin \theta = 10 \sin \frac{4\pi}{3} = 10\left(-\frac{\sqrt{3}}{2}\right) = -5\sqrt{3}$

$(x, y) = (-5, -5\sqrt{3})$

19. $(r, \theta) = \left(2, -\frac{\pi}{4}\right)$

$x = r \cos \theta = 2 \cos\left(-\frac{\pi}{4}\right) = 2\left(\frac{\sqrt{2}}{2}\right) = \sqrt{2}$

$y = r \sin \theta = 2 \sin\left(-\frac{\pi}{4}\right) = 2\left(-\frac{\sqrt{2}}{2}\right) = -\sqrt{2}$

$(x, y) = (\sqrt{2}, -\sqrt{2})$

21. $(x, y) = (4, 0)$ is on the polar axis.

$(r, \theta) = (4, 0)$

23. $(x, y) = (-2, 0)$ is on the ray opposite the polar axis.

$(r, 0) = (-2, 0)$ or $(2, \pi)$.

25. $(x, y) = (2, 2)$ is in quadrant I.

$r = \sqrt{x^2 + y^2} = \sqrt{2^2 + 2^2} = \sqrt{8} = 2\sqrt{2}$

$\tan \theta = \frac{y}{x} = \frac{2}{2} = 1$

$\theta = \frac{\pi}{4}$

$(r, \theta) = \left(2\sqrt{2}, \frac{\pi}{4}\right)$

27. $(x, y) = (-2, 2)$ is in quadrant II.

$r = \sqrt{x^2 + y^2} = \sqrt{(-2)^2 + 2^2} = \sqrt{8} = 2\sqrt{2}$

$\tan \theta = \frac{y}{x} = \frac{2}{-2} = -1$

$\theta = \frac{3\pi}{4}$

$(r, \theta) = \left(2\sqrt{2}, \frac{3\pi}{4}\right)$

29. $(x, y) = (1, -\sqrt{3})$ is in quadrant IV.

$r = \sqrt{x^2 + y^2} = \sqrt{1^2 + (-\sqrt{3})^2} = \sqrt{4} = 2$

$\tan \theta = \frac{y}{x} = \frac{-\sqrt{3}}{1} = -\sqrt{3}$

$\theta = -\frac{\pi}{3}$

$(r, \theta) = \left(2, -\frac{\pi}{3}\right)$

31. $(x, y) = (3, -\sqrt{3})$ is in quadrant IV.

$r = \sqrt{x^2 + y^2} = \sqrt{3^2 + (-\sqrt{3})^2} = \sqrt{12} = 2\sqrt{3}$

$\tan \theta = \frac{-\sqrt{3}}{3}$

$\theta = -\frac{\pi}{6}$

$(r, \theta) = \left(2\sqrt{3}, -\frac{\pi}{6}\right)$ or $\left(2\sqrt{3}, \frac{11\pi}{6}\right)$

33. $r = 2$

r is 2 for every value of θ.

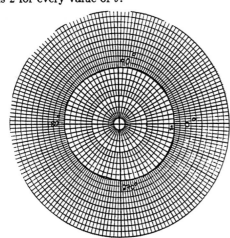

35. $\theta = \frac{\pi}{3}$

θ is $\frac{\pi}{3}$ for every value of r

Note: it is usual to allow negative values of r; then the graph is a complete line.

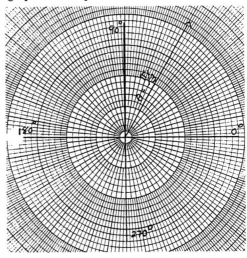

37. $r = |\theta|$

θ	0	$\pm\frac{\pi}{6}$	$\pm\frac{\pi}{3}$	$\pm\frac{\pi}{2}$	$\pm\frac{2\pi}{3}$	$\pm\frac{5\pi}{6}$	$\pm\pi$	$\pm\frac{7\pi}{6}$
r	0	.52	1.05	1.57	2.09	2.62	3.14	4.19

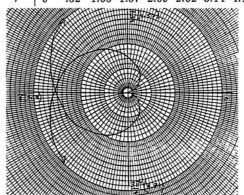

39. $r = 2(1 - \cos\theta)$ is a cardioid.

θ	0°	30°	60°	90°	120°	150°	180°
r	0	.27	1	2	3	3.7	4

θ	210°	240°	270°	300°	330°
r	3.7	3	2	1	.27

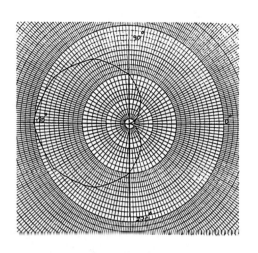

41. $r = 2 + \cos\theta$ is a limaçon without a dent.

θ	0°	30°	60°	90°	120°	150°	180°
r	3	2.9	2.5	2	1.5	1.1	1

θ	210°	240°	270°	300°	330°
r	1.1	1.5	2	2.5	2.9

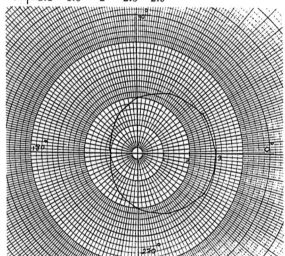

43. $r = 3\cos 2\theta$ is a four-leaved rose.

θ	0°	30°	45°	60°	90°	120°	135°
r	3	1.5	0	−1.5	−3	−1.5	0

θ	150°	180°	210°	225°	240°	270°	300°
r	1.5	3	1.5	0	−1.5	−3	−1.5

θ	315°	330°	360°
r	0	1.5	3

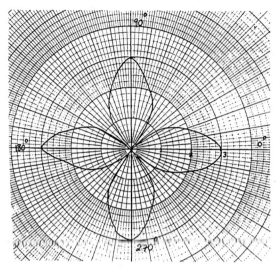

45. $r = \sin 3\theta$ is a three-leaved rose.

θ	0°	15°	30°	45°	60°	75°	90°
r	0	.7	1	.7	0	−.7	−1

θ	105°	120°	135°	150°	165°	180°	195°
r	−.7	0	.7	1	.7	0	−.7

θ	210°	225°	240°	255°	270°	285°	300°
r	−1	−.7	0	.7	1	.7	0

θ	315°	330°	345°
r	−.7	−1	−.7

47. $r = \sin 4\theta$

θ	0°	15°	22.5°	30°	45°	60°	67.5°
r	0	.87	1	.87	0	−.87	−1

θ	75°	90°	105°	112.5°	120°	135°	150°
r	−..87	0	.87	1	.87	0	−.87

312

θ	157.5°	165°	180°
r	-1	$-.87$	0

Continue the pattern the table suggests.

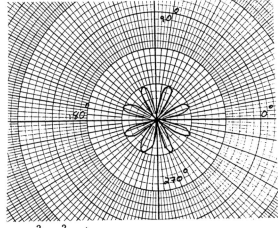

49.
$$x^2 + y^2 = 4$$
$$r^2 = 4$$
$$r = 2$$

51.
$$y = x^2$$
$$r \sin \theta = (r \cos \theta)^2 = r^2 \cos^2 \theta$$

Because the pole is on the graph we may divide by r.

$$\sin \theta = r \cos^2 \theta$$
$$r = \frac{\sin \theta}{\cos^2 \theta} = \tan \theta \sec \theta$$

53.
$$\tan \theta = 2$$
$$\frac{y}{x} = 2$$
$$y = 2x$$

Because the pole is on the graph, the equation is valid for $x = 0$ as well.

55.
$$r = \cos 2\theta$$
$$r = \cos^2 \theta - \sin^2 \theta$$

The pole is on the graph so we may multiply by r^2.

$$r^3 = (r \cos \theta)^2 - (r \sin \theta)^2$$
$$r^3 = x^2 - y^2$$
$$(r^2)^3 = (x^2 - y^2)^2$$
$$(x^2 + y^2)^3 = (x^2 - y^2)^2$$

The graph of $(x^2 + y^2)^{3/2} = x^2 - y^2$ consists of only two of the four leaves.

57. a)
$$r = \frac{5}{3 \sin \theta - 2 \cos \theta}$$
$$3r \sin \theta - 2r \cos \theta = 5$$
$$3y - 2x = 5$$

A line.

b)
$$r = 4 \cos \theta - 6 \sin \theta$$

The pole is on the graph so we may multiply by r.

$$r^2 = 4r \cos \theta - 6r \sin \theta$$
$$x^2 + y^2 = 4x - 6y$$
$$x^2 - 4x + 4 + y^2 + 6y + 9 = 4 + 9$$
$$(x - 2)^2 + (y + 3)^2 = 13$$

A circle.

59. $r = 4$
A circle of radius 4.

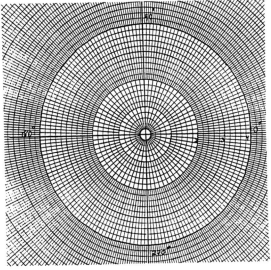

61. $r = 2(1 - \sin \theta)$
A cardioid.

θ	0°	30°	60°	90°	120°	150°	180°
r	2	1	.27	0	.27	1	2

θ	210°	240°	270°	300°	330°	360°
r	3	3.73	4	3.73	3	2

313

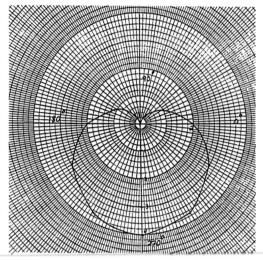

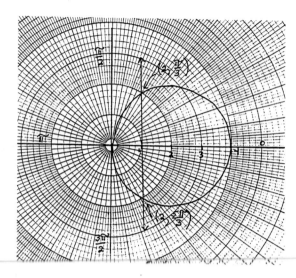

63. $r^2 = \sin 2\theta$

$r = \pm\sqrt{\sin 2\theta}$

θ	0°	15°	30°	45°	60°	75°	90°
r	0	±.7	±.9	±1	±.9	±.7	0

Do not choose 90° < θ < 180°, since $\sin 2\theta < 0$.

θ	195°	210°	225°	240°	255°	270°
r	±.7	±.9	±1	±.9	±.7	0

Do not choose 270° < θ < 360°, since $\sin 2\theta < 0$.

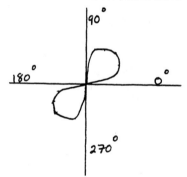

b) $r_1 = 2\sqrt{3}\sin\theta$, a circle

$r_2 = 2(1 + \cos\theta)$, a cardioid

θ	0°	30°	60°	90°	120°	150°	180°
r_1	0	1.7	3	3.5	3	1.7	0
r_2	4	3.7	3	2	1	.3	0

θ	210°	240°	270°	300°	330°
r_1	−1.7	−3	−3.5	−3	−1.7
r_2	.3	1	2	2	3.7

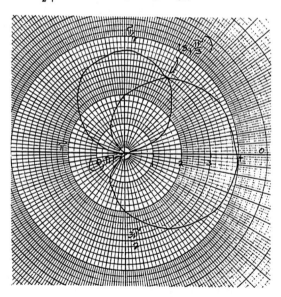

65. a) $r = 4\cos\theta$, a circle

θ	0°	30°	60°	90°	120°	150°	180°
r	4	3.46	2	0	−2	−3.46	−4

θ	210°	240°	270°	300°	330°
r	−3.4	−2	0	2	2.46

$r\cos\theta = 1$, the line $x = 1$.

67.

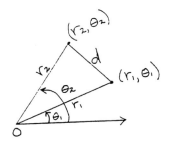

By the law of cosines,

$$d^2 = r_1{}^2 + r_2{}^2 - 2r_1r_2\cos(\theta_2 - \theta_1)$$
$$d = \sqrt{r_1{}^2 + r_2{}^2 - 2r_1r_2\cos(\theta_2 - \theta_1)}$$

With $r_1 = 4$, $\theta_1 = \frac{2\pi}{3}$, $r_2 = 8$, $\theta_2 = \frac{\pi}{6}$,

$$d = \sqrt{4^2 + 8^2 - 2(4)(8)\cos\left(-\frac{\pi}{2}\right)}$$
$$= \sqrt{80 - 64(0)}$$
$$= \sqrt{80}$$
$$= 4\sqrt{5}$$

69.

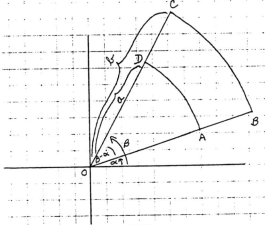

area of polar rectangle ABCD

$$= \text{area of sector } OBC - \text{area of sector } OAD$$
$$= \tfrac{1}{2}b^2(\beta - \alpha) - \tfrac{1}{2}a^2(\beta - \alpha)$$
$$= \tfrac{1}{2}(\beta - \alpha)(b^2 - a^2)$$
$$= \tfrac{1}{2}(\beta - \alpha)(b + a)(b - a)$$

71.

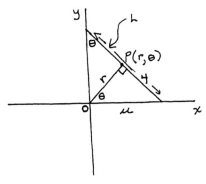

Since $\cos\theta = \frac{r}{u}$ and $\sin\theta = \frac{u}{4}$,

$r = u\cos\theta$ and $u = 4\sin\theta$.

Therefore,

$r = (4\sin\theta)\cos\theta = 2\sin 2\theta$

The equation defines a four-leaved rose, as shown in Example C.

73. Press $\boxed{\text{MODE}}$

Select *Param* and press $\boxed{\text{ENTER}}$

a) Press $\boxed{\text{Y=}}$

Enter $(5 + 4 \boxed{\text{SIN}} \boxed{\text{X|T}}) \boxed{\text{COS}} \boxed{\text{X|T}}$ as X_{1T}.

Enter $(5 + 4 \boxed{\text{SIN}} \boxed{\text{X|T}}) \boxed{\text{SIN}} \boxed{\text{X|T}}$ as Y_{1T}.

Press $\boxed{\text{GRAPH}}$

b) Press $\boxed{\text{Y=}}$

Replace the coefficient 4 with 5 in both X_{1T} and Y_{1T}.

Press $\boxed{\text{GRAPH}}$

c) Press $\boxed{\text{Y=}}$

Replace the term 5 with 3 in both X_{1T} and Y_{1T}.

Press $\boxed{\text{GRAPH}}$.

75. Press $\boxed{\text{MODE}}$
Select *Param* and press $\boxed{\text{ENTER}}$

Press $\boxed{\text{RANGE}}$

Set TSTEP to .01

Press $\boxed{\text{Y=}}$

Enter as X_{1T}:

$6 \boxed{\text{COS}} \boxed{\text{X|T}} ((\boxed{\text{COS}} 4 \boxed{\text{X|T}}) \wedge 4 + \boxed{\text{SIN}} 3 \boxed{\text{X|T}})$

Enter as Y_{1T}:

$6 \boxed{\text{SIN}} \boxed{\text{X|T}} ((\boxed{\text{COS}} 4 \boxed{\text{X|T}}) \wedge 4 + \boxed{\text{SIN}} 3 \boxed{\text{X|T}})$

Press $\boxed{\text{GRAPH}}$

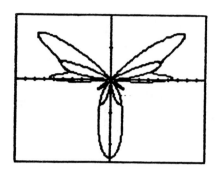

1. $x = 4$
$r \cos \theta = 4$

3. $x = -3$
$r \cos \theta = -3$

5. $r = \dfrac{6}{\cos \theta}$
$r \cos \theta = 6$
$x = 6$

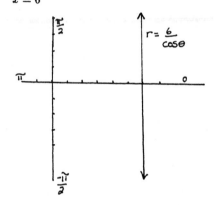

7.

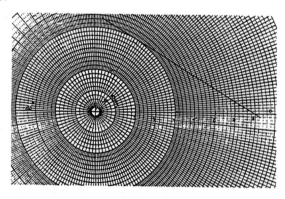

$$r = \frac{4}{\cos\left(\theta - \frac{\pi}{3}\right)}$$
$$d = 4$$
$$\theta_0 = \frac{\pi}{3}$$
$$r = \frac{4}{\cos \theta \, \cos \frac{\pi}{3} + \sin \theta \, \sin \frac{\pi}{3}}$$
$$= \frac{4}{\frac{1}{2} \cos \theta + \frac{\sqrt{3}}{2} \sin \theta}$$
$$= \frac{8}{\cos \theta + \sqrt{3} \sin \theta}$$
$$8 = r \cos \theta + \sqrt{3} \sin \theta = x + \sqrt{3}y$$
$$x + \sqrt{3}y = 8$$

9. $r = \dfrac{5}{\cos\left(\theta + \frac{\pi}{4}\right)}$
$$d = 5$$
$$\theta_0 = -\frac{\pi}{4}$$

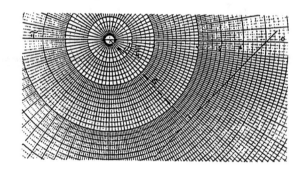

$$r = \frac{5}{\cos \theta \, \cos \frac{\pi}{4} - \sin \theta \, \sin \frac{\pi}{4}}$$
$$= \frac{5}{\frac{\sqrt{2}}{2} \cos \theta - \frac{\sqrt{2}}{2} \sin \theta}$$
$$\frac{\sqrt{2}}{2} r(\cos \theta - \sin \theta) = 5$$
$$r \cos \theta - r \sin \theta = \frac{10}{\sqrt{2}}$$
$$x - y = 5\sqrt{2}$$

11. $r = 2a \cos(\theta - \theta_0)$ with $\theta_0 = 0$; $a = 4$
$$r = 8 \cos \theta$$

To find the xy-equation, multiply by r:
$$r^2 = 8r \cos \theta$$
$$x^2 + y^2 = 8x$$

316

13. $r = 2a \cos(\theta - \theta_0)$ with $\theta_0 = \frac{\pi}{3}$; $a = 5$

$$r = 10 \cos\left(\theta - \frac{\pi}{3}\right)$$

Find the xy-equation:

$$r = 10\left[\cos\theta\cos\frac{\pi}{3} + \sin\theta\sin\frac{\pi}{3}\right]$$
$$= 10\left[\frac{1}{2}\cos\theta + \frac{\sqrt{3}}{2}\sin\theta\right]$$
$$r^2 = 5r\cos\theta + 5\sqrt{3}r\sin\theta$$
$$x^2 + y^2 = 5x + 5\sqrt{3}y$$

15. $r = \dfrac{4}{1 + \frac{2}{3}\cos\theta}$

Since $e = \frac{2}{3} < 1$, this is an ellipse.

$$ed = \frac{2}{3}d = 4$$
$$d = \frac{12}{2} = 6$$

The directrix is perpendicular to the polar axis, to the right of the focus. Its equation is $x = 6$.

17. $r = \dfrac{5}{2 + 4\cos\theta} = \dfrac{\frac{5}{2}}{1 + 2\cos\theta}$

Since $e = 2 > 1$, this is a hyperbola.

$$ed = 2d = \frac{5}{2}$$
$$d = \frac{5}{4}$$

The directrix is perpendicular to the polar axis, to the right of the focus. Its equation is $x = \frac{5}{4}$.

19. $r = \dfrac{7}{1 - \cos\theta} = \dfrac{1 \cdot 7}{1 - 1\cos\theta}$

Since $e = 1$, this is a parabola. $d = 7$.

The directrix is perpendicular to the polar axis, to the left of the focus. Its equation is $x = -7$.

21. $r = \dfrac{\frac{1}{2}}{\frac{3}{2} - \cos\theta} = \dfrac{\frac{2}{3} \cdot \frac{1}{2}}{1 - \frac{2}{3}\cos\theta}$

Since $e = \frac{2}{3} < 1$, this is an ellipse. $d = \frac{1}{2}$.

The directrix is perpendicular to the polar axis, to the left of the focus. Its equation is $x = -\frac{1}{2}$.

23. $r = \dfrac{5}{1 + \sin\theta} = \dfrac{1 \cdot 5}{1 + 1\sin\theta}$

Since $e = 1$, this is a parabola. $d = 5$.

The directrix is parallel to the polar axis, above the focus. Its equation is $y = 5$.

25. $r = \dfrac{6}{2 - \sin\theta} = \dfrac{\frac{1}{2} \cdot 6}{1 - \frac{1}{2}\sin\theta}$

Since $e = \frac{1}{2} < 1$, this is an ellipse. $d = 6$.

The directrix is parallel to the polar axis, below the focus. Its equation is $y = -6$.

27. $r = \dfrac{4}{2 + \frac{5}{2}\sin\theta} = \dfrac{2}{1 + \frac{5}{4}\sin\theta}$

Since $e = \frac{5}{4} > 1$, this is a hyperbola.

$$ed = \frac{5}{4}d = 2$$
$$d = \frac{8}{5}$$

The directrix is parallel to the polar axis and above the focus. Its equation is $y = \frac{8}{5}$.

29. $r = \dfrac{4}{1 + \frac{2}{3}\cos\theta}$

Since $e = \frac{2}{3} < 1$, this is an ellipse.

θ	0°	30°	60°	90°	120°	150°	180°
r	2.4	2.53	3	4	6	9.46	12

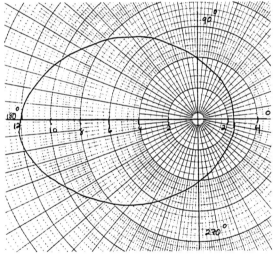

31. $r = \dfrac{5}{1 + \sin\theta}$

Since $e = 1$, this is a parabola. The directrix is parallel to the polar axis and above the focus; the parabola opens downward.

θ	0°	30°	60°	90°	120°	150°	180°
r	5	3.3	2.7	2.5	2.7	3.3	5

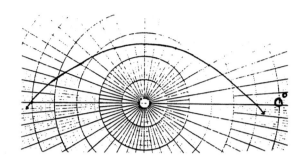

33. $r = \dfrac{18}{2 + 3\cos\theta} = \dfrac{9}{1 + \frac{3}{2}\cos\theta}$

Since $e = \frac{3}{2} > 1$, this is a hyperbola.

θ	0°	30°	60°	90°	120°	150°	180°
r	3.6	3.9	5.1	9	36	−30	−18

θ	210°	240°	270°	300°	330°
r	−30	36	9	5.1	3.9

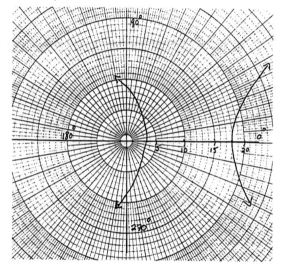

35. a) $y = 3$

$r\sin\theta = 3$

$r = \dfrac{3}{\sin\theta}$

$r = 3\csc\theta$

b) $x^2 + y^2 = 9$

$r^2 = 9$

$r = 3$

c) $(x + 9)^2 + y^2 = 81$

$x^2 + y^2 + 18x + 81 = 81$

$r^2 + 18r\cos\theta = 0$

The pole lies on the result so we may divide by r.

$r + 18\cos\theta = 0$

$r = -18\cos\theta$

d) $(x - 3)^2 + (y - 3)^2 = 18$

$x^2 - 6x + 9 + y^2 - 6y + 9 = 18$

$x^2 + y^2 - 6x - 6y = 0$

$r^2 - 6r\cos\theta - 6r\sin\theta = 0$

The pole lies on the result so we may divide by r.

$r - 6\cos\theta - 6\sin\theta = 0$

$r = 6\sqrt{2}\left(\dfrac{1}{\sqrt{2}}\cos\theta + \dfrac{1}{\sqrt{2}}\sin\theta\right)$

$= 6\sqrt{2}\left(\cos\theta\cos\dfrac{\pi}{4} + \sin\theta\sin\dfrac{\pi}{4}\right)$

$r = 6\sqrt{2}\cos\left(\theta - \dfrac{\pi}{4}\right)$

Another way to view the problem is to realize the equation defines a circle through the pole with center at $(3, 3)$ and radius $\sqrt{18} = 3\sqrt{2}$. Since the center is at $(3\sqrt{2}, \pi/4)$ in polar coordinates, the polar equation is that given above.

37. a) $y^2 = 4(x + 1)$

$x^2 + y^2 = x^2 + 4x + 4$

$r^2 = (x + 2)^2$

$r = x + 2$

$r = r\cos\theta + 2$

$r - r\cos\theta = 2$

$r = \dfrac{2}{1 - \cos\theta}$

b) $x^2 = -8(y - 2)$

$x^2 + y^2 = y^2 - 8y + 16$

$r^2 = (y - 4)^2$

$r = -(y - 4)$

$r = -(r\sin\theta - 4)$

$r + r\sin\theta = 4$

$r = \dfrac{4}{1 + \sin\theta}$

39. $0 = \cos^2\theta + \sin\theta\cos\theta - 6\sin^2\theta$

$0 = r^2\cos^2\theta + r^2\sin\theta\cos\theta - 6r^2\sin^2\theta$

$0 = x^2 + xy - 6y^2$

$0 = (x + 3y)(x - 2y)$

The pole lies on the graph so we may multiply by r^2.

41. $r = \dfrac{8}{2 + \cos \theta} = \dfrac{\frac{1}{2}(8)}{1 + \frac{1}{2}\cos \theta}$

$e = \frac{1}{2}$ and $d = 8$

The endpoints of the major diameter are on the polar axis, at $(0°, \frac{8}{3})$ and $(180°, 8)$. The length of the major diameter is

$$8 + \frac{8}{3} = \frac{32}{3}$$

The minor diameter is

$$\frac{32}{3}\sqrt{1 - e^2} = \frac{32}{3}\sqrt{1 - \left(\frac{1}{2}\right)^2} = \frac{32}{3}\sqrt{\frac{3}{4}} = \frac{16\sqrt{3}}{3}$$

43. Since the focus is at the pole, the latus rectum is the distance from

$$\left(\frac{8}{2 + \cos 90°}, 90°\right) = (4, 90°)$$

to

$$\left(\frac{8}{2 + \cos 270°}, 270°\right) = (4, 270°)$$

The distance is $4 + 4 = 8$.

45. $r = 6 + 3\cos \theta$

a) This is a limaçon with $a = 6$ and $b = 3$. Because $a \geq 2b$, there is no dent.

θ	0°	30°	60°	90°	120°	150°	180°
r	9	8.6	7.5	6	4.5	3.4	3

θ	210°	240°	270°	300°	330°
r	3.4	4.5	6	7.5	8.6

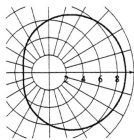

b) A chord through the pole will touch the limaçon at two points, (r_1, θ_1) and $(r_2, \theta_1 + \pi)$. Since these points lie on a straight line through the pole, the distance between them is

$$r_1 + r_2 = (2a + a\cos \theta_1) + [2a + a\cos(\theta_1 + \pi)]$$
$$= 2a + a\cos \theta_1 + 2a - a\cos \theta_1$$
$$= 2a$$

47. Press $\boxed{\text{MODE}}$ PARAM $\boxed{\text{ENTER}}$

a) Press $\boxed{\text{Y=}}$

Enter 5 $\boxed{\text{COS}}$ $\boxed{\text{X|T}}$ $\div$ $\boxed{\text{COS}}$ ($\boxed{\text{X|T}}$ $- 1$) as X_{1T}

Enter 5 $\boxed{\text{SIN}}$ $\boxed{\text{X|T}}$ $\div$ $\boxed{\text{COS}}$ ($\boxed{\text{X|T}}$ $- 1$) as Y_{1T}

Press $\boxed{\text{GRAPH}}$.

b) Press $\boxed{\text{Y=}}$

Replace the number 1 with 2.5 in X_{1T} and Y_{1T}.

Press $\boxed{\text{GRAPH}}$.

c) Press $\boxed{\text{Y=}}$

Enter 10 $\boxed{\text{COS}}$ $\boxed{\text{X|T}}$ $\div$ $\boxed{\text{COS}}$ ($\boxed{\text{X|T}}$ $- 2.5$) as X_{1T}

Enter 10 $\boxed{\text{SIN}}$ $\boxed{\text{X|T}}$ $\div$ $\boxed{\text{COS}}$ ($\boxed{\text{X|T}}$ $- 2.5$) as Y_{1T}

Press $\boxed{\text{GRAPH}}$.

d) Press $\boxed{\text{Y=}}$.

Replace 2.5 with 4 in both X_{1T} and Y_{1T}.

Press $\boxed{\text{GRAPH}}$

Chapter 13 Review Problem Set

1. False

$$4p = 6$$
$$2p = 3$$

It is 3 units from the focus to the directrix.

2. True

3. True

4. False. The eccentricity is $\dfrac{\sqrt{34}}{3}$.

5. True

6. True

7. False. For example, if B = 2, the graph is

$$(x + y)^2 = 1$$
$$x + y = \pm 1$$

consisting of two parallel lines.

8. True

9. False. $m = \dfrac{-6}{3} = -2$

10. True

11. a) $x^2 + y^2 = \dfrac{10}{3}$ circle (iii)

b) $\dfrac{(x + 2)^2}{12} + \dfrac{(y - 1)^2}{3} = 1$ horizontal ellipse (v)

c) $-\dfrac{(x + 2)^2}{4} + \dfrac{(y - 1)^2}{1} = 1$ vertical hyperbola (viii)

d) $\dfrac{(x+2)^2}{4} - \dfrac{(y-1)^2}{4} = 1$ horizontal hyperbola (iv)

e) $-\dfrac{(x-1)^2}{5} - \dfrac{y^2}{20} = 1$ the empty set (vii)

f) $y - 2 = \pm 2$ parallel lines (vi)

g) $y - 2 = \pm 2x$ intersecting lines (xi)

h) $y = 2x^2 - x - 6$ vertical parabola (ii)

i) $(x+4)^2 + (y-2)^2 = 36$ circle (iii)

j) $2x = 3y^2 + 6y - 4$ horizontal parabola (ix)

k) $3x - 5y = 0$ single line (x)

12. $(x+1)^2 = -12(y-2) = 4(-3)(y-2)$

The vertex is at $(-1,2)$ and the focus is 3 units below, at $(-1,-1)$.

13. Since this is a horizontal parabola with vertex at the origin, the equation is of the form $y^2 = 4px$. Since $p = 5$, $y^2 = 20x$.

14. Since this is a vertical parabola with vertex at the origin, the equation is of the form $y = kx^2$. With $x = -2$ and $y = -4$ we get $-4 = 4k$, $k = -1$. Therefore $y = -x^2$.

15. Since the directrix is parallel to the x-axis and below the vertex, this is a vertical parabola opening upwards. The equation will be of the form

$$(x-3)^2 = 4p(y-5)$$

p is the distance from the directrix to the vertex.

$$p = 5 - 1 = 4$$

Therefore,

$$(x-3)^2 = 16(y-5)$$

16. Place the reflector with its vertex at the origin, opening upward. The equation of the parabola is $x^2 = 4py$. Since it passes through the point $(3,1)$,

$$9 = 4p$$

$$p = 2.25$$

The rays are focused at a point 2.25 feet above the vertex.

17. $25(x+2)^2 + 9(y-1)^2 = 225$

$$\dfrac{(x+2)^2}{9} + \dfrac{(y-1)^2}{25} = 1$$

$a^2 = 25$, $a = 5$

$b^2 = 9$, $b = 3$

$c^2 = a^2 - b^2 = 16$, $c = 4$

The vertices lie 5 units away from the center on the major diameter, which is vertical. Since the center is $(-2,1)$, the vertices are $(-2,6)$ and $(-2,-4)$.

The foci lie 4 units from the center on the major diameter, at $(-2,5)$ and $(-2,-3)$.

18. The major diameter is 8, so $a = 4$. The minor diameter is 4, so $b = 2$. The equation of the ellipse is

$$\dfrac{x^2}{16} + \dfrac{y^2}{4} = 1$$

19. $a = 6$

$a^2 = 36$

$e = \dfrac{c}{a}$

$\dfrac{1}{2} = \dfrac{c}{6}$

$c = 3$

$c^2 = 9$

$b^2 = a^2 - c^2 = 27$

Since this is a vertical ellipse, a^2 goes with y^2.

$$\dfrac{x^2}{27} + \dfrac{y^2}{36} = 1$$

20. $c = 2\sqrt{5}$

$a^2 - b^2 = c^2 = 20$

Substituting $a = b + 2$ gives

$$(b+2)^2 - b^2 = 20$$
$$b^2 + 4b + 4 - b^2 = 20$$
$$4b = 16$$
$$b = 4$$
$$a = 6$$

Since this is horizontal ellipse, a^2 goes with x^2.

$$\dfrac{x^2}{36} + \dfrac{y^2}{16} = 1$$

21. $a = 8$, $a^2 = 64$

$b = 6$, $b^2 = 36$

$$\dfrac{x^2}{36} + \dfrac{y^2}{64} = 1$$

When $y = 6$,

$$\dfrac{x^2}{36} + \dfrac{36}{64} = 1$$

$$\dfrac{x^2}{36} = \dfrac{28}{64}$$

$$x^2 = \dfrac{28 \cdot 36}{64} = \dfrac{7 \cdot 9}{4}$$

$$x = \dfrac{3}{2}\sqrt{7}$$

The width of the ellipse when $y = 6$ is

$$2\left(\dfrac{3}{2}\sqrt{7}\right) = 3\sqrt{7} \approx 7.94 \text{ feet}$$

22. $25x^2 - 4y^2 = 100$

Find the vertices by setting $y = 0$:

$$25x^2 = 100$$
$$x^2 = 4$$
$$x = \pm 2$$

The vertices are at the points $(\pm 2, 0)$. The equations of the asymptotes are

$$5x \pm 2y = 0$$
$$y = \pm \frac{5}{2}x$$

23.
$$a = 6$$
$$a^2 = 36$$
$$e = \frac{c}{a}$$
$$\frac{3}{2} = \frac{c}{6}$$
$$c = 9$$
$$c^2 = 81$$
$$b^2 = c^2 - a^2 = 45$$

Since this is a vertical hyperbola, a^2 goes with y^2.
$$\frac{y^2}{36} - \frac{x^2}{45} = 1$$

24. $\frac{x^2}{16} - \frac{y^2}{b^2} = 1$

Because the point $(5, \frac{3}{2})$ is on the hyperbola,

$$\frac{25}{16} - \frac{\frac{9}{4}}{b^2} = 1$$
$$\frac{9}{16} = \frac{9}{4b^2}$$
$$4b^2 = 16$$
$$b^2 = 4$$

The equation of the hyperbola is
$$\frac{x^2}{16} - \frac{y^2}{4} = 1$$

25.
$$a = 5$$
$$5y = 2x$$
$$y = \frac{2}{5}x$$

The slope of the asymptote is $\frac{2}{5}$.
$$\frac{b}{a} = \frac{2}{5}$$
$$\frac{b}{5} = \frac{2}{5}$$
$$b = 2$$
$$b^2 = 4$$
$$a^2 = 25$$

This is a horizontal hyperbola.
$$\frac{x^2}{25} - \frac{y^2}{4} = 1$$

26. $(x+1)^2 = 12(y - 2)$

This is a vertical parabola with vertex at $(-1, 2)$.

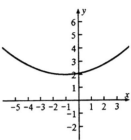

27. $4(x-4)^2 - (y+1)^2 = 16$

$$\frac{(x-4)^2}{4} - \frac{(y+1)^2}{16} = 1$$

This is a horizontal hyperbola with center at $(4, -1)$. The vertices are 2 units away from the center on the horizontal axis.

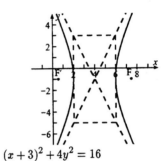

28. $(x+3)^2 + 4y^2 = 16$

$$\frac{(x+3)^2}{16} + \frac{y^2}{4} = 1$$

This is a horizontal ellipse with center at $(-3, 0)$. The vertices are 4 units left and right of the center.

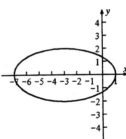

29.
$$x^2 + y^2 - 14x + 2y + 25 = 0$$
$$x^2 - 14x + 49 + y^2 + 2y + 1 = -25 + 49 + 1$$
$$(x-7)^2 + (y+1)^2 = 25$$

Circle with center at $(7, -1)$ and radius 5.

30.
$$9x^2 + 4y^2 - 90x + 16y + 205 = 0$$
$$9(x^2 - 10x \quad) + 4(y^2 + 4y + \quad) = -205$$
$$9(x^2 - 10x + 25) + 4(y^2 + 4y + 4) = -205 + 225 + 16$$
$$9(x-5)^2 + 4(y+2)^2 = 36$$
$$\frac{(x-5)^2}{4} + \frac{(y+2)^2}{9} = 1$$

We have a vertical ellipse with center at $(5, -2)$, $a = 3$, $b = 2$.

31.
$$9x^2 - 4y^2 - 54x + 16y + 65 = 0$$
$$9(x^2 - 6x + \quad) - 4(y^2 - 4y + \quad) = -65$$
$$9(x^2 - 6x + 9) - 4(y^2 - 4y + 4) = -65 + 81 - 16$$
$$9(x-3)^2 - 4(y-2)^2 = 0$$
$$4(y-2)^2 = 9(x-3)^2$$
$$2(y-2) = \pm 3(x-3)$$

Two straight line that intersect at $(3, 2)$.

32.
$$7x^2 - 4\sqrt{3}xy + 3y^2 = 36$$
$$x = u \cos 60° - v \sin 60° = \frac{1}{2}u - \frac{\sqrt{3}}{2}v$$
$$y = u \sin 60° + v \cos 60° = \frac{\sqrt{3}}{2}u + \frac{1}{2}v$$
$$7\left(\frac{1}{2}u - \frac{\sqrt{3}}{2}v\right)^2 - 4\sqrt{3}\left(\frac{1}{2}u - \frac{\sqrt{3}}{2}v\right)\left(\frac{\sqrt{3}}{2}u + \frac{1}{2}v\right)$$
$$+ 3\left(\frac{\sqrt{3}}{2}u + \frac{1}{2}v\right)^2 = 36$$
$$7\frac{u^2 - 2\sqrt{3}uv + 3v^2}{4} - 4\sqrt{3}\frac{\sqrt{3}u^2 - 2uv - \sqrt{3}v^2}{4}$$
$$3\frac{3u^2 + 2\sqrt{3}uv + v^2}{4} = 36$$
$$7u^2 - 14\sqrt{3}uv + 21v^2 - 12u^2 + 8\sqrt{3}uv + 12v^2$$
$$+ 9u^2 + 6\sqrt{3}uv + 3v^2 = 144$$
$$4u^2 + 36v^2 = 144$$
$$\frac{u^2}{36} + \frac{v^2}{4} = 1$$

Ellipse with center $(0, 0)$, $a = 6$, $b = 2$.

33.

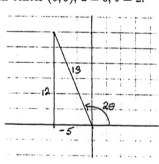

$$3x^2 + 12xy + 8y^2 = 12$$
$$A = 3, \ B = 12, \ C = 8$$
$$\cot 2\theta = \frac{A - C}{B} = \frac{-5}{12}$$
$$\cos 2\theta = -\frac{5}{13}$$
$$\cos \theta = \sqrt{\frac{1 + \cos 2\theta}{2}} = \sqrt{\frac{1 - \frac{5}{13}}{2}} = \sqrt{\frac{8}{26}} = \frac{2}{\sqrt{13}}$$
$$\sin \theta = \sqrt{\frac{1 - \cos 2\theta}{2}} = \sqrt{\frac{1 + \frac{5}{13}}{2}} = \sqrt{\frac{18}{26}} = \frac{3}{\sqrt{13}}$$
$$x = u \cos \theta - v \sin \theta = \frac{1}{\sqrt{13}}(2u - 3v)$$
$$y = u \sin \theta + v \cos \theta = \frac{1}{\sqrt{13}}(3u + 2v)$$

Substitute for x and y:
$$3 \cdot \frac{1}{13}(2u - 3v)^2 + 12 \cdot \frac{1}{13}(2u - 3v)(3u + 2v)$$
$$+ 8 \cdot \frac{1}{13}(3u + 2v)^2 = 12$$
$$3(4u^2 - 12uv + 9v^2) + 12(6u^2 - 5uv - 6v^2)$$
$$+ 8(9u^2 + 12uv + 4v^2) = 156$$
$$12u^2 - 36uv + 27v^2 + 72u^2 - 60uv - 72v^2$$
$$+ 72u^2 + 96uv + 32v^2 = 156$$
$$156u^2 - 13v^2 = 156$$
$$u^2 - \frac{v^2}{12} = 1$$
$$a^2 = 1, \ b^2 = 12$$
$$c^2 = a^2 + b^2 = 13, \ c = \sqrt{13}$$

The distance between the foci is $2c = 2\sqrt{13}$.

34.
$$x = 4t - 5 \qquad\qquad y = -3t + 2$$
$$3x = 12t - 15 \qquad\quad 4y = -12t + 8$$

Adding the last two equations gives
$$3x + 4y = -7$$

35.
$$x = 4 \cos t \qquad\qquad y = 3 \sin t$$
$$\left(\frac{x}{4}\right)^2 = \cos^2 t \qquad\quad \left(\frac{y}{3}\right)^2 = \sin^2 t$$
$$\frac{x^2}{16} + \frac{y^2}{9} = \cos^2 t + \sin^2 t = 1$$

36.
$$x = 2t$$
$$t = \frac{1}{2}x$$
$$y = 12t^2 + 2t - 4$$
$$= 12\left(\frac{1}{2}x\right)^2 + 2\left(\frac{1}{2}x\right) - 4$$
$$= 3x^2 + x - 4$$

37. $x = 4 \cot t$, $y = 4 \sin^2 t$

θ	1°	30°	60°	90°	120°	150°	179°
x	229	6.9	2.3	0	−2.3	−6.9	−229
y	.001	1	3	4	3	1	.001

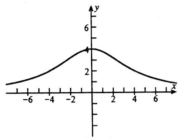

38. a) Cartesian equation: $x^2 + y^2 = 4$

b) Polar equation: $r = 2$

c) Parametric equation:
$x = r \cos \theta = 2 \cos \theta$, $y = r \sin \theta = 2 \sin \theta$

39. a) $x = r \cos \theta = 2 \cos \frac{2\pi}{3} = 2\left(-\frac{1}{2}\right) = -1$

$y = r \sin \theta = 2 \sin \frac{2\pi}{3} = 2\left(\frac{\sqrt{3}}{2}\right) = \sqrt{3}$

$(x, y) = (-1, \sqrt{3})$

b) $x = r \cos \theta = 4 \cos\left(-\frac{\pi}{2}\right) = 4(0) = 0$

$y = r \sin \theta = 4 \sin\left(-\frac{\pi}{2}\right) = 4(-1) = -4$

$(x, y) = (0, -4)$

c) $x = r \cos \theta = -10 \cos 240° = -10\left(-\frac{1}{2}\right) = 5$

$y = r \sin \theta = -10 \sin 240° = -10\left(-\frac{\sqrt{3}}{2}\right) = 5\sqrt{3}$

$(x, y) = (5, 5\sqrt{3})$

40. a) $(x, y) = (-4, 0)$. $(r, \theta) = (4, \pi)$

b) $(x, y) = (-4\sqrt{2}, 4\sqrt{2})$ is in quadrant II

$r = 4\sqrt{2}(\sqrt{2}) = 8$

$\cos \theta = \frac{-4\sqrt{2}}{8} = -\frac{\sqrt{2}}{2}$. $\theta_0 = \frac{\pi}{4}$.

$(r, \theta) = \left(8, \frac{3\pi}{4}\right)$

c) $(x, y) = (3, -3\sqrt{3})$ is in quadrant IV

$r = \sqrt{3^2 + (-3\sqrt{3})^2} = \sqrt{9 + 27} = \sqrt{36} = 6$

$\cos \theta = \frac{3}{6} = \frac{1}{2}$. $\theta_0 = \frac{\pi}{3}$

$(r, \theta) = \left(6, \frac{5\pi}{3}\right)$

41.

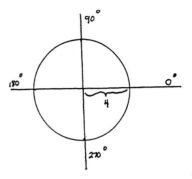

42.

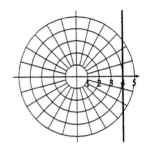

43. $r = 4 \sin \theta$

θ	0°	30°	60°	90°	120°	150°	180°
r	0	2	3.46	4	3.46	2	0

θ	210°	240°	270°	300°	330°
r	−2	−3.46	−4	−3.46	−2

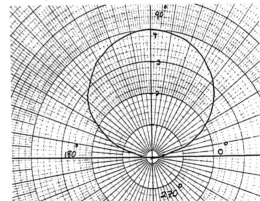

44. $r = 4 \cos 3\theta$

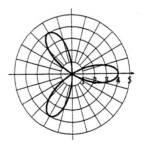

45. $r = \dfrac{2}{1+\sin\theta}$

This is a vertical parabola facing downwards.

θ	0°	30°	60°	90°	120°	150°	180°
r	2	1.33	1.07	1	1.07	1.33	2

θ	210°	240°	270°	300°	330°
r	4	14.9	und	14.9	4

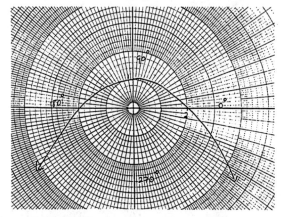

46. $r = 2(1+\sin\theta)$ is a cardioid.

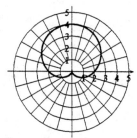

47. $\qquad x^2 + y^2 + 4x - 2y = 0$

$$r^2 + 4r\cos\theta - 2r\sin\theta = 0$$

The pole lies on the graph so we may divide by r.

$$r + 4\cos\theta - 2\sin\theta = 0$$

$$r = 2\sin\theta - 4\cos\theta$$

48. $\qquad r = \dfrac{2\sin 2\theta}{\cos^3\theta - \sin^3\theta}$

$$r(\cos^3\theta - \sin^3\theta) = 2\sin 2\theta = 4\sin\theta\cos\theta$$

The pole lies on the graph so we may multiply by r^2.

$$(r\cos\theta)^3 - (r\sin\theta)^3 = 4(r\sin\theta)(r\cos\theta)$$

$$x^3 - y^3 = 4xy$$

49. a) $r\cos\theta = -4$

$\qquad\qquad x = -4;$ $\qquad\qquad$ vertical line (vii)

b) $\qquad r = -4\cos\theta$ $\qquad\qquad$ circle (iv)

c) $\qquad r = \dfrac{3}{1+\cos\theta}$, $e = 1$ $\quad$ parabola (i)

d) $\qquad r = \dfrac{2}{1+4\cos\theta}$, $e = 4$ $\quad$ hyperbola (ii)

e) $\qquad r = \dfrac{5}{\cos\left(\theta - \frac{\pi}{4}\right)}$ $\qquad$ nonvertical line (vi)

f) $\qquad r = \dfrac{3}{2+3\cos\theta} = \dfrac{\frac{3}{2}}{1+\frac{3}{2}\cos\theta}$

$\qquad\qquad e = \dfrac{3}{2}$ $\qquad\qquad$ hyperbola (v)

g) $\qquad r = \dfrac{4}{4-\cos\theta} = \dfrac{1}{1-\frac{1}{4}\cos\theta}$

$\qquad\qquad e = \dfrac{1}{4}$ $\qquad\qquad$ ellipse (iii)

h) $\qquad r = 2\cos\theta - 4\sin\theta$

$\qquad r^2 = 2r\cos\theta - 4r\sin\theta$

$\quad x^2 + y^2 = 2x - 4y$ $\qquad$ circle (iv)

50. a) $|r| = \left|\dfrac{1}{1-\cos\theta}\right|$

has a minimum of $\frac{1}{2}$ when $\cos\theta = -1$

b) $|r| = |3 - 2\sin\theta|$

has a minimum of 1 when $\sin\theta = 1$.